SYMMETRIES AND LAPLACIANS

Introduction to Harmonic Analysis, Group Representations and Applications

DAVID GURARIE

Mathematics Department
Case Western Reserve University

DOVER PUBLICATIONS, INC.
Mineola, New York

Bibliographical Note

This Dover edition, first published in 2008, is an unabridged republication of the work originally published in 1992 by North-Holland Publishing Company, Amsterdam, as Volume 174 in the *North-Holland Mathematics Studies* series.

Library of Congress Cataloging-in-Publication Data

Gurarie, David.
Symmetries and Laplacians : introduction to harmonic analysis, group representations, and applications / David Gurarie. — Dover ed.
p. cm.
Originally published: Amsterdam ; New York : North-Holland, 1992.
Includes index.
ISBN-13: 978-0-486-46288-2
ISBN-10: 0-486-46288-9
1. Harmonic analysis. 2. Representations of groups. I. Title.

QA403.G87 2008
515' .2433—dc22

2007030271

www.doverpublications.com

In the memory of my parents, whom I owe so much

to Valentina

Eli and Mark

TABLE OF CONTENTS

Introduction

Throughout this book word Group will be synonymous with transformations and symmetries. Groups are omnipresent in the mathematical and (possibly) physical universe. In Mathematics symmetries play several important functions. On the one hand they allow to reduce the "number of variables", and often render problem soluble (like reducing the order, or separating variables in differential equations). On the other hand they allow to analyze and synthesize complex objects in terms of simple (elemental) blocks/constituents (eigenvalues of matrices and operators, expansion in Fourier modes, harmonic analysis in general).

The prime example came from the very onset of the subject, the Galois theory of algebraic equations. The thrust of this early work was precisely to relate the symmetry (Galois) group of the equation to its solvability[1] in radicals. Then Sophus Lie set up the task to develop the "Galois theory" for differential equations. He naturally came up with the concept of a continuous *Lie group* and its infinitesimal versions *Lie algebra.* Since its inception in the mid XIX-century the Lie theory rapidly grew and spread across the broad range of subjects, to occupy its present central position at the crossroads of Geometry, Analysis, Differential equations, Classical mechanics and Physics in general. Symmetries in physical systems[2] are inextricably linked to conservation laws (via Noether's Theorem), and the Galois chain repeats: "symmetries of hamiltonians" $\Rightarrow$ "conserved integrals" $\Rightarrow$ "solvability (via symmetry-reductions)".

Our book sets more limited goals. The main objective is to introduce the reader to a wide range of concepts, ideas, results and techniques, that evolve around symmetry-groups, representations, and Laplacians. We strived to stress the diversity and versatility of the subject, but at the same time to unravel its common roots.

More specifically, our main interest lies in geometrical objects and structures $\{X\}$, discrete or continuous, that possess sufficiently large symmetry-group G, like regular graphs (Platonic solids); lattices; symmetric Riemannian manifolds, and other. All such objects have a natural *Laplacian* Δ, a linear operator on functions over X, invariant under the group action. There are many problems associated with Laplacians on X. Typically one is interested in certain "continuous or discrete-time" evolutions, on X, random walks, diffusion processes, wave-propagation. All these problems require to

[1]Thence came the terminology of *solvable groups* (abelian extensions of abelian groups), as the abelian ones were directly associated with solvability.

[2]Symmetries acquire yet greater significance in the murky world of subatomic physics, where nothing could be observed, measured, sensed and compared to the everyday experiences of the macro-world. Here symmetries and conservation laws provide the only guiding light, and Lie groups enter in the very formulation of fundamental models of particles, fields and forces.

compute certain "functions of Δ", $L = f(\Delta)$, like powers $\{\Delta^n\}$; the semigroup, generated by Δ $\{e^{-t\Delta}\}$; function $\{\cos(t\sqrt{\Delta})\}$, etc. To explain and motivate the problems and the methods, based on symmetries, we shall start with a few simple prototypical examples.

Fibonacci sequence is defined by a 2-term recurrence relation: $\boxed{x_{n+1} = x_n + x_{n-1}}$ (or more general, $x_{n+1} = ax_n + bx_{n-1}$), and the two initial values $\{x_0; x_1\}$. One is asked to find all $\{x_n\}$. To solve the problem, one first converts the 2-term numerical recurrence to a one-term recurrence for vectors: $y_n = \begin{bmatrix} x_{n-1} \\ x_n \end{bmatrix}$, $\boxed{y_{n+1} = Ly_n}$ with a Frobenius-type matrix $L = \begin{bmatrix} 0 & 1 \\ a & b \end{bmatrix}$. Clearly,

$$y_n = L^n[y_0], \tag{1}$$

and we need to compute all iterates $\{L^n\}$. The latter could be accomplished by diagonalizing matrix L, i.e. solving the eigenvalue problem: finding *spectrum* $\{\lambda_1; \lambda_2\}$, and the matrix of eigenvectors $U = \{\psi_1; \psi_2\}$. The latter serves to diagonalize L, $L = U\Lambda U^{-1}$, where $\Lambda = \mathrm{diag}\{\lambda_1; \lambda_2\}$. Once L is brought into the diagonal form, its iterates $\{L^n = U\Lambda^n U^{-1}\}$ come right away, as well as more general "function of L", $f(L) = Uf(\Lambda)U^{-1}$. So we immediately get an explicit solution [3].

Of course, no group-analysis was needed here, as the "eigenvalue problem" had an easy direct solution. However, similar "finite" problems could quickly become intractable, as we shall demonstrate in our next example.

Random walk on graphs Γ. The process consists of random jumps from any site (vertex) of the graph to any one of its nearest neighbors with equal probabilities. Starting with the initial position at $\{x_0\}$, or the initial probability distribution $P_0(x)$ $(x \in \Gamma)$, one is asked to find the probability distribution after the n time-steps, $\{P_n(x)\}$. Once again vector $\{P_n(x)\}$ is completely determined by the initial state P_0 and the transition-matrix L, which could be called the "Laplacian" of the graph, that sends any vector/function $\psi(x)$ on Γ into

$$L\psi(x) = \frac{1}{m}\sum_y \psi(y),$$

sum over all m nearest vertices $\{y\}$ of x. Since the jumps are *statistically independent* (the process has no memory of the past, when deciding to go from $\{x\}$ to $\{y\}$), and *stationary* in time (the "jump-rules" remains the same for all times $n = 0; 1; \ldots$), we get a sequence of iterates,

[3] The classical Fibonacci numbers are given by $x_n = c_1(\frac{1+\sqrt{5}}{2})^n + c_2(\frac{1-\sqrt{5}}{2})^n$, the generalized ones, $x_n = c_1{\lambda_1}^n + c_2{\lambda_2}^n$; where coefficients, $\{c_1; c_2\}$ depend on the initial data $\{x_0; x_1\}$.

$$P_n = L^n[P_0], \text{ at time } n. \tag{2}$$

To compute powers $\{L^n\}$ we need once again to diagonalize L, i.e. to solve the eigenvalue problem. But this time the difficulty increases drastically with the size of Γ, if one tries a direct approach (imagine diagonalizing a 20×20 "dodecahedral" matrix!). There is not much to do in the general case, so we turn to a special, but important, class of *regular graphs* (like polygons, Platonic solids, a "Knight on a periodic chessboard", etc.). Regular graphs Γ are characterized by the abundance of symmetries. One can associate with any such Γ group G of vertex-transformations, that preserve all links (the incidence relations) between vertices. For regular Γ, group G takes any vertex x into any other[4], so Γ becomes a *homogeneous space of* G, $\Gamma = H\backslash G$, where H denotes the stabilizer subgroup of some $x_0 \in \Gamma$, $H = \{g: x_0{}^g = x_0\}$. For instance, the regular m-gone is itself a finite abelian group $\mathbb{Z}_m = \{\text{integers modulo } m\}$ (its symmetries are made of all finite rotations and reflections), whereas Platonic solids (cube, octahedron, dodecahedron, etc.) are homogeneous spaces $H\backslash G$ of some well-known symmetric or alternating groups (see §1.1 and §3.2).

Given a homogeneous space X, group G acts on X by translations, $g: x \to x^g$, and this action gives rise to the *regular representation* $R = R^X$ of G on the function-space $\mathcal{L} = \mathcal{L}(X)$ over X,

$$R_g f(x) = f(x^g). \tag{3}$$

One easily verifies that map $g \to R_g$, from G into operators on $\mathcal{L}$, takes group multiplication into the product of operators, in other words defines a group-representation. Since Laplacian L commutes with all operators $\{R_g\}$, group G leaves eigenspaces $\{E_\lambda\}$ of L in $\mathcal{L}$ invariant, and we have to analyze the resulting "reduced representations" of G on eigenspaces $\{E_\lambda\}$. Among all representations of the group one distinguishes the minimal ones, called *irreducible*, the ones that do not allow further reduction to smaller invariant subspaces (they play the role of "joint eigenvalues" for a family of operators $\{R_g: g \in G\}$). The typical problems in representation theory include:
(i) *characterization all irreducible* $\{\pi\text{'s}\}$ *for a given group* G (analysis-problem); and
(ii) *decomposition of an arbitrary (natural) representation* R *into the direct sum (or integral,* to be explained below) *of irreducibles* (synthesis problem).

The harmonic analysis on X suggests a possible approach to the spectral problem

[4] In fact, in many cases G takes any connected pair of vertices $\{x; y\}$ to any other such pair. Such graphs Γ's could be called 2-point symmetric spaces, by analogy with their continuous counterparts.

for operator L. Namely, we need

(I) to study the structure of the symmetry-group G;

(II) to determine its elemental (irreducible) representations $\{\pi\}$;

(III) to interpret Laplacian L in terms of group-elements of G, to be able to assign operators $\{\pi_L\}$ to it for various irreducible $\{\pi\}$

(IV) finally, to decompose the regular representation into the direct sum of *irreducible components:* $R^X = \bigoplus_j \pi^j$.

The latter means that space $\mathcal{L}(X)$ is split into the direct sum of subspaces $\bigoplus_j E_j$, invariant under all operators $\{R_g: g \in G\}$, irreducible relative to the G-action. Such decomposition, generalizes the notion of spectral/resolution (diagonalization) of a single operator.

Steps (I-IV), when successfully accomplished, would lead to the desired spectral resolution of L. Namely, the reduced Laplacian $L \mid E_j \sim \pi_L$ becomes scalar[5], $\{\pi_L = \lambda_\pi(L)$; for each irreducible $\pi\}$. So the "*eigenvalue spectrum* $\{\lambda_\pi(L)\}$" could be identified the "*representation spectrum* $\{\pi^j\}$ of R", the "*eigenspaces*" being "*irreducible components*" of R. Once L is diagonalized we immediately obtain all iterates $\{L^n\}$, and more general functions $\{f(L)\}$, needed to solve (3).

The details of the scheme could be found in §2.3 (Finite Fourier transform) for polygons, and in §3.3 for Laplacians on Platonic solids.

After a brief excursion through the finite/discrete cases (graphs) we shall turn now to a completely different geometric setup: smooth Riemannian manifolds, Lie groups and their quotients, symmetric spaces X. All these objects will be introduced and studied at length in the due course (chapters 1,4,5,6,7). Here for the sake of introduction we consider 3 familiar examples: the Euclidian space $\mathbb{R}^n$, the n-sphere $S^n \subset \mathbb{R}^{n+1}$, and the hyperbolic (Poincare-Lobachevski) plane[6] $\mathbb{H}^2 = \{z = x+iy: y \geq 0\}$.

To motivate the continuous case we shall mention 3 typical problems in the integral geometry with some physical context. The setup for all 3 is basically the same. We are given a solid body D of constant (or variable) density $\rho(x)$. In the first case one considers all lines $\{\gamma\}$, passing through D, and computes integrals along γ: $\hat{\rho}(\gamma) = \int_\gamma \rho ds$. In the

[5]There is a general reason, why Laplacian L goes into the scalar operator π_L under any irreducible π, known as Schur's lemma.

[6]the latter could be also interpreted, as a continuous limit of properly scaled in space and time random walks on trees.

second solid D (assumed to be convex, symmetric) is illuminated from all possible directions, and we measure the area of the shadows, cast by D. In the third case (somewhat simpler than the second) we take all cross-sectional areas of D by the family of planes, passing through its center. In all 3 cases one would like to recover either density ρ (case 1), or the geometric shape of D (case 2 and 3) from such data. The first case is essentially the celebrated X-ray transform, used in tomography. The 2-nd and 3-rd give somewhat different versions of the Radon transform on the 2-sphere. Indeed, a (convex) body D can be represented as a graph of function $f(x)$ on the 2-sphere, $S^2 = \{|x| = 1\}$. The cross-sectional areas of D are obtained by integrating $\frac{1}{2}f^2$ along all great circles $\{\gamma\}$ in S^2. So in all cases we are dealing with certain integral transforms of functions, either on the Euclidian space, or the sphere.

All 3 spaces, $\mathbb{R}^n$, S^n, $\mathbb{H}$ have the natural Riemannian metric (§1.1), and possess large symmetry-group G of distance-preserving transformations (*isometries*). The Euclidian isometries consists of rigid motions: group $\mathbb{E}_n$, generated by all translations and rotations in $\mathbb{R}^n$; the n-sphere S^n has the *orthogonal* symmetry-group $SO(n+1)$, made of all $(n+1)\times(n+1)$ - orthogonal matrices in $\mathbb{R}^{n+1} \supset S^n$; symmetries of $\mathbb{H}$ coincide with the ubiquitous *unimodular* group $SL_2(\mathbb{R})$ - all 2×2 real matrices of $\det g = 1$, acting by the fractional-linear transformations, $g: z \to \frac{az+b}{cz+d}$. In fact, all 3 examples represent the so called *symmetric spaces*, quotients $K\backslash G$ of group G modulo the maximal compact subgroup K, that fixes a particular point in X, $K = \{g: x_0{}^g = x_0\}$. In the above examples: $K = SO(n)$ for $\mathbb{R}^n$ and S^n, while the hyperbolic $K = SO(2)$.

The analog of the discrete-time random-walk on space X is played either by *spherical means:*

$$M_r f(x) = \frac{1}{vol(S_r(x))}\int_{d(x;y) = r} f(y)dS(y),$$

(average value of function f over the sphere of radius r, centered at $\{x\}$), or by more general integral kernel,

$$u(x) \to K[u] = \int K(x;y)u(y)dy.$$

Transition-operator[7] K takes any distribution $p_0(y)$ on X into $p_1(x) = K[p_0]$, at a single time-step. As before we are interested in iterates $\{K^n\}$, and once again have to face the "diagonalization problem" for K. In general, the problem is hardly tractable, so we turn to a special class of integral kernels, $K(x;y)$, that commute with the G-action,

[7]Strictly speaking, density $K(x;y)dy$ represents transitional probabilities to jump from point $\{y\}$ to $\{x\}$, provided $\int K(x;y)dy = 1$.

(3) on X. Any G-invariant operator has kernel K, that depends only on the distance[8] $r = d(x;y)$, between x and y, $K(x;y) = K(r)$. Spherical means give one such example, the Radon and X-ray transforms also obey this condition, but the foremost of all G-invariant operators is the natural Laplacian[9] Δ (the Laplace-Beltrami operator) on X. Associated to Δ there are various evolution processes on X (continuous analogs of the "discrete-time random walks"), described by partial differential equations (pde). One of them is the standard *heat-diffusion problem,*

$$u_t - \Delta u = 0;\ u(0) = f; \tag{4}$$

whose formal solution of (4) is given by the *heat-semigroup* of Δ,

$$u = e^{t\Delta}[f].$$

Another important model is the wave equation: $u_{tt} - \Delta u = ...$, whose solutions can also be represented by "functions of Δ", $\{u = e^{\pm it\sqrt{-\Delta}}$; or $\cos t\sqrt{-\Delta}$; $\sin t\sqrt{-\Delta}\}$.

Since Δ commutes with the G-action on X, all its functions $\{f(\Delta)\}$ do the same, including the heat and wave-propagators (another names for the "transition-matrices", $K = e^{t\Delta}$; $e^{it\sqrt{\Delta}}$). But any such K has a radially symmetric kernel $K = K(r)$. So a pde-problem for K, is often reduced to an ordinary differential equation (see chapters 2;4;5). Those in many cases, including $\mathbb{R}^n$, S^n, $\mathbb{H}$, could be solved explicitly in terms of the well-known special functions. Once again we are lead to analyze the G-action on X, and the resulting regular representation $R_g f(x) = f(x^g)$, on suitable function-spaces over X. So spectral decomposition of any radial operator K is reduced to the harmonic analysis of representation R. The latter involves among other a good understanding of the symmetry-group G itself, its "elemental (irreducible) representations" $\{\pi\}$, and the nature of the direct sum and/or integral decomposition,

$$R = \bigoplus_j \pi^j;\ \text{or}\ \oint \pi^s d\mu(s). \tag{5}$$

To give some feeling of the issues involved, let us take the Euclidian space $\mathbb{R}^n$, as an abelian group with translations only. Irreducible representations of $\mathbb{R}^n$ are all 1-

[8]This result is a direct consequence of the "double-transitive" action of G on rank-one symmetric spaces X, i.e. any pair $\{x;y\} \subset X$, is taken any other equidistant pair.

[9]In fact, any spherically symmetric kernel K on X (in either one of 3 examples) is given by a "function of the Laplacian", $K = f(\Delta)$. This result, far from obvious, comes from the harmonic analysis of rank-one symmetric spaces (chapters 4-5). It has to do with the "multiplicity-free spectrum" of regular representation R, equivalently, with the fact that the commutator of R (all operators, that commute with it) forms a commutative algebra.

"Functions of operators" $\{f(L)\}$ will be introduced and discussed in chapter 2 and Appendix A via spectral resolution of operator L.

dimensional (characters!), they consist of the family of exponentials $\{e^{ix\cdot\xi}: \xi \in \mathbb{R}^n\}$. Decomposition (5), amounts to the Fourier-transform of $f(x)$, expansion in the Fourier-series or integral. In chapter 2 we explore in detail various aspects of the commutative Fourier analysis and apply it to differential equations.

The relevant analysis on the n-sphere, however, has a very different flavor. Here one has to start with "irreducible representations" $\{\pi\}$ of the orthogonal group $SO(n+1)$, that would play the role of exponentials $\{e^{ix\cdot\xi}\}$ on $\mathbb{R}^n$. This, by itself a challenging task, requires to develop a fair amount of the linear and multilinear (tensorial) algebra, as well as the relevant structure theory of Lie groups and algebras (chapters 4-5). Irreducibles $\{\pi\}$ of $SO(n+1)$ turned out to be finite dimensional, and the decomposition of R has purely "discrete spectrum"[10],

$$R = \bigoplus_{k=0}^{\infty} \pi^k.$$

The corresponding irreducible subspaces $\{\mathcal{H}_k\}$ of $L^2(S^n)$, called *spherical harmonics*, have many remarkable features. Aside of being eigenspaces of the spherical Laplacian, they proved to be closely connected to harmonic polynomials on $\mathbb{R}^{n+1}$, and the classical Legendre and Jacobi functions.

Our last example $\mathbb{H}^2$ reveals even more striking difference. SL_2 is a *simple, not compact* group (in fact, the "smallest" of them). Unlike $\mathbb{R}^n$ or $SO(n)$, its irreducible unitary representations are infinite-dimensional, they make up certain continuous and discrete families $\{\pi^s: s \in \mathbb{R}\}$ and $\{\pi^m: m \in \mathbb{Z}\}$. The decomposition problem for regular representations $\{R^X\}$ on homogeneous spaces, strongly depends on the "geometric nature" of quotient $X = H\backslash G$. In some cases (Poincare half-plane $\mathbb{H}$) it becomes purely continuous (direct integral), $R = \oint \pi^s d\mu(s)$; in other cases (compact quotients $\Gamma\backslash G$, Γ - discrete subgroup of G) - purely discrete (direct sum) $R = \bigoplus_{0}^{\infty} \pi^{s_j}$. It could also be a combination of both, like the regular representation R on the entire group G! The relevant harmonic analysis becomes quite involved (see chapter 7). Its most spectacular application to the spectral theory of Laplacians appears in the context of quotients $X = \Gamma\backslash G$, modulo discrete subgroup $\Gamma \subset G$. It turns out that "spectrum of such X" (either Laplacian Δ_X, or representation R^X) is intimately connected with some fine "arithmetic properties" of Γ, via the celebrated Selberg-trace Theorem (a noncommutative version of the "Poisson summation" on $\mathbb{R}^n$). As an off-short we

[10]Both results are general and hold for arbitrary *compact* (finite in size) groups (chapter 3). But specific examples, like $SO(n)$ will carry a great deal more structure and information about $\{\pi\}$ and R.

establish some interesting links between the "*eigenvalue spectrum* $\{\lambda_k\}$ of Δ_X", and the "*geometric length-spectrum of X*" (length of all closed geodesics in X). We elaborate some aspects of spectral theory on compact quotients $\Gamma\backslash G$ in §7.6-7.7.

Our discussion of the "continuous cases" vs. "discrete cases", although very different in technical terms, clearly demonstrates similarities in the basic procedures. Namely,

I) investigation of the group-structure of G, and its "elemental" irreducible representations $\{\pi\}$;

II) study of G-invariant objects "Laplacians" on G, or on its quotients $X = H\backslash G$;

III) decomposition of natural G-actions, regular representations R^X, R^G; its connection to Laplacians L on G and X;

IV) application of parts (I)-(III), particularly, the correspondence between "*spectrum of L*" and "*spectrum of R*" to various "*functions of L*", like iterates $\{L^n\}$, semigroups $\{e^{-tL}; e^{-t\sqrt{L}}\}$, and solutions of the associated differential equations.

The development in the book largely will largely follow the general scheme (I-IV), although not always step by step, as in the model examples. We may or may not start with a simple model problem. Often a significant effort has to be spent first to study the relevant groups, and their representation theory: problems of irreducibility and decomposition (cf. chapters 5-7). As we mentioned our main interest lies in the natural G-actions and representations, regular R^G; R^X, and their generalizations called *induced* representations. The latter extend G-actions on scalar function $\{f(x)\}$, to vector-functions and sections of vector-bundles over X (so G-action combines translations in the X-space with "twisting" in fiber-spaces). The decomposition of the regular and induced representations makes up the content of the so called *Plancherel Theorem.*

Although we follow steps (I-IV) throughout the book their relative "size" and "weight", as well as the amount and depth of applications, varies from chapter to chapter. Thus chapter 2 is largely devoted to *applications* of the commutative Fourier transform, while chapters 5 or 7, deal mostly with the *analysis* of groups and representations.

An introductory chapter 1 brings in the principal players: basic examples of groups and geometric structures (§1.1). Then we proceed to develop some fundamental concepts and results of the representation theory (§§1.2-3), and the Lie theory (§1.4). In chapters 2-7 we take on the main themes of the book with many facets of the general scheme (I-IV). Each chapter has its own "favorite" Laplacians and symmetry-groups:

- standard Laplacian Δ on commutative groups $\mathbb{R}^n;\mathbb{T}^n$ and certain domains $D \subset \mathbb{R}^n$, in chapter 2

- polyhedral Laplacians and the relevant finite groups in chapter 3

- the spherical Laplacian Δ on S^2 and S^n, whose theory is based on Lie groups $SU(2); SO(3)$, in chapter 4

- Laplacians on more general compact Lie groups and symmetric spaces in chapter 5

- the harmonic oscillator $H = -\Delta + |x|^2$, and the Heisenberg group in chapter 6

- Laplacians on the Poincare plane $\mathbb{H}$, SL_2, and compact Riemann surfaces $\Gamma\backslash\mathbb{H}$ in chapter 7.

Chapter 8 stands somewhat aside from the mainstream, as it takes on the subject of "*symmetries in nonlinear(!) hamiltonian systems*", and their role in *integrability*. But the very end of chapter 8 (§8.5) brings us back to the main issues in group-representations. We discuss the quantization problem of classical hamiltonians, particularly, the "*quantized Kepler problem*" - the *hydrogen atom*, one of the best studied models in quantum mechanics (see [LL]). The "Laplacian" pops in here in a quite interesting and unexpected form. A remarkable feature of the Kepler (hydrogen) hamiltonian, Schrödinger operator $H = -\Delta - \frac{1}{|x|}$ in $\mathbb{R}^3$, is that the "discrete (negative) part of H" is equivalent to the inverse Laplacian Δ^{-1} on the 3-sphere.

Along with the main themes (I-IV) a number of side issues and problems come to the discussion in different places. The reader will learn, for instance, why the geodesic flow is integrable on S^2 (chapter 8), and ergodic on compact negative-curved Riemann surfaces X (chapter 7); what group-representations have to do with a topological problem of "counting linearly independent vector fields on spheres" (chapter 3); how Lie groups and representations explain some peculiar properties and relations of special functions (Legendre, Jacobi, Hermite, etc.).

The book was designed as an introduction to harmonic analysis and group representations for graduate students in Mathematics and applications, or anybody (non-expert), interested in the subject, who would like to gain a broad perspective, but also learn some basic techniques and ideas. In the words of H. Weyl "... *it is primarily meant for the humble, who want to learn as new the things set forth therein, rather than for the proud and learned who are already familiar with the subject and merely look for quick and exact information...*"

The material of the book is based on the lectures and seminars, given by the author over the passed few years at UC Irvine, Caltech, and CWRU. Student comments and suggestions were helpful in preparing the manuscript.

Our goal was to cover a wide range of topics, rather than to delve deeply into any particular one. The exposition is largely based on examples and applications, which either precede or follow the general theory. Some are important on their own, others serve to elucidate and motivate general concepts and statements. Of course, if the general approach seemed conceptually easy and directly leads to the main point, we do not hesitate to bring it forth (like the "Peter-Weyl theory" of chapter 3). But we never engage in "abstract" studies for their own sake.

In order to keep the minimal prerequisites, and to shorten the background preparations, the book often appeals to intuition, example and analogy, rather than formal derivations. So the reader versed to some degree in the basic Riemannian geometry, functional analysis and operator theory, should be able to go through most of the topics without difficulty. The less prepared reader would be granted a paragraph (or two) of a footnote/appendix style explanations, to enable him to grasp quickly a new concept (or idea), and to follow the rest. Certain parts of the book (sections/paragraphs) are addressed to a better prepared reader, without detracting from the main themes.

As we wanted not to rely heavily on the standard (4-8 semester) staple of real analysis, algebra, topology, the book provides a number of 'shortcuts'. So some basic concepts in algebra, geometry, topology are introduces "on fly", as the need arises. We were somewhat more patient and systematic with the analysis of operators and differential equations: chapter 2 could serve as a brief introduction to PDE's, mostly from the classical standpoint (cf. [CH]; [WW]). We also provided 3 appendices: on spectral decomposition of self-adjoint operators (A); on integral operators (B), and on basic Riemannian geometry (C).

Finally, to cover a sizable material in a moderate-size volume required a departure from certain standards of mathematical exposition. We found it impossible (and undesirable) to try to maintain a uniform level of rigor and detail throughout the text. So some results are provided with fairly complete arguments, others are only outlined, relegated to problems, or just stated. The role of formal proofs is in general downplayed.

The book contains sufficient material for a 1 or 2-semester course in the Harmonic analysis and Group representations. The instructor could make several choices, and follow different path in selecting topics. Aside of chapter 1, that provides a general core and background, all other parts are relatively independent. So the reader could start at practically any place in the book, going back and forth, as deemed necessary. The only exception are chapter 4 and 5, which should proceed in their natural order. In each chapter we marked with * more advanced topics, that could be omitted in the first reading.

In writing an introductory text on a fairly broad subject, one inevitably has to make certain choices, and put aside some important topics. Our selection and style reflected largely on the author's personal experience and prejudices, rather than anything else.

Among a few important topics, left outside the scope, let us mention the representation theory of infinite-dimensional Lie groups and algebras (Kac-Moody), which was actively pursued over the passed 20 years, and recently came to the focus in connection with the String theory (monographs [Kac] and [GSW] present the mathematical and the physical view on the subject). Another important topic is related to symmetries of differential equations, dynamical systems, integrable hamiltonians. Although we do touch upon integrability in chapter 8, our analysis is limited to finite-dimensional systems. The exciting developments in the field of "infinite integrable hamiltonians" over the past 30 years were also left out (see [Per]; [Olv] for references).

The book was composed on the EXP-2 word processing system, with an additional help of the Microsoft Windows PBRUSH-graphics, responsible for the figures (the author takes entire responsibility for errors, misprints, omissions). After a somewhat bumpy initiation to the world of the modern information technology, the author had a highly rewarding experience working with both programs. Both became the most indispensable tools in the arduous enterprise of writing and organizing the manuscript.

He thought he saw a Garden-Door
That opened with a key
He looked again, and found it was
A Double Rule of Three:
"And all its mystery", he said,
"Is clear as day to me !"

He thought he saw an Argument
That proved he was the Pope
He look again, and found it was
A Bar of Mottled Soap.
"A fact so dread," he faintly said,
"Extinguishes all hope !"

Lewis Carroll, "*The Mad Gardener's Song*"

Chapter 1. Basics of representation theory.

§1.1. Groups and group actions.

We introduce basic examples of discrete and continuous transformation groups; classical matrix Lie groups; isometries of symmetric spaces; rigid motions of the Euclidian, spherical and hyperbolic geometry, as well as symmetries of regular polyhedra.

1.1. Geometric transformations. Groups, discrete and continuous, typically arise as symmetries of "geometric" structures of different kinds, which includes both discrete objects (graphs, polyhedra), and the continuous, like manifolds, symmetric spaces. Important examples of discrete groups include:

i) *permutations of a finite sets* $\Lambda = \{1;...n\}$, $G(\Lambda) = \mathbb{W}_n$;

ii) *isomorphisms of graphs, lattices, regular polyhedra*;

iii) *matrix groups over finite fields.*

Some of them will be described at the end of the section (example 1.1).

Throughout this book we shall be mostly interested in the continuous (Lie) groups. The latter typically arise as transformations (linear or nonlinear) of vector spaces, or manifolds $\mathcal{M}$, equipped with certain geometric structure. Such transformations $\phi{:}\mathcal{M}\to\mathcal{M}$ preserve the structure (or transform it in a prescribed manner), in other words they represent *symmetries* of $\mathcal{M}$. For instance, *classical (matrix) Lie groups* over vector spaces $\mathbb{R}^n$, $\mathbb{C}^n$, or *quaternionic*[1] $\mathbb{Q}^n$, consists of linear transformations, that preserve certain bilinear/quadratic forms, like the *general linear group* made of all non-singular $n \times n$ - matrices, $GL_n = \{A{:}\ det\,A \neq 0\}$, its subgroup $SL_n = \{A{:}\ det\,A = 1\}$, called *special linear (*or *unimodular) group*, as well as their numerous subgroups (orthogonal, unitary, symplectic etc.).

Lie groups also arise naturally as *isometries* of certain Riemannian (pseudo-Riemannian) manifolds, or more general *conformal transformations*[2]. The foremost cases are *symmetric spaces*, Riemannian manifolds which possess large isometry groups. Here we shall briefly discuss 3 basic examples of symmetric spaces (flat, spherical and hyperbolic), and the related symmetry groups.

a) Euclidian space $\mathbb{R}^n$, *respectively Minkowski* $\mathbb{M}^n$, equipped with either positive-

---

[1] *Quaternions* $\mathbb{Q} = \{\xi = a + b\boldsymbol{i} + c\boldsymbol{j} + d\boldsymbol{k}{:}\ a,b,c,d \in \mathbb{R}\}$ form a noncommutative field (division algebra), generated by 3 imaginary units: $\boldsymbol{i}^2 = \boldsymbol{j}^2 = \boldsymbol{k}^2 = -1$; $\boldsymbol{ij}=-\boldsymbol{ji}=\boldsymbol{k}$; $\boldsymbol{jk} = -\boldsymbol{kj}=\boldsymbol{i}$; $\boldsymbol{ki} = -\boldsymbol{ik} = \boldsymbol{j}$. They can be also represented by complex 2-vectors: $\xi = z + w\boldsymbol{k}$ $(z, w \in \mathbb{C})$, where $\boldsymbol{k}^2 = -1$, with the multiplication rule: $\boldsymbol{k}w = \overline{w}\,\boldsymbol{k}$. Quaternions along with $\mathbb{R}$ and $\mathbb{C}$ are known to form the only 3 possible division algebras over reals (see J3.1 for further details).

definite product: $x \cdot y = \sum x_j y_j$, or indefinite product[3]: $\langle x \mid y \rangle = x_0 y_0 - \sum_1^{n-1} x_j y_j$. The isometries of the Euclidian space form a group $\mathbb{E}_n$ of *rigid motions* of $\mathbb{R}^n$, generated by all *translations* $\{a: x \to x + a\}$, and *rotations (orthogonal matrices)* $\{U: {}^tUU = I\}$. So each transformation $\phi \in \mathbb{E}_n$ is of the form $\phi(x) = Ux + b$. The proof is outlined in problem 1. The tricky part is to show that any rigid motion ϕ is *affine*, $\phi(x) = Ax + b$, with some matrix A. Then orthogonality of A follows fairly straightforward. Thus group $\mathbb{E}_n$ becomes a subgroup of a larger *affine group*,

$$Aff_n = \{\phi(x) = Ax + b: A \in GL_n;\ b \in \mathbb{R}^n\}.$$

The $\mathbb{E}_n$-linear factors $\{A\}$ are either general orthogonal matrices, $A \in O(n)$, or special orthogonal, $A \in SO(n) = \{U: det\, U = 1\}$, if ϕ preserves the orientation.

In Minkowski space $\mathbb{M}^n$ the role of the orthogonal affine factors is taken by the *Lorentz (pseudoorthogonal)* group $SO(1; n-1)$, which consists of matrices B, that preserve the indefinite product $\langle x \mid y \rangle = Jx \cdot y$, where $J = \begin{pmatrix} 1 & & \\ & -1 & \\ & & -1 \end{pmatrix}$ - matrix of Minkowski form,

$$\{A: \langle Ax \mid Ay \rangle = \langle x \mid y \rangle\}.$$

So symmetries of the Minkowski space form the *Poincare group* $\mathbb{P}_n$ of Special Relativity. The latter is generated by all translations and all Lorentz transformations,

$$\phi(x) = Ax + b;\ b \in \mathbb{R}^n;\ A \in SO(1; n).$$

Let us notice that in all three cases (affine, Euclidian, Poincare) elements ϕ are

---

[2]We recall that a *Riemannian manifold* $\mathcal{M}$ carries a positive definite metric $g = \sum g_{ij} dx^i dx^j$, i.e. an inner product $\langle \xi \mid \xi \rangle_g = \sum g_{ij} \xi_i \xi_j > 0$, on tangent spaces of $\mathcal{M}$ $\{\xi \in T_x\}$. *Pseudo-riemannian* refers to the indefinite metric g, usually of the type $(+; -; -; ...)$. Any diffeomorphism (change of variables) $x = \phi(y)$, on a Riemannian (pseudo-Riemannian) manifold $\mathcal{M}$ transforms the metric: $g \to \tilde{g} = g^\phi = \phi'^*(g \circ \phi)\phi'$, the new entries being

$$\tilde{g}_{km}(y) = \sum (g_{ij} \circ \phi) \frac{\partial x^i\, \partial x^j}{\partial y^k \partial y^m};\ \text{where } \phi' = \left(\frac{\partial x^i}{\partial y^k}\right) \text{ denotes the Jacobian (matrix) of } \phi.$$

Map ϕ defines an *isometry* of $\mathcal{M}$, if the transformed metric g^ϕ is equal to g. Such maps ϕ obviously preserve the *length of any path* $\gamma = \{\gamma(t): 0 \le t \le T\}$,

$$L[\gamma] = \int_0^T \sum g_{ij}(\gamma(t))\, \dot{\gamma}_i \dot{\gamma}_j\, dt,$$

hence the distance between points, $d(x; y) = min\{L[\gamma]: \gamma(0) = x;\ \gamma(T) = y\}$. Conversely, for any distance preserving diffeomorphism $\phi: \mathcal{M} \to \mathcal{M}$, the differential ϕ'_x (Jacobian matrix of ϕ), considered as the map of tangent spaces, $\phi': T_x \to T_y$ $(y = \phi(x))$, preserves the metric (norm) on T_x, $\langle \phi'_x(\xi) \mid \phi'_x(\xi) \rangle = \langle \xi \mid \xi \rangle$, for any tangent vector ξ. Isometries of $\mathcal{M}$ clearly form a group. *Conformal maps* ϕ do not preserve metric, but multiply it with a scalar (conformal) factor $\rho(x)$, i.e. $g^\phi(x) = \rho(x) g$. So they form a larger symmetry group of $\mathcal{M}$.

[3]In Special Relativity $\mathbb{R}^4 = \{x_0; ...; x_3\}$ represents a simplified version of space-time, where x_0 serves as time variable, while $(x_1; x_2; x_3)$ represent space coordinates.

identified with pairs $(A;b)$, and the group multiplication takes the form

$$(A;b)\cdot(A';b') = (AA';Ab'+b). \tag{1.1}$$

It is easy to check that translations $\{b\}$ form a normal subgroup $H \simeq \mathbb{R}^n$ of G, while linear factors $\{A\}$, or $\{U\}$, form a subgroup $K \simeq \boldsymbol{GL}_n$, or $\boldsymbol{SO}(n)$; $\boldsymbol{SO}(1;n)$. So group G is decomposed into a *semidirect product*[4], $G = H \triangleright K$. The representation theory of semidirect products will be analyzed in chapters 3 and 6.

Two other examples of geometric symmetries arise on the *sphere* and the *hyperbolic (Poincare-Lobachevski) space*, the prototypes of the *spherical and hyperbolic geometries*. They also serve as the simplest prototypes of symmetric spaces of the compact and non-compact (hyperbolic) type.

b) *Sphere*: $S^{n-1} = \{x:\|x\|^2 = 1\}$ in $\mathbb{R}^n$ with the natural (Euclidian) metric has the isometry group $G = \boldsymbol{SO}(n)$ - all orthogonal transformations (rotations) in $\mathbb{R}^n$. In fact, S^{n-1} can be identified with the quotient (*coset*) space of group $G = \boldsymbol{SO}(n)$, modulo the *stabilizer (isotropy subgroup)*[5] $K = \boldsymbol{SO}(n-1)$ of a fixed point x_0 in S^{n-1} (the North Pole!), $S^n \simeq G/K$. In other words (definition of stabilizer), $K = \{U: U(x_0) = x_0\}$.

c) *Hyperbolic space* $\mathbb{H}$, can be realized either as Poincare-Lobachevski complex half-plane $\{z = x+iy: y>0\}$, with metric

$$ds^2 = \frac{dx^2+dy^2}{y^2} = \frac{dz\,d\bar{z}}{y^2},$$

or as complex disk $\mathbb{D} = \{|z|<1\}$ with metric, $ds^2 = \frac{dz\,d\bar{z}}{(1-|z|^2)^2}$. Both spaces are related by the *Möbius transformation,*

$$\sigma: z \to z^\sigma = i\frac{z+i}{z-i} = w.$$

One can easily check that σ takes $\mathbb{H}$ onto $\mathbb{D}$, and transforms their Poincare metrics, one into the other. The isometry group of $\mathbb{H}$ in both realizations is made of the fractional-linear transformations:

$$\phi: z \to w = \frac{az+c}{bz+d}. \tag{1.2}$$

---

[4]A *semidirect product* $G = H \triangleright U$, of groups H and U, where U acts by automorphisms on H, $u: x \to x^u$, $(x \in H; u \in U)$, consists of all pairs $g = (x,u)$ with the multiplication rule

$$(x,u)\cdot(y,v) = (x\cdot y^{u^{-1}}; uv).$$

It is easily seen that $H = \{(x;e)\}$ forms a normal subgroup of G, while $K = \{(e;u)\}$ a subgroup, so that $H \bigcap K = \{e\}$, and the whole group $G = H\cdot K$.

[5]Stabilizer (isotropy subgroup) of point x in a G-space X, consist of all elements $\{g \in G\}$, that leave x fixed, $K_x = \{g: x^g = x\}$.

In the half-plane case matrices $\left\{A = \begin{bmatrix} a & c \\ b & d \end{bmatrix} : ad - bc = 1\right\}$ belong to $SL_2(\mathbb{R})$ - the *real unimodular group*. In the disk realization they are given by complex matrices

$$\left\{A = \begin{bmatrix} \alpha & \beta \\ \bar{\beta} & \bar{\alpha} \end{bmatrix} : \det A = |\alpha|^2 - |\beta|^2 = 1\right\},$$

in other words matrices A that preserve the indefinite hermitian form: $J(z) = |z_1|^2 - |z_2|^2$, on $\mathbb{C}^2 = \{(z_1; z_2)\}$. This group is called *conformal* and denoted $SU(1;1)$, by analogy with the standard *unitary group* $SU(2)$, that preserves the positive hermitian product, $|z|^2 = |z_1|^2 + |z_2|^2$. Once again the hyperbolic space coincides with the quotient of $G = SL_2$ (or $SU(1;1)$), modulo stabilizer K of a fixed point $z_0 = i \in \mathbb{H}$. The stabilizer K is easily seen to coincide with an orthogonal group in $\mathbb{R}^2$, $K = \left\{A : \frac{ai + c}{bi + d} = i\right\} = SO(2)$, so $\mathbb{H} \simeq SL_2/SO(2)$.

Remark: The hyperbolic nature of the Poincare-Lobachevski geometry stems from its close connection to a hyperboloid $\Gamma = \{z^2 - (x^2+y^2) = 1\}$ in $\mathbb{R}^3$. The former represents an orbit of the Lorentz group $SO(2;1)$, acting by linear transformations of $\mathbb{R}^3 \simeq \mathbb{M}^3$. So the Lorentz metric $ds^2 = dx^2+dy^2 - dz^2$, restricted on Γ, remains invariant under $SO(2;1)$. Furthermore, its restriction, $ds^2 \mid \Gamma$ becomes positive-definite (since all Γ-tangent vectors $\xi = (x; y; z)$ are *space-like*, $x^2+y^2 - z^2 > 0$!). So Γ turns into a Riemannian manifold with a large symmetry group (symmetric space). In fact, $\Gamma \simeq SO(2;1)/SO(2)$, where $SO(2)$ acts as a stabilizer of the vertex (0;0;1) of Γ.

The connection between Γ and $\mathbb{D}$ is established via a *stereographic map* Φ (see fig.1). We parametrize Γ and $\mathbb{D}$ by polar coordinates in the xy-plane,

$$\{(r;\theta): z^2 = 1 + r^2\} \text{ for } \Gamma, \text{ and } \{(\rho,\theta)\} \text{ for } \mathbb{D}.$$

Then $\Phi: \rho \to r = \frac{2\rho}{1-\rho^2}$. So the metric on Γ:

$$ds^2 = \frac{dr^2}{1+r^2} + r^2 d\theta^2,$$

is taken into the Poincare metric in $\mathbb{D}$,

$$ds^2 = 4\frac{d\rho^2 + \rho^2 d\theta^2}{(1 - \rho^2)^2}.$$

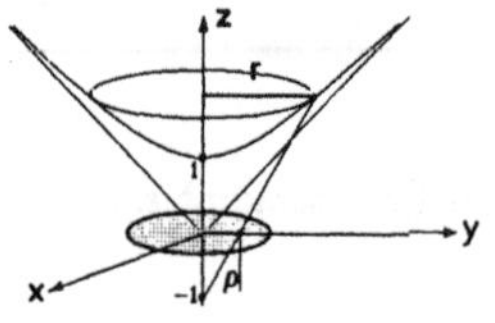

Fig. 1. *Stereographic map Φ takes a hyperboloid Γ: $z^2 - (x^2 + y^2) = 1$, in $\mathbb{R}^3 \simeq \mathbb{M}^3$, into the unit disk $\mathbb{D}$: $\rho^2 = x^2 + y^2 < 1$, in the xy-plane, and transforms the natural ($SO(2;1)$)-invariant metric on* Γ into the Poincare metric on $\mathbb{D}$.

The relation between Γ, $\mathbb{D}$ and $\mathbb{H}$ suggests that their groups of isometries should be identical. Indeed, we shall see (chapters 5,7) that $SL_2(\mathbb{R}) \simeq SU(1;1)$ makes a two-fold cover of $SO(2;1)$ (problem 5).

Thus we have exhibited 3 classes of manifolds with rich symmetry groups: Euclidian/Minkowski spaces, spheres and the hyperbolic plane. In all 3 cases group G acts *transitively* on $\mathcal{M}$ (each point is moved to any other by an element of G). Hence space $\mathcal{M}$ is identified with the quotient G/K, where K denotes a stabilizer of a point $x_0 \in \mathcal{M}$, namely

$\mathbb{R}^n \simeq \mathbb{E}_n/SO(n)$; $K = SO(n)$ - stabilizer of $\{0\}$;

$\mathbb{M}^n \simeq \mathbb{P}_n/SO(1;n-1)$ - stabilizer of $\{0\}$ in $\mathbb{P}_n$

$S^{n-1} \simeq SO(n)/SO(n-1)$ - stabilizer of $x_0 = (1;0;\ ...\ )$

$\mathbb{H} \simeq SL_2(\mathbb{R})/SO(2)$; or $\mathbb{D} = SU(1;1)/SO(2)$; $K = SO(2)$- stabilizer of $\{i\}$ (or $\{0\}$)

In fact in all three cases group G acts in a stronger *double-transitive* manner on $\mathcal{M}$, meaning that any pair of points $\{x;y\}$ can be transformed into any other equidistant pair $\{x';y'\}$, $d(x;y) = d(x';y')$. This fact has important implications for the analysis on such manifolds.

Let us remark that all the Riemannian examples above (except Minkowski) belong to a wide class of *symmetric spaces*. These manifolds can be generally described as quotients of Lie groups modulo maximal compact subgroups, $\mathcal{M} \simeq G/K$, and we shall see many more examples in subsequent sections.

1.2. Finite groups. The easiest to describe are *commutative finite group.* The complete list includes *cyclic groups:* $\mathbb{Z}_n \simeq \mathbb{Z}/n\mathbb{Z}$, which could also be written in the complex form: $\{e^{ik\omega}: 0 \le k \le n-1\}$, ($\omega = \frac{2\pi}{n}$ - the n-th primitive root of unity), and their direct sums: $G = \mathbb{Z}_{n_1} \times ... \times \mathbb{Z}_{n_m}$.

2) *Symmetries of regular polyhedra and finite subgroups of* $SO(3)$. We shall start with regular polygons in the plane.

i) *Symmetries of the regular n-gon* consist of rotations by angles $\{\frac{2\pi}{n}k\colon k = 0;1...\} \simeq \mathbb{Z}_n$ - a finite subgroup of all planar rotations $SO(2)$, or the larger *dihedral group,* $\mathbb{D}_n = \mathbb{Z}_n \triangleright \mathbb{Z}_2 \subset O(2)$ - a semidirect product of rotations and a reflection about any symmetry axis. $\mathbb{D}_n$ can be thought of as the symmetry group of the *dihedral* Δ_n, a solid in $\mathbb{R}^3$, built of 2 pyramids based on a regular polygon Ω_n in $\mathbb{R}^2$ and a pair of opposite vertices that project onto the center of Ω_n (see figure). Then $\mathbb{Z}_n$ implements axisymmetric rotations of the dihedral in the plane, while a generator of $\mathbb{Z}_2$ flips 2 opposite (spatial) vertices.

In 3-D one also has 5 Platonic perfect solids, each one with a symmetry group of orthogonal transformations in $\mathbb{R}^3$ that preserve its vertices:

ii) *tetrahedral symmetries*: $\mathbb{A}_4$ – *alternating group* (even permutations) of order 4. The generators of $\mathbb{A}_4$ are cyclic permutations of 3 vertices in each tetrahedral face, that leave the opposite vertex fixed.

iii) *cubo-octahedral symmetries*: the symmetric (permutation) group $\mathbb{W}_4$. Cube contains two opposite regular inscribed tetrahedra (fig.2), so its group contains an $\mathbb{A}_4$ plus an element σ that transposes both tetrahedra. Hence $G_{cube} = \mathbb{A}_4 \bigcup \sigma\mathbb{A}_4 = \mathbb{W}_4$.

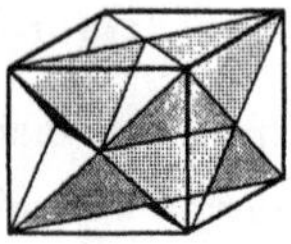

Fig.2 shows cube with 2 opposit inscribed tetrahedra.

iv) *Icoso-dodecahedral group*: $\mathbb{A}_5$ - alternating group of order 5 [Cox]. Indeed, a dodecahedron contains 5 inscribed tetrahedra $\{T_1;...;T_5\}$ (the set of 20 vertices is evenly split into 5 quadruples). A rotation of order 3 about any pair of opposit verteces leaves a pair of tetrahedra (say $T_1;T_2$) fixed and cyclicly permutes the remaining triple $(T_3;T_4;T_5)$. Obviously, even permutations, $\{(123);(234);(345); ...\}$ generate $\mathbb{A}_5$!

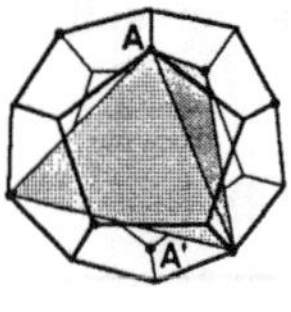

Fig.3 demonstrates one of 5 regular tetrahedrons, inscribed inside the dodecahedron. Any pair of opposit vertices $(A; A')$ selects a pair of tetrahedra, and any rotation of order 3 about the AA$'$-aixis, leaves the pair fixed, and cyclicly permutes the remaining triple.

We shall see now that (i-iv) completely describe all finite subgroups of the orthogonal group $SO(3)$.

Classification Theorem 1: *Any finite subgroup G of* $SO(3)$ *coincides with one of the above polyhedral groups (i)-(iv).*

The proof involves several steps.

1) We observe that any rotation $U \in SO(3)$ has two fixed points on the unit sphere $S^2 = \{\|x\| = 1\}$, indeed the eigenvalues of an orthogonal U in $\mathbb{R}^2$ are $\lambda = e^{\pm i\beta}; 1$!

2) Take set X of all G-fixed points on S^2. Set X is G–invariant, hence splits it into the union of G–orbits: $X = \omega_1 \bigcup \dots \bigcup \omega_m$. For each point $x \in X$ we consider its *stabilizer* $G_x = \{U: Ux = x\}$, and call $|G_x| = n_x$ - the *degree* of $\{x\}$. Obviously, all points on the same orbit have equal degrees (stabilizers are conjugate!), so $n_x = n(\omega_j) = n_j$, for $x \in \omega_j$.

3) Next we count the number of pairs $\{(x;U): x \in X; U \in G_x\}$ (the "total degree of G") in two ways: $\#\{$fixed points of all $U \in G \backslash \{e\}\}$ = "sum of degrees of all $\{x\}$". This yields,

$$2(|G|-1) = \sum_{x \in X} (|G_x| - 1) = \sum_1^m |\omega_j|(n_j - 1). \tag{1.3}$$

4) Dividing both sides of (1.3) by $|G|$ we derive the equation relating the order of G to degrees of its fixed points:

$$\boxed{2\left(1 - \frac{1}{|G|}\right) = \sum_1^m 1 - \frac{1}{n_j}} \tag{1.4}$$

The rest of analysis closely resembles a classification of Platonic solids. Namely, one can show that m in (1.4) could take on two values m = 2;3 only! Each case is analyzed separately.

5) In case $\boxed{m = 2}$, one easily shows $n_1 = n_2 = n = |G|$. So G has two 1-point orbits of degree n, which implies $G = \mathbb{Z}_n$- cyclic!

6) In case $\boxed{m = 3}$ there are 3 possible subcases:

Case $\boxed{n_1 = n_2 = 2;\ n_3 = n}$. Here G has two "n-point" orbits of deg = 2 and a "2-point" orbit of deg = n, which implies $G = \mathbf{D}_n$- the dihedral group!

Case $\boxed{n_1 = 2;\ n_2 = 3}$ yields 3 possibilities for n_3

$$n_3 = \begin{cases} 3;\ |G| = 12;\ G_{tetr}\ \text{(3 orbits of degree: 2;3;3)} \\ 4;\ |G| = 24;\ G_{cube}\ \text{(3 orbits of degree: 2;3;4)} \\ 5;\ |G| = 60;\ G_{dodec}\ \text{(3 orbits of degree:2;3;5)} \end{cases}$$

In a similar vein one can describe symmetries of regular polytopes in higher dimensions [Cox].

3. Other interesting classes arise as *automorphisms of finite groups,* and *regular* graphs. Among them we shall mention *finite groups of Lie type*: $G = GL_n(\mathbb{F})$, and $SL_n(\mathbb{F})$, made of $n \times n$-matrices with entries in a finite field $\mathbb{F}$. There many similarities

in the analysis and representation theory of classical and finite-type Lie groups, but our attention will be focused mostly on the continuous case.

1.3. Compact groups. These are *topological groups* with compact space G. The most important examples include *torus* $\mathbb{T}^n = \{(t_1;...t_n): t_j \in [0;1]\} \simeq \mathbb{R}^n/\mathbb{Z}^n$; the *classical compact Lie groups*, like orthogonal - $SO(n)$; unitary - $SU(n)$; symplectic - $Sp(n)$, et al. The general theory of the compact and finite groups will be developed in chapter 3. The classical compact Lie theory will be covered in detail in chapters 4-5.

Other interesting examples arise as matrix groups over *p-adic numbers* (p-prime): $\mathbb{Q}_p$, or integers $\mathbb{Z}_p$, i.e. the closure of rationals $\mathbb{Q}$, or integers $\mathbb{Z}$, in the p-adic norm: $\|a\|_p = p^{-n}$. Here n denotes the largest power (positive or negative) of prime p in fraction $a = p^n a'$. The corresponding *p-adic Lie groups*: $GL_n(\mathbb{Q}_p)$; $SL_n(\mathbb{Q}_p)$, and their compact subgroups $GL_n(\mathbb{Z}_p)$; $SL_n(\mathbb{Z}_p)$, consists of all $\mathbb{Q}_p(\mathbb{Z}_p)$-valued $n \times n$ matrices (respectively matrices of $det = 1$). These groups find applications in the number theory and algebraic geometry, but we shall not venture into the subject of the p-adic analysis (see [GGP]; [JL]).

1.4. Lie groups form the most important and interesting class, which plays the fundamental role in many areas of Mathematics and Physics: analysis, differential equations, geometry; classical, quantum, statistical mechanics. In general, Lie groups are defined as *manifolds* with smooth (differential) group structure: multiplication and inversion operations. A brief introduction to the Lie theory is provided in §1.4. The main bulk of the book will deal with the analysis and representations of Lie groups, emphasizing both general aspects of the Lie group theory as well as many specific examples and applications.

Lie groups could be divided into two large and distinct classes: *solvable and nilpotent* (a subclass of solvable); *simple and semisimple*. The definitions of both classes will be given in §1.4, and the detailed analysis conducted in chapter 6 (nilpotent and solvable groups), and chapters 4,5,7 (simple and semisimple groups). Although both classes differ substantially in their structure, there are abundant connections, and parallels in the harmonic analysis and representation theory.

The first class is exemplified by

i) *1-D affine group*: $Aff_1 = \{\phi_{a,b}: x \to ax + b\}$- all *affine transformations* of $\mathbb{R}$. This group can be also realized by 2×2 matrices of the form $\left\{A = \begin{pmatrix} a & b \\ & 1 \end{pmatrix}; b \in \mathbb{R}; a \in \mathbb{R}^*\right\}$.

ii) the celebrated *Heisenberg group* $\mathbb{H}_1$ (and its higher-D cousins $\mathbb{H}_n$), realized by 3×3 matrices

$$\mathbb{H}_1 = \left\{ A = \begin{pmatrix} 1 & a & c \\ & 1 & b \\ & & 1 \end{pmatrix} : a, b, c \in \mathbb{R} \right\}. \tag{1.5}$$

iii) The group B_n of upper/lower triangular matrices in GL_n (called often the *Borel subgroup*),

$$B_3 = \left\{ A = \begin{pmatrix} \lambda_1 & a & c \\ & \lambda_2 & b \\ & & \lambda_3 \end{pmatrix} : \lambda_j; a, b, c \in \mathbb{R}, \text{ or } \mathbb{C} \right\}.$$

Groups Aff_1 and B_3 *are solvable*, whereas $\mathbb{H}_1$ - nilpotent.

In §1.4 we shall see that any Lie group can be decomposed into a *semidirect product* $B \triangleright H$, of the solvable normal subgroup B, and a semisimple subgroup H. Any *semisimple* group H in turn can be decomposed into the direct product of *simple groups.* The latter were completely classified in celebrated works of E. Cartan. They comprise 4 series of *classical Lie groups*, listed below, and a few exceptional groups.

Many classical Lie groups arise as linear transformations of vector spaces over $\mathbb{R}$, $\mathbb{C}$ or quaternions $\mathbb{Q}$, that preserve certain bilinear/quadratic forms, in other words as subgroups of GL_n or SL_n. These include,

i) *Orthogonal groups:* $O(n)$, $SO(n)$, preserve Euclidian inner product $x \cdot y$, $O(n) = \{U : Ux \cdot Uy = x \cdot y\}$, and $SO(n) = \{U \in O(n) : \det U = 1\}$. There are also *indefinite orthogonal groups* $O(p;q)$ and $SO(p;q)$, that preserve *indefinite products:*

$$\langle x \mid y \rangle = \sum_1^p x_j y_j - \sum_{p+1}^{p+q} x_j y_j \text{ in } \mathbb{R}^{p+q}, \text{ or } \mathbb{C}^{p+q},$$

so that

$$\langle Ux \mid Uy \rangle = \langle x \mid y \rangle, \text{ for all } x, y.$$

The principal difference between the *definite-type* and *indefinite-type* groups in the real case is that the former are *compact*, while the latter are not, as exemplified by the $SO(n+1)$ and the Lorentz $SO(1;n)$, mentioned earlier. But in the complex case the difference disappears, so $SO(p;q) \simeq SO(p+q)$, for p, q.

ii) *Unitary groups*: $U(n)$, $U(k;m)$, preserve definite (or indefinite) hermitian inner product in $\mathbb{C}^n$, $z \cdot \overline{w}$, or

$$\langle z \mid w \rangle = \sum_1^k z_j \overline{w}_j - \sum_{k+1}^{k+m} z_j \overline{w}_j; \text{ in } \mathbb{C}^{k+m}.$$

A particular example is the conformal group $SU(1;1)$.

iii) *Symplectic groups: $Sp(n)$* consist of all $2n \times 2n$ matrices that preserves a skew-symmetric form $J = \begin{bmatrix} & I \\ -I & \end{bmatrix}$,

$$\langle z \mid z' \rangle = x \cdot y' - x' \cdot y = Jz \cdot z'; \; z = (x, y).$$

In all cases the corresponding group consists of matrices, that satisfy

$$U^*JU = J,$$

where U^* denotes the transpose (or hermitian adjoint) of U, and J is either identity (for the orthogonal and unitary groups), or the symmetric form I_{km} with k pluses and m minuses on the main (for the indefinite orthogonal and unitary groups), or the skew symmetric form J in the symplectic case.

The above list contains most of the classical examples, but it does not represent their classification scheme (see chapter 5), our emphasis was mainly on *construction*. The reader will find (problems 5,6), that 3 series overlap, particularly in low dimensions. More examples of this nature will come in §1.4 and subsequent chapters.

1.5. Discrete groups. Those typically arise as discrete subgroups of continuous (Lie) groups, like *lattices* $\Gamma \simeq \mathbb{Z}^n$ in $\mathbb{R}^n$, or their noncommutative counterparts, *lattices* $\Gamma \subset \mathbf{GL}_n$; $\mathbf{SL}_n$, and other Lie groups. Important examples of the noncommutative lattices are

i) the *unimodular group* $\Gamma = \mathbf{SL}_n(\mathbb{Z})$, and some related subgroups of $\mathbf{SL}_n$

ii) *discrete Heisenberg group*: matrices (1.5) with integer $a, b, c \in \mathbb{Z}$.

iii) discrete subgroups Γ of the Euclidian motion group $\mathbb{E}_n$.

The latter are called *crystallographic groups*, as they describe all possible crystalline arrangements in Euclidian spaces. They were thoroughly investigated, and classified in dimensions 2 and 3. As $\mathbb{E}_n$ itself such groups are decomposed into a semidirect product, $\Gamma = \Lambda \triangleright U$, where $\Lambda \simeq \mathbb{Z}^n$ forms a lattice in $\mathbb{R}^n$, while a finite group $U \subset \mathbf{GL}_n$ acts by automorphisms (linear transformations) of Λ.

In 2-D the complete list contains 17 groups, whose U-components could take only 10 possible values: $\{\mathbb{Z}_m;\mathbb{D}_m: m = 1;2;3;4;6\}$. The complete list in 3-D includes 219 nonisomorphic and 230 nonconjugate crystallographic groups. It was worked out by Fedorov, Schoenfliess and Barlow at the turn of the last century. As a step towards classification one needs all finite orthogonal symmetries, derived in Theorem 1. For further details and relates issues we refer to [BH]; [HC]; [Sch].

Problems and Exercises:

1. Show that any rigid motion on $\mathbb{R}^n$ (Euclidian or Minkowski) is linear, $\phi(x) = Ax + a$, for some $a \in \mathbb{R}^n$, and matrix A. Follow steps:

   i) The Jacobian $\phi'(x)$ is an orthogonal matrix, i.e. $\partial_i\phi \cdot \partial_j\phi = \delta_{ij}$

   ii) Differentiate the orthogonality relation (i) in the k-th variable, $\partial_k(\ldots)$, then dot-multiply with $\partial_m\phi$. Show the resulting 4-tensor $t_{ik}^{mj} = \langle \partial_m\phi \;\; \partial_{ik}^2\phi \;\; \partial_j\phi \rangle$ to be symmetric in both row-indexes: $t_{ik} = t_{ki}$; $t^{mj} = t^{jm}$; and antisymmetric in column-indexes (changing $i \leftrightarrow j$ or $k \leftrightarrow m$ changes its sign).

   iii) Show that any such t must be at once symmetric and antisymmetric in the pair of indexes $\{ij\}$. Hence $t = 0$, for all quadruples $ijkm$, which implies $\partial^2\phi = 0$, i.e. ϕ is linear!

2. The isometry group of S^{n-1} consist of rotations $\{U \in SO(n)\}$. Hint: any isometry of S^{n-1} extends to an isometry of $\mathbb{R}^n \backslash \{0\} \simeq S^{n-1} \times \mathbb{R}_+$. But the latter are linear by problem 1!.

3. Show that fractional-linear transformations (1.2) of SL_2 on $\mathbb{H}$, or $SU(1;1)$ on $\mathbb{D}$, are isometries.

4. Show that all finite subgroups of $SO(3)$ are either polyhedral, or polygonal groups, described in Example 1.1.

5. Establish and find explicit form of the isomorphism $SL_2(\mathbb{R}) \to SU(1;1)$, using their fractional-linear actions on $\mathbb{H}$ and $\mathbb{D}$, and the *Caley transformation* ϕ: $z \to \frac{z-i}{z+i}$, from $\mathbb{H}$ to $\mathbb{D}$. Notice that both are subgroups of the larger complex unimodular group $SL_2(\mathbb{C})$, and can be obtained by conjugation, $u \to \sigma^{-1}u\sigma$, with a Caley element $\sigma \in SL_2(\mathbb{C})$. Find σ!

6. Show that the symplectic group $Sp(1)$ coincides with SL_2.

7. Show that complex groups $SO(p;q)$ and $SO(p+q)$ are isomorphic (Hint: the indefinite product $\langle z \mid w \rangle$ in $\mathbb{C}^{p+q}$ is equivalent to the definite product $z \cdot w$ by conjugation with a complex diagonal matrix).

§1.2. Regular and induced representations; Haar measure and Convolution algebras.

We introduce two basic concepts of regular and induced representations, discuss continuity and unitarity and develop some basic algebraic constructions: direct sum, direct integral, tensor product. In the process we introduce the (invariant) *Haar measure* on groups and homogeneous (coset) spaces $H\backslash G$, define *convolution (group) algebras* $L^1(G)$, and find the links between representations of the group and those of its group-algebra.

2.1. Regular representations: In the last section we have described some examples of transformation groups acting on various geometric structures (graphs, manifolds, symmetric spaces), including the action of group on itself by the right/left multiplication (*translation*),

$$g\colon x \to xg;\ or\ x \to g^{-1}x.$$

With any such action of group G on space X (denoted by $x \to x^g$), we can associate a *linear G-action* on functions over X, $\{f(x)\} = \mathfrak{C} = \mathfrak{C}(X)$, i.e. a homomorphism of G into linear operators over $\mathfrak{C}$,

$$\boxed{R_g\colon f(x) \to f(x^g);\ f \in \mathfrak{C}(X)}. \tag{2.1}$$

In particular, the right/left translations, $x \to xg$; $x \to g^{-1}x$, on G give

$$\boxed{R_g f(x) = f(g^{-1}x)\ \text{(left), or}\ f(xg)\ \text{(right), on}\ \mathfrak{C}(G)} \tag{2.2}$$

Homomorphisms $g \to T_g$, of group G into linear transformations/matrices on a vector space $\mathcal{V}$, are called *representations*, and formulae (2.1)-(2.2) provide important examples of *regular representations* of G on X, or on itself. In the analysis of group actions and associated representations on various function-spaces, it is often necessary to "integrate" over G or X. So we need some measures on X and G.

2.2. The Haar measure. It turns out that all discrete and continuous (locally compact) groups G, as well as large classes of *homogeneous spaces*[6] $X = H\backslash G$ have an *invariant measure*,

$$d\mu(x^g) = d\mu(x),\ \text{for all}\ x \in X,\ g \in G,$$

called the *Haar measure*. The general proof will be outlined below. More important, however, will be to compute the Haar measure in a suitably chosen coordinate system on group G or space X. Here we shall list a few examples of Haar measures on groups and homogeneous spaces (many more will appear throughout the book).

---

[6] homogeneous (quotient) space $X = H\backslash G$, or G/H, consists of all right/left cosets $\{x = Hg_0 : g_0 \in G\}$, or $\{x = g_0H : g_0 \in G\}$. Group G acts on X by right/left translations, $g\colon x \to x^g = (Hg_0)g$, or $x \to g^{-1}(x)$. So X could be viewed as a G-space, with a transitive G-action, and subgroup H coincides with the stabilizer of a fixed point, $H = \{g : x_0{}^g = x_0\}$.

a) finite/discrete group G has $d\mu(x) = \sum_y \delta_y(x)$- sum of δ-functions over all $y \in G$.

b) for commutative groups $\mathbb{R}^n;\mathbb{T}^n$, $d\mu$ coincides with the standard Lebesgue measure (volume element), $dx = dx_1 dx_2 ... dx_n$. We have shown in §1.1 that $\mathbb{R}^n$ could be regarded as a homogeneous space of the Euclidian motion group $G = \mathbb{E}_n$. The Lebesgue measure is clearly invariant under all translations and rotations on $\mathbb{R}^n$, hence it also forms a G-invariant measure on homogeneous space $\mathbb{R}^n = \mathbb{E}_n/SO(n)$.

c) on compact Lie groups $SO(n)$, $SU(n)$, the Haar measure can be explicitly calculated in terms of suitably chosen coordinates, like Euler angles (chapters 3-4). The same could be done on non-compact (Lorentz, conformal, etc.) groups, like $SO(1;n)$ or $SU(1;n)$, by a combination of spherical and hyperbolic angles.

d) the hyperbolic (Poincare-Lobachevski) half plane $\mathbb{H}$ is a homogeneous space of $G = SL_2(\mathbb{R})$. One can check (problem 1), that measure $d\mu = \frac{dx\,dy}{y^2}$ is G-invariant.

e) $GL_n(\mathbb{R})$ has a natural set of coordinates, matrix entries $\{x_{jk}\}_{jk=1}^n$ of g. One can show (problem 6) the invariant measure on GL_n

$$d\mu(g) = \frac{dx}{(\det x)^n} = \frac{1}{(\det x)^n} \prod_{jk} dx_{jk}.$$

For more examples see problems 5-8.

Existence Theorem: i) *On any compact (and more general locally compact) group there exists a Borel measure $d\mu(x)$, positive on all open subsets, finite on all compact subsets and invariant under all right (or left) translations: $\mu(Ex) = \mu(E)$, for all subsets $E \subset G$ and elements $x \in G$.*

ii) A right and left Haar measures: $d_r x$; $d_l x$ are unique up to a constant factor.

iii) On a compact group G the left and right Haar measures are equal, furthermore $d\mu$ is invariant under the group inversion and conjugation,

$$d\mu(x^{-1}) = d\mu(x),\ d\mu(g^{-1}xg) = d\mu(x).$$

Let us briefly outline the proof.

Existence: To construct a right-invariant measure $d\mu$, we pick a small neighborhood U of the identity $e \in G$ to serve as a gauge. Given a (open, closed) subset $E \subset G$, we choose optimal (least) covers of E and G by translates of U,

$$E \subset \bigcup_{j=1}^m Ux_j,\quad G = \bigcup_{j=1}^n Uy_j.$$

The ratio $\frac{m}{n}$ represents an approximate relative size of E in G in the "U-gauge".

Taking limit $\frac{m(E;U)}{n(G;U)}$, as $U \to \{e\}$, (by standard arguments such limit always exists!), one gets an honest Borel measure on G. The limiting measure, $d\mu$, thus constructed is easily seen to be right-invariant. Also $\mu(E) > 0$ for any open E, and $\mu(G) < \infty$ for a compact G, since G is covered by finitely many translates of E.

Uniqueness: If $d\tilde{\mu}$ is another right-invariant measure, then for small U, $m(E,U) \approx \frac{\tilde{m}(E)}{\tilde{m}(U)}$, while $n(G;U) \approx \frac{\tilde{m}(G)}{\tilde{m}(U)}$, whence $\mu(E) = \lim_{U \to \{e\}} \frac{m}{n} = Const\ \tilde{m}(E)$, for all subsets E.

iii) To show that the right and the left-invariant Haar measures on a compact G are equal, we take a left translate, $d\tilde{\mu}(x) = d\mu(ax)$, of a right-invariant measure $d\mu(x)$. Of course, $d\tilde{\mu}$ is also right-invariant. Hence $d\tilde{\mu} = \rho(a)d\mu$, by the uniqueness of $d\mu$. The map $a \to \rho(a)$ is a homomorphism of G into the multiplicative group $\mathbb{R}_+$. But there are no such nontrivial *continuous* homomorphisms on a compact group (continuous functions on compact sets are always bounded, whereas nontrivial homomorphisms ρ: $G \to \mathbb{R}_+$ are unbounded!). Hence, $\rho = 1$, and $d\mu$ is also right-invariant.

Two other invariance properties of $d\mu$ easily follow now. Indeed, change of variable $x \to x^{-1}$ takes a right-invariant measure into a left-invariant measure, the conjugate-invariance immediately follows from the bi-invariance: $d\mu(axb) = d\mu(x)$ for all a, b. This completes the proof.

Remark: In many cases (e.g. Lie groups) the right and left invariant measures are absolutely continuous one relative to the other. The corresponding density $\frac{d_l(g)}{d_r(g)} = \Delta(g)$ is a *character* (homomorphism into the numbers) of G with positive real values

$$\Delta: G \to \mathbb{R}_+;\ \Delta(gh) = \Delta(g)\Delta(h),$$

called the *modular function.* Groups with equal right and left Haar measures, $\Delta(g) = 1$, are called *unimodular.* Large classes of groups are known to be unimodular, for instance, all compact groups, semisimple and nilpotent Lie groups. Nonunimodular examples include: affine groups and more general Borel groups of all upper/lower triangular matrices (problem 8).

The invariant (Haar) measure, topology/metric, and the differential structure on group G allow to introduce a variety of function-spaces. The most important among those are

- $\mathcal{C}(G)$ - continuous (bounded) functions $\{f(x)\}$ on G, with norm

$$\| f \|_\infty = \sup_{x \in G} | f(x) |,$$

and its subspaces $\mathfrak{C}_0$ - functions vanishing at $\{\infty\}$, and $\mathfrak{C}_c$ - compactly supported functions;

• L^p-spaces on G with respect to the Haar measure dx,

$$\{f(x): \|f\|_p = \left(\int |f(x)|^p dx\right)^{\frac{1}{p}} < \infty\};\ 1 \le p \le \infty,$$

in particular Hilbert space $L^2(G)$ with the standard inner product

$$\langle f_1 | f_2 \rangle \gg \int f_1 \bar{f}_2 \, dx.$$

• For Lie groups G (*smooth manifold*) one can consider a variety of *differentiable function-spaces:* $\mathfrak{C}^m(G)$ - *"m-smooth functions"*, $\mathfrak{C}^\infty(G)$ - *infinitely smooth functions*; their subspaces of functions vanishing at $\{\infty\}$, or compactly supported $\mathfrak{C}_0^m$; $\mathfrak{C}_c^m$, with norm

$$\|f\|_m = \sum \sup_{x \in G} |\partial^\alpha f(x)|,$$

sum over all "partial derivatives" of order $|\alpha| \le m$ (the meaning of "derivatives") on Lie groups will be discussed later in §1.4).

Then there are other versions of smooth functions, defined via L^2-norms on G, instead of sup-norm, called *Sobolev spaces* $\{\mathfrak{H}_m(G): m \in \mathbb{Z}\}$, or $\{\mathfrak{H}_s(G): s \in \mathbb{R}\}$ (the latter exploit 'fractional powers' of the Laplacian on G). We shall discuss them at a proper place.

Finally, associated to smooth functions are various spaces of *distributions* on G (see §2.2). Let us remark that all function-spaces $\mathcal{V} = \mathfrak{C}$, $\mathfrak{C}^m$, L^p, etc., are invariant under right/left translations, $f \to R_u f(x) = f(xu)$, or $R_u f = f(u^{-1}x)$.

2.3. Convolution: The group multiplication extends to function spaces on G and renders them an algebraic structure. The resulting product is called *convolution* of functions, it is defined by

$$\boxed{(f_1 * f_2)(x) = \int_G f_1(xy^{-1}) f_2(y)\, dy = \int_G f_1(y)\, f_2(y^{-1}x)\, dy} \tag{2.3}$$

integration with respect to the Haar measure dy on G. On finite/discrete groups integration in (2.3) is replaced by the sum,

$$f * h = \sum_{y \in G} f(xy^{-1})\, h(y).$$

The meaning of *convolution,* as an extension of the group multiplication to functions, becomes transparent here, as each group element $\{x\}$ is identified with the delta-function δ_x, so that

$$\delta_x * \delta_y = \delta_{xy}; \text{ for all } x, y \in G.$$

For continuous groups G convolution-integral (2.3) is well defined on compactly supported functions ($\mathfrak{C}_c$;$\mathfrak{C}_c^m$), then it extends to larger classes by the standard density arguments[7]. Spaces closed under convolution forms *convolution/group algebras.* One important example is space $L^1(G)$ of all integrable functions on G. Here,

$$\|f*h\|_1 \le \|f\|_1 \|h\|_1; \text{ for all pairs } f, h \in L^1. \tag{2.4}$$

A similar estimate holds for a convolution of L^1 and L^p-functions,

$$\|f*h\|_p \le \|f\|_1 \|h\|_p; \text{ for all pairs } f \in L^1,\ h \in L^p. \tag{2.5}$$

In the standard terminology L^1 is a *Banach algebra*, while other L^p are *modules* over L^1.

Estimates (2.4)-(2.5) follows from the well known *Minkowski inequality*,

$$\left(\int \left| \int F(x;y)\,dy \right|^p dx\right)^{1/p} \le \int \left(\int |F(x;y)|^p dx \right)^{1/p} dy, \text{ for any } F(x,y), \tag{2.6}$$

applied to function $F = f(xy^{-1})\,h(y)$. The inequality (2.6) has an obvious interpretation, any function $F(x;y)$ $(x \in X; y \in Y)$ defines a map $F: Y \to L^p(dx)$, then (2.6) claims: $\| \int F(y)\,dy \| \le \int \|F(y)\|\,dy$, where $\|\ \|$ means the $L^p(dx)$-norm. Other examples of convolution-algebras include spaces of continuous and differentiable functions: $\mathfrak{C}$, $\mathfrak{C}^m$, $\mathfrak{C}^\infty$, as well as bounded (Borel) measures on G, $\mathcal{M}(G) \supset L^1(G)$. Condition (2.5) can be stated as *boundedness* of the convolution (bilinear) map,

$$L^1 * L^p \to L^p, \text{ or } L^1 * L^p \subset L^p,\ 1 \le p \le \infty.$$

It is also easy to see that $L^2 * L^2 \subset L^\infty$, and $\|f*h\|_\infty \le \|f\|_2 \|h\|_2$ (Cauchy-Schwartz). Such results can be extended to many other triples: $L^p * L^q \subset L^r$, and other "scales of function-spaces" by *interpolation* discussed in chapter 2 (J2.2).

2.4. Continuity and unitarity. The study of group representations combines two aspects: *algebraic*, which reflects purely algebraic properties of groups and operators, and *analytic (topological)*, which involves the geometric structure of manifolds $\mathcal{M}$ (differentiable, complex, etc.), as well as topological properties of the relevant vector (function) spaces and operators over $\mathcal{M}$.

In the topological context (locally compact/Lie groups G, Banach or Hilbert spaces $\mathcal{V}$) one needs to impose some continuity assumptions on T. One possibility would

[7] Spaces $\mathfrak{C}_c$ and $\mathfrak{C}_c^m$ are dense in all L^p, $1 \le p < \infty$!

be to require continuity of the map $g \to T_g$ in the *operator (norm) topology*[8] on space $\mathfrak{B}(\mathcal{V})$, of all bounded operators on $\mathcal{V}$. This notion, however, turns out to be too restrictive, (see problem 2, and the author's paper [9], JFA, 35, 1980).

A weaker notion of continuity asks vector functions $\{T_g\xi : G \to \mathcal{V}\}$, or the *matrix entries* of T, $\{t_{\xi\eta}(g) = \langle T_g\xi \mid \eta\rangle : \xi \in \mathcal{V}; \eta \in \mathcal{V}^*\text{-dual space}\}$ to be continuous. It is called *strong continuity*. Strong continuity holds in most natural examples, including regular representations (2.1), (2.2), and (2.8) below, in different function-spaces: $\mathfrak{C}$; $\mathfrak{C}^m$; $L^p(G)$, etc. (both on G and on homogeneous spaces $X = H\backslash G$, G modulo a closed subgroup H). Let us remark that quotient-spaces X of topological groups often inherit basic structures of G, like metric, differential structure, and in some cases G-invariant (*Haar*) measure.

An important class of group representations consist of *unitary representations* in Hilbert (inner product) spaces: ,

$$\langle T_g\xi \mid T_g\eta\rangle = \langle \xi \mid \eta\rangle, \text{ or } \|T_g\xi\| = \|\xi\|, \ \textit{for all } \xi,\eta \in \mathcal{H},\ g \in G.$$

The natural example is furnished by the regular representation R on square-integrable functions $\mathcal{V} = L^2(X)$, provided space X has a G-invariant *Haar measure*, $d(x^g) = dx$, for all $g \in G$. The *Haar measure* was shown to exist on any locally compact group G, but quotient spaces $X = H\backslash G$ may or may not have it, as we shall demonstrate. The invariant measure always exists, if subgroup H is compact, e.g. on symmetric spaces. However, in any case a homogeneous space (manifold) X comes equipped with some measure (volume element) dx. Group elements $\{g \in G\}$, acting on X transform dx into another measure dx^g, which will differ from dx by a *density factor*: $\alpha(x;g) = \frac{dx^g}{dx}$. In case of the smooth (Riemannian) manifolds and G-actions (e.g. symmetric space X), α represents the Jacobian determinant of the map $\phi : x \to x^g$,

$$\alpha(x;g) = \det \phi'(x). \tag{2.7}$$

We can still construct a unitary representation of G on $L^2(X;dx)$ by combining the group (translational) action on X with multiplication by $\sqrt{\alpha}$,

$$\boxed{T_g f(x) = \sqrt{\alpha(x;g)}\, f(x^g)}. \tag{2.8}$$

Function $\alpha(x;g)$ is easily seen to obey the so called *cocycle condition*,

---

[8]We remind the reader that the space of all bounded linear operators in a normed vector space $\mathcal{V}$ is equipped with an *operator-norm*: $\|A\| = \sup\{\|A\xi\| : \text{all } \|\xi\| \le 1\}$, or $\sup\{\langle A\xi \ \eta\rangle : \text{all } \xi;\eta\}$ in the Hilbert space setup.

[9]We have proved that *any norm-continuous irreducible T must be finite dimensional,* which automatically excludes large and important classes of ∞-D representations.

$$\alpha(x;g_1g_2)=\alpha(x;g_1)\,\alpha(x^{g_1};g_2);\ \text{all } g_1;g_2\in G, \tag{2.9}$$

which yields the representation property for operators T_g (2.8),

$$T_{g_1g_2}=T_{g_1}T_{g_2},\ \text{for all } g_1;g_2 \text{ in } G.$$

Obviously operators $\{T_g\}$ preserve the L^2-norm of f, so we get once again a unitarity representation of G on $L^2(X)$!

2.5. Induced representations. Regular representations (2.2)-(2.1) on G and quotients $H\backslash G$, as well as more general representation (2.8) are obtained by the process called *induction*. In general, *induction* allows to construct representations of group G, starting from representations of its subgroup H. An alternative definition is based on the notion of *cocycle* $\sigma(x;g)$ on the homogeneous space $X=H\backslash G$, i.e. a function

$$\sigma: X\times G\to \{\text{numbers}\} \text{ or } \{\text{operators}\},$$

that satisfies the *cocycle condition* (2.9)

$$\sigma(x;g_1g_2)=\sigma(x;g_1)\,\sigma(x^{g_1};g_2);\ \text{for all } g_1;g_2\in G,\ x\in X.$$

The *induced representation*, $T=ind(\sigma\mid H;G)$, acts on the space of ($\mathfrak{V}$-valued) functions[10] on X,

$$\mathfrak{C}=\mathfrak{C}(X;\mathfrak{V})=\{f(x): X\to\mathfrak{V}\},$$

and is given by

$$\boxed{T_gf(x)=\sigma(x;g)[f(x^g)]}\ f\in\mathfrak{C}. \tag{2.10}$$

We leave to the reader, as an easy exercise, to check that (2.10) does give a representation of G. Furthermore, if σ is continuous/unitary and X has a Haar measure dx, then T_g also becomes a continuous/unitary representation[11] of G on space $L^2(X;\mathfrak{V})$, of square-integrable $\mathfrak{V}$-valued functions on X (problem 9). Let us remark that the cocycle condition (2.9) for elements of stabilizer $H=G_x=\{g: x^g=x\}$, yields a representation $\sigma(h)=\sigma(x;h)$ of H on $\mathfrak{V}$, $\sigma(h_1h_2)=\sigma(h_1)\sigma(h_2)$. So one could start with a representation σ of H and construct a cocycle $\sigma(x;g)$! This alternative way to define *ind* will be discussed in §3.2.

Regular representations R on G and quotients $X=H\backslash G$, are induced by the trivial cocycle σ,

$$R=ind(1\mid H;G).$$

[10] In the topological (Lie) setup one takes a suitable function-space on X, e.g. *continuous, differentiable, L^2-functions*, etc.

[11] In the absence of invariant measure on X we can modify operators T_g (2.10) by a scalar factor $\sqrt{\alpha(x;g)}$ (2.7) to make T unitary. Obviously, the product of two cocycles σ and $\sqrt{\alpha}$ makes another cocycle!

We shall see the induction procedure to appear throughout this book in various places and contexts, in the representation theory of finite, compact, simple, nilpotent, solvable groups alike. The role of induction will be twofold: on the one hand it extends the notion of regular representations on G and X to vector functions and *sections of vector bundles*[12], on the other hand many *irreducible representations* of groups will be constructed via induction.

Here we shall illustrate the induced action by two examples of homogeneous spaces $H\backslash G$, without G-invariant measure (see problem 4), and construct the relevant regular/induced *unitary* representations.

1) Affine group $G=\{g=(a;b): x \to ax+b=x^g\}$ of §1.1 acts transitively on $\mathbb{R} \simeq H\backslash G$, where H denotes the stabilizer of $\{0\}$, $H=\{(a;0)\} \simeq \mathbb{R}^*$ (multiplicative group of reals). The "$ax+b$" action on $\mathbb{R}$ has no G-invariant measure. Indeed, the translational invariance ($\{b\}$-part) requires dμ to be Lebesgue, but multiplication with the $\{a\}$-factors dilate it, d$\mu \to |a|\, d\mu$. However, we get the cocycle

$$\alpha(x;g)=\frac{dx^g}{dx}=|a|, \tag{2.11}$$

hence a unitary action of G on $L^2(\mathbb{R})$,

$$T_g f(x)=|a|^{1/2} f(ax+b).$$

The representation theory of affine groups will be discussed in chapter 6 (§6.1).

2) Group $G=SL_2(\mathbb{R})$ and its subgroup of upper-triangular matrices $H=\left\{\begin{pmatrix} a & b \\ & a^{-1}\end{pmatrix}\right\}$ yield the quotient-space $X \simeq \mathbb{R}$, with G acting by the fractional linear transformations,

$$g: x \to \frac{ax+b}{cx+d}, \text{ for } g=\begin{pmatrix} a & c \\ b & d\end{pmatrix}. \tag{2.12}$$

Indeed, cosets $x \in H\backslash G$ can be parametrized by matrices $\left\{\gamma_x=\begin{pmatrix} 1 & \\ x & 1\end{pmatrix}: x \in \mathbb{R}\right\}$ - *coset representatives*, via the Gauss factorization: *(almost) any element* $g=\begin{pmatrix} a & c \\ b & d\end{pmatrix}$, $(d \neq 0)$ *factors into the product,*

$$\begin{pmatrix} a & c \\ b & d\end{pmatrix}=\begin{pmatrix} d^{-1} & c \\ & d\end{pmatrix}\begin{pmatrix} 1 & \\ b/d & 1\end{pmatrix}=h\,\gamma.$$

To find the cocycle we take the product $\gamma_x g$, and factor out the h-term,

$$\gamma_x \cdot g=\begin{pmatrix} a & b \\ ax+b & cx+d\end{pmatrix}=h \cdot \gamma,$$

where

$$h=h(x;g)=\begin{pmatrix} (cx+d)^{-1} & b \\ & cx+d\end{pmatrix}; \text{ and } \gamma=\gamma_{x^g}=\begin{pmatrix} 1 & \\ \frac{ax+b}{cx+d} & 1\end{pmatrix}.$$

[12] A vector bundle $\mathfrak{B}$ consists of the base manifold X, stuffed with *fibers* (linear vector spaces) $\{\mathcal{V}_x\}$ associated to points of $x \in X$. The natural example is the tangent bundle, made of all tangent spaces $\{T_x: x \in X\}$ of a smooth base-manifold.

The density factor α is computed from the transformation law (2.12) to be

$$\alpha(x;g) = |cx+d|^{-2}. \tag{2.13}$$

So a unitary representation of SL_2 on space $L^2(\mathbb{R})$ takes the form,

$$\boxed{T_g f(x) = |cx+d|^{-1} f(\tfrac{ax+b}{cx+d})}.$$

More general representations of $SL_2(\mathbb{R})$ on $L^2(\mathbb{R})$ are obtained by taking complex powers of the cocycle,

$$\boxed{T_g^s f(x) = |cx+d|^{-1+is} f(\tfrac{ax+c}{bx+d})}. \tag{2.14}$$

These are so called ***principal series representations*** of SL_2, that will be studied in detail in chapter 7.

2.6. Representations of group algebras. Any group representation T_g gives rise to a representation of a suitable group-algebra $\mathcal{L} = \{f\}$, of functions on G, by integrating T against f,

$$\boxed{T_f = \int_G f(g) T_g \, dg} \tag{2.15}$$

It is easy to check that map $f \to T_f$ satisfies

$$T_{f*h} = T_f T_h, \text{ for any pair } f, h \in \mathcal{L}.$$

Conversely, any algebra-representation $f \to T_f$, subject to a minor technical assumption (span$\{T_f(\mathcal{V}) : f \in \mathcal{L}\}$ is dense in $\mathcal{V}$), can be shown to come from a group representation $g \to T_g$ via (2.15). As usually, in the topological context one has to give a meaning to integration (2.15) and to find a proper class of functions $\{f\}$ on G. If T is continuous (which is always the case), and f - compactly supported, integral T_f obviously exists and yields a bounded operator,

$$\|T_f\| \leq \int |f(x)| \|T_x\| dx. \tag{2.16}$$

Estimate (2.16) also indicates the class of functions, for which T_f can be extended, as a bounded operator. The latter depends on the behavior of function $w(x) = \|T_x\|$ on G. Clearly T_f extends to the space (algebra) of functions, integrable with *weight* w, $\{f: \int |f(x)| w(x) dx < \infty\}$. In particular, unitary (or more general bounded) representations, $\|T_g\| \leq C$, yield bounded operators $\{T_f\}$ for all f in the group algebra $\mathcal{L} = L^1(G)$, indeed, $\|T_f\| \leq C \|f\|_1$.

If T is a left regular representation (2.2) on G, then operators (2.15) are nothing but left convolutions with f,

$$R_f[h] = f*h.$$

The main advantage of passing from the group representation T_g to its integrated form (2.15) has to do with nice (smoothing/regularizing) properties of operators $\{T_f\}$. To wit the "unitary/bounded *group operators*" $\{T_g\}$ could yield a better class of "algebra operators" $\{T_f\}$: *compact, Hilbert-Schmidt,* or *trace-class.* The latter often depends on the regularity properties of $\mathcal{L} = \mathfrak{C}^m(G);\mathfrak{C}^\infty(G)$, the "smoother" $\{f\text{'s}\}$ one takes the "better" $\{T_f\}$ results.

A simple illustration is furnished by convolutions with $f \in \mathfrak{C}^m(G)$ (or $\mathfrak{C}^\infty$). It is easy to check (problem 11), that any such operator

$$R_f: L^p \to \mathfrak{C}^m\ (\mathfrak{C}^\infty),\ \text{i.e. } \mathfrak{C}^m * L^p \subset \mathfrak{C}^m,\ \text{for all } p, m.$$

In other words, convolutions with m-smooth $\{f\}$ transform L^p-singular functions $\{h\}$ into m-smooth functions. Smoothing properties of convolution operators with nice (smooth) functions are well known and frequently utilized techniques in harmonic analysis and differential equations, that go under the name of *regularization methods.* We shall use them extensively in subsequent sections, but here we just give a simple illustration of *smoothing* by convolutions in problem 10.

2.7. Algebraic constructions. We shall conclude this section with a few *algebraic constructions* of representations.

• *Direct sum* of $\{T^1; T^2;...\}$, acting in spaces $\mathcal{V}_1; \mathcal{V}_2...$, is defined in the obvious way by taking the direct sum of spaces $\mathcal{V}_1 \oplus \mathcal{V}_2 \oplus ...$, and setting,

$$(T^1 \oplus T^2 \oplus\ ...)_g(\xi_1;\xi_2;...) = T^1_g\xi_1 \oplus T^2_g\xi_2 \oplus ...;\ \xi_j \in \mathcal{V}_j. \tag{2.17}$$

There is a continuous version of (2.17) called the *direct integral decomposition.* Given a family of representations $\{T^\lambda: \lambda \in \Lambda\}$, in spaces $\{\mathcal{H}^\lambda\}$ (assumed to be regularly embedded in the embient space $\mathcal{H}$), and a measure $d\mu$ on Λ, we define *direct integral space:* $\mathcal{H} = \oint \mathcal{H}^\lambda d\mu(\lambda)$, to consist of all L^2-vector valued functions on Λ, that for each $\lambda \in \Lambda$, take on values in $\mathcal{H}_\lambda$: $F = f(\lambda) \in \mathcal{H}^\lambda$. We then define direct-integral operators,

$$T_g = \oint T^\lambda_g\, d\mu(\lambda),$$

to map vector-functions $f(\lambda) \to T^\lambda[f(\lambda)]$.

• *Tensor (Kronecker) product* $T^1 \otimes T^2$ is defined in the *tensor-product space*[1] $\mathcal{V}_1 \otimes \mathcal{V}_2$:

$$(T^1 \otimes T^2)_g(\xi \otimes \eta) = T^1_g\xi \otimes T^2_g\eta.$$

An analytic version requires a proper ***topological product*** $\mathcal{V}_1 \otimes \mathcal{V}_2$. Thus for regular representations R^j $(j = 1,2)$ in spaces $\mathcal{V}^j = L^2(X_j)$, the product space $\mathcal{V} = \mathcal{V}_1 \otimes \mathcal{V}_2 \simeq L^2(X_1 \times X_2)$, and G acts on $\mathcal{V}$ by

$$R_g f(x;y) = f(x^g;y^g).$$

• ***The contragredient*** (dual) representation $\tilde{T}_g = T^*_{g^{-1}}$, in the dual space $\mathcal{V}^*$ to $\mathcal{V}$. Here T^* denotes the dual (adjoint) operator to T.

• ***The adjoint representation:*** $Ad = Ad^T$ in the space $\mathcal{L}(\mathcal{V})$-all linear transformations on $\mathcal{V}$, or its subspaces, $\mathrm{Ad}_g(A) = T_g^{-1} A T_g$, $A \in \mathcal{L}(\mathcal{V})$. Space $\mathcal{L}(\mathcal{V})$ is naturally identified with the tensor product of $\mathcal{V} \otimes \mathcal{V}'$ ($\mathcal{V}$ by its dual $\mathcal{V}'$), where tensor $\xi \otimes \ell \rightarrow$"rank-one operator $\langle \ell | ... \rangle \xi$". Then one can show that $Ad^T = T \otimes \tilde{T}$ - tensor product of T and its contragredient. A natural example of Ad is the action of GL_n by conjugations on the space of all matrices, $g: A \rightarrow g^{-1} A g$; $g \in GL_n$, $A \in Mat_n$. More generally, the adjoint action appears in the context of Lie groups acting on their Lie algebras, and will be explained in §1.4.

[13]Tensor product of 2 vector spaces $\mathcal{W} = \mathcal{V}_1 \otimes \mathcal{V}_2$ is spanned by linear combinations of elemental products $\{\xi \otimes \eta : \xi \in \mathcal{V}_1; \eta \in \mathcal{V}_2\}$, subject to the natural bilinear rules: $(\xi + \eta) \otimes \zeta = \xi \otimes \zeta + \eta \otimes \zeta$, and similarly for $\xi \otimes (\eta + \zeta)$. The easiest way to think of $\mathcal{W}$ is in terms of function-spaces: $\mathcal{V}_1 \simeq \mathbb{C}(X)$, $\mathcal{V}_2 \simeq \mathbb{C}(Y)$ (for finite-D spaces $\mathcal{V}$, sets X and Y are finite!). Then $\mathcal{W}$ is also a function-space $\mathbb{C}(X \times Y)$, where elemental tensors $\{\xi \otimes \eta\}$ correspond to product-functions $\{\xi(x)\eta(y)\}$. Such correspondence extends to some infinite function-spaces, e.g. $L^2(X) \otimes L^2(Y) \simeq L^2(X \times Y)$.

Problems and Exercises:

1. Show that the fractional-linear action (2.12) of group $G = SL_2(\mathbb{R})$ on the Poincare-Lobachevski half-plane **H** leaves measure $d\mu = y^{-2}dxdy$, invariant and that $d\mu$ is the only G-invariant measure of **H**.

2. Show that the regular representation of $G = \mathbb{R}^n$ ($\mathbb{T}^n$) is strongly continuous on various function spaces over $\mathbb{R}^n$ ($\mathbb{T}^n$):

   (i) $\mathfrak{C}_0 = \{$all continuous functions vanishing at $\infty\}$ with norm $\|f\| = max|f(x)|$;

   (ii) $\mathfrak{C}_0^m = \{m$-times continuously differentiable functions, vanishing at $\infty\}$, with norm $\|f\|_m = \sum_{|\alpha| \leq m} max|\partial^\alpha f(x)|$;

   (iii) $\mathfrak{C}_0^\infty = \bigcap_m \mathfrak{C}_0^m$; with a sequence of seminorms $\{\|f\|_m: m=0;1;...\}$

   (iv) Lebesgue L^p-spaces, $1 \leq p < \infty$, with $\|f\|_p = \left(\int |f(x)|^p dx\right)^{1/p}$.

   Hint: Verify continuity of the map $a \to R_a f$, for compactly supported functions f in $\mathfrak{C}_0$ or $\mathfrak{C}_0^m$ (uniform continuity!), then utilize the standard functional analysis: *density of compactly supported f's in* $\mathfrak{C}_0$ and L^p, and the *uniform boundedness* of the family of translations $\{R_a: a \in \mathbb{R}^n\}$, $\|R_a\| = 1$!

   The above argument applies to other groups G and homogeneous spaces X, and their regular representations R on $\mathfrak{C}(X)$; $L^p(X)$, etc. as long as transformations $g: x \to x^g$, are continuous and measure preserving, i.e. X has a G-invariant Haar measure.

   (v) The regular representation R on G or $X = H \backslash G$ is not *uniformly continuous*, as $\|R_g - I\| = 2$, for all g in the vicinity of $\{e\}$.

3. Check that any *cocycle* $\alpha(x;g)$ defines a representation $T = T^\alpha$ via formula (2.8). Show that the product, $\alpha\beta$, and a power α^s of cocycles, are also cocycles, hence T^s of (2.14) is also a representation.

4. A cocycle α is *trivial*, if $\alpha(x,g) = \dfrac{\beta(x^g)}{\beta(x)}$, for some function $\beta(x)$.

   (i) Show that cocycles (2.11) and (2.13) are nontrivial.

   (ii) A trivial cocycle yields a Haar measure on X.

   Thus in both examples (affine and SL_2) the quotient $H \backslash G$ has no Haar measure.

5. Show that the Haar measure on a semidirect product $G = H \rhd K$, with compact group K (e.g. Euclidian motions $\mathbf{E}_n = \mathbb{R}^n \rhd SO(n)$), is equal to $dg = dh\,dk$ - the product of Haar measures (Hint: Haar measure is invariant under any compact group of automorphisms).

6. Show that the Haar measure on $SL_2(\mathbb{R}) = \left\{g = \begin{pmatrix} x & y \\ z & v \end{pmatrix}; \ detg = 1\right\}$, is given by

$$\boxed{dg = \frac{dx\,dy\,dz}{x}}.$$

   Use the Haar measure on $GL_2 = SL_2 \times \mathbb{R}^*$, write any $g \in GL_2$ as

$$\begin{pmatrix} a & c \\ b & d \end{pmatrix} = \lambda \begin{pmatrix} x & y \\ z & v \end{pmatrix}, \quad v = \frac{1+yz}{x},$$

   and compute the Jacobian

$$\frac{\partial(a;b;c;d)}{\partial(x;y;z;\lambda)} = \frac{1}{\lambda x}!$$

7*. Show that the Haar measure on $SL_n(\mathbb{R})$ parametrized by its entries $\{g = (x_{jk}): j+k > 2$ (excluding x_{11}!)$\}$ is equal to

$$\boxed{dg = \frac{1}{M_{11}(g)} \prod_{j+k>2} dx_{jk}};$$

where M_{11} denotes the 1,1-minor of g.

Formulae of problems 5-6 look elegant, but found little practical use in harmonic analysis on SL_n. The latter exploits a different set of coordinate, based on certain factorizations of SL_n (Cartan, Iwasawa), described in chapter 7.

8. a) Find the left and right Haar measures on the affine group $G = \{ax+b\}$ over $\mathbb{R}$, or $\mathbb{C}$, and compute the modular factor: $\Delta(a;b) = 1/a$.

b) Do the same problem for the Borel group of all upper-triangular matrices

$$B_n = \left\{ g = \begin{bmatrix} a & * & * \\ & \ddots & * \\ & & c \end{bmatrix} : det g \neq 0 \right\}.$$

9. Show that formula (2.10) defines a representation T of G. The induced representation is continuous/unitary, provided σ is continuous/unitary and X has a G-invariant measure. Another simple (functorial) property of ind,

$$ind(\sigma_1 \oplus \sigma_2) = ind(\sigma_1) \oplus ind(\sigma_2).$$

10. Consider convolutions on the unit circle $G = \mathbb{T}$,

$$R_f[h] = f*h = \int_0^{2\pi} f(t-\theta)\, h(\theta)\, d\theta,$$

on L^2-space ($h \in L^2$), and show

i) L^1-functions f yield *compact* operators R_f

ii) L^2-functions f yield *Hilbert-Schmidt* operators R_f

iii) $\mathcal{C}^m$-smooth functions ($m \geq 2$) yield trace-class operators.

Hint: use Fourier expansion of $f = \sum a_m e^{imt}$; and some basic facts of Fourier analysis on $\mathbb{T}$: (a) $f \in L^1 \Rightarrow \{a_k\} \to 0$, as $k \to \infty$; (b) $f \in L^2 \Leftrightarrow \{a_k\} \in \ell^2$; and (c) $f \in \mathcal{C}^m \Rightarrow a_k = O(k^{-m})$, as $k \to \infty$.

We remind the reader that *compactness* for diagonalizable (self-adjoint) operators means that their eigenvalues $\lambda_k \to 0$, *Hilbert-Schmidt* means $\sum |\lambda_k|^2 < \infty$, and *trace-class* is equivalent to $\sum |\lambda_k| < \infty$ (see Appendix B).

11. Show that convolutions $f*h$ with $f \in \mathcal{C}^m$ on groups $\mathbb{R}^n$, $\mathbb{T}^n$, take any L^p-space back into $\mathcal{C}^m$ (interchange integration $\int ...dy$, and differentiation ∂_x^α). The same argument works on any Lie group and its homogeneous spaces, when integration and differentiation are G-invariant!

§1.3. Irreducibility and decomposition.

Section 1.3 studies *irreducibility* and *decomposition* of representations. We establish basic irreducibility tests: Schur's Lemma and Burnside's Theorem, discuss direct sum/integral decompositions, equivalence, intertwining spaces, and outline some basic problems of the representation theory.

3.1. Irreducibility. One of the main goals in the representation theory is to analyze the structure of any representation T in terms of its simple constituents, called *irreducible representations.* In this section we shall discuss some general features of irreducibility and decomposition.

A representations T of group/algebra in space $\mathcal{V}$ is called *irreducible*, if operators $\{T_g\}$ have no joint nontrivial *invariant subspaces*[14]: $\mathcal{V}_0 \subset \mathcal{V}$;

$$T_g \mid \mathcal{V}_0 \subset \mathcal{V}_0, \text{ for all } g \in G, \Rightarrow \mathcal{V}_0 = \{0\} \text{ or } \mathcal{V}.$$

An invariant subspace $\mathcal{V}_0 \subset \mathcal{V}$ reduces T, so one can talk about a *subrepresentation* (restriction) $T^0 = T \mid \mathcal{V}_0$, and the *factor-representations* on the quotient-spaces $\mathcal{V}/\mathcal{V}_0$. Irreducible representations play the role of elementary building blocks, in the decomposition of an arbitrary T. In this respect they resemble *eigenvalues* of linear operators, or *characters* (joint eigenvalues) of commuting families of operators. Two general results provide useful irreducibility tests: *Schur's Lemma* and *Burnside (von Neumann) Theorem.* Each has a purely "algebraic" (finite-dimensional) version, and a "topological" (unitary) version. Since the arguments are very similar in both cases, we shall discuss them in parallel. Given a representation T, we introduce its *commutator algebra:*

$Com(T)=\{A\text{: } AT_g = T_gA, \text{ for all } g \in G\}$ - all operators that commute with T.

Schur's Lemma: (i) *An irreducible finite-D representation T in complex space $\mathcal{V}$ has scalar commutator algebra, $Com(T) = \{\lambda I\}$.*

(ii) A unitary representation T is irreducible, iff $Com(T)$ is scalar.

Proof: (i) If $Com(T)$ has a nonscalar operator Q, then any (nontrivial) eigensubspace $E_\lambda = \{\xi : Q\xi = \lambda\xi\}$ of Q, is obviously invariant under T, which means reducibility.

(ii) The unitary version requires some basic facts of the spectral theory of unitary and selfadjoint operators, described in Appendix A. Let us remark that algebra

---

[14]In the topological context (Banach/Hilbert spaces $\mathcal{V}$) one usually considers *closed invariant subspaces* $\{\mathcal{V}_0\}$.

$Com(T)$ is closed under conjugation $Q \to Q^*$ (adjoint operator to Q). So a nontrivial algebra $Com(T) \neq \{\lambda I\}$ has a nonscalar selfadjoint operator Q.

The role of the eigenspaces in part (i) will be played now by the *spectral subspaces*[15] $\{E(\Delta)\}$ of Q. Each subspace $E(\Delta)$ is known to be invariant under the commutator algebra of Q, since the corresponding spectral projections $\{P(\Delta)\}$ are approximated by "functions (polynomials) of Q". The argument is easily completed now. A nontrivial $Q = Q^*$ in $Com(T)$ would yield a nontrivial T-invariant spectral subspace $E(\Delta)$. So irreducibility of T implies $\mathrm{Com}(T) = \{\lambda I\}$. Conversely, if T has an invariant subspace $\mathcal{V}_0$, then the orthogonal projection P: $\mathcal{V} \to \mathcal{V}_0$, gives a nonscalar element of $Com(T)$, QED.

Let us remark that for general (nonunitary) T, Schur's Lemma gives only a necessary condition of irreducibility. It is easy to find examples of *reducible* T with scalar commutator, for instance, all upper-triangular matrices in $\mathbb{C}^n$ (problem 3).

An immediate corollary of Schur's Lemma is

Corollary: *An irreducible finite-D or unitary representation of a commutative group/algebra is one-dimensional,* $T_g = \chi(g)$ *- a character of G.*

Remark: There is a version of Schur's Lemma for representations in real spaces $\mathcal{V}$. Here irreducibility of T does not always imply $\mathrm{Com}(T)$ to be scalar. But the commutator algebra must contain nonsingular matrices only, hence it makes a *division algebra.* There are only 3 such beasts: reals $\mathbb{R}$ (the corresponding T are called *real-type*); complex numbers $\mathbb{C}$ (*complex-type* T), and quaternions $\mathbb{Q}$ (*quaternionic-type* T). All 3 cases can be characterized in terms of complexification of T, i.e. the family of operators $\{T^c(\xi + i\eta) = T\xi + i(T\eta) : \xi, \eta \in \mathcal{V}\}$. Then "real type" yields an irreducible representation $T^C = T \otimes \mathbb{C}$; "complex-type" - a pair of inequivalent (conjugate) representations: $T^C \simeq T_1 \oplus \overline{T}_1$; while quaternionic-type results in a pair of equivalent representations: $T^C \simeq T \oplus T$.

A simple illustration of this statement is given by 3 algebras, $\mathbb{R};\mathbb{C};\mathbb{Q}$, themselves,

[15] *Spectrum* of a selfadjoint operator Q in Hilbert space $\mathcal{H}$, consists of $\{\lambda \in \mathbb{R}$: so that $(\lambda I - Q)$ is not boundedly invertible$\}$. Spec(Q) is always a closed subset of $\mathbb{R}$, and operators with 1-pointed spectrum, spec$(Q) = \{\lambda\}$, are known to be scalar, $Q = \lambda I$. Furthermore, any closed subset $\Delta \subset \mathbb{R}$ has an associated *spectral subspace* $E(\Delta)$ and the corresponding *spectral projection,* $P(\Delta)$: $\mathcal{H} \to E(\Delta)$. Spaces $\{E(\Delta)\}$ are invariant under Q, and have $spec(Q \mid E(\Delta)) \subset \Delta \bigcap \Sigma$. For operators with discrete spectrum $\{\lambda_k\}$, space $E(\Delta) = \oplus E(\lambda_j)$, consists of all eigensubspaces with $\lambda_j \in \Delta$. An easy way to obtain $\{E(\lambda)\}$ is to use the *canonical model* of a selfadjoint operator Q, namely, a multiplication: $f(\lambda) \to \lambda f(\lambda)$, on the space of square-integrable (scalar or vector) functions on $\Sigma = spec(Q)$, $f \in L^2(\Sigma; d\mu)$. Then $E(\Delta)$ consists of all L^2-functions vanishing outside Δ (see Appendix A).

when realized by matrices (problem 11),

$$\mathbb{R} = \{[a];\ a \in \mathbb{R}\};\ \mathbb{C} = \left\{\begin{bmatrix} a & b \\ -b & a \end{bmatrix}: a,b \in \mathbb{R}\right\};\ \text{and}\ \mathbb{Q} = \left\{\begin{bmatrix} a & b \\ -\bar{b} & \bar{a} \end{bmatrix}: a,b \in \mathbb{C}\right\}. \tag{3.1}$$

Burnside (von Neumann) Theorem: *(i) A finite-D representation T in space $\mathcal{V}$ is irreducible if and only if the algebra $\mathcal{A} = \mathcal{A}(T)$, spanned by T, coincides with the full operator/matrix algebra $\mathcal{L}(\mathcal{V}) \simeq Mat_d$, where $d = d(T) = dim\mathcal{V}$, called the degree of T.*

(ii) A unitary (infinite-D) representation T in Hilbert space $\mathcal{H}$ is irreducible, if and only if the closure of $\mathcal{A}(T)$ in the weak operator topology[16] *coincides with the algebra of all bounded operators $\mathfrak{B}(\mathcal{H})$.*

Sufficiency follows from the irreducibility of the full matrix/operator algebra $Mat_n/\mathfrak{B}(\mathcal{H})$. The proof of necessity: *"irreducibility"* ⇒ *"full-operator algebra"*, exploits a "bootstrap" argument (see problems 7,8). In the ∞-D case one could not expect the purely algebraic (Burnside's) version of the Theorem to hold, as there are many transitive operator-algebras, different from $\mathfrak{B}(\mathcal{H})$, like all finite-rank operators $\mathfrak{F}$, or compact operators $\mathfrak{K}$ in $\mathcal{H}$. But the *weak (operator) closure* of both coincides with $\mathfrak{B}(\mathcal{H})$! A Theorem of von Neumann extends the algebraic result to *symmetric transitive subalgebras* $\mathcal{A} \subset \mathfrak{B}(\mathcal{H})$, by combining "bootstrap" with some standard functional analysis, to show that $\mathcal{A}$ must be weakly dense in $\mathfrak{B}(\mathcal{H})$.

3.2. Decomposition. If a finite-D representation T on space $\mathcal{H}$ is reducible, one can always find a minimal invariant subspace $\mathcal{H}_0$, so that $T \mid \mathcal{H}_0$ becomes irreducible. Hence, any such representation can be brought into a "*quasi-triangular form*" with irreducible diagonal blocks. But for unitary T each invariant subspace $\mathcal{H}_0$ has also an T-invariant orthogonal complement $\mathcal{H}_1 = \mathcal{H} \ominus \mathcal{H}_0$. Therefore, *quasi-triangular* reduction for unitary T amounts to "*quasi-digonalization*" or *complete decomposition.*

Proposition 1: *A finite-dimensional unitary representation can be decomposed into the direct orthogonal sum of irreducible components: $T = \bigoplus_j T^j$, i.e. $\mathcal{H} = \bigoplus_j \mathcal{H}_j$- direct sum of invariant subspaces, and $T^j = T \mid \mathcal{H}_j$.*

In other words a unitary finite-D representation is *completely reducible.* Thus irreducible representations provide the "building blocks" of which any T is made. In this sense Proposition 1 extends the well-known *diagonalization* of a unitary/self-adjoint

[16] *Weakly dense* algebra $\mathcal{A}$ means that any operator $B \in \mathfrak{B}(\mathcal{H})$ can be approximated by operators $\{A \in \mathcal{A}\}$, relative to a sequence of *seminorms* $\{\langle A\xi \mid \eta\rangle: \xi, \eta \in \mathcal{H}\}$, equivalently, by all trace-class operators K, identified with functionals on $\mathfrak{B}(\mathcal{H})$, via the natural pairing K: $A \to tr(KA)$. In other words the distance between operators $\{A; B\}$ is measured by n-tuples of vectors $\{\xi_j; \eta_j: 1 \le j \le m\}$, or trace-class operators $\{K\}$: $\{\langle (A-B)\xi_j \mid \eta_j\rangle\}_j$ or $\{tr\, K(A-B)\}$.

operator (Appendix A). There exists a continuous version of Proposition 1, called the *direct integral decomposition* of a unitary representation in ∞-D case. We briefly introduced it in the previous section and will discussed later in §6.1.

Let us remark that the decomposition of Proposition 1 is not unique in general, due to possible multiplicities of irreducible components of T, the same way as multiple eigenvectors of a matrix are not unique. We shall describe a canonical (unique!) decomposition of T into the direct sum of *multiples of irreducibles*, $\pi \otimes m$, where m $m(\pi;T)$ denotes the multiplicity of π in T.

Theorem 2 (primary decomposition): *A unitary finite-D representation T in space $\mathcal{H}$ is uniquely decomposed into the direct sum of multiples of irreducible representations, i.e. $\mathcal{H} = \bigoplus_k \mathcal{H}_k$ - direct sum of invariant subspaces, so that $T/\mathcal{H}_k \simeq \pi^k \otimes m_k$ - a multiple of an irreducible π^k, and*

$$\boxed{T \simeq \bigoplus_k \pi^k \otimes m_k} \tag{3.2}$$

with $m_k = m(\pi^k;T)$ - multiplicity of π^k in T.

Multiples of irreducibles $\{\pi^k\}$ are often called *primary components* of T, spaces $\{\mathcal{H}_k\}$ are *primary subspaces,* and (3.2) is referred to as a *primary decomposition* of T. The main difference between Theorem 2 and Proposition 1 is that the primary subspaces $\mathcal{H}_j = \mathcal{H}(\pi^j)$ are uniquely determined, whereas within each $\mathcal{H}_j$ there are many ways to break $T \mid \mathcal{H}_j$ into the sum of irreducibles. So the primary subspaces represent a noncommutative analog of the *eigenspaces* of a linear operator.

There are different ways to establish the primary decomposition. Some depend on specific structures of groups in question. For instance, for finite and compact groups G (3.2) could be derived via *irreducible characters* of G and the associated *primary projections* (see chapters 3-5). Another method is based on *central projections* of the commutator algebra $Com(T)$. Here we shall exploits matrix entries of T: given a pair of vectors $\xi,\eta \in \mathcal{H}$, we define the matrix entry $t_{\xi\eta}(g) \overset{def}{=} \langle T_g\xi \mid \eta\rangle$ defines a function on G. For each vector $\xi \in \mathcal{H}$ we introduce a *cyclic subspace*, $\mathcal{V}(\xi) = Span\{T_g\xi\colon g \in G\}$. Obviously, $\mathcal{V}(\xi)$ is invariant under T. Any vector $\eta \in \mathcal{H}$ defines a linear map, $Q_\eta\colon \xi \to \langle T_g\xi \mid \eta\rangle$, that takes $\mathcal{V}(\xi)$ into a subspace $\mathcal{F}$ of functions on G (matrix-entries), invariant under right translations, $f(xg) \in \mathcal{F}$, for any $f \in \mathcal{F}$. In fact, Q_η *intertwines* the subrepresentation $T \mid \mathcal{V}(\xi)$ and the regular representation $R \mid \mathcal{F}$,

$$R_g Q_\eta = Q_\eta T_g, \text{ for all } g \in G.$$

This shows, in particular, that any irreducible T can be embedded into the regular representation on a suitable function space. We say that vector $\xi \in \mathcal{H}$ *transforms according to an irreducible* π, if all matrix-entries $\{t_{\xi\eta}(x): \eta \in \mathcal{H}\}$ belong to the linear span $\mathcal{F}(\pi)$ of matrix-entries of π, a R-invariant subspace of $\mathcal{L}(G)$. An equivalent statement: all irreducible constituents of $T \mid \mathcal{V}(\xi)$ are copies of π, so $T \mid \mathcal{V}(\xi) \simeq \pi \otimes m$. We claim that $T \mid \mathcal{H}(\xi) \simeq \pi$ has multiplicity 1. Indeed, by Schur's Lemma one can show that the only *cyclic subrepresentations* of $S = \pi \otimes m$ are copies of π. Now we are able to identify primary subspaces $\{\mathcal{H}(\pi)\}$ of Theorem 2: $\mathcal{H}(\pi) = \{\xi \in \mathcal{H}$: transformed according to $\pi\}$. Spaces $\mathcal{H}(\pi)$ are obviously T-invariant, their intersection $\mathcal{H}(\pi) \bigcap \mathcal{H}(\pi') = 0$, if $\pi \neq \pi'$, and their sum, $\bigoplus_{\pi} \mathcal{H}(\pi)$, spans $\mathcal{H}$. Thus we get a unique (canonical!) primary decomposition of T, QED.

Proposition 1 and Theorem 2 naturally lead to the following problems:

(I) *Construct and classify all irreducible representations of the group/algebra G.*

(II) *Decompose the given representation (regular, tensor-product, etc.) into the sum/integral of irreducibles.*

These are indeed two basic questions in representation theory, that will be addressed at length in chapters 3-7. We shall construct, classify and analyze irreducible representations for many different classes of groups:

- finite and general compact groups (chapter 3);

- classical compact Lie groups $SO(3); SU(2)$ (chapter 4), as well as their higher-D analogs: $SU(n); SO(n), \ldots$ (chapter 5);

- semidirect products, including the Heisenberg group: $\mathbb{H}_n \simeq \mathbb{R} \triangleright \mathbb{C}^n$, the Euclidian motion groups: $\mathbb{E}_n = \mathbb{R}^n \triangleright SO(n)$, and affine groups (chapter 6);

- the celebrated SL_2 in its various incarnations: conformal $SU(1;1)$; Lorentz $SO(1;2)$ and $SO(1;3)$, and also the Poincare group: $\mathbb{P}_4 \simeq \mathbb{R}^4 \triangleright SO(1;3)$ in chapter 7.

Once the list of irreducible components of G is found (usually denoted by $\hat{G}$, and called the *dual object* of G), one can turn to the decomposition Problem (II) for specific classes of representations. The most interesting examples include *regular and induced representation* R on G, or homogeneous spaces $X = H \backslash G$. We shall see that in the compact case (X or G), the decomposition is always "discrete",

$$\boxed{R \simeq \bigoplus_{\pi \in \hat{G}} \pi \otimes m(\pi)} \text{ - direct sum of irreducibles.}$$

For non-compact groups/spaces the decomposition is typically "continuous" (direct integral),

$$\boxed{R \simeq \oint_{\hat{G}} \pi^{\lambda} \otimes m_{\lambda} d\mu(\lambda)},$$

or more generally a combination of the "discrete" and the "continuous" parts, as is the case with $SL_2(\mathbb{R})$ (chapter 7). The resulting measure $d\mu$ on $\hat{G}$ (whether discrete or continuous) is called the ***Plancherel measure*** of R. It will be one of principal objects of our study.

The classification problem could easily become "unmanageable", if one insists on "*all irreducible representations*" of G, since any T can be deformed in infinitely many ways by conjugations with invertible operators,

$$Q: T_g \rightarrow T'_g = Q^{-1} T_g Q.$$

Such deformations, $T \rightarrow T'$, however, retain many essential features of T (e.g. spectrum), and could be regarded trivial. Thus we are lead to the notion of *equivalence*: a pair of representations, T and S in spaces $\mathcal{H}$ and $\mathcal{W}$, are called *equivalent*, if they are conjugated (*intertwined*) by an invertible linear map Q: $\mathcal{H} \rightarrow \mathcal{W}$,

$$\boxed{S_g Q = Q T_g} \text{ or } \boxed{S_g = Q T_g Q^{-1}} \text{ for all } g \in G.$$

The space of all *intertwining operators* is denoted by $Int(T;S)$, its dimension (if finite) is called an *intertwining (branching) number* $m(T;S)$. Operators that intertwine T with itself form the *commutator algebra* of T, $Int(T;T) = Com(T)$, and other spaces $Int(S;T)$ acquire the structure of $\{Com(T);Com(S)\}$ - *bimodule*, i.e. $Com(T)$ acts on $Int(S;T)$ by the left multiplication while $Com(S)$ by the right one.

Problem (I) usually refers to *classification of equivalence classes* of irreducible representations, rather than individual $\{T$'s$\}$. In some cases, however, equivalent representations T and S may be given in different realizations. Then one is interested in specific intertwining operators that implement the equivalence.

Let us mention yet another useful application of Schur's Lemma.

Proposition 3: *For any pair of irreducible representations T and S, space $Int(T;S)$ is either 1-D, if T and S are equivalent, or $\{0\}$. So the intertwining number $m(T;S) = 1$ or 0.*

Indeed, for any $Q \in Int(T;S)$, its range $\mathcal{R}(Q)$ and kernel $\mathcal{N}(Q)$ are invariant subspaces of S and T. So space Int of an irreducible finite-D pair $(T;S)$ consists of invertible operators Q. But $Q_1 Q_2^{-1} \in Com(S)$ for any pair Q_1, Q_2 in $Int(T;S)$, so by Schur's

Lemma, $Q_1Q_2^{-1}=\lambda I$, i.e. $Q_1=\lambda Q_2$. The unitary version requires a slight modification. We observe that two spaces $Int(T;S)$ and $Int(S;T)$ are naturally conjugate one to the other by taking the adjoint $Q\leftrightarrow Q^*$. By Schur's Lemma $Q^*Q=\lambda I$, hence operator $\lambda^{-1/2}Q$ is unitary and invertible, and the rest of the argument proceeds as above.

Our final result in this section gives irreducible unitary representations of the direct product group $G=H\times K$ in terms of its factors.

Theorem 4: *Each irreducible representation π of $G=H\times K$ is decomposed into the tensor product*[17] *of two irreducible representations of H and K, $\pi=\sigma\otimes\gamma$, where $\sigma\in\widehat{H}$, $\gamma\in\widehat{K}$. Conversely, the product of two irreducibles $\sigma\in\widehat{H}$ and $\gamma\in\widehat{K}$, is an irreducible representation of G, $\pi=\sigma\otimes\gamma\in\widehat{G}$. So the dual object of $G=H\times K$ also factors into the product,*

$$\boxed{\widehat{G}\simeq\widehat{H}\times\widehat{K}}$$

Proof: The second statement follows directly from the Burnside's Theorem and the known facts of multilinear algebra about tensor-products of operator/matrix algebras (problem 10),

$$Mat_n\otimes Mat_m\simeq Mat_{nm};\tag{3.3}$$

or in terms of operator algebras

$$\mathfrak{B}(\mathcal{H})\otimes\mathfrak{B}(\mathcal{V})\simeq\mathfrak{B}(\mathcal{H}\otimes\mathcal{V}).\tag{3.4}$$

To prove the first statement we take an irreducible $\pi\in\hat{G}$ in space $\mathcal{H}$, restrict it on the first factor $\pi\mid H$, and write the primary decomposition of Theorem 2,

$$\pi\mid H\simeq\bigoplus_k\sigma_k\otimes m_k;$$

where $\sigma_k\in\widehat{H}$, and m_k- the multiplicity of σ_k in $\pi\mid H$. So the representation space $\mathcal{H}\simeq\bigoplus_k\mathcal{H}_k$ - direct sum of primary subspaces of H. Since each $\mathcal{H}_k$ is invariant under the commutator $Com(\pi\mid H)$, hence under K, it follows that $\mathcal{H}$ has only one primary component, $\mathcal{H}\simeq\mathcal{H}_k$, and $\pi\mid \mathrm{H}\simeq\sigma\otimes\mathrm{I}_m$. In other words, space $\mathcal{H}$ is a direct sum of $\{m\}$ copies of the *"irreducible space" $\mathcal{V}$ of σ*,

$$\mathcal{H}\simeq\underbrace{\mathcal{V}\oplus\ldots\oplus\mathcal{V}}_{m\text{-times}}\simeq\mathcal{V}\otimes\mathbf{C}^m.$$

Any operator A in $\mathcal{H}$ can be represented by a block-matrix $A=(a_{ij})$, $a_{ij}\in Mat(\mathcal{V}_i;\mathcal{V}_j)$. We are interested in operators $A\in Com(\pi\mid H)$. By Proposition 3 all "matrix entries" $\{a_{ij}\}$ of such A must be scalars: $a_{ij}=\gamma_{ij}I$; $\gamma_{ij}\in\mathbf{C}$. When restricted on subgroup K, matrices $(\gamma_{ij}(k))$ clearly yield a representation of K. So space $\mathcal{H}$ is decomposed into the product $\mathcal{V}\otimes\mathbf{C}^d$, as well as operators $\{\pi_g\}$, $g=(h;k)\in G$, $\pi_g=\sigma_h\otimes\gamma_k$. Obviously, for π

[17] The reader should not confuse the *tensor product* $\sigma\otimes\gamma$ of Theorem 4 with the *Kronecker product* of §1.2. The latter is obtained, when the tensor product $\sigma\otimes\gamma$ of the product-group $G\times G$ is restricted on the diagonal subgroup $\{(g;g)\}\simeq G\subset G\times G$.

to be irreducible, γ must also be irreducible, which completes the proof!

Problems and Exercises.

1. Show that the natural actions of GL_n and $U(n)$ on $\mathbf{C}^n$ are irreducible.

2. Show that an eigenspace E_λ or a root subspace $E_\lambda^m = \{\xi: (T-\lambda I)^m\xi = 0\}$ of a matrix T in $\mathbf{C}^n$ are invariant under $Com(T)$.

3. Show: the group B_n of all upper/lower triangular matrices in $\mathbf{R}^n$; $\mathbf{C}^n$ has trivial (scalar) commutator, though B_n is clearly reducible (This exercise shows Schur's Lemma to provide only a necessary condition for general non-unitary representations!).

4. A symmetric group $\mathbf{W}_n$ (permutations of n objects) acting on $X = \{1,2,...n\}$ induces the regular representation R on the function-space: $\mathcal{C}(X) \simeq \mathbf{C}^n$, $R_s f(j) = f(j^s)$.

   (i) Show that $\mathcal{C}(X)$ is the sum of two invariant subspaces: $\mathcal{H}_0 = \{\lambda f_0\}$ - spanned by the constant function $f_0(j) = 1$, and $\mathcal{H}_1 = \Re f$: $\sum f(j) = 0\}$.

   (ii) Show that subrepresentations $\pi^0 = R\,|\,\mathcal{H}_0$-trivial, and $\pi^1 = R\,|\,\mathcal{H}_1$-irreducible (Use Schur's Lemma for π^1, represent any operator Q on $\mathcal{C}(X)$ by a kernel/matrix $[q_{jk}]$, $Qf(j) = \sum q_{jk}f_k$).

5*.(i) Show that the adjoint representation $\mathrm{Ad}_g(A) = g^{-1}Ag$, of the group $G = SL_n$ on the space of all traceless matrices: $\mathcal{M}_0 = \{A \in Mat_n:\ trA = 0\}$ is irreducible (Hint: any $Q \in Com(Ad)$ is uniquely determined by its values on the subspace $\mathfrak{D} \subset \mathcal{M}_0$ of diagonal matrices, moreover $Q:\mathfrak{D}\to\mathfrak{D}$, commutes with the natural action of $\mathbf{W}_n$ on $\mathfrak{D}$. Use problem 3!).

   (ii) Prove the same result for the special unitary group $SU(n)$.

6. Show that a pair of matrices: $A = \begin{bmatrix}\lambda_1 & & \\ & \ddots & \\ & & \lambda_n\end{bmatrix}$ (diagonal), $B = \begin{bmatrix} & 1 & \\ & & 1 \\ 1 & & \end{bmatrix}$ (cyclic permutation), generate the full matrix algebra in $\mathbf{C}^n$, provided all $\{\lambda_j\}$ are different.

7. Let $\mathcal{A}$ be an operator-algebra in space $\mathcal{V}$, vector $\xi \in \mathcal{V}$. We denote by $\mathfrak{J} = \mathfrak{J}(\xi)$ the *annihilator* of ξ in $\mathcal{A}$, $\mathfrak{J} = \{A: A\xi = 0\}$ - a *left ideal* $\mathcal{A}$, $(\mathcal{A}\mathfrak{J} \subset \mathfrak{J})$, and by $\mathcal{V}_0 = \mathcal{V}_0(\xi)$ - a *cyclic subspace* of ξ, $\mathcal{V}_0 = \{A\xi: A \in \mathcal{A}\}$.

   (i) Show that $\mathcal{V}_0$ is $\mathcal{A}$-invariant, and the restriction of $\mathcal{A}$ on $\mathcal{V}_0$ is equivalent to the regular representation of $\mathcal{A}$ on factor-space $\mathcal{A}/\mathfrak{J}$, $R_A(B) = AB\,(mod\ \mathfrak{J})$.

   (ii) If $\mathfrak{J}(\xi) \subset \mathfrak{J}(\eta)$, then there exists a linear map Q: $\mathcal{V}_0(\xi)\to\mathcal{V}_0(\eta)$, that intertwines two representations of $\mathcal{A}$ on $\mathcal{V}_0(\xi)$ and $\mathcal{V}_0(\eta)$. (Set: $Q[A\xi] = A\eta$, for all $A \in \mathcal{A}$, and check that Q is well defined!).

8. Prove the Burnside's Theorem by the "bootstrap argument", as outlined below.

   It is convenient to use the following terminology

   • an operator-algebra $\mathcal{A}$ is *transitive* (irreducible!), if for any vectors $\xi \neq 0$ and η there exists $A \in \mathcal{A}$, that takes $A: \xi\to\eta$;

   • algebra $\mathcal{A}$ is *double-transitive*, if for any pair $(\xi_1;\xi_2)$ of linearly independent vectors and any pair $(\eta_1;\eta_2)$ there exists $A \in \mathcal{A}$, so that $A\xi_1 = \eta_1$, $A\xi_2 = \eta_2$.

   • Similarly, a k-*transitive* $\mathcal{A}$ takes any linearly independent k-tuple $(\xi_1;...\xi_k)$ into any k-tuple $(\eta_1;...\eta_k)$, i.e. $A\xi_j = \eta_j$, $j = 1;...k$.

Clearly, a n-transitive algebra in n-space coincides with the full matrix algebra!

The idea of "bootstrap" is to show that "*transitivity*"$\Rightarrow$"*2-transitivity*"$\Rightarrow$..."*k-transitivity*".

Step 1°: Show "transitive$\Rightarrow$2-transitive".

For any vector $\xi \in \mathcal{V}$ denote by $\mathfrak{J} = \mathfrak{J}(\xi)$ the ***annihilator*** of ξ in $\mathcal{A}$ (problem 6). Given an operator-algebra $\mathcal{A}$ and a left ideal $\mathfrak{J}$, consider ***$\mathfrak{J}$-cyclic subspaces*** $\mathcal{V}_0 = \mathcal{V}_0(\eta) = \{B\eta: B \in \mathfrak{J}\}$, $\eta \in \mathcal{V}$. Each of them is $\mathcal{A}$-invariant, $\mathcal{A}\mathfrak{J} \subset \mathfrak{J}$. So for transitive (irreducible) $\mathcal{A}$, each space $\mathcal{V}_0$ must be either 0 or the entire $\mathcal{V}$.

Pick a linearly independent pair $(\xi_1;\xi_2)$, and consider the ideal $\mathfrak{J} = \mathfrak{J}(\xi_1)$ of ξ_1, and the $\mathfrak{J}$-cyclic subspace, $\mathcal{V}_0 = \mathcal{V}_0(\xi_2)$. Use problem 6 to show: if $\mathcal{V}_0 = \{0\}$, i.e. $\mathfrak{J}(\xi_1) \subset \mathfrak{J}(\xi_2)$, then map Q: $\xi_1 \to \xi_2$, extends to an intertwining operator $Q \in \mathrm{Com}(\mathcal{A})$, impossible for irreducible $\mathcal{A}$ (Schur's Lemma)!

Conclude, that $\mathcal{V}_0 = \mathcal{V}$, i.e. for any $\eta \in \mathcal{V}$ there exists $A \in \mathcal{A}$, so that $A\xi_1 = 0$, $A\xi_2 = \eta$. Show that the latter implies double-transitivity of $\mathcal{A}$.

Step 2°. Demonstrate in a similar fashion that "k-transitive$\Rightarrow$(k+1)-transitive", and completes the argument of Burnside's Theorem.

The same "bootstrap" applies to transitive algebras $\mathcal{A}$ in (∞-D) Hilbert spaces, to show that $\mathcal{A}$ is k–transitive for all k! Then fairly standard functional analysis implies that $\mathcal{A}$ is ***weakly closed*** in $\mathfrak{B}(\mathcal{H})$ (von Neumann).

9. Prove: (i) the commutator algebra of a primary representation $T \simeq \pi \otimes m$, $Com(T) \simeq Mat_m$, and any $Q \in Com(T)$ factors into the product $I_d \otimes Q$, in the tensor-product decomposition of space $\mathcal{V}_T \simeq \mathcal{V}_\pi \otimes \mathbb{C}^m$ (m = multiplicity of π in T!).

   (ii) If representation $T \simeq \oplus\, \pi^j \otimes m_j$ - is completely reducible (direct sum of primary components), then the algebra spanned by T, $\mathcal{A}(T) \simeq \oplus\, Mat_{d_j} \otimes I_{m_j}$ ($d_j = deg\,\pi^j$; m_j = multiplicity of π^j in T), while the commutator algebra
$$Com(T) \simeq \bigoplus_j I_{d_j} \otimes Mat_{m_j}.$$
   (iii) The ***intertwining number*** of a pair of representations $m(T;S)$ is defined as $dim\,Int(T;S)$. Show, if $T \simeq \oplus\, \pi \otimes m(\pi;T)$, and $S \simeq \oplus\, \pi \otimes m(\pi;S)$, then
$$m(T;S) = \sum_\pi m(\pi;T)m(\pi;S).$$

10. Show that the tensor product of matrix/operator algebras is equivalent to the full matrix algebra on the tensor-product space: $Mat_n \otimes Mat_m \simeq Mat_{n\cdot m}$; and $\mathfrak{B}(\mathcal{H}) \otimes \mathfrak{B}(\mathcal{V}) \simeq \mathfrak{B}(\mathcal{H} \otimes \mathcal{V})$ (Use rank-one operators and vectors).

11. Check formulae (3.1) for $\mathbb{C}$ and $\mathbb{Q}$ in terms of real/complex matrices. Show that complexification of $\mathbb{C}$ yields the direct sum of two conjugate representations: $z \to (z;\bar{z})$, $z \in \mathbb{C}$; while complexified $\mathbb{Q}$ yields two copies of $\mathbb{Q}$ itself.

§1.4. Lie groups and algebras; The infinitesimal method.

Here we shall introduce two principal players in the Lie theory: *Lie algebras* and *Lie groups*. Although different in appearance both are intimately linked through the *exponential* map, and its inverse *log*. Two maps allow to express the local group structure (multiplication) through the Lie bracket (Campbell-Hausdorff formula). We discuss the basic classes of Lie algebras: solvable; nilpotent; simple, semisimple, and state the decomposition result (Levi-Malcev). The section is concluded with an infinitesimal method, that allows to reduce group representations to those of its Lie algebra, and thus provides a powerful tool in their analysis.

4.1. Lie groups and Lie algebras. We remind the reader that *Lie groups* are smooth manifolds G with *differential,* (sometimes *analytic,* or *algebraic) group structure.* This means that the group operations: $(x,y)\to xy$ and $x\to x^{-1}$, are differentiable (analytic or algebraic). The most common examples of Lie groups are *subgroups of matrices,* $G\subset Mat_n$, like the *general and special linear groups:* GL_n and SL_n; *orthogonal, unitary, symplectic,* and other groups described in §1.1. Let us remark, however, that not all Lie groups can be realized by finite-D matrices. One such example is a *simply connected cover* of SL_2, $G=\widetilde{SL}_2$ (examples in 4.2 below). The class of Lie groups admits all natural operations, performed with groups in general, like taking closed subgroups, factor-groups, direct and semidirect products, etc. All those result in new Lie groups.

Lie algebras can be viewed as "linearizations" of *Lie groups,* although formal definition does not allude to an underlying group structure. Formally, *Lie algebra* $\mathfrak{G}$ is defined as a vector space equipped with the *Lie bracket* (product) $[X;Y]$, a bilinear map $\mathfrak{G}\times\mathfrak{G}\to\mathfrak{G}$, that satisfies:

(i) *skew symmetry:* $[X;Y]=-[Y;X]$ for all X,Y.

(*ii*) *Jacobi identity:* $[X;[Y;Z]]+[Y;[Z;X]]+[Z;[X;Y]]=0$, for all triples X,Y,Z.

We can rephrase (i-ii) is in terms of the so called *adjoint action* of Lie algebra on itself,

$$\mathrm{ad}_X(Y)=[X;Y];\ \mathrm{ad}_X:\mathfrak{G}\to\mathfrak{G}. \tag{4.1}$$

Then,

(i) $\mathrm{ad}_X(Y)=-\mathrm{ad}_Y(X)$, for all X,Y

(*ii*) $\mathrm{ad}_{[X;Y]}=[\mathrm{ad}_X;\mathrm{ad}_Y]=\mathrm{ad}_X\mathrm{ad}_Y-\mathrm{ad}_Y\mathrm{ad}_X$,

In particular, the Jacobi identity means that map, $X \to \mathrm{ad}_X$, is a *representation* of the Lie algebra $\mathfrak{G}$ by linear transformations over $\mathfrak{G}$. It can also be written as

$$\mathrm{ad}_X[Y;Z] = [\mathrm{ad}_X(Y);Z] + [Y;\mathrm{ad}_X(Z)], \text{ for all } Y,Z, \tag{4.2}$$

which means that operators $\{\mathrm{ad}_X\}$ are *derivations* of the Lie algebra. The examples of Lie algebras are abundant.

• *Any associative algebra (e.g. matrix/operator algebra* Mat_n*) turns into the Lie algebra with respect to the commutator bracket:* $[x,y] = xy - yx$. Considered as a Lie algebra it is usually denoted $\mathbf{gl}(n)$.

• *Any vector space* $\mathcal{A}$ *with the trivial bracket,* $[x;y] = 0$, *gives a commutative Lie algebra.*

• Important examples of Lie algebras are given by *vector fields on a manifold* $\{X = \sum a_i(x)\partial_i\}$ (in local coordinates $\{x_1;...x_n\}$), *with the standard Lie bracket of vector fields,*

$$[X;Y] = XY - YX = \sum_i \Big(\sum_j a_j\partial_j b_i - b_j\partial_j a_i\Big)\partial_i = \sum_i (\partial_X[b_i] - \partial_Y[a_i])\partial_i; \tag{4.3}$$

Here $\partial_X;\partial_Y$ denote directional (Lie) derivatives along fields X,Y. Those include the algebra $\mathfrak{D}(\mathcal{M})$ of all vector fields on $\mathcal{M}$, or its subalgebras of fields, preserving certain structures on $\mathcal{M}$: volume, symplectic (chapter 8), etc.

Many examples arise as *Lie subalgebras* of Mat_n, and will be examined below. There is a close relationship between Lie groups and algebras.

Theorem 1: (i) *Any Lie group* G *has an associated Lie algebra* $\mathfrak{G}$, *which consists of infinitesimal generators of all one-parameter subgroups of* G.

(ii) Conversely, given a Lie algebra $\mathfrak{G}$ *one can construct a "local" Lie group*[18] G *with algebra* $\mathfrak{G}$.

(iii) A local Lie group can be extended to a global group in many different ways. Namely, there exists a unique simply-connected extension $\tilde{G}$, *and any other Lie group* G *with Lie algebra* $\mathfrak{G}$ *is isomorphic to a factor-group* $\tilde{G}/\Gamma$, $\tilde{G}$ *modulo a discrete subgroup of the center* $\Gamma \subset Z(\tilde{G})$.

Theorem 1 summarizes the basic facts of the *Lie theory*. We shall not provide complete details of the proof, but rather demonstrate different statements in several

[18] By *local Lie group* we mean an open neighborhood of the identity $\{e\}$ in a Lie group, i.e. a manifold, where the group operations: $x \cdot y$, and x^{-1}, are defined only locally.

important cases. The easiest case to consider are *matrix groups and algebras,* i.e. subgroups and subalgebras of Mat_n. We shall start with *infinitesimal generators.* It is well known that any (continuous) one-parameter subgroup $\{U_t\}$ of Mat_n is *generated* by a matrix A, in the sense that

$$U_t = e^{tA};\ \text{where } A = \frac{d}{dt}U_t\Big|_{t=0}. \tag{4.4}$$

Such A is naturally to call a *generator* of $\{U_t\}$. Let now G be a closed (Lie) subgroup of Mat_n.

Proposition 2: *(i) The set of infinitesimal generators $\{A\}$ of G coincides with the tangent space $T_e(G)$ at the identity.*

(ii) The commutator of two generator $A, B \in T_e(G)$ is also a generator, $[A;B] \in T_e(G)$, so tangent space forms a Lie subalgebra $\mathfrak{G} \subset Mat_n$.

Proof: (i) Obviously, generators $\{A\} \subset T_e$, and the latter forms a subspace of Mat_n. To show that $\{A\}$ is a subspace in T_e we pick two subgroups of G $\{U_t = e^{tA}\}$, $\{V_t = e^{tB}\}$, generated by $A, B \in T_e$ according to (4.4), then a subgroup generated by $A + B$, can be approximated by an "amalgamation" of U, V,

$$e^{t(A+B)} = \lim_{n\to\infty}\left(exp(\tfrac{t}{n}A)\,exp(\tfrac{t}{n}B)\right)^n, \tag{4.5}$$

hence $\{e^{t(A+B)}\} \subset G$ (problem 2). So $A+B \in T_e$, and clearly $\lambda A \in T_e$ for any scalar λ. Conversely, if matrix A is tangent to G, i.e. tangent to a curve $g(t) \subset G$, $A = g'(0)$, then we can easily construct a 1-parameter subgroup tangent to A, by setting,

$$U_t = \lim_{n\to\infty}(g(t/n))^n. \tag{4.6}$$

The generator of U_t is A (problem 2), thus we show that generators $\{A\}$ span the entire space T_e!

(ii) It suffices to exhibit $[A;B]$ as a tangent vector of some curve in G. Once again we take a pair of subgroups $U_t = e^{tA}$, $V_t = e^{tB}$, and consider the *group commutator* $g(t) = U_t^{-1}V_t^{-1}U_tV_t$. Expanding U and V in the Taylor series, multiplying and collecting terms we get

$$g(t) = I + t^2[A;B] + \sqcap(t^3).$$

Neglecting the cubic and higher order terms the commutator becomes the tangent vector at $\{e = I\}$,

$$[A;B] = \lim_{t\to 0}\frac{g(t)-I}{t^2};\ \text{QED.}$$

We have shown that the Lie algebra of a matrix group G can be identified with the "set of infinitesimal generators" = "tangent space of G at $\{e\}$". A similar description could be given for all (non-matrix) Lie groups, namely $\mathfrak{G} \simeq T_e(G)$. Lie algebra $\mathfrak{G}$ can also be characterized, as *the set (subalgebra) of left/right invariant vector fields on G.*

Group G acts on itself by left/right translations g: $x \to g^{-1}x$ (or $x \to xg$), gives rise to left/right invariant vector fields, $\{X = \sum a_j(x)\partial_j$ - in local coordinates $(x_1;...x_n)$ on $G\}$. *Left-invariance* of X means that the transformed field[19]: $X^g(x) = g'[X] \circ g^{-1} = X$, where g' denotes the Jacobian map of the left translate g: $x \to gx$. Left-invariant vector fields form a subalgebra $\mathfrak{G}$ in the algebra of all vector fields $\mathfrak{V}(G)$ with the Lie bracket (4.3). Any left-invariant vector field X can be identified with its value at $\{e\}$, a tangent vector $\xi \in T_e$, $X(g) = g'[\xi]$, for any point $g \in G$. So $\mathfrak{G} \simeq T_e \simeq$ {left/right invariant fields on g}.

4.2. The *exp* and *log* maps. The correspondence between Lie groups and Lie algebras is given by two maps: *exponential and log*,

$$exp: \mathfrak{G} \to G; \quad log: G \to \mathfrak{G}.$$

Map *exp*: $\mathfrak{G} \simeq T_e \to G$, sends a tangent vector/line $\{t\xi\}$ into a one-parameter subgroup $\{exp\, t\xi\}$ of G, tangent to ξ at $\{e\}$[20]. Map $\phi(\xi) = exp(\xi)$ is a local diffeomorphism (since its Jacobian $\phi'_e = I$!), and its inverse is called *log*. So *exp* takes a neighborhood of $\{0\}$ in $\mathfrak{G}$ onto a neighborhood of identity $\{e\}$ in G, while *log* does vice versa. The easiest way to introduce *exp* and *log* is for matrix groups and algebras. Here two maps are given by their Taylor-expansions,

$$\boxed{exp X = \sum_0^\infty \frac{1}{k!} X^k; \quad log(I+Y) = \sum_1^\infty \frac{(-1)^{k-1}}{k} Y^k} \tag{4.7}$$

Exponential map on the full matrix algebra $Mat_n = gl_n$ takes it onto the full linear group $GL_n = \{x \in Mat_n: \ det(x) \neq 0\}$. Clearly, *exp* on Mat_n is a local diffeomorphism (near $X = 0$), but globally it may not be 1-1. For instance, all complex matrices $\{X\}$ with purely imaginary integer eigenvalues $\{\lambda_j = 2\pi i m\}$, are taken to the identity by *exp*. We shall list now a few basic and familiar properties of *exp* and *log*.

---

[19] We remind the reader that a diffeomorphism (coordinate change) $y = \phi(x)$ on a manifold transforms vector field $X = \sum a_j \partial_j$ into a field $Y = X^\phi = (\phi'[X]) \circ \phi^{-1}$, with components $b_i = \sum \phi'_{ij} a_j$, where ϕ' denotes a Jacobian (matrix) of ϕ.

[20] Such notion of exp arises in differential geometry and is closely connected with *geodesics* and *parallel transport* (Appendix C). On Lie groups the parallel transport is furnished by the left/right group multiplication, $x \to g^{-1}x$, $g \in G$. The latter gives rise a notion of parallel field (along a curve $\gamma(t)$), the geodesic path γ (whose tangent field $\gamma'(t)$ is parallel), and exp:$t\xi \in T_a \to \gamma_\xi(t)$ - a geodesic initiated at $\{a\}$ in the direction ξ!

Lemma 3: (i) *If matrices A and B commute, $AB = BA$, then $e^{(A+B)} = e^A e^B$, in particular $\{e^{tA}\}_{t \in \mathbb{R}}$ is a one-parameter subgroup of Mat_n.*

(ii) $det(e^A) = e^{trA}$.

(iii) *exp preserves complex conjugation, transposition and taking adjoint matrix:*

$$exp(\bar{A}) = \overline{exp\, A};\ exp({}^{\mathrm{T}}A) = {}^{\mathrm{T}}expA;\ exp(A^*) = (expA)^*.$$

The proof is fairly straightforward (problem 1). Using Lemma 3 we are able to characterize Lie algebras of many classical Lie groups (orthogonal, unitary, symplectic, etc.), listed in §1.1. For instance, the *exp*-map takes a Lie algebra of $n \times n$ *hermitian antisymmetric matrices,* denoted $\mathfrak{u}(n)$, or *real antisymmetric matrices,* denoted $\mathfrak{o}(n)$, into the *unitary and orthogonal groups:* $U(n)$ and $O(n)$. *Special* (traceless) *Lie algebras* $\{X: trX = 0\}$ are taken into *special* $\{detg = 1\}$ *Lie groups,* like

- $\mathfrak{sl}(n) = \{A:\mathrm{tr}A = 0\}$ (all traceless matrices) goes into SL_n
- $\mathfrak{so}(n) = \{{}^{\mathrm{T}}A = -A;\ tr\,A = 0\}$ (orthogonal traceless) goes into $SO(n)$
- $\mathfrak{su}(n) = \{A^* = -A;\ tr\,A = 0\}$ (unitary traceless) goes into $SU(n)$.

In fact, in all cases the *exp* map is locally onto, since by Lemma 3 generators of groups SL_n; $U(n)$; $O(n)$, etc., must traceless, unitary, orthogonal.

We have already shown that Lie algebras are obtained as tangent spaces at $\{e\}$ of Lie groups G, or as infinitesimal generators of one-parameter subgroups of G. Going in the opposite direction one would like to construct Lie group G, starting from Lie algebra $\mathfrak{G}$. As we stated earlier (Theorem 1) the correspondence $\mathfrak{G} \to G$ is not 1-1, as many Lie groups could have the same Lie algebra. Although all such groups are locally isomorphic, their global structure may be different.

The *local group-structure,* however, is completely determined by the *Lie algebra bracket,* via the so called *Campbell-Hausdorff-Dynkin formula*

$$\boxed{log(e^A e^B) = A + B + \tfrac{1}{2}[A;B] + \sum_{m=3}^{\infty} \frac{(-1)^{m-1}}{m} \sum_{p_1 q_1 \dots q_m} \frac{1}{p_1 + \dots + q_m}[A^{p_1} B^{q_1} \dots B^{q_m}]} \tag{4.8}$$

Here the inner sum extends over all tuples of integers $\{p_1, q_1; \dots\ p_m, q_m\}$ with the property: $p_j + q_j \geq 1$, and $[A^{p_1}[B^{q_1} \dots B^{q_m}]]$ denotes the iterated commutator $[A[A \dots [B \dots [B \dots]]]]$. The Campbell-Hausdorff series can be shown to converge for all Lie algebra elements (matrices) of norm $\|A\|; \|B\| < \frac{1}{4}$, and thus defines a local group structure in a neighborhood of $\{0\}$ in $\mathfrak{G}$. Formula (4.8) explains, why the exponential sends any Lie subalgebra $\mathfrak{G} \subset Mat_n$ into a Lie group. Indeed,

$$\exp A \exp B = \exp(A+B+\tfrac{1}{2}[A;B]+...),$$

so $\exp \mathfrak{G}$ is closed under the matrix multiplication.

The local Lie group uniquely extends to a *global simply connected Lie group* $\tilde{G}$ (not necessarily matrix group) with the Lie algebra $\mathfrak{G}$. Any other Lie group G with algebra $\mathfrak{G}$ is then isomorphic to $\tilde{G}/\Gamma$, where Γ is a discrete subgroup of the center of $\tilde{G}$, identified with the *fundamental group*[21] of G. So all locally isomorphic Lie groups $\{G\}$ with the given Lie algebra are in 1-1 correspondence with discrete subgroups $\Gamma \subset Z(\widetilde{G})$.

We shall illustrate the relations between Lie algebras $\mathfrak{G}$ and groups G; $\widetilde{G}$ and Γ with a few simple examples.

1) The easiest of the sort is the abelian Lie algebra $\mathfrak{G} = \mathbb{R}^n$, which gives rise to several Lie groups:

$$G = \mathbb{R}^n; \text{ or } G = \mathbb{T}^n = \mathbb{R}^n/\,\mathbb{Z}^n \text{ or } G = \mathbb{T}^m \oplus \mathbb{R}^{n-m}, \text{ for any } m = 1,2,...$$

corresponding to various discrete subgroups $\Gamma \simeq \mathbb{Z}^n$, or $\mathbb{Z}^m$, of G.

2) Another example is the unitary group $SU(2)$ (simply connected), which forms a 2-fold cover of $SO(3)$ (chapter 3). All other (higher-D) orthogonal groups $SO(n)$ also have 2-fold simply connected covers, called spinor groups $Spin(n)$, $SO(n) \simeq Spin(n)/\mathbb{Z}_2$.

3) Group $G = SL_2(\mathbb{R})$ is not simply connected, and its simply connected cover $\widetilde{G}$ cannot be realized as a matrix group!

Indeed, group G is topologically equivalent to the product $\mathbf{R}^2 \times \mathbf{T}^1$ (this follows from realization of the Poincare-Lobachevski half-plane $\mathbf{H}$ as quotient $SL_2/SO(2)$, and could also be established by the Iwasawa decomposition, as explained in §7.1, chapter 7). So the

---

[21]We remind the reader that a *fundamental group* of manifold $\mathcal{M}$ consists of the *homotopy equivalence classes* of closed path (loops) in $\mathcal{M}$. Two loops $\gamma_1;\gamma_2$ in $\mathcal{M}$ are equivalent, if γ_1 can be continuously deformed into γ_2 inside $\mathcal{M}$. Any pair $\{\gamma_1;\gamma_2\}$ of closed path (loops) passing through a fixed point $x_0 \in \mathcal{M}$ can be formally *multiplied* by combining them into a single path $\gamma_1 \circ \gamma_2$ ("γ_1" followed by "γ_2"), and any γ can be inverted, by traversing it in the opposite direction. Such multiplication and inversion are easily verified to respect the (homotopy) equivalence on the space of loops passing through a fixed point. So equivalence classes of $\{\gamma\}$ acquire the group structure, the identity being made of all path contractible to $\{x_0\}$. The resulting *fundamental* group of $\mathcal{M}$, $\Gamma = \pi_1(\mathcal{M})$, is a discrete group, that carries some important topological information about the manifold. It acts freely (by the so called *deck transformations*) on a simply connected cover $\widetilde{\mathcal{M}}$ of $\mathcal{M}$, so the latter is isomorphic to a quotient-space $\widetilde{\mathcal{M}}/\Gamma$. The fundamental group of a Lie group G is identified with a normal discrete subgroup $\Gamma \subset \widetilde{G}$, hence Γ belongs to the center of $\widetilde{G}$ (problem 8), and $G \simeq \widetilde{G}\,/\Gamma$.

fundamental group Γ of SL_2 is isomorphic to $\mathbb{Z}$, and its simply connected cover $\tilde{G}$ has *center* $Z \simeq \mathbb{Z}$. If SL_2 were realized by matrices we would get a finite-D *faithful* (one-to-one) representation of SL_2 in $\mathbb{C}^n$. The detailed analysis of finite-D representations of SL_2 in chapter 4 shows that any such T has $\{\det T_g = 1$; for all $g \in G\}$, and the same holds for representations of $\tilde{G}$ (in other words T maps $SL_2 \to SL_n$!). Furthermore, any representation T of G or $\tilde{G}$ is completely reducible ($\simeq$ "direct sum of irreducibles π"). Since center of any group is represented by scalars $\{\lambda I\}$ under an irreducible $\pi \in \hat{G}$ (Schur's Lemma), and for $\tilde{SL}_2$ these scalars must be m-th roots of the identity $\{\omega: \omega^m = 1\}$ (due to $\det \pi_g = 1$), it follows that center $Z(\tilde{G})$ is mapped by T into a finite group of matrices with diagonal entries $\{\omega\}$. But $Z \simeq \mathbb{Z}$ is infinite, so some $g \in Z$ must go to $\{I\}$, which means T can not be exact. In other words, $\tilde{SL}_2$ can not be realized as matrix group!

Let us make a few comments concerning the *exp*-log correspondence: $\mathfrak{G} \leftrightarrow G$. It clearly, preserves all general (functorial) properties of groups and algebras. So

- *closed (connected) subgroups* of $G \leftrightarrow$ *subalgebras* of $\mathfrak{G}$,

- *normal subgroups* $H \subset G \leftrightarrow$ *ideals* $\mathfrak{H} \subset \mathfrak{G}$ - subalgebras invariance under the adjoint action of $\mathfrak{G}$: $[\mathfrak{G};\mathfrak{H}] \subset \mathfrak{H}$, or $\mathrm{ad}_X(\mathfrak{H}) \subset \mathfrak{H}$, for all $X \in \mathfrak{G}\}$.

- *other "natural subgroups"*, like *center* $Z = Z(G)$, commutant $G_1 = [G;G]$ (a subgroup, generated by all *group commutators* $\{g^{-1}h^{-1}gh: g,h \in G\}$), etc., have proper counterparts in the Lie algebra: *central ideal* $\mathfrak{Z} = \mathfrak{Z}(\mathfrak{G})$, *commutant*

$$\mathfrak{G}_1 = [\mathfrak{G};\mathfrak{G}] = Span\{[X;Y]: X,Y \in \mathfrak{G}\}, \text{ etc.}$$

Also any decomposition of group in the *direct, semidirect product,* yields a similar decomposition of its Lie algebra.

4.3. The adjoint action. Lie group acts on its Lie algebra by the so called *adjoint action* Ad_g. First let us observe that G acts on itself by the group conjugation $g: h6 \to g^{-1}hg$. The latter can be lifted to the Lie algebra via exponential map:

$$g^{-1}(\exp A)g = \exp(g^{-1}Ag), \text{ for } A \in \mathfrak{G}. \tag{4.9}$$

We call the resulting action (4.9) of G on $\mathfrak{G}$ the *adjoint representation* Ad_g. Of course, for matrix groups G, the adjoint action is nothing but a conjugation with g,

$$\mathrm{Ad}_g(A) = g^{-1}A\,g.$$

In general, adjoint action represents G by automorphisms of its Lie algebra,

$$\mathrm{Ad}_g[A;B] = [\mathrm{Ad}_g(A);\mathrm{Ad}_g(B)].$$

We shall see that the *adjoint action* of group G by automorphisms of its Lie algebra corresponds to the *adjoint action* (4.1) of $\mathfrak{G}$ on itself by derivations, $\mathrm{ad}_A(B) = [A;B]$. In fact derivations $\{\mathrm{ad}_A\}$ represent *infinitesimal generators* of one-parameter subgroups of automorphisms $\{\mathrm{Ad}_{\exp tA}\}$, as will be explained in Theorem 5 below.

4.4. Basic classification of Lie groups and algebras; Examples. On the algebraic level the realm of Lie algebras (it is easier to start with them, as one avoids many topological complication) contains two large and distinct classes:

I) *solvable and nilpotent algebras* (the latter make up a subclass of solvables);

II) *simple and semisimple algebras.*

We shall briefly describe both of them. The definition of the former is given in terms of the *upper derived series* of repeated commutators,

$$\mathfrak{G} \supset \mathfrak{G}' = [\mathfrak{G};\mathfrak{G}] \supset \mathfrak{G}'' = [\mathfrak{G}';\mathfrak{G}'] \supset \dots \supset \mathfrak{G}^{(k)} = [\mathfrak{G}^{(k-1)};\mathfrak{G}^{(k-1)}] \supset \dots \tag{4.10}$$

Here each subsequent term $\mathfrak{G}^{(k)}$ is *derived* via *commutators,* $\mathrm{Span}\{[X;Y];\ X,Y \in \mathfrak{G}^{(k-1)}\}$ of the preceding term $\mathfrak{G}^{(k-1)}$.

Algebra $\mathfrak{G}$ is *solvable* (of step n), if its derived series terminates at $\{0\}$, $\mathfrak{G}^{(n+1)} = \{0\}$ (similar notion applies to groups with *group-commutators* $\{g^{-1}h^{-1}gh\}$ in place of Lie brackets $[X;Y]$). A commutator $\mathfrak{G}'$ of Lie algebra is the smallest ideal $\mathfrak{H} \subset \mathfrak{G}$, with the commutative factor-algebra $\mathfrak{G}/\mathfrak{H}$, and one can check that all $\{\mathfrak{G}^{(k)}\}$ are also ideals of $\mathfrak{G}$. So solvable algebras/groups are obtained by subsequent abelian extensions of abelian algebras[22]. *Nilpotent algebras* form a subclass of solvable, they are defined in a similar manner with a *(smaller) lower derived series,*

$$\mathfrak{G} \supset \mathfrak{G}_1 = [\mathfrak{G};\mathfrak{G}] \supset \mathfrak{G}_2 = [\mathfrak{G};\mathfrak{G}_1] \supset \dots \supset [\mathfrak{G};\mathfrak{G}_k] \supset \dots \tag{4.11}$$

So each subsequent term is the commutator of the entire algebra $\mathfrak{G}$ with the preceding term. Once again the series is required to terminate at $\{0\}$, $\mathfrak{G}_{n+1} = [\mathfrak{G};\mathfrak{G}_n] = \{0\}$ *(step-n nilpotent).* In particular, $\mathfrak{G}_n$ belongs to the center of $\mathfrak{G}$, and the entire algebra is obtained by a sequence of *central extensions.*

The natural examples of solvable and nilpotent algebras are made of upper-

[22]The terminology came from the Galois theory, where *solvability* of the Galois group of an algebraic equation implies that the equation can be solved in radicals. Radicals correspond to cyclic subgroups of quotients $G^{(k)}/G^{(k-1)}$.

triangular matrices, respectively matrices with {0} on the main diagonal (problem 5). Other examples include the affine group Aff_1, realized by the upper-triangular matrices of the form $\left\{\begin{bmatrix} a & b \\ & 1 \end{bmatrix}\right\}$ (a step-one solvable group), and the Heisenberg groups $\mathbb{H}_1$ and $\mathbb{H}_n$ (step-one nilpotent groups) (see §1.1 and chapter 6).

Next we turn to *semisimple* and *simple* algebras. The former are characterized by the property that the commutator $[\mathfrak{G};\mathfrak{G}] = \mathfrak{G}$ (in other they have no commutative quotients), the latter (*simple)* have no ideals at all (hence no nontrivial quotients)[23]. These properties have important implications for the structure and analysis of simple and semisimple groups and algebras (see chapters 4-6). At this point we shall mention only two basic results:

i) *any semisimple algebra is a direct sum of simple ones,* $\mathfrak{G} \simeq \bigoplus_j \mathfrak{G}_j$; in other words *the adjoint action* (4.1) *on* $\mathfrak{G}$ *is completely reducible;*

ii) *all simple Lie algebras can be completely classified; they consist of the classical series: unimodular, orthogonal, symplectic, and a few exceptional examples.*

Here we shall list the basic *classical simple Lie algebras,* which correspond to *classical Lie groups* of §1.1.

Examples: 1) Lie algebra $\mathfrak{gl}(n) = Mat_n$, of the *general linear group* GL_n; and its subalgebra $\mathfrak{sl}(n) = \{A: trA = 0\}$, which corresponds to a *special linear group* SL_n.

Other classical groups (§1.1) preserve certain quadratic/bilinear forms J on $\mathbb{R}^n$; $\mathbb{C}^n$, so that

$$^{\mathrm{T}}gJg = J;\ \text{or}\ g^*Jg = J;\ \text{for group elements}\ \{g\}. \tag{4.12}$$

The corresponding condition for the Lie algebra elements $\{X\}$ becomes

$$^{\mathrm{T}}X\,J + J\,X = 0;\ \text{or}\ X^*J + J\,X = 0,$$

which easily derived from (4.12) by differentiating $g = exp\,tX$, at $t = 0$. Thus we get

2) $\mathfrak{so}(n)$ - algebra of all $n\times n$ real/complex antisymmetric traceless matrices $\{^{\mathrm{T}}X = -X;\ \mathrm{tr}\,X = 0\}$, which corresponds to the orthogonal group $SO(n)$. In the real case "traceless" condition holds automatically for all skew-symmetric matrices.

3) $\mathfrak{u}(n)$ and $\mathfrak{su}(n)$ - algebras of all $n\times n$ complex (hermitian) antisymmetric matrices, $\{X^* = -X\}$, and traceless antisymmetric matrices $\{tr\,X = 0\}$, correspond to to the *unitary*, or *special unitary groups,* $U(n)$ and $SU(n)$. In a similar fashion one can

[23]We can phrase both definitions in terms of the adjoint action: *simplicity means irreducibility* of ad_X *on* $\mathfrak{G}$, *while semisimple implies that operators* $\{\mathrm{ad}_X(\mathfrak{G}): X \in \mathfrak{G}\}$ *span entire space* $\mathfrak{G}$.

describe Lie algebras $\mathfrak{so}(p;q)$; $\mathfrak{su}(p;q)$ of $(p;q)$-indefinite orthogonal and unitary groups $SO(p;q)$ and $SU(p;q)$, *symplectic algebra* $\mathfrak{sp}(n)$.

For instance, matrices $X \in \mathfrak{so}(p;q)$ have block structure $\left\{ X = \begin{pmatrix} a & c \\ b & d \end{pmatrix} \right\}$ with $a \in \mathfrak{so}(p)$; $d \in \mathfrak{so}(q)$, and $c = {}^{T}b \in Mat_{p \times q}$, and similar representations could be found for $\mathfrak{su}(p;q)$, $\mathfrak{sp}(n)$ (problem 6). So *vector space* $\mathfrak{so}(p;q)$ is isomorphic to $\mathfrak{so}(p) \times \mathfrak{so}(q) \times Mat_{p \times q}$, which yields the dimension of Lie group $SO(p;q)$.

We shall see more examples in subsequent chapters 4-6. Two classes: *solvable* and *semisimple,* provide the building blocks for arbitrary Lie algebras $\mathfrak{L}$. The following general result will be stated here without proof.

Levi-Malcev decomposition Theorem: *Any finite-dimensional Lie algebra* $\mathfrak{L}$ *has the maximal solvable ideal* $\mathfrak{R}$, *called radical, and a semisimple subalgebra* $\mathfrak{G}$, *so that* $\mathfrak{L}$ *is decomposed into a semidirect product:* $\mathfrak{L} = \mathfrak{R} \oplus \mathfrak{G}$. *The radical* $\mathfrak{R}$ *is a characteristic ideal of* $\mathfrak{L}$ *(invariant under all automorphisms of* $\mathfrak{L}$*), and obviously unique, while any two semisimple factors* $\mathfrak{G}$; $\mathfrak{G}'$ *differ by an automorphism of* $\mathfrak{L}$, $\mathfrak{G}' = \alpha(\mathfrak{G})$, $\alpha \in Aut\,\mathfrak{L}$.

Let us recall that a *semidirect product* $\mathfrak{L} = \mathfrak{R} \oplus_s \mathfrak{G}$, is defined for a pair of Lie algebras $\{\mathfrak{R};\mathfrak{G}\}$, where $\mathfrak{G}$ acts by *derivations* on $\mathfrak{R}$,

$\mathfrak{G} \ni x \rightarrow \alpha = \alpha_x \in Der(\mathfrak{R})$, where $\alpha([a;b]) = [\alpha(a);b] + [a;\alpha(b)]$; for all $a, b \in \mathfrak{R}$.

By definition, $\mathfrak{L}$ consists of all pairs $\{(a;x)\}$ with the Lie bracket (problem 6),

$$[(a;x);(b;y)] = (a + \alpha_x(b); x + y). \tag{4.13}$$

The natural examples of semidirect products include the *Euclidian motion* and the *Poincare algebras:* $\mathfrak{E}_n = \mathbb{R}^n \oplus \mathfrak{so}(n)$ and $\mathfrak{P}_n = \mathbb{M}^n \oplus \mathfrak{so}(1;n-1)$. In both cases the radical $\mathfrak{R} = \mathbb{R}^n$ (or $\mathbb{M}^n$) is commutative, and a (simple) factor $\mathfrak{G} \simeq \mathfrak{so}(n)$ or $\mathfrak{so}(1;n-1)$ (orthogonal/pseudo-orthogonal) acts on $\mathfrak{R}$ by linear transformations. Another important example, discussed in chapter 6, is a semidirect product of the Heisenberg and symplectic Lie algebras: $\mathfrak{H}_n \oplus \mathfrak{sp}(n)$.

Further details and references on the general Lie theory could be found in [Che]; [Jac]; [Kir]; [Ser]; [Hel].

4.5. The infinitesimal method. The correspondence between Lie groups and Lie algebras extends to their representations. It is often convenient to replace Lie group representations by those of its Lie algebra, as the latter are much easier to analyze (study irreducibility; decomposition, etc.). We shall conclude this section with a general result that serves the basis of the *infinitesimal method* in representation theory. It will require a notion of the *generator* of a 1-parameter group of unitary (bounded) operators U_t in an ∞-D space $\mathcal{H}$.

Proposition 4: *(i) A one-parameter strongly continuous group* $\{U_t\}$ *of matrices, or bounded operators in space* $\mathcal{H}$ *has a generator A, defined on a dense subspace of "smooth vectors"* $\{\xi\}$ *(i.e. vectors which yield smooth vector-functions* $\xi(t) = U_t\xi$*),*

$$A\xi = \frac{dU_t\xi}{dt}\Big|_{t=0}.$$

Conversely, any selfadjoint (bounded and unbounded) operator A (or more general A, subject to some technical constraints) generates a 1-parameter group, denoted by $U_t\xi = e^{tA}\xi$.

(ii) *Generators of orthogonal/unitary groups* $\{U_t\}$ *are skew-symmetric,* $A^* = -A$.

The algebraic (finite-D) version exploits the well known feature of any $\{U_t\}$: differentiability to any order (even real analyticity)! The same holds for a *uniformly continuous* group U_t in Hilbert/Banach spaces. In both cases, $U_t = e^{tA}$, with a bounded operator[24] A (problem 4). The strongly continuous (unitary) case is more subtle, as not all vectors $\xi \in \mathcal{H}$ have differentiable "tails" $\{\xi(t) = U_t\xi\}$, in general. However, any U has *smooth vectors* $\{\xi : \xi(t) \in \mathcal{C}^1\}$ (or $\mathcal{C}^m$; $\mathcal{C}^\infty$; real analytic), which form dense subspaces in $\mathcal{H}$: $\mathcal{H} \supset \mathcal{H}_1 \supset \dots \mathcal{H}_m \supset \dots \mathcal{H}_\infty \supset \mathcal{A}$. Indeed, smooth vectors can be obtained by *mollifying (smoothing) operators* of J1.2,

$$U_f\xi \overset{def}{=} \int f(t)U_t\xi\,dt;\ f \in \mathcal{C}_0^m, \text{ or } \mathcal{C}_0^\infty.$$

It is easy to check that operators $\{U_f\}$ send $\mathcal{H}$ into the m-smooth subspace $\mathcal{H}_m$.

The density of the embedding $\mathcal{H}_m \subset \mathcal{H}$ is verified by observing that m-smooth compactly supported functions $\{f \in \mathcal{C}_0^m\}$ approximate any L^1-function on $\mathbb{R}$, as well as distribution δ (in the weak sense!). If $f_n \to \delta$, then images $U_{f_n}[\xi]$ are easily seen to approach any $\xi \in \mathcal{H}$. Thus we get a densely defined (closed) generator: $A = \frac{d}{i\,dt}U_t$ on $\mathcal{H}_1 \subset \mathcal{H}$. For unitary $\{U\}$ operator A is selfadjoint, $A^* = A$, so its exponential e^{itA} (as well as any other function "$f(A)$") can be defined via spectral resolution (Appendix A),

$$e^{itA} = \int e^{i\lambda t}\,dE(\lambda) = U_t.$$

Smooth vectors form a *domain* $\mathfrak{D}(A) = \mathcal{H}_1$ of a *closed, skew-adjoint* (in the unitary case)

---

[24]This fact is a generalization of differentiability of continuous one-parameter groups of numbers, or *characters* $\{\chi(t) : t \in \mathbb{R}\} \subset \mathbb{R}$ or $\mathbb{C}$. Such χ is shown to be differentiable, by smoothing it out via convolution with nice $f \in \mathcal{C}_c^r(\mathbb{R})$, i.e. $\chi(t) = \frac{1}{c}(\chi * f)(t)$, where $c = \int f(s)\chi(s)\,ds \neq 0$, and the RHS is obviously as smooth as f! Now it follows immediately, that $\chi(t) = e^{at}$, for $a \in \mathbb{R}$ or $\mathbb{C}$.

unbounded operator A. Space $\mathfrak{D}(A)$ is closed under a *stronger (graph) norm*,

$\|\xi\|_1 = \|\xi\| + \|A\xi\|$, or norm $\|(I+|A|)\xi\|$, where $|A| = (A^*A)^{1/2}$- the modulus of A.

In fact, any such A defines a scale of densely embedded ("Sobolev") spaces

$$\mathcal{H} \supset \mathcal{H}_1 \supset \dots \supset \mathcal{H}_m \supset \dots = \mathcal{H}_\infty;$$

where $\mathcal{H}_m = (I + |A|)^{-m}\mathcal{H} =$ domain of $A^m = \{m\text{-smooth vectors in } \mathcal{H}\}$. The intersection $\mathcal{H}_\infty = \bigcap_{m \geq 0} \mathcal{H}_m$- consists of ∞-smooth vectors, and is also dense in $\mathcal{H}$ (moreover, $\mathcal{H}_\infty$ contains yet another dense subspace $\mathcal{A}$ of *analytic vectors!*). All subspaces are invariant under the group U_t, and of course, $U_t \mid \mathcal{H}_m$ is strongly continuous in the m-norm.

Let us illustrate the general concepts of *smooth vectors; closedness; domains; "strong" vs. "uniform"* continuity of operator-groups with a few simple examples.

4.6. Examples: 1) *Multiplication operator*, $Af(x) = xf(x)$, on L^2-spaces: $L^2(\mathbb{R})$; $L^2([a;b])$; or more general $L^2(\mathbb{R}; dm)$, generates a unitary group: $U_t f = e^{itx} f(x)$, also acting by multiplications.

i) On finite intervals $[a;b]$, or for measures dm of "finite support", the group is *uniformly (norm) continuous*, and generator A is *bounded*, $\|A\| \leq C$, if supp $\subset [-C;C]$. The two conditions (norm-continuity and boundedness) always come together. Furthermore, *all vectors* $\{f\}$ in L^2 are "∞-smooth" (even real analytic), so $\mathcal{H} = \mathcal{H}_\infty = \mathcal{A} = L^2$.

ii) On infinite intervals (e.g. $\mathbb{R}$), A is no more bounded, and U_t is only *strongly continuous:* $\|U_t f - f\| \to 0$, as $t \to 0$, for any $f \in L^2$. Not all vectors are smooth, however, the space of "1-smooth vectors":

$$\mathcal{H}_1 = \mathfrak{D}(A) = \left\{f : \|f\|_1 = \|(1+|x|)f(x)\|_{L^2} < \infty\right\},$$

similarly "m-smooth vectors":

$$\mathcal{H}_m = \mathfrak{D}(A^m) = \left\{f : \|f\|_m = \|(1+|x|)^m f(x)\|_{L^2} < \infty\right\}, \text{ etc.}$$

Vector-function $e^{itx} f(x)$ could be differentiated at $t = 0$, only for 1-smooth vectors $\{f\}$, to get $Af = xf$!

2) The translation-group $\mathbb{R}$: $U_t f(x) = f(x-t)$, acing on L^2 (or other L^p-spaces) is *strongly continuous*. Its generator $A = \frac{d}{dx}$ is an unbounded operator with a dense domain made of "1-smooth" vectors/functions, $f \in L^2$, so that f' is also L^2,

$\mathcal{H}_1 = \left\{f : \|f\|_1 = \|f\|_{L^2} + \|f'\|_{L^2}\right\}$ - the so called *first-Sobolev space*.

Similarly, "*m-smooth vectors*" coincide with "*m-smooth (Sobolev) functions*" (see §2.2),

$$\mathcal{H}_m = \left\{ f : \|f\|_m = \|f\|_{L^2} + \dots + \left\| f^{(m)} \right\|_{L^2} \right\} \text{ - } m\text{-th Sobolev space}^{25}.$$

The last example explains the terminology "*smooth vectors*".

It could be extended to more general one-parameter group actions, e.g. *flows on manifolds*, ϕ_t: $x \to y(x;t)$ on $\mathcal{M}$. Any such flow is *generated* by a vector field $X = \sum a_j \partial_j$, as a solution of an ordinary differential system: $\dot{y} = X(y)$; $y(0) = x$ (so we could write: $\phi_t = \exp tX$). Associated to any flow ϕ_t is a 1-parameter group of linear transformation on function-spaces over $\mathcal{M}$, e.g. $L^2(\mathcal{M})$,

$$U_t f(x) = f(\phi_t(x)); \text{ or more general } V_t f(x) = \alpha(x;t) f \circ \phi_t(x),$$

where α satisfies the cocycle condition: $\alpha(x;\, t{+}s) = \alpha(x;t)\,\alpha(\phi_t(x);s)$. The generator of group U_t, is the vector-field X considered as a *1-st order differential operator* $\sum a_j \partial_j$ on functions over $\mathcal{M}$ (it plays the role of $D = \frac{d}{dx}$ in example 2). *Smooth vectors* consist of functions differentiable along the flow. In the second case, the generator is the 1-st order operator of the form: $A = X + q$, where q is a multiplication operator with function $q(x) = \alpha_t(x;0)$ (∂_t- derivative of α at $t = 0$). If function q is sufficiently smooth, one can check that "*m-smooth vectors*" of A are the same as for X.

Once the notion of *generator* of one-parameter groups was made clear, we can proceed to more general Lie group representations.

Theorem 5: *Any representation $g \to T_g$ of Lie group G in space $\mathcal{H}$ gives rise to a representation $X \to T_X$ of its Lie algebra $\mathfrak{G}$:*

$$T_X = \frac{d}{dt}\Big|_{t=0} T_{\exp tX};\; X \in \mathfrak{G}. \tag{4.14}$$

In the ∞-D case generators $\{T_X\}$ are unbounded operators defined a joint dense core of "smooth vectors" $\mathcal{H}_1 = \{\xi : \langle T_g \xi \mid \eta \rangle \in C^1(G);$ for all $\eta \in \mathcal{H}\}$. Furthermore, for unitary T operators *$\{T_X\}$ are skew-symmetric. Conversely, a representation of Lie algebra $\mathfrak{G}$ by (closed, unbounded) generators $\{T_X\}$ can be exponentiated (lifted) to a representation of its simply connected Lie group,*

$$T_{\exp X} = \exp T_X. \tag{4.15}$$

Proof: On the purely algebraic level (finite-D, uniformly continuous case) the result becomes a simple consequence of the *exp* and *log* maps (defined via Taylor expansions)

[25] One can show (§2.2) that the m-th Sobolev norm is equivalent to $\|(1+|D|)^m f\|_{L^2}$; where $|D| = (D^*D)^{1/2}$ is the *modulus* of the differentiation operator $D = \frac{d}{dx}$.

and the Campbell-Hausdorff formula (4.2). On the functional-analytic (unitary) level some technical complications appear, as generators $\{T_X\}$ could be unbounded (densely defined) operators, each with its own domain $\mathfrak{D}(T_X)$. However, all of them have a joint dense *core* of *1-smooth vectors* $\mathcal{H}_1$. As in the one-parameter case, $\mathcal{H}_1$ contains the range $\{\xi = T_f(\eta): \eta \in \mathcal{H}\}$ of any *mollifier* T_f with $f \in C_c^1(G)$. In fact, there exists the whole scale of *m-smooth spaces:* $\mathcal{H} \supset \mathcal{H}_1 \supset ...\mathcal{H}_m \supset ...\mathcal{H}_\infty \supset \mathcal{A}$ (analytic vectors), with $\mathcal{H}_m \supset \{T_f(\eta): f \in C_c^m(G)\}$. All of them are dense and invariant under the group action $\{T_g\}$ (for analytic vectors this result is due to L.Gårding). On each space $\mathcal{H}_m$ one can define the m-fold products and powers of generators $\{T_{X_1}...T_{X_m}: X_j \in \mathfrak{G}\}$, furthermore on analytic vectors $\{\xi \in \mathcal{A}\}$ the Taylor series: $\sum_0^\infty \frac{1}{m!} T_X^m(\xi)$, converges. Thus we get the group action on analytic vectors $\{T_g \mid \mathcal{A}: g = exp\, X\}$, which then extends rough the whole space $\mathcal{H}$. An alternative way to obtain the group representation in the unitary case (skew-symmetric generators) is via *spectral resolution:*

$$T_X = \int \lambda\, dE(\lambda),\ E(\lambda)\text{-spectral measure,}$$

whence the corresponding unitary group:

$$exp(t\, T_X) = T_{exp tX} = \int e^{it\lambda} dE(\lambda),$$

as in Proposition 4. This completes the proof.

Let us remark that many natural properties of representations, including irreducibility, decomposition[26], etc. can be transferred from Lie algebra $\mathfrak{G}$ to Lie group G. However, in the absence of *simple connectivity* of G local representation $\{T_{exp\, X}\}$ given by (4.15) may not be extendible through the whole group. Indeed, only "half" of irreducible representations of simply connected group $SU(2)$ extend through the representations of $SO(3) \simeq SU(2)/\{\pm I\}$ (chapter 4), and a similar pattern holds for all other non-simply-connected groups. Clearly, the dual object of any factor-group G/H consists of all $\pi \in \hat{G}$, that annihilate H, $\pi \mid H = 1$.

[26] with some obvious qualification: for any invariant subspace of the Lie algebra generators $\{T_X\}$ we take its closure in the $\mathcal{H}$-norm. So *irreducibility of* $\{T_X\}$ means any $\{T_X\}$-invariant subspace $\mathcal{H}'$ is dense in $\mathcal{H}$.

Problems and Exercises:

1. Prove Lemma 3 (Hint: some statements, like (ii), are easy to check first for *diagonalizable matrices* $A = U^{-1}DU$ (D - diagonal, $U \in GL_n$), and then to use density of diagonalizable $\{A\}$ in Mat_n!).

2. (i) Prove formula (4.5) for matrix subgroups, using the Taylor expansion of $\{exp\, tA\}$. It is known as Trotter product formula (4.5), and has numerous extensions and applications to groups (and semigroups) of operators in spaces of finite and infinite dimensions.

   (ii) Show that formula (4.6) defines a 1-parameter group with generator A.

3. Show that subspaces sl_n; $so(n)$ and $su(n)$ of $gl_n = Mat_n$ form Lie algebras.

4. A (strongly continuous) one-parameter group of unitary Hilbert-space operators $\{U_t\}$ has a selfadjoint generator $A = A^*$, i.e. $U_t = exp(iAt)$. Show U_t is uniformly (norm) continuous iff operator A is bounded. Steps:

   i) Use spectral decomposition of a selfadjoint (not necessarily bounded!) operator: A is unitarily equivalent to a multiplication, $x: f \to xf(x)$, in $L^2(\mathbb{R}; d\mu)$ or in the direct sum $\bigoplus_j L^2(\mathbb{R}; d\mu_j)$.

   ii) Show: A is bounded iff $supp\{d\mu\}$ or $\overline{\bigcup supp\{d\mu_j\}}$ is compact in $\mathbb{R}$.

   iii) Unitary group $\{U_t\colon f \to e^{itx}f(x)\}$ in $L^2(d\mu)$ is norm-continuous iff $supp\{d\mu\}$ is compact.

5. Show that all upper triangular matrices over $\mathbb{R}$ or $\mathbb{C}$ form a solvable Lie algebra $\mathfrak{B}_n$. Find the derived series for $\mathfrak{B}_n$, and the corresponding Lie group. Do the same for the the algebra $\mathfrak{U}_n$ of all upper triangular matrices with 0 on the main diagonal (show that $\mathfrak{U}$ is nilpotent, find its lower derived series, and its Lie group).

6. Find block-structure (as explained in example 3) of Lie algebras: $so(n)$, $su(n)$, $so(p;q)$, $su(p;q)$, $sp(n)$, and use it to compute dimensions of the corresponding Lie groups.

7. Show that (4.13) defines Lie bracket on the semidirect product $\mathfrak{R} \oplus \mathfrak{G}$ of Lie algebras.

8. Show that a normal discrete subgroup Γ of a Lie group G belongs in the center $Z(G)$ (Hint: conjugate each $\gamma \in \Gamma$ by elements $g \in G$, $\gamma \to g^{-1}\gamma g$, and observe that all $\{g\}$ in a vicinity of $\{e\}$ fix γ).

Chapter 2. Commutative Harmonic Analysis.

Schur's Lemma implies that irreducible representations of a commutative group G consists of 1-D *characters*, i.e. homomorphisms $\{\chi: G \to \mathbb{C}^*$ (multiplicative complex numbers)$\}$, or $\{\chi: G \to \mathbb{T} = \{e^{i\theta}\}$ - *unitary characters*$\}$. We shall mostly deal with the latter. The characters can be multiplied and inverted:

$$(\chi_1; \chi_2) \to \chi_1\chi_2;\ \chi \to \chi^{-1} = \frac{1}{\chi(x)};$$

so they also form a commutative group, called the *dual group* $\widehat{G}$. Dual group of a locally compact G, is itself locally compact [Pon], hence carries an invariant (Haar) measure. Three commutative groups lie at the heart of the classical harmonic analysis: $\mathbb{R}^n$; $\mathbb{Z}^n \subset \mathbb{R}^n$ and $\mathbb{T}^n \simeq \mathbb{R}^n/\mathbb{Z}^n$. The characters of all three consist of exponentials

$$\chi_\xi(x) = e^{i\xi \cdot x}.$$

In the case of $\mathbb{R}^n$, parameter ξ runs over $\mathbb{R}^n$, so the group becomes isomorphic to its dual. For $\mathbb{Z}^n$, identified with an integral lattice $\{m = (m_1;...m_n)\}$ in $\mathbb{R}^n$, characters

$$\chi = \chi_\theta(m) = e^{i\theta \cdot m},$$

are labeled by the *dual torus*, $\theta = (\theta_1;...\theta_n)$, $0 \le \theta \le 2\pi$, so $\widehat{\mathbb{Z}}^n \simeq \mathbb{T}^n$. The characters of torus $\mathbb{T}^n = \{(\theta_1;...\theta_n)\text{: } 0 \le \theta_j \le 2\pi\}$, or more general $\{0 \le \theta_j \le T_j\}$, are in turn labeled by the lattice points,

$$\chi = \chi_m(\theta) \quad e^{im \cdot \theta};\ m \in \mathbb{Z}^n,$$

so $\widehat{\mathbb{T}}^n \simeq \mathbb{Z}^n$. Let remark that the mutual duality of groups $\mathbb{T}^n$ and $\mathbb{Z}^n$ exemplifies the general *Pontrjagin Duality principle* for locally compact Abelian groups [Pon]: *the second dual group (characters of* $\widehat{G}$*) is isomorphic to* G *itself, via the map:* $a \to \widehat{a}(\chi) = \chi(a)$, $a \in G$ (elements $a \in G$ define characters on $\widehat{G}$). So the classification problem (I) of chapter 1 for commutative groups G amounts to characterization of $\widehat{G}$, while the decomposition problem (II) for the regular representation R will lead to study the *Fourier transform* $\mathcal{F}$ on L^2-spaces and other spaces. This marks the starting point of the *Fourier analysis*. **Chapter 2** gives the basic overview of the Commutative Fourier analysis, and some of its applications, the main emphasis being on differential equations.

§2.1. Fourier transform: Inversion and Plancherel Formula.

In this section we shall introduce the Fourier transform/series expansion on $\mathbf{R}^n$; $\mathbf{Z}^n$; $\mathbf{T}^n$, discuss their basic properties (with respect to translation, differentiation, dilation, rotations. Then we establish the key result of the Fourier analysis, Plancherel/Inversion formula, in 2 different ways. One exploits Gaussians and Schwartz functions, another Poisson summation.

1.1. ***Fourier transform*** $\mathfrak{F}$ on group $G = \mathbb{R}^n; \mathbb{T}^n$ is defined by integrating functions $\{f\}$ against characters $\{\chi = e^{i\xi \cdot x}\}$,

$$\boxed{\mathfrak{F}: f(x) \to \hat{f}(\xi) = \int_{\mathbb{R}^n} f(x)\, e^{-i\xi \cdot x}\, dx} \tag{1.1}$$

while on $\mathbb{Z}^n$ integration becomes summation,

$$\boxed{\mathfrak{F}: f = \{f_m\} \to \hat{f}(\theta) = \sum_m f_m e^{-im \cdot \theta}\, dx} \tag{1.2}$$

Sometimes we shall indicate the "x to ξ" transform by $\mathfrak{F}_{x \to \xi}$. From the definition one could easily check a few basic properties of $\mathfrak{F}$.

1) *Convolution→product*, $\mathfrak{F}: f * g \to \hat{f}(\xi)\hat{g}(\xi)$

2) *Translation→multiplication with character*, $\mathfrak{F}: R_a f = f(x-a) \to e^{ia \cdot \xi}\hat{f}(\xi)$

3) *Differentiation→multiplication*, $\mathfrak{F}: \partial_j f \to i\xi_j \hat{f}(\xi)$; hence $\mathfrak{F}: \partial^\alpha f \to (i\xi)^\alpha \hat{f}$

Here $\{\partial_j\}$ and $\{\partial^\alpha : \alpha = (\alpha_1; \ldots \alpha_n)\}$ denote partial differentiations in x. The latter relation is easily verified via integration by parts:

$$\int (\partial^\alpha f)\, g = \int f\, (-\partial)^\alpha g; \text{ with } g = e^{-ix \cdot \xi}.$$

We introduce an *involution* on functions $\{f(x)\}$ on G, as a linear map,

$$*: f(x) \to f^*(x) = \overline{f(-x)},$$

with the obvious properties: $(f * g)^* = g^* * f^*$; $(f^*)^* = f$. Clearly,

4) *Involution→complex conjugation*, $\mathfrak{F}: f^*(x) \to \overline{\hat{f}(\xi)}$.

5)A linear change of variable, $A: x \to Ax$ $(A \in GL_n)$ on $\mathbb{R}^n$, defines a transformation of functions,

$$f \to f^A(x) = f \circ A = f(Ax).$$

Fourier transform takes such linear change into its inverse transpose, precisely

$$\mathfrak{F}: f^A \to \frac{1}{\det A} \hat{f} \circ ({}^t A^{-1}).$$

Finally, relation (3) ***differentiation→multiplication***, can be reversed into

6) ***multiplication→differentiation***, $\mathfrak{F}: x_j f(x) \to \frac{1}{i}\partial_{\xi_j}\widehat{f}(\xi); \Rightarrow x^\alpha f \to (\frac{1}{i}\partial_\xi)^\alpha \widehat{f}$

As we mentioned earlier characters $\{\chi\}$ describe all irreducible representations of the commutative group G. On the other hand they can be interpreted as *joint eigenfunctions* of all *translations* $\{R_a: a \in G\}$, respectively, of their *infinitesimal generators* $\{\partial_j\}$ (on $\mathbb{R}^n$; $\mathbb{T}^n$), as well as *convolution operators:* $R_f[g] = (f * g)(x)$. Formulae (1-3) can be rephrased now in the language of the representation theory by saying: *map* $\mathfrak{F}$ *diagonalizes regular representation* R *of* G,

$$\mathfrak{F} R \mathfrak{F}^{-1} \simeq \text{"direct sum/integral of characters"}.$$

Precisely, any function f (in a suitable space) can be expanded into the direct sum/integral of characters $\chi = e^{ix \cdot \xi}$,

$$f(x) = \sum b_\xi e^{ix \cdot \xi}; \text{ or } f(x) = \int b(\xi) e^{ix \cdot \xi} d\xi, \tag{1.3}$$

with "Fourier coefficients/transform" $b(\xi) = \widehat{f}(\xi)$, and the regular representation turns into multiplication on the space of "Fourier coefficients",

$$R_a: \widehat{f}(\xi) \to e^{ia \cdot \xi} \widehat{f}(\xi).$$

Later on (chapters 3-7) we shall establish various noncommutative versions of this result.

1.2. Plancherel and Inversion formula. The fundamental property of exponentials $\{e^{i\xi \cdot x}\}$ is their *orthogonality* and *completeness.* In case of compact G (e.g. $\mathbb{T}^n$), orthogonality can be understood literally[1],

$$\boxed{\langle \chi_m \mid \chi_k \rangle = \int_{\mathbb{T}^n} e^{i(m-k) \cdot x} dx = (2\pi)^n \delta(m-k)} \tag{1.4}$$

as all $\{\chi_m\}$ are L^2-functions on G. Completeness means that any f can be approximated, hence uniquely expanded into the Fourier series. An elementary proof for $\mathbb{T}^n$ exploits the *Cezaro means* (problem 3).

More general explanation has to do with *compactness* of convolution-operators, $R_f: u(x) \to f * u(x)$, in L^2 (L^p)-spaces on $\mathbb{T}^n$ (Appendix B). Compact self-adjoint ($f = f^*$) operators, are well known to have a complete orthogonal system of eigenvectors, and exponentials $\{\chi_m = e^{ix \cdot m}\}$ are precisely the eigenvectors of R_f, $f * e^{ix \cdot \xi} = \widehat{f}(m) e^{ix \cdot \xi}$.

---

[1]The same holds on an arbitrary compact group G (see [HR]; [Loo]), namely,

$$\langle \chi_m \mid \chi_k \rangle = \int_G \chi_m \bar{\chi}_k dx = |G| \delta_{km}; \; |G| = vol(G); \; \delta_{km}\text{- Kronecker } \delta.$$

Similar arguments apply to all compact groups (chapter 3).

For non-compact groups $\mathbb{R}^n$; $\mathbb{Z}^n$ *orthogonality* should be understood in a generalized (distributional) sense,

$$\boxed{\langle e^{i\xi\cdot x} \mid e^{i\eta\cdot x}\rangle = \int e^{i(\xi-\eta)\cdot x}dx = (2\pi)^n\delta(\xi-\eta)} \tag{1.5}$$

In other words, the role of *Kronecker* $\delta_{mk} = \delta(m-k)$, being played by the *Dirac δ-function.* The integral (1.5) is strictly speaking divergent, but it makes sense as a *distribution.* Namely, for any nice (testing) function $f(\eta)$ on $\mathbb{R}^n$,

$$\int\int e^{ix\cdot(\xi-\eta)}f(\eta)\,d\eta\,dx = \int e^{ix\cdot\xi}\widehat{f}(x)\,dx = (2\pi)^n f(\xi).$$

We shall see that in such form orthogonality is equivalent to the *Inversion/Plancherel formula* for $\mathfrak{F}$. Namely the transform,

$$\mathfrak{F}'\colon h(\eta)\to\check{h}(x) = \langle h \mid e^{-i\eta\cdot x}\rangle = \int h(\eta)e^{i\eta\cdot x}d\eta, \tag{1.6}$$

takes $\widehat{f}$ back into Const $\times f$. Therefore, (1.6) defines the *inverse Fourier transform* of h. The precise result is usually stated for a suitable class of functions on G. We shall treat separately compact $(\mathbb{T}^n;\mathbb{Z}^n)$-cases and non-compact $\mathbb{R}^n$- case.

1.3. Compact case. A suitable class of functions on torus, called *Fourier algebra,* consists of absolutely convergent Fourier series,

$$\mathcal{A} = \{f(x) = \sum a_k e^{ix\cdot k} \colon \sum |a_k| < \infty\}.$$

Clearly, algebra $\mathcal{A}\subset C(\mathbb{T}^n)$, is made of continuous functions on $\mathbb{T}^n$, and contains sufficiently many smooth functions, $\mathcal{A}\supset C^m$, for $m>n$. The latter is easy verified using differentiation formula 3). Indeed, Fourier coefficients of a partial derivative $\partial^\alpha f = \sum (ik)^\alpha a_k e^{ix\cdot k}$, by (iii). So if $f\in C^m$, then its Fourier coefficients $a_k = O(|k|^{-m})$, which guarantees convergence of $\sum |a_k|$.

Uniqueness and Plancherel Theorem on $\mathbb{T}^n$: (i) *Any function $f\in\mathcal{A}(\mathbb{T}^n)$ is uniquely determined by its Fourier coefficients,*

$$\{a_k\}\leftrightarrow f(x) = \sum a_k e^{ix\cdot k}.$$

(ii) *For any function $f = \sum a_k e^{ix\cdot k}\in L^2$,*

$$\boxed{\|f\|^2 = (2\pi)^n\sum |a_k|^2} \tag{1.7}$$

In other words $\mathfrak{F}\colon L^2(\mathbb{T}^n)\to\ell^2(\mathbb{Z}^n)$, is a unitary map. Both results are fairly obvious corollaries of the completeness and orthogonality[2] of $\{e^{ix\cdot k}: k\in\mathbb{Z}^n\}$ (problem 3).

1.4. Continuous case $\mathbb{R}^n$. Here a suitable class consists of all functions $\{f\}$, integrable along with their $\mathfrak{F}$-transforms,

$$\{f\in L^1: \hat{f}\in L^1\} = L^1\cap\mathcal{A}.$$

As above, $\mathcal{A}$ denotes the *Fourier algebra* $\{\hat{f}(x): f\in L^1\}$, of absolutely convergent Fourier integrals. As with $\mathbb{T}^n$ algebra $\mathcal{A}$ is shown to consist of continuous functions, decaying at $\{\infty\}$, a consequence of the *Riemann-Lebesgue Theorem* (problem 1 of §2.2). Furthermore, $\mathcal{A}\supset\mathcal{C}_c^m$ - compactly supported m-smooth functions[3], so

$$\mathcal{C}_c^m\subset\mathcal{A}\subset\mathcal{C}_0, \text{ for } m>n.$$

Inversion/Plancherel Theorem on $\mathbf{R}^n$: (I) *Any function $f\in\mathcal{A}\cap L^1$, can be uniquely recovered from $\hat{f}$ by the Inverse Fourier transform*

$$\boxed{\mathfrak{F}': \hat{f}\to f(x) = \frac{1}{(2\pi)^n}\int\hat{f}(\xi)e^{ix\cdot\xi}d\xi} \tag{1.8}$$

(II) *transformation $\mathfrak{F}$, normalized by factor $\frac{1}{(2\pi)^{n/2}}$, becomes a unitary map from $L^2(\mathbb{R}^n)$ onto $L^2(\mathbb{R}^n)$, i.e.*

$$\boxed{\|\hat{f}\|^2 = (2\pi)^n\|f\|^2, \textit{ for all } f\in L^2} \tag{1.9}$$

The Fourier transform of an L^2-function in (1.9) should be understood in the "square-mean" sense, i.e. by approximating f with nice (L^1; compactly supported) functions. Let us make a few comments:

• the general inversion formula (1.8) is equivalent to a special case $x=0$ (and is often stated in that form),

$$f(0) = \int_{\hat{G}}\hat{f}(\xi)\,d\xi,$$

multiplication of $\hat{f}(\xi)$ by $e^{ix\cdot\xi}$ corresponds to a shift $f(0)\to f(x)$, under the inverse transform $\mathfrak{F}': \hat{f}\to f$.

---

[2]Indeed, for any complete orthogonal system (basis) $\{\psi_k\}$ in Hilbert space $\mathcal{H}$, any $f\in\mathcal{H}$, is uniquely decomposed into a series $f=\sum a_k\psi_k$, and $\|f\|^2 = \sum|a_k|^2\|\psi_k\|^2$ (Parceval identity). Clearly, factor $(2\pi)^n$ in (1.9) represents L^2-norms of Fourier harmonics $\{\psi_k = e^{ix\cdot k}\}$.

[3]The inclusion $\mathcal{C}_c^m\subset\mathcal{A}$ follows from the differentiation formula (3): $\mathfrak{F}[\partial^\alpha f] = (i\xi)^\alpha\hat{f}$ whence $\hat{f}(\xi) = O(|\xi|^{-m})$, for any $f\in\mathcal{C}_c^m$, which yields absolute convergence of the integral $\int|\hat{f}(\xi)|\,d\xi$.

• Plancherel formula (1.9) follows from the Inversion (1.8) applied to a function:

$$g(x) = \int f(x+y)\overline{f(y)}dy = f*f^*. \tag{1.10}$$

Indeed, from (i)-(ii) it follows that $\widehat{g}(\xi) = \left|\widehat{f}(\xi)\right|^2$, while $g(0) = \|f\|^2$. So inverting $\widehat{g}$ via (1.8), we get the Plancherel formula (1.9) for f.

• The result holds for any locally compact Abelian group G ([Loo]; [HR]). Here $\mathcal{F}$ transforms functions $\{f(x)\}$ on G into $\{\widehat{f}(\chi)\}$ on $\widehat{G}$, takes $\mathcal{A}\bigcap L^1(G)\to\mathcal{A}\bigcap L^1(\widehat{G})$, and defines a unitary map: $L^2(G)\to L^2(\widehat{G})$.

We shall present two arguments based on very different, but equally important ideas. One of them exploits the *Gaussian* and *Schwartz functions*, another a "discrete (lattice) approximation" of $\mathbb{R}^n$, and the so called *Poisson summation formula.*

The Gaussian method. Obviously, it suffices to prove (1.9) for a dense class of functions in $\mathcal{A}\bigcap L^1$ or L^2. One possible choice are infinitely smooth $\{f\}$, rapidly decaying at $\{\infty\}$ with all their derivatives

$$|x|^{-m}\partial^\alpha f(x)\to 0, \text{ as } x\to\infty, \text{ for all } m \text{ and } \alpha,$$

equipped with a suitable system of seminorms, e.g.

$$\|f\|_{\alpha\beta} = \max_x \left|x^\alpha f^{(\beta)}(x)\right|.$$

These are so called *Schwartz functions* on $\mathbf{R}^n$. The class of Schwartz functions and distributions is discussed in detail in §2.2. Here we shall only remark that the Schwartz class is sufficiently rich. It contains, all C^{∞}- compactly supported $\{f\}$; the *Gaussian function:* $G = G_t(x) = e^{-tx^2}$ ($x^2 = x\cdot x = |x|^2$), and all products: $G\times$ "polynomials".

The latter class, $\mathcal{P} = \Re f = p(x)e^{-x^2/2}$: polynomial $p(x)\}$, will be used in our proof of inversion Theorem. Let us remark that space $\mathcal{P}$ is dense in $C_0(\mathbf{R}^n)$, by the Weierstrass "uniform approximation" principle, as well as in other spaces: L^2, L^p, $\mathcal{A}\bigcap L^1$, $\mathcal{I}$, with appropriate norms. The inversion for the Gaussian itself is verified directly (problem 5):

$$\mathcal{F}: G_t \to \hat{G}_t = (2\pi t)^{n/2} e^{-\xi^2/4t},$$

i.e. Gaussian is Fourier-transformed into another (differently scaled) Gaussian, whence follows $G(x) = \mathcal{F}^{-1}[\hat{G}]$. Then it remains to apply the differentiation and multiplication rules 3)-6), to get Fourier-inversion for any product-function, $f = p_n(x)G$ in $\mathcal{P}$. The proof is completed by the standard density argument: once Plancherel formula is established for a dense subspace $\mathcal{P}\subset L^2$, it follows for all $f\in L^2$.

1.5. Poisson Summation Formula: *Let functions f and $\hat{f}$ be integrable and* *continuous on* $\mathbb{R}^n$ *(e.g.* $f \in \mathcal{A} \cap L^1$*). Then for any period* $T > 0$,

$$\boxed{\sum_{k \in \mathbb{Z}^n} f(Tk) = \frac{1}{T^n} \sum_{m \in \mathbb{Z}^n} \widehat{f}(\tfrac{2\pi}{T}m)} \tag{1.11}$$

Similar result holds for an arbitrary lattice: $\Gamma = \{m = \sum m_j t_j\}$, *spanned by vectors* $\{t_1;...t_n\} \subset \mathbb{R}^n$, *and its dual lattice,* $\Gamma' = \{\gamma = \sum k_j a_j\}$, *where* $a_i \cdot t_j = 2\pi\delta_{ij}$. *Namely,*

$$\boxed{\sum_{m \in \Gamma} f(m) = \frac{1}{vol(\mathcal{M})} \sum_{\gamma \in \Gamma'} \widehat{f}(\gamma)} \tag{1.12}$$

where $\mathcal{M} = \mathbb{R}^n/\Gamma \simeq \mathbb{T}^n$ *denotes the fundamental domain of* Γ.

The proof is fairly easy and straightforward. We define a new Γ-periodic function:

$$F(x) = \sum_{k \in \Gamma} f(x+k).$$

Series obviously converges and gives a continuous function on torus $\mathbb{R}^n/\Gamma \simeq \mathbb{T}^n$. Function F is expanded in the Fourier series,

$$F(x) = \sum_{m \in \Gamma'} a_m e^{im \cdot x};$$

with coefficients $\{a_m\}$ on the dual lattice Γ':

$$a_m = \frac{1}{vol(\mathbb{R}^n/\Gamma)} \int F(x) exp(im \cdot x) dx.$$

Next the *torus inversion formula* applies,

$$F(0) = \sum_{k \in \Gamma} f(k) = \sum_{m \in \Gamma'} a_m = \sum_m \frac{1}{vol} \int \sum_k f(x+k) exp(-im \cdot x) = \frac{1}{vol} \sum_m \widehat{f}(m),$$

and the result follows.

Remark: Formula (1.11) has many different meanings, interpretations, and applications. One of them is in terms of distribution: $\delta_\Gamma = \sum_{m \in \Gamma} \delta(x-m)$ on $\mathbb{R}^n$, "δ-function of Γ". The LHS of (1.12) pairs δ_Γ to a nice *test-function* $f(x)$, while the RHS defines the *Fourier transform* of $\phi = \delta_\Gamma$. By definition, $\langle \widehat{\phi} \mid \widehat{f} \rangle = \langle \phi \mid f \rangle$, for any distribution ϕ and a test-function f. So (1.11)-(1.12) reads,

$$\widehat{\delta}_\Gamma(x) = \mathcal{F}(\sum \delta_k) = \sum_{k \in \Gamma} e^{-ik \cdot \xi} = \frac{1}{vol(M)} \sum_{m \in \Gamma'} \delta_m(\xi) = \frac{1}{vol(M)} \delta_{\Gamma'};$$

in other words "*δ-function of Γ is Fourier-transformed into* $\frac{1}{vol} \times$ "*δ-function of Γ'*".

In such s form Poisson summation turns into a special case of the general result for arbitrary Abelian locally compact groups: *if Γ is a closed subgroup of G, and Γ' denotes the annihilator of Γ in $\widehat{G}$, $\Gamma' = \{\chi : \chi \mid \Gamma = 1\}$, then δ-function of Γ (understood as integral $\langle \delta_\Gamma \mid f \rangle = \int_\Gamma f dy$, dy-Haar measure on Γ), is Fourier transformed into the δ-function of Γ', $\mathfrak{F}: \delta_\Gamma \to Const\, \delta_{\Gamma'}$.* The key idea is that averaging over Γ, sends functions $f(x)$ on G into functions on the factor-group G/Γ,

$$f(x) \to \tilde{f}(x+\Gamma) = \int_\Gamma f(x+y)dy = \delta_\Gamma * f.$$

The dual group to G/Γ is identified with Γ', and the Fourier transform takes the projection map: $f \to \delta_\Gamma * f$ (from G to G/Γ), into the restriction map: $\widehat{f} \to \widehat{f} \mid \Gamma'$.

Poisson summation immediately yields the inversion/Plancherel formula on $\mathbb{R}^n$. We let $T \to \infty$ in (1.11), and remember that $f(x) \to 0$ at $\{\infty\}$. The LHS of (1.11) clearly goes to $f(0)$, while the RHS converges to

$$\frac{1}{(2\pi)^n} \int \widehat{f}(\xi) d\xi, \text{ so } f(0) = \frac{1}{(2\pi)^n} \int \widehat{f}(\xi) d\xi, \text{ QED.}$$

Further applications of the Poisson summation will appear in §2.4 and chapter 7.

Problems and Exercises:

1. Show that any continuous character on $\mathbf{R}$ (a homomorphism $\chi: \mathbf{R} \to \mathbf{C}^*$) is of the form $\chi(a) = e^{za}$, for some $z \in \mathbf{C}$ (prove it first for rational $a \in \mathbf{Q}$, then use continuity of χ!). Deduce the character formulae (1.0) for $\mathbf{R}^n$; $\mathbf{Z}^n$; $\mathbf{T}^n$.

2. i) If a sequence of functions $\{F_n\}$ on $\mathbb{R}^n$; $\mathbf{T}^n$ approximates δ-function: $\langle F_n \mid h \rangle \to h(0)$, as $n \to \infty$, for any continuous h, then the corresponding convolution kernels: $F_n * f(x) \to f(x)$, as $n \to \infty$, for all $\{f\}$ in $\mathcal{C}$ and L^p, $1 \le p < \infty$ (such sequences are called *approximate unities*).

   ii) Any sequence of positive functions $\{F_n\}$ with the properties:
   $$\text{(a) } \int F_n dx = 1; \text{ (b) } \int_{|x| \le \epsilon} F_n dx \to 0, \text{ as } n \to \infty, \text{ for any } \epsilon \;\; 0,$$
   forms an approximate unity.

3. *Completeness of characters* $\{\chi_m = e^{im \cdot x}\}$ on $\mathbf{T}^n$: any $f \in \mathcal{C}$, or L^2 is approximated by trigonometric polynomials,
   $$p_N = \sum_{|m| \le N} b_m e^{im \cdot x}.$$
   Hint: It suffices to check the result for continuous $\{f\}$ with the standard $\| \ \|_\infty$. Note, that partial sums,
   $$s_n(x; f) = \sum_{|k| \le n} a_k e^{ik \cdot x},$$
   may not converge to $f \in \mathcal{C}$ uniformly, in general. However, certain *means* of s_n do, like *Cesaro*,
   $$c_n(x) = \frac{1}{n+1} \sum_{k \le n} s_k(x).$$
   i) Show that both sums s_n and c_n are given by convolutions,
   $$s_n(f) = S_n * f; c_n(f) = F_n * f,$$
   with kernels, which in 1-D case take the form,
   $$S_n(\theta) = \frac{\sin(n+\frac{1}{2})\theta}{\sin\frac{1}{2}\theta}; \text{ and } F_n(\theta) = \sum_{|k| \le n} (1 - \frac{k}{n+1}) e^{ik\theta} = \frac{\sin^2\frac{n\theta}{2}}{\sin^2\frac{\theta}{2}},$$
   called the *Fejer kernel.*

   ii) Show that the Fejer kernel forms an approximate unity (problem 2), hence $F_n * f(x) \to f(x)$, uniformly in $\mathcal{C}$, as well as in any L^p-norm, $1 \le p < \infty$.

4. Establish the decay rates of Fourier transform of m-smooth functions $\mathcal{C}^m$ and $\mathcal{H}_m$ on $\mathbf{T}^n$and $\mathbb{R}^n$: estimates (1.11); (1.0); (1.0).

5. **Fourier integrals** on $\mathbb{R}$ are often evaluated by the techniques of complex analysis: calculus of residues, branch-cuts (if necessary), changing path of integration.

   Use complex analysis to evaluate the following integrals:

   i) Gaussians: $G_t = \mathcal{F}(e^{-tx^2/2}) = \frac{1}{\sqrt{2\pi t}} e^{-\xi^2/2t} = G_t$ (*in* 1−*D*); and $(2\pi t)^{-n/2} e^{-\xi^2/2t}$ (*in* *n*−*D*).

   ii) $\mathcal{F}(e^{-t|x|}) = \frac{t}{\pi(\xi^2 + t^2)} = P_t$; and $\mathcal{F}(\frac{t}{x^2 + t^2}) = e^{-t|x|}$.

   iii) $\mathcal{F}(\frac{\sinh ax}{\sinh bx}) = 2\pi \frac{\sin\frac{\pi a}{b}}{\cosh\frac{\pi x}{b} + \cos\frac{\pi a}{b}}$, on $\mathbb{R}$; $0 < a < b$ (calculus of residues).

   Integrals (ii-iii) are evaluated by the residue calculus. They represent the *Poisson kernels* (§2.2) in the half-plane and a strip.

The following 3 Fourier integrals appear in the harmonic analysis on group $SL_2(\mathbf{R})$ - the celebrated Selberg-trace formula (§7.6 of chapter 7):

iv) $\int_0^\infty \frac{\lambda^{is}}{1+2\lambda\cos\theta+\lambda^2}d\lambda = \frac{\pi \sinh s\theta}{\sin\theta \sinh\pi s}$; $-\pi<\theta<\pi$

v) $\int_0^\infty \frac{\lambda^{is}}{1-2\lambda\cos\theta+\lambda^2}d\lambda = \frac{\pi \sinh s(\theta-\pi)}{\sin\theta \sinh\pi s}$; $0<\theta<\pi$;

vi) $\int_0^\infty \frac{\lambda^{is}}{1-2\lambda\cos\theta+\lambda^2}d\lambda = -\frac{\pi \sinh s(\theta+\pi)}{\sin\theta \sinh\pi s}$; $-\pi<\theta<0$.

6. *Show that the Fourier transform of the radial function* $f = f(|x|)$ in $\mathbf{R}^n$ is also radial, find the convolution of two radial functions $f(r)$, $g(r)$ in polar coordinates.

7. Show that the Gaussian $G_t(x)$ and the Poisson kernel P_t form convolution-semigroups:

$$G_t*G_s = G_{t+s}, \text{ and } P_t*P_s = P_{s+t}.$$

Use the "product→convolution" property of $\mathfrak{F}$, and explicit form of $\widehat{G}$ and $\widehat{P}$).

§2.2. Fourier transform on function-spaces.

In this section we shall briefly discuss the basic properties of Fourier transform on various function-spaces, including L^p-spaces, Sobolev spaces, distributions and analytic functions. The material will provide some background for a subsequent study of differential equations in §2.4 and other parts of the book.

2.1. L^p- spaces and interpolation. Plancherel formula implies unitarity of the normalized Fourier transform $\mathfrak{F}':f\to(2\pi)^{-n/2}\hat{f}$ on L^2,

$$\|f\|_{L^2}=\|\hat{f}\|_{L^2}.$$

On the other hand Riemann-Lebesgue Lemma means that $\mathfrak{F}:L^1\to L^\infty$ boundedly,

$$\|\hat{f}\|_\infty\leq\|f\|_1.$$

Such estimates can be extended to other L^p-spaces by the *Riesz interpolation Theorem*. Namely, $\mathfrak{F}:L^p\to L^q$, where $q=\frac{p}{p-1}$, is the dual Hölder index, furthermore

$$\boxed{\|\hat{f}\|_q\leq C(p)\|f\|_p}\ \textit{(Hausdorff-Young inequality)}.$$

The exact constant in the Hausdorff-Young was calculated by Beckner in 70's,

$$C(p)=\left(\frac{p^{1/p}q^{1/q}}{2\pi}\right)^{n/2}, \tag{2.1}$$

and was shown to be attained on the Gaussian function $f=e^{-x^2/2}$. Interpolation provides a powerful tool in estimating linear (bilinear, multi-linear) operators in $\{L^p\}$ and other scales of spaces. The simplest linear version states,

Riesz-Thorin Theorem: *if operator T maps boundedly $L^{p_1}(X)\to L^{q_1}(Y)$, and $L^{p_2}(X)\to L^{q_2}(Y)$,*

$$\|Tf\|_{q_1}\leq C_1\|f\|_{p_1},\ \textit{and}\ \|Tf\|_{q_2}\leq C_2\|f\|_{p_2},$$

then for any pair of indices:

$$\frac{1}{p}=\frac{1-s}{p_1}+\frac{s}{p_2};\ \frac{1}{q}=\frac{1-s}{q_1}+\frac{s}{q_2},$$

between $(p_1;q_1)$ and $(p_2;q_2)$, operator T is bounded from $L^p(X)$ to $L^q(Y)$, and its (p,q)-norm is estimated by

$$\|T\|_{L^pL^q}\leq C_1^{1-s}C_2^s.$$

So function $\phi(\frac{1}{p};\frac{1}{q})=\log\|T\|_{L^pL^q}$ is convex in the plane of reciprocal parameters $\{\frac{1}{p};\frac{1}{q}\}$.

The Hausdorff-Young inequality arises via interpolation between pairs of Hölder indices $(1;\infty)$ and $(2;2)$ for $\mathfrak{F}$. Of course, interpolation does not yield the best constant

(2.1), as $C_1 = C_2 = 1 \Rightarrow C(p;q) = 1$, for all p.

The interpolation also applies to convolutions of L^p-spaces. This time we deal with triples of indices $(\frac{1}{p};\frac{1}{q};\frac{1}{r})$, so that $L^p * L^q \subset L^r$, in other words, $(f;g) \to f*g$, maps boundedly $L^p \times L^q \to L^r$. The general principle for bilinear T remains the same, as for linear operators. So if T maps boundedly $L^{p_1} \times L^{q_1} \to L^{r_1}$, and $L^{p_2} \times L^{q_2} \to L^{r_2}$, with bounds $C_1; C_2$, then for any interpolating triple:

$$\frac{1}{p} = \frac{s}{p_1} + \frac{s-1}{p_2};\ \frac{1}{q} = \frac{s}{q_1} + \frac{s-1}{q_2};\ \frac{1}{r} = \frac{s}{r_1} + \frac{s-1}{r_2};$$

T maps $L^p \times L^q \to L^r$, with the norm estimate

$$\|T\|_{L^p \times L^q; L^r} \leq C_1^{1-s} C_2^s. \tag{2.2}$$

To apply (2.2) to convolutions of L^p-spaces we observe that $L^1 * L^p \subset L^p$ (in other words operators $\{R_f : f \in L^1\}$ are bounded in L^p by $\|f\|_1$). On the other hand $L^2 * L^2 \subset L^\infty$. Interpolating between the resulting triples $(1;\frac{1}{p};\frac{1}{p})$ and $(\frac{1}{2};\frac{1}{2};0)$, we get the *Young's inequality,*

$$\boxed{\|f*h\|_r \leq \|f\|_p \|h\|_q;\ \text{for all triples } \frac{1}{p} + \frac{1}{q} = 1 + \frac{1}{r}} \tag{2.3}$$

Young's inequality holds for all groups $G = \mathbb{R}^n;\ \mathbb{T}^n;\ \mathbb{Z}^n$. Moreover, for compact groups (like $\mathbb{T}^n$) the range of (2.3) can be extended to $\{(p,q,r): \frac{1}{r} \geq \frac{1}{p} + \frac{1}{q} - 1\}$, due to the obvious inclusions: $L^p \subset L^q$, for all $1 \leq q < p \leq \infty$.

The proof of the Riesz interpolation Theorem is based on a version of the maximum principle for holomorphic functions in the strip, called 3-line Lemma (see problems 2,3,4).

Let us remark, that so far we didn't use any special features of $\mathbb{R}^n$; $\mathbb{T}^n$ or $\mathbb{Z}^n$, so all results about L^p-spaces, which involve interpolation (like Young and Hausdorff-Young inequalities) extend to arbitrary locally compact Abelian groups [HR].

2.2. Smooth functions and Sobolev spaces. Riemann-Lebesgue Theorem (problem 1) implies that continuous functions on $\mathbb{T}^n$ (or continuous, compactly supported $\{f\}$ on $\mathbb{R}^n$) have bounded, decaying at $\{\infty\}$ Fourier transform. Thence one could easily show (problem 5) that m-smooth functions $f \in C^m(\mathbb{T}^n)$, or $C_0^m(\mathbb{R}^n)$, have their Fourier transform/coefficients $\{\hat{f}(k)\}$; $\{\hat{f}(\xi)\}$ decay at a polynomial rate:

$$|\hat{f}(\xi)| \leq C|\xi|^{-m}, \text{ with constant } C = \sum_{|\alpha| \leq m} \|\partial f\|_{L^1}. \tag{2.4}$$

Conversely, if Fourier transform $\hat{f}(\xi)$ decays as $|\xi|^{-p}$, with $p > m+n$, then $f(x)$ is at least m-smooth. Both results are easily verified via the differentiation formula: $(\widehat{\partial f}) = (i\xi)^\alpha \hat{f}(\xi)$, and the obvious relation: $\hat{f} = O(|\xi|^{-n-\epsilon}) \Rightarrow \hat{f} \in L^1$. Hence, by

Riemann-Lebesgue Theorem, $f = \mathfrak{F}^{-1}(\widehat{f}) \in \mathfrak{C}_0$.

Thus Fourier transform relates the "*degree of smoothness* of f" to the "*decay rate* of $\widehat{f}$ at $\{\infty\}$". More precise results, however, could be stated in terms of L^2-norms, rather than uniform (L^∞). The main disadvantage of smooth functions from the standpoint of harmonic analysis, is that $\mathfrak{C}^m$ does not allow a simple characterization in terms of Fourier transform/coefficients, which could done for L^2-functions. Thus we are naturally lead to the notion of *Sobolev spaces.*

Sobolev space $\mathcal{H}_m$, of order m, consists of functions $\{u\}$, which belong to L^2 with all their derivatives $\{u^{(\alpha)} \in L^2 \colon |\alpha| \leq m\}$. It is equipped with the norm

$$\|u\|_m = \Big\{ \sum_{|\alpha| \leq m} \|\partial^\alpha u\|_{L^2}^2 \Big\}^{\frac{1}{2}}. \tag{2.5}$$

The Plancherel Theorem and the differentiation rule of §2.1 give an explicit characterization of $\mathcal{H}_m$ in terms of Fourier transform/coefficients $\{\widehat{u}(\xi)\}$ of u,

$$\|u\|_m^2 = \sum_{|\alpha| \leq m} \sum_{k \in \mathbb{Z}^n} |k^\alpha \widehat{u}_k|^2; \text{ or } \sum_{|\alpha| \leq m} \int_{\mathbb{R}^n} |\xi^\alpha \widehat{u}(\xi)|^2 d\xi, \tag{2.6}$$

- sum of *weighted* L^2-norms with polynomial weights $w(\xi) = \xi^\alpha$.

It is easy to check that the sum of monomials $\{k^\alpha\}$ in (2.6) (respectively partials $\{\partial^\alpha\}$ in (2.5)), could be replaced by powers of $(1+|k|^2)^m$, respectively $(1-\Delta)^m$, where $\Delta = \nabla \cdot \nabla = \sum \partial_i^2$ is the Laplacian on $\mathbb{R}^n$. So norm (2.6) is equivalent to

$$\|u\|_m^2 \simeq \int (1+|\xi|^2)^m |\widehat{u}(\xi)|^2 = \left\|(1-\Delta)^{m/2} u\right\|_{L^2}.$$

In other words $\mathfrak{F}$ takes Sobolev space $\mathcal{H}_m$ onto the *weighted* L^2*-space*, $L^2(w_m d\xi)$ (or $\ell^2(w_m)$ in the $\mathbb{T}^n$-case), with polynomial weight $w_m(\xi) = (1 + |\xi|^2)^m$, and

$$\left\|(1-\Delta)^{m/2} u\right\|_{L^2(dx)} \simeq \left\|(1+|\xi|^2)^{m/2} \widehat{u}\right\|_{L^2(d\xi)}.$$

This allows to extend the notion of *Sobolev norms* and *derivatives (powers of Laplacian)* from integers to all real (fractional; negative) values of m. By definition,

$$(1-\Delta)^{-s} u = \mathfrak{F}^{-1}(1+|\xi|^2)^{-s/2} \mathfrak{F} u = \int (1+|\xi|^2)^{-s/2} \widehat{u}(\xi) e^{ix \cdot \xi} d\xi. \tag{2.7}$$

Fractional Laplacian (2.7) are given by convolution kernels, $B_s * u$, with the so called *Bessel potential*

$$B_s = K_\nu(|x|) = \mathfrak{F}^{-1}[(1+|\xi|^2)^{-s/2}],$$

where $K_\nu(r)$*-modified Bessel function* (3-rd kind) of order $\nu = \frac{n-s}{2}$ (problem 3).

Thus we get a scale of Sobolev spaces $\{\mathcal{H}_s : s \geq 0\}$,

$$\mathcal{H}_s = (1-\Delta)^{-s/2}[L^2] = B_s*L^2 = \mathcal{F}^{-1}[L^2(w_s d\xi)], \text{ where } w_s = (1+|\xi|^2)^s.$$

The above definition also extends for negative s by duality. Namely, we define space $\mathcal{H}_{-s}$ as the dual space (of distributions) to $\mathcal{H}_s$, via the pairing,

$$\langle F \mid f\rangle = \int f(x)F(x)dx = \int \hat{f}(\xi)\hat{F}(\xi)d\xi,\ f \in \mathcal{H}_s;\ F \in \mathcal{H}_{-s}. \tag{2.8}$$

We shall list a few basic properties of Sobolev spaces.

Theorem 1: i) *Spaces $\{\mathcal{H}_s\}$ form a continuous scale of embedded Hilbert spaces:*

$$\boxed{\mathcal{D} \supset \mathcal{H}_{-\infty} = \bigcup_t \mathcal{H}_{-t} \supset \ldots\mathcal{H}_{-s} \supset \ldots\mathcal{H}_0 = L^2 \supset \ldots\ \mathcal{H}_s \supset \ldots\mathcal{H}_\infty = \bigcap_t \mathcal{H}_t = \mathcal{C}^\infty}$$

with $\mathcal{H}_{-s}$ = "dual to $\mathcal{H}_s$ via the natural pairing (2.8)".

ii) *On torus $\mathbf{T}^n$ (or more general compact manifolds, domains) the embedding $\mathcal{H}_s \subset \mathcal{H}_t$ is compact for all $t < s$.*

iii) *Sobolev embedding: spaces $\{\mathcal{H}_s\}$ are related to $\{\mathcal{C}^m\}$ in the following way:*

$$\mathcal{H}_m \subset \mathcal{C}, \text{ for } m > n/2,$$

more generally,

$$\mathcal{C}^{m+1} \subset \mathcal{H}_s \subset \mathcal{C}^{m-n/2}, \text{ for any } m \le s < m+1,$$

Proof: (i-ii) Each $\mathcal{H}_t$ is mapped onto $\mathcal{H}_s$ by a fractional Laplacian $(1-\Delta)^{-m}$, $m = \frac{s-t}{2}$, which is Fourier-equivalent to a multiplication of ℓ^2-sequences of coefficients by a decaying sequence $\{a_k = (1+|k|^2)^{-m/2}\}$, clearly a compact operator.

(iii) Multiplying and dividing a sequence of Fourier coefficients of $f \in \mathcal{H}_s$ by a polynomial weight $(1+|k|^2)^{m/2}$, and applying the Cauchy-Schwartz inequality, we get

$$\sum |a_k| = \sum |(1+|k|^2)^{m/2}a_k|(1+|k|^2)^{-m/2} \le \left\|\{(1+|k|^2)^{m/2}a_k\}\right\|_{\ell^2} \sum (1+|k|^2)^{-m} < \infty,$$

which shows that the Fourier series of $u(x) = \sum a_k e^{ik\cdot x}$ is absolutely convergent, whence follows continuity.

Remark 1: The last statement of the Theorem extends to embedding of Sobolev spaces $\mathcal{H}_s$ into the L^p-scale. The general result (Sobolev-Hardy-Littlewood Theorem) claims: *space $\mathcal{H}_s \subset L^p$ - continuously embedded, i.e.*

$$\|f\|_{L^p} \le \left\|(1-\Delta)^{s/2} f\right\|_s; \tag{2.9}$$

for all $\frac{1}{p} = \frac{1}{2} - \frac{s}{n}$ (on $\mathbf{R}^n$), and $\frac{1}{p} \ge \frac{1}{2} - \frac{s}{n}$ (on $\mathbf{T}^n$). The argument exploits interpolation for Bessel potentials $B_s \in L^q$, with $q = \frac{n-s}{n}$. Convolutions, $u \to B_s * u$, with such B_s shifts any L^p-space into L^r, where

$$\frac{1}{r}=\frac{1}{p}-(1-\frac{1}{q})=\frac{1}{p}-\frac{s}{n}.$$

Consequently, for negative s, $\mathfrak{D}_{m+1}\supset\mathcal{H}_{-s}\supset\mathfrak{D}_{m-n/2}$. Therefore,

$$\mathcal{C}^\infty=\bigcap_{s>0}\mathcal{H}_s;\quad \mathfrak{D}=\bigcup_{s<0}\mathcal{H}_s;$$

as claimed in (i).

Remark 2: Sobolev spaces provide a natural framework to analyze solutions of differential equations, as outlined in §2.3. This often requires Sobolev spaces of both integral and fractional order, as well as Sobolev spaces on manifolds and domains $\Omega\subset\mathbb{R}^n$. Let us also notice that spaces $\mathcal{H}_m$ constitute "m-smooth vectors" of regular representation R on $L^2(\mathbb{R}^n)$, or $L^2(\mathbb{T}^n)$, in the terminology of §1.4.

2.3. Distributions $\{F\}$ are defined as continuous linear functionals on suitable *spaces of test-functions* $\{u\}$, like $\mathcal{C}^m$, $\mathcal{C}^\infty$, etc. One usually writes them as formal integrals

$$\langle F\mid u\rangle=\int F(x)u(x)dx. \tag{2.10}$$

So for the class of bounded continuous functions, $f\in\mathcal{C}(X)$, all functionals are given by finite Borel measures $\mu\in\mathcal{M}(X)$, $\langle\mu\mid u\rangle=\int u(x)d\mu(x)$. Test-spaces $\mathcal{C}^m$, $\mathcal{C}^\infty$ are equipped with a sequence of seminorms,

$$\Big\{\|u\|_m=\sum_{|\alpha|=m}max\,|\,\partial^\alpha u(x)\,|\,;m=1;2;...\},$$

therefore the corresponding distribution spaces, called $\mathfrak{D}^m$, $\mathfrak{D}^\infty$, consist of functionals satisfying

$$|\langle F\mid u\rangle|\le C\sum_{j=0}^m\|u\|_j. \tag{2.11}$$

Using (2.11) one can show

Proposition: *Any distribution* $F\in\mathfrak{D}^m$ $(m=0;1;...\infty)$, *is equal to a finite sum of derivatives of Borel measures,*

$$F=\sum_{|\alpha|\le m}\partial^\alpha(\mu_\alpha), \tag{2.12}$$

Integer m is called the *(differential) order* of F. Derivatives of measures in (2.12) are defined via pairing to smooth functions:

$$\langle\partial^\alpha\mu\mid f\rangle=(-1)^\alpha\int\partial^\alpha f(x)d\mu(x).$$

The proof of (2.12) is fairly straightforward. Indeed, the m*–jet* map

$$f(x)\to\{f^{(\alpha)}(x):\,|\alpha|\le m\},$$

takes $\mathcal{C}^m(\mathbb{R})$ into the direct sum:

$$\mathcal{C}_N=\bigoplus_{|\alpha|\le m}\mathcal{C}(\mathbb{R}^n)\text{ - }N\text{-component vector functions on }\mathbb{R}^n;\ N=\binom{n+m}{m}.$$

Distribution F defines a bounded linear functional (in the sup-norm of $\mathfrak{C}_N$) on the image $\mathfrak{C}^m \subset \mathfrak{C}_N$, which can be extended through the entire space $\mathfrak{C}$. But any linear functional on vector-valued continuous functions, $\mathfrak{C}_N$, is given by an N-tuple of Borel measures $\{d\mu_\alpha : |\alpha| \le m\}$, QED.

Let remark that all natural linear operations on test-functions can be transferred to distributions by duality. For instance, differentiation is introduced via

$$\langle \partial^\alpha F \mid u\rangle = \langle F \mid (-\partial)^\alpha u\rangle,$$

by analogy with the standard "integration by parts" formula for test-functions. Clearly, $\partial^\alpha F \in \mathfrak{D}^\infty$, for any $F \in \mathfrak{D}^\infty$, in fact $\partial^\alpha : \mathfrak{D}^m \to \mathfrak{D}^{m+|\alpha|}$. Thus we get the following chain of test-function spaces along with their distribution-spaces

$$\boxed{\mathfrak{C}^\infty = \bigcap_0^\infty \mathfrak{C}^k \subset \ldots \subset \mathfrak{C}^m \subset \ldots \subset \mathfrak{C} \subset L^1 \subset \mathcal{M} = \mathfrak{D}_0 \subset \ldots \subset \mathfrak{D}_m \subset \ldots \subset \mathfrak{D} = \bigcup_0^\infty \mathfrak{D}_k} \quad (2.13)$$

Examples of distributions include

(i) δ-functions and their derivatives: $\delta_z(x) = \delta(x-z)$, or

$$\langle \delta^{(\alpha)}; u\rangle = (-1)^{|\alpha|} u^{(\alpha)}(0),$$

as well as their linear combinations, e.g. $\delta_\Lambda = \sum \delta(x-m)$ (Λ- lattice in $\mathbb{R}^n$);

(ii) δ-functions of hyper-surfaces and submanifolds: $\langle \delta_\Sigma \mid u\rangle = \int_\Sigma u\,dS$, where dS is the natural surface area element on Σ, like the spherical distribution $\delta_S = \int_S \ldots dS$, ($S = S_r$- sphere of radius r in $\mathbb{R}^n$), that appears in the *wave propagator* (fundamental solution of the wave equation);

(iii) Certain singular integrals, like *Hilbert transform*,

$$\langle H \mid u\rangle = P.V.\int \frac{u(x)}{x}dx \text{ (principal-value integral)} = \lim_{\epsilon\to 0}\int_{|x|\ge\epsilon} \frac{u(x)}{x}dx,$$

as well as its higher-D analogs, called Calderon-Zygmund integrals, can be interpreted as distributions ([Zy];[St];[SW]).

(iv) Distributions often arise as limits of families of regular functions[4], e.g.

$$\lim_{\epsilon\to 0}\frac{1}{x+i\epsilon} = H + i\delta(x).$$

Another linear operation on functions, the change of variable, $f \to f\circ\phi$, can also

[4]Such approximating families $\{f_\epsilon(x)\}$, called *regularizations*, exist for any $F \in \mathfrak{D}$. The easiest way to construct a regularization is to take a $\mathfrak{C}^\infty$- smooth approximate unity $\{h_\epsilon \to \delta\}$, e.g. $h_\epsilon = \epsilon^{-n} h(\frac{x}{\epsilon})$, with $h \in \mathfrak{C}_c^\infty$ ($h \ge 0$, and $\int h = 1$), and to convolve $h_\epsilon * F = f_\epsilon$ (problem 1).

be transferred to distributions (problem 2),

$$\langle F\circ\phi \mid u\rangle \overset{def}{=} \langle F \mid (\frac{1}{|\phi'|}u)\circ\phi^{-1}\rangle = \int F(x)(\frac{1}{|\phi'|}u)\circ\phi^{-1}(x)dx.$$

Distribution spaces on torus can be characterized by their Fourier coefficients, the same way as smooth functions. For any $F \in \mathfrak{D}^m$ we define

$$b_k = \langle F \mid e^{-ix\cdot k}\rangle = \int_{\mathbf{T}^n} F(x)e^{-ik\cdot x}.$$

Fourier coefficients of continuous functions, $u(x) = \sum a_k e^{ik\cdot x}$, are bounded and go to 0, as $k\to\infty$, those of $\mathcal{C}^m$-functions decay at a polynomial rate, $a_k = \mathcal{O}(|k|^{-m})$. Since, $\langle F \mid u\rangle = \sum a_k b_k$ (an extension of Plancherel's Theorem to pairs "function-distributions"), it is natural to expect $\{b_k\}$ to increase polynomially. Indeed, if $F \in \mathfrak{D}_m$

$$\boxed{|b_k| = |\langle F \mid e^{ik\cdot x}\rangle| \le C\, max\,|\partial^\alpha(e^{ik\cdot x})| = \mathcal{O}(|k|^m)} \qquad (2.14)$$

for any F of order m. We shall illustrate (2.14) by the example of δ-function and its derivatives,

$$\delta = \sum e^{ik\cdot x}, \text{ so all } b_k = 1;$$

$$\delta^{(\alpha)} = \sum (ik)^\alpha\, e^{ik\cdot x}, \text{ so } b_k = (ik)^\alpha.$$

The latter extends to all (integer and fractional) powers of the Laplacian:

$$(-\Delta)^s\delta = \sum |k|^{2s} e^{ik\cdot x}, \text{ or } (1-\Delta)^s\delta = \sum (1+|k|^2)^s e^{ik\cdot x}.$$

Hence both belong to space $\mathfrak{D}^m$, for $m > 2s$.

Remark 3: On $\mathbb{R}^n$ test-function and distribution spaces are defined similar to the compact (torus) case, but additional constraints are imposed at $\{\infty\}$. If no such constraints are imposed (spaces $\mathcal{C}^m$ and $\mathcal{C}^\infty$) one gets *compactly supported distributions* $\mathfrak{D}$. Conversely, *compactly supported $\mathcal{C}$-spaces* ($\mathcal{C}_c^m$; $\mathcal{C}_c^\infty$) give all distributions (with unlimited support). The $\mathbb{R}^n$-Sobolev spaces $\mathcal{H}_m$ satisfy all the properties (i-iii), except compactness of the embedding $\mathcal{H}_t \subset \mathcal{H}_s$. To show for instance, the embedding: $\mathcal{H}_m \subset \mathcal{C}_0$, for $m > n/2$, we replace Fourier sums by Fourier integrals,

$$\int |\hat f(\xi)|\, d\xi = \int (1+|\xi|^2)^{m/2}\hat f(1+|\xi|^2)^{-m/2} \le \left\|(1+|\xi|^2)^{m/2}\hat f\right\|_{L^2} \int (1+|\xi|^2)^{-m}\, d\xi < \infty,$$

If $\hat f \in L^1$, then $f = \mathfrak{F}^{-1}(\hat f) \in \mathcal{C}_0$. As in the torus case *smoothness* of f can be linked to the rate of decay of $\hat f$ at ∞,

$$f \in \mathcal{C}_c^m \Rightarrow \hat f(\xi) = O(|\xi|^{-m}); \text{ or } f \in \mathcal{H}_m \Rightarrow (1+|\xi|^m)\hat f \in L^2.$$

Conversely, sufficient decay rate $\mathcal{O}(|\xi|^{-m})$ of $\hat f$ guarantees $(m-\frac{n}{2})$-smoothness of f.

2.4. Schwartz functions and tempered distributions. The most convenient classes of functions and distributions combine both features (smoothness and the decay rate ay $\{\infty\}$). These are so called *Schwartz functions* $\mathcal{S}$ and *tempered distributions* $\mathcal{S}'$. Space $\mathcal{S}$ consists of $\mathbb{C}^\infty$-functions, that decay faster than any polynomial, $\mathcal{S} = \{f: x^\alpha f^{(\beta)}(x) \to 0, \text{as } x \to \infty\}$, along with all their derivatives. It is equipped with a sequence of semi-norms,

$$\|f\|_{\alpha\beta} = \left\|x^\alpha \partial^\beta f\right\|_{L^\infty}; \text{ or equivalently } \|f\|_{pq} = \left\|(1+|x|^2)^p(1-\Delta)^q f\right\|_{L^\infty}. \tag{2.15}$$

Let us that the choice of *sup*-norm in (2.15) is not essential, it could be replaced with any other L^p-norm. Also the order of two operations (differentiation and multiplication) could be interchanged (problem 4). So space $\mathcal{S}$ can be characterized by semi-norms,

$$\left\|(1-\Delta)^p(1+|x|^2)^p f\right\|_{L^2} < \infty. \tag{2.16}$$

Space $\mathcal{S}$ is sufficiently rich, it contains $\{\mathbb{C}_c^\infty\}$; the Gaussian $\{e^{-tx^2}\}$; $\{\text{polynomials}\} \times \{e^{-tx^2}\}$, etc. The important features of the Schwartz spaces are due to the fact that $\mathcal{F}: \mathcal{S} \to \mathcal{S}$, represents an isomorphism of $\mathcal{S}$, i.e. semi-norms (2.15) or (2.16) of f and $\widehat{f}$ are equivalent. *Schwartz (tempered) distributions* $\mathcal{S}'$ are linear functionals on $\mathcal{S}$. So distribution $F \in \mathcal{S}'$, if it satisfies,

$$|\langle F \mid u\rangle| \leq C \sum \|u\|_{pq}; \text{ for any Schwartz } u.$$

This class includes all examples we encountered so far, like the lattice δ-function, $\delta_\Lambda = \sum \delta(x-k)$; its transform; $\hat{\delta}_\Lambda = \sum e^{ikx}$; δ-functions of hyper-surfaces, or submanifolds in $\mathbb{R}^n$; their derivatives, etc. We shall list a few other examples of Schwartz distributions and their transforms (see problem 7,8):

i) Dirac-δ is transformed into constant 1, hence distribution $\delta \in \mathcal{H}_{-s}$, for $s > n/C$.

ii) δ-function of subspace $E \subset \mathbb{R}^n$, $\delta_E \to (2\pi)^{n-k}\delta_{E'}$; where E' is the orthogonal complement of E, $k = dim E$;

iii) $F = \delta_S$ (δ-function of the r-sphere S) $\to \widehat{F}(\xi) = c_n(r|\xi|)^{-n/2+1} J_{n/2-1}(r|\xi|)$ - the Bessel function of order $(\frac{n}{2}-1)$;

iv) fractional derivatives of Dirac δ: $(-\Delta)^{-s}\delta$, or $(1-\Delta)^{-s}\delta$ turn into $|\xi|^{-2s}$ *(Riesz potential)*, and $(1+|\xi|^2)^{-s/2}$ *(Bessel) potentials* (problem 9).

v) δ-function of the light cone $C = \{z = c\sqrt{x^2+y^2}\}$ in $\mathbb{R}^3$ is Fourier-transformed into

$$\hat{\delta}_C(\xi;\eta;\zeta) = \frac{1}{\sqrt{\xi^2+\eta^2}} H(c\zeta - \sqrt{\xi^2+\eta^2}),$$

where H is the Heaviside function. So the Fourier transform of δ_C is supported in the dual cone $C' = \{\xi : \xi \cdot x \geq 0\}$. As with $\mathcal{S}$ the Fourier transform maps $\mathcal{S}' \to \mathcal{S}'$ isomorphically.

2.5. Analytic functions and the Paley-Wiener Theorem. After the Fourier analysis of L^p-spaces and smooth functions $\{\mathbb{C}^m; \mathcal{H}_m\}$, we turn our attention to exponentially decaying functions $L^p{}_\alpha = \{f : f(x)e^{\alpha|x|} \in L^p\}$, and yet smaller spaces of compactly supported functions. Clearly, in both cases the Fourier transform $\hat{f}(\xi)$ extends from the real space $\mathbb{R}^n$ to a complex domain $\{\zeta = \xi + i\eta\} \subset \mathbb{C}^n$. In the former case the domain of $\hat{f}$ is a tubular neighborhood of the real space $\mathbb{R}^n$, $\Omega_\alpha = \{\xi + i\eta : |\eta| \leq \alpha\}$, while in the latter (compactly supported) case $\hat{f}$ extends analytically throughout the entire $\mathbb{C}^n$-space, but the support of $f(x)$, $B_\rho = \{|x| \leq \rho\}$ could be directly linked to the exponential rate of increase of function $\hat{f}(\xi + i\eta)$ along the imaginary space. Precisely,

Paley-Wiener Theorem ([PW]): *An L^2-function $f(x)$ on $\mathbb{R}^n$ is supported in the ball B_ρ of radius ρ, iff its Fourier transform $\hat{f}(\xi + i\eta)$ belong to space $L^2(d\xi)$ for each imaginary $i\eta$, furthermore,*

$$\left\| \hat{f}(\xi + i\eta) \right\|_{L^2(d\xi)} \leq e^{\rho\eta} \| f \|_{L^2}. \tag{2.17}$$

The necessity of (2.17) is fairly straightforward. The proof of sufficiency is more involved. To give the reader a flavor of the arguments we shall briefly outline the 1-D case.

The proof exploits the so called *L^2-Hardy spaces*, made of analytic functions $\{F(z)\}$ in the half-plane $\mathbb{H} = \{\Im z \geq 0\}$, square-integrable along lines $\{\Im z = y_0\}$, parallel to the real axis, such that all L^2-norms are uniformly bounded,

$$\int_{-\infty}^{\infty} |F(x+iy)|^2 dx \leq C; \text{ for all } y > 0.$$

Any Hardy function $F(z)$ is obtained by Fourier-transforming an L^2-function $f(t)$, supported on the positive half-line $[0;\infty)$ (problem 6):

$$F(z) = \int_0^{\infty} f(t) e^{iz \cdot t} dt, \; z \in \mathbb{H}. \tag{2.18}$$

A similar result holds for functions $\{F(z)\}$, holomorphic in the lower half-plane, the corresponding $\{f(t)\}$ being supported on the negative half-line $(-\infty; 0]$. For exponentially increasing functions $\{F(z)\}$ (2.17) the support of $\{f(t)\}$ gets shifted by $\mp\rho$ to the left/right, an easy consequence of the Fourier-translation formula of J2.1:

$$e^{iz\rho} F(z) = \mathcal{F}_{t \to z}[f(t+\rho)],$$

and the fact that $\{e^{iz\rho} F(z)\}$ belongs to the Hardy L^2-class. Therefore, analytic function $F(z)$, exponentially bounded in both the upper and the lower half-planes, has its inverse $\mathcal{F}$-transform inside the interval $[-\rho;\rho]$, QED.

Problems and Exercises:

1. **Riemann-Lebesgue Theorem:** *Fourier transform $\hat{f}(\xi)$ of any L^1-function f on $\mathbb{R}^n$ is continuous and goes to 0, as $\xi \to \infty$, i.e. $\hat{f} \in \mathfrak{C}_0$.*
Hint: Check the result directly for indicator functions $\{\chi_B\}$ of rectangular boxes: $B=[a_1;b_1]\times\ldots\times[a_n;b_n]$. Then use the standard density argument: any $f \in L^1$ is approximated by linear combinations of χ_B.

2. **Three line Lemma:** *If function $F(z)$ is holomorphic in a complex strip $0<\Re ez<1$, and has maximal values on both sides of the strip:*
$$M_0 = \sup_t |F(it)|;\ M_1 = \sup_t |F(1+it)|,$$
then F is bounded inside the strip and
$$|F(s+it)| \le M_0^{1-s} M_1^s, \text{ for all } z=s+it.$$
1°. Truncate $F(z)$ by a family of exponentials $\{M_0^{1-z} M_1^z e^{(z^2-1)/n}\}$, to ensure that
(i) product $G_n(z) = M_0^{1-z} M_1^z e^{(z^2-1)/n} F(z)$, goes to 0, as $z=s+it\to\infty$, and
(ii) maximal values of G_n on both sides of the strip ≤ 1.
2°. Use the maximum principle for holomorphic functions to show $|G_n(z)| \le 1$, for all z. Observe, $G_n(z) \to F(z)$, as $n\to\infty$!

3. Prove the Riesz interpolation Theorem,
$$\left|\int (Tf)(x)\, g(x)\, dx\right| \le M_0^{1-s} M_1^s \|f\|_{p(s)} \|g\|_{q'(s)};\ q' = \frac{q}{q-1},$$
for a dense subset of step-functions functions $\{f;g\}$, i.e. linear combinations of the indicator functions $\{\chi_j = \chi_{\Omega_j}\}$ of subsets $\{\Omega \subset X\}$.
1°. Represent: $f = \sum_j a_j \chi_j;\ \ g = \sum_k b_k \chi_k$; with coefficients
$$a_j = r_j e^{i\theta_j};\ b_k = \rho_k e^{i\phi_k}.$$
2°. It is convenient to use reciprocal Hölder indices:
$$\alpha = \tfrac{1}{p};\ \beta = \tfrac{1}{q};\ \alpha_0 = \tfrac{1}{p_0};\ \beta_0 = \tfrac{1}{q_0};\ \alpha_1,\ \beta_1;\ \alpha(z) = (1-z)\alpha_0 + z\alpha_1, \text{ etc.}$$
Introduce two families of functions (one-parameter deformations of $f;g$) depending on complex z,
$$f_z = \sum r_j^{\alpha(z)/\alpha} e^{i\theta_j} \chi_j;\ g_z = \sum \rho_k^{[1-\beta(z)]/(1-\beta)} e^{i\phi_k} \chi_k,$$
where α,β in the denominators of powers of $\{r_j\}$, $\{\rho_k\}$ mean α(s) and β(s) for some real $0<s<1$ (without loss of generality assume $\|f\|_\alpha = 1$; $\|g\|_{1-\beta} = 1$).
3°. Consider a holomorphic function $F(z) = \int T(f_z) g_z dx$, and estimate it on both sides of the strip: $0<\Re ez<1$.
For $\boxed{z=it}$ express $F(it) = \int T(\tilde{f})\tilde{g}dx$, in terms of new functions
$$\tilde{f} = \sum r_j^{\alpha_0/\alpha} e^{i(\theta_j+\ldots)} \chi_j;\ \tilde{g} = \sum \rho_k^{(1-\beta_0)/(1-\beta)} e^{i(\phi_k+\ldots)} \chi_k.$$
Apply the $(\alpha_0;\beta_0)$-estimate to the pair $\{\tilde{f};\tilde{g}\}$, to show
$$\left|\int T(\tilde{f})\, \tilde{g} dx\right| \le \ldots \le M_0 \|\tilde{f}\|_{1-\alpha_0} \|\tilde{g}\|_{1-\beta_0} = M_0.$$
Similarly, on the other side of the strip $\boxed{z=1+it}$
$$|F(1+it)| = \left|\int T(f_{1+it}) g_{1+it} dx\right| \le M_1 \|f\|_\alpha \|g\|_{1-\beta} = M_1.$$
Conclude the proof by the Three Line Lemma (problem 2).

4. Outline the proof of the Interpolation Theorem (specifically the construction of holomorphic function $F(z)$), for a bilinear map T.

5. Compute L^p-norm of the Gaussian and check the Beckner estimate of the Young-Hausdorff constant (2.1).

6. **Hardy classes on the half-plane** [Hof] are defined by (2.18) with an L^2-norm (or, more generally, L^p-norm, $1 < p < \infty$).

i) Hardy functions are known to have the (limiting) boundary values in L^2 (L^p):

$$F(x+i\epsilon) \to F(x), \text{ as } \epsilon \to 0, \text{ and } \| F(x) \| \leq C.$$

This holds for yet more general *harmonic functions* $u(z)$ in $\{\Im z > 0\}$ (solutions of the Laplaces equation: $\Delta u = \partial^2_{z\bar{z}} u = 0$, while $F(z)$ solves only the Cauchy-Riemann equation: $\partial_{\bar{z}} F = 0$!), as a consequence of the Poisson representation of u (§2.2),

$$u(x+iy) = (P_y * u)(x) = \int P_y(x-\tau) u(\tau) d\tau,$$

with the Poisson kernel

$$P_y(x) = \frac{y}{\pi(x^2+y^2)}.$$

The latter forms an approximate unity $\{P_y\}$, as $y \to 0$, i.e. $P_y \to \delta$, and $P_y * u \to u$, for all $u \in L^p$ (see problems 2,5,7 of §2.1 and problem 2 of §2.2).

ii) Furthermore, analytic (Hardy) functions $F(z)$ are represented through their boundary value by the Cauchy integral:

$$F(z) = \frac{1}{2\pi i} \int_{-\infty}^{\infty} \frac{1}{z-s} F(s) ds = K_y * F,$$

a convolution of the Cauchy kernel $K_y(x) = \frac{1}{2\pi i(x+iy)}$, whose real and imaginary parts are

$$\Re K = P_y(x) = \frac{y}{2\pi(x^2+y^2)} \to \tfrac{1}{2}\delta(x) \text{ - Poisson kernel;}$$

$$\Im K = H_y(x) = \frac{x}{2\pi(x^2+y^2)} \to \tfrac{1}{2} H(x);$$

where H denotes a distribution $P.V.(\frac{1}{x})$, called *Hilbert transform* (see §2.2, and [SW]).

iii) The Fourier transforms of both distributions, Hilbert H and Dirac δ, on $\mathbb{R}$ are easy to compute: $\hat{\delta} = 1$; $\hat{H}(t) = \operatorname{sgn} t$ (§2.2).

iv) Use (i-iii) to show that, *any Hardy function $F(z)$ in $\mathbb{H}$ is given by Fourier transforming an L^2-function $f(t)$ on the half-line* (2.18).

The latter provides a crucial step in the proof of the Paley-Wiener Theorem.

7. *Smooth regularization of distributions.* Take a smooth compactly supported function $h(x)$, consider a family of dilations: $h_\epsilon(x) = \epsilon^{-n} h(x/\epsilon)$, and show that convolutions $\{h_\epsilon * F\}$ yield a family of regular functions, that converge to F: $\langle f_\epsilon \mid u \rangle \to \langle F \mid u \rangle$, for any test-function u.

8. a) Show that $(\delta \circ \phi) = \sum_{y \in \Gamma_0} \frac{1}{|\phi'(y)|} \delta(x-y)$, for the Dirac delta on $\mathbb{R}$, sum over all zeros $\Gamma_0 = \{y\}$ of function ϕ.

b) If ϕ: $\mathbb{R}^n \to \mathbb{R}$, then $(\delta \circ \phi) = \int_{\phi(x)=0} \frac{dS}{|\nabla \phi|}$, where dS is the Euclidian surface element on the level set $\{x : \phi(x) = 0\}$.

9. *Bessel and Riesz potentials* are formally defined as Fourier transforms of tempered distributions: $\hat{B}_s(\xi) = (1+|\xi|^2)^{-s/2}$, and $\hat{R}_s(\xi) = |\xi|^{-s}$ on $\mathbb{R}^n$.

i) Use polar coordinates and the Bessel-function representation of $\hat{\delta}_S = J_\nu(r...)$ (on the r-sphere) to compute $B_s(|x|)$ in terms of the modified Bessel function K_ν. A *Hankel-type integral* relates Bessel functions *J and K* of the 1-st and 3-rd kind:

$$\int_0^\infty \frac{t^{\nu+1} J_\nu(rt)}{(t^2+1)^{\mu+1}} dt = \frac{1}{\Gamma(\mu+1)} \left(\frac{r}{2}\right)^\mu K_{\nu-\mu}(r).$$

In our case $\nu = \frac{n}{2} - 1$; $\mu = \frac{s}{2} - 1$.

ii) Check that Fourier transform takes homogeneous functions/distributions of degree s: $f(tx) = t^{-s} f(x)$, $t > 0$, into homogeneous functions of degree $(s-n)$: $\hat{f}(t\xi) = t^{s-n}\hat{f}(\xi)$. Apply it to the Riesz potential to show, $R_s(|x|) = C(s)|x|^{s-n}$. Compute constant $C(s)$.

10. Show that the L^∞-norm in the definition of Schwartz functions $\mathcal{S}$ can be replaced with any other L^p-norm. Show also that operations $(1+|x|^2)$ and $(1-\Delta)$ can be interchanged in the definition of $\mathcal{S}$. (Hint: apply a version of the Hardy-Littlewood-Sobolev inequality (2.17) for *L^p-Sobolev spaces:* $\mathcal{H}_s^p = (1-\Delta)^{-s/2} L^p$, to show that $\mathcal{H}_s^p$ is embedded in L^q, for $\frac{1}{q} = \frac{1}{p} - \frac{s}{n}$, and

$$\|f\|_{L^q} \leq \left\|(1-\Delta)^{s/2} f\right\|_{L^p}.$$

§2.3. Some applications of Fourier analysis.

Here we picked up a few selected topics and applications of the commutative Fourier analysis: Central limit Theorem of probability; the Heisenberg uncertainty principle; Finite Fourier transforms, Bochner's Theorem and the Mellin transform. The latter has application to special functions and differential equations in $\mathbf{R}^n$, and also in the harmonic analysis on SL_2 (chapter 7).

3.1. The uncertainty principle. The Heisenberg *uncertainty principle* of the Quantum mechanics will be discussed in §6.2 of chapter 6. It has a simple Fourier-analytic formulation:

$$\boxed{\left(\int |\psi|^2 dx\right)^2 \leq \frac{4}{n^2}\left(\int |x\psi|^2 dx\right)\left(\int |\xi\widehat{\psi}|^2 d\xi\right)} \tag{3.1}$$

for any function ψ in $\mathbb{R}^n$, such that ψ and $\hat{\psi}$ are both square-integrable with weights $|x|^2$ and $|\xi|^2$ respectively.

Rephrasing (3.1) in terms of the derivative (gradient) $\partial\psi$, we get

$$\left(\int |\psi|^2 dx\right)^2 \leq \frac{4}{n^2}\left(\int |x\psi|^2 dx\right)\left(\int |\partial\psi|^2 d\xi\right). \tag{3.2}$$

In this form it becomes an easy consequence of the standard integration by parts formula,

$$\int x\psi\psi' = \int x(\tfrac{1}{2}|\psi|^2)' = -\tfrac{1}{2}\int |\psi|^2, \text{ (1-D); or} \int \nabla\psi\cdot x\psi = -\int \tfrac{1}{2}|\psi|^2(\nabla\cdot x) = -\tfrac{n}{2}\int |\psi|^2, \text{ (}n\text{-D)},$$

and the Cauchy-Schwartz inequality applied to the LHS,

$$\left|\int x\psi\cdot\nabla\psi\right| \leq \left(\int |x\psi|^2\right)^{1/2}\left(\int |\nabla\psi|^2\right)^{1/2}.$$

Relation (3.2) implies that one can simultaneously localize functions ψ and $\hat{\psi}$ at $\{0\}$ maintaining the L^2-norm $\|\psi\|_{L^2} = 1$ (LHS), in other words, localizing $\text{supp}(\psi)$ in $x-$space will "stretch" its Fourier transform $\hat{\psi}$, hence increase gradient, and vice versa. In quantum mechanics operators $x:\psi\to\psi$, and $\partial:\psi\to\partial\psi$, represent the "position and momentum" of the quantum particle, while integrals

$$\left(\int |x\psi|^2 dx\right) \text{ and } \left(\int |\partial\psi|^2 d\xi\right)$$

estimate errors in their measurements. So the Heisenberg principle prevents simultaneous precise determination of both operators to any degree of accuracy.

3.2. Finite Fourier transform. Here we shall briefly discuss the Fourier transform on finite commutative groups. The simplest of them is a cyclic group $\mathbb{Z}_n$. Its characters coincide with the n-th roots of unity $\{\chi_\ell(j) = \omega^j\}$, where $\omega = exp(\frac{2\pi i}{n})$. So the dual group $\hat{\mathbb{Z}}_n \simeq \mathbb{Z}_n$, and the Fourier transform on $\mathbb{Z}_n$ becomes

$$\mathfrak{F}: f(j) \to \hat{f}(\ell) = \sum_{0 \le j \le n-1} f(j)\omega^{\,j\ell}. \tag{3.3}$$

Characters $\{\chi_\ell = (\omega^{\ell j})\}$ form an orthogonal system of eigenvectors of the translation operator $R: f(j) \to f(j+1)$, on space $L^2(\mathbb{Z}_n) \simeq \mathbb{C}^n$, with norms

$$\langle \chi \mid \chi \rangle = \|\chi\|^2 = n.$$

The inverse transform takes the form,

$$\mathfrak{F}^{-1}: b(\ell) \to \check{b}(j) = \frac{1}{n} \sum_{0 \le \ell \le n-1} b(\ell)\omega^{j\ell},$$

while the Plancherel formula,

$$\|f\|^2 = \sum |f(j)|^2 = \frac{1}{n}\sum |\hat{f}(\ell)|^2 = \|\hat{f}\|^2,$$

both being obvious consequences of the orthogonality of $\{\chi_\ell\}$.

We shall study now the *circulant matrix* A, that appears in many applications of finite groups (in linear algebra, probability, etc.),

$$A = \begin{bmatrix} a_1 & a_2 & \cdots & a_n \\ a_2 & a_3 & \cdots & a_1 \\ \cdots & \cdots & \cdots & \cdots \\ a_n & a_1 & \cdots & a_{n-1} \end{bmatrix}$$

each subsequent row of A is obtained by a cyclic permutation of the n-tuple $(a_1; ... a_n)$. Notice that matrix A is a convolution with function $a(k)$ on $\mathbb{Z}_n$,

$$Af = a * f.$$

So Fourier transform (3.3) applies to diagonalize A (find its eigenvalues and eigenvectors), compute inverse, determinant, etc.

A specific model of such A is given by the *random walk* on $\mathbb{Z}_n$, where $a_2 = a_n = \frac{1}{2}$ (probabilities to jump from point $\{1\}$ to its neighbors $\{2\}$ and $\{n\}$ in a unit time), and the rest $a_j = 0$. More generally, we consider a *stationary stochastic matrix*[5]: $a_j \ge 0$; $\sum a_j = 1$ (a_j measures probability to jump from cite $\{1\}$ to $\{j\}$).

Characters $\{\chi_\ell\}$ form a complete set of eigenfunctions of A with eigenvalues,

$$\lambda_\ell = \hat{a}(\ell) = \sum a_j \omega^{j\ell},$$

- the ℓ-th Fourier coefficient of function $\{a(j)\}$, $\ell = 0; 1; ... n-1$. Diagonalizing A, via $\mathfrak{F}$,

[5] Stationarity means that all rows of A are permutations of a single row, so the process is independent of the starting point: probability of "$i \to j$"-jump depends only on the difference $i - j \in \mathbb{Z}_n$: $a_{ij} = a(i-j)$.

allows to compute $\det A$; A^{-1}, all iterates A^m, and yet more general "functions of A", $B = f(A)$. Indeed, any such B is itself a convolution with function $b(j)$, whose Fourier transform, $\widehat{b} = f \circ \widehat{a}$, so

$$b(j) = \mathfrak{F}^{-1}[f \circ \widehat{a}] = \tfrac{1}{n}\sum_{\ell} f \circ \widehat{a}(\ell)\,\omega^{j\ell}.$$

As an application to random walk on $\mathbb{Z}_n$, we immediately find the probability distribution of the process after m steps (units of time), i.e. find entries of the m-th iterate A^m. If $p^{(m)}_{j;\,j+k}$ denotes the probability of jump from j to $j+k$ after m steps, then

$$p^{(m)}_{j;\,j+k} = \mathfrak{F}^{-1}[\lambda^m_{\ell}].$$

Hence,

$$p^{(m)}_{j;\,j} = \tfrac{1}{n}\sum_{\ell} \cos^m(\tfrac{2\pi\ell}{n});\ \text{and}\ p^{(m)}_{j;\,j+k} = \tfrac{1}{n}\sum_{\ell} \cos^m(\tfrac{2\pi\ell}{n})\cos(\tfrac{2\pi\ell k}{n}).$$

Remark: More general finite groups G are known to decompose into the direct product of cyclic groups, $G \simeq \prod \mathbb{Z}_n$. Hence, $\hat{G} = \prod \hat{\mathbb{Z}}_n \simeq G$, and the Fourier transform reduces to the Fourier transforms (1.3) on cyclic components of G. One interesting application of such finite Fourier transform is the random walk on the n-cube (problem 9).

3.3. Central Limit Theorem. Another interesting application of the Fourier Analysis is to *random walks* on groups $\mathbb{R}$, $\mathbb{Z}$, or more generally, to *sums of independent random variables*, $S_n = X_1 + X_2 + \ldots + X_n$. Here we shall establish the *Central Limit Theorem*, that describes asymptotic distribution of S_n, as $n \to \infty$.

Any real random variable X defines a probability measure (distribution), $d\mu_X$ on $\mathbb{R}$. Independence of two random variables X, Y means that their joint distribution $d\mu_{X,Y}(x,y)$ on $\mathbb{R}^2$ is the product $d\mu_X(x)d\mu_Y(y)$. So the sum of two independent random variables $X+Y$ has distribution,

$$d\mu_{X+Y} = d\mu_X * d\mu_Y \tag{3.4}$$

- convolution of two probability measures[6] on $\mathbb{R}$. Another simple transformation of X, scaling $X \to \alpha X$, results in the dilation of the measure

$$d\mu_X \to d\mu_X(\tfrac{x}{a}). \tag{3.5}$$

We consider the sum of n *independent identically distributed random variables,*

$$S_n = X_1 + \ldots + X_n,$$

with equal probability distributions,

$$d\mu_{X_j} = d\mu,$$

and ask about the asymptotic behavior of sums $\{S_n\}$ at large n.

[6]Convolution of L^1-functions $f*g$ on $\mathbb{R}^n$ (or any group G) can be extended to bounded measures $\{dm\}$, and other (compactly supported) distributions/generalized functions (see §2.2), via pairing to continuous test-functions $\{f\}$.

A typical example will be random walk on the lattice $\mathbb{Z}$. Here variable X takes on values ± 1 (jumps from k to $k \pm 1$) with equal probabilities $\frac{1}{2}$, so $d\mu = \frac{1}{2}(\delta_1 + \delta_{-1})$. Then the sum of n variables, S_n, measures the position of the walk after n time-steps, relative to the initial point.

It turns out that the limiting (large n) behavior of sums $\{S_n\}$ carries some universal features, independent of a particular distribution $d\mu$. In what follows, we always talk about convergence of random variables $\{Y_n\}$ in the *distributional (weak) sense*, namely,

$$Y_n \to Y, \text{ if } \int f d\mu_{Y_n} \to \int f d\mu_Y, \text{ for any continuous } f(x) \text{ on } \mathbb{R}.$$

Theorem 1: *Let $\{X_n\}_1^\infty$ be a sequence of independent identically distributed random variables with mean: $E = \int x d\mu(x)$, and the dispersion*

$$\sigma = \int (x-E)^2 d\mu = \int x^2 d\mu - E^2 > 0.$$

Then

(i) $S_n = \sum_{j=1}^n X_j$ has no limiting distribution, as $n \to \infty$.

(ii) $\frac{1}{n} S_n \to \delta_E$

(iii) $\frac{1}{\sqrt{n}} \sum_1^n (X_j - E) \to N(E;\sigma) = \frac{1}{\sqrt{2\pi\sigma}} e^{-x^2/2\sigma}$ - normal distribution with dispersion σ.

Rephrasing Theorem 1 we can say that the mean value of sum: $E(S_n) \approx nE$, increases linearly with n, while its dispersion $\sigma(S_n - nE) \sim \sigma\sqrt{n}$, grows as $\sqrt{n}$.

The proof exploits the notion of the characteristic function of variable X,

$$\phi_X(\xi) \stackrel{\text{def}}{=} \int_\Omega e^{i\xi X} dP = \int e^{ix\xi} d\mu,$$

the Fourier transform of measure μ. The characteristic function has the following properties:

A) $\phi(\xi)$ is continuous and positive-definite[7],

$$\sum_{jk} \phi(\xi_j - \xi_k) a_j \bar{a}_k \geq 0, \tag{3.6}$$

for all tuples $\{\xi_1; \dots \xi_n\}$ in $\mathbb{R}$ and $\{a_1; \dots a_n\}$ in $\mathbb{C}$; also

$$\phi(0) = 1; \ |\phi(\xi)| \leq 1.$$

B) if measure $d\mu$ has all moments to order k, i.e.

[7]The converse result is also true (see Bochner's Theorem below). is also true. Namely (problem 12): any positive-definite continuous function on group $\mathbb{R}$; $\mathbb{R}^n$, or more general commutative G, is the Fourier transform of a positive (probability) measure.

$$\int |x^j|\, d\mu < \infty, \ j=0;1;...k,$$

then function $\phi(\xi)$ if k-smooth, and its derivatives: $\phi'(0) = iE$; $\phi''(0) = -\sigma - E^2$.

C) For any sequence of random variable $\{Y_k\}$ measures $d\mu_{Y_n} \to d\mu_Y$, iff $\phi_{Y_n}(\xi) \to \phi_Y(\xi)$, uniformly on compact sets of **R**.

The first two properties follow directly from the definition of $\mathcal{F}$ and positivity of measures $d\mu$. The last one relies on the fact that families of exponential $\{e^{ix\xi}\}$ with ξ varying over compact sets D, approximate all continuous functions $\{f\}$ on compacts in $\{x\}$. So the convergence of characteristic functions is equivalent to the weak convergence of measures.

We consider characteristic functions $\{\phi_n\}$ of S_n. By (3.4)

$$\phi_n(\xi) = \mathcal{F}(\mu * \mu * ... * \mu) = \phi(\xi)^n,$$

where $\phi(\xi) = \hat{\mu}(\xi)$ - the characteristic function of a single variable distribution $d\mu$. Property (A) shows that $\phi(\xi)^n$ does not converge to a continuous function, unless $\phi \equiv 1$, i.e. $\mu = \delta$. This proves (i). Next we apply (3.4) and (3.5) to the "ergodic mean" $\frac{1}{n}S_n$,

$$\phi_{S_n/n}(\xi) = \phi(\xi/n)^n.$$

Taking the Taylor expansion of $\phi(\xi)$ at $\{0\}$, we get

$$\{1 + iE\tfrac{\xi}{n} + O(n^{-2})\}^n \to e^{i\xi E} = \mathcal{F}(\delta_E), \text{ as } n \to \infty.$$

This implies by (C), $d\mu_{S_n/n} \to \delta_E$, as claimed in (ii). Finally, Fourier transforming distribution-measures $d\mu_{Y_n}$ of variable $Y_n = \frac{1}{\sqrt{n}}(S_n - nE)$, we get by (3.4)-(3.5), $\phi\left(\frac{\xi}{\sqrt{n}}\right)^n$, where ϕ denotes the shifted measure $d\mu(x+E)$, with the zero mean,

$$\int x\, d\mu(x+E) = 0.$$

Taking the Taylor expansion of $\phi(\xi)$ at 0 we get by (B)

$$\{1 - \tfrac{1}{2}\sigma\tfrac{\xi^2}{n} + O(n^{-3})\}^n \to e^{-\sigma\xi^2/2} = \mathcal{F}\left(\frac{1}{\sqrt{2\pi\sigma}}e^{-x^2/2\sigma}\right) \text{ - the Gaussian.}$$

Sequence of measures $\{d\mu_{Y_n}\}$ converges, as above, to the normal distribution, QED.

We shall illustrate the foregoing argument with an example of the random walk on the lattice $a\mathbb{Z}$ (problem 8). Then measure $d\mu = \frac{1}{2}(\delta_a + \delta_{-a})$, has characteristic function

$$\phi = \cos a\xi = \{1 - \tfrac{a^2}{2}\xi^2 + ...\},$$

with mean: $\int x\, d\mu = 0,$

and dispersion: $\sigma = \int x^2 d\mu = a^2.$

Hence, the limit:

$$\left(1 - \frac{a^2\xi^2}{2n} + ...\right)^n \to e^{-a^2\xi^2/2}.$$

3.4. Positive-definite functions and Bochner's Theorem. In the previous part we encountered Fourier transforms of positive (probability) measures $\phi(\xi) = \int e^{-ix\cdot\xi} d\mu(x)$, which are positive-definite on $\mathbb{R}$ (or $\mathbb{R}^n$) in the sense of definition (6). It turns out that all positive-definite functions are obtained in this way.

Bochner's Theorem: *Any continuous positive-definite function* $\phi(\xi)$ *on* $\mathbb{R}^n$ *is the Fourier transform of a positive (probability) measure,*

$$\phi(\xi) = \int e^{ix\cdot\xi} d\mu(x).$$

Two direct arguments, outlined in problem 5, utilize spectral theory of self-adjoint operators, and regularization of distributions. There is yet another fairly general argument, based on the *convexity* and *extreme points.* It applies to all locally compact groups (and even more *positive functionals* on Banach algebras).

We observe that the set of all continuous positive-definite functions $\{\phi\}$ is a convex cone $\mathfrak{K}$ in space $\mathfrak{C}(\mathbb{R}^n)$, $a\phi + b\psi$ is positive-definite, along with $\{\phi;\psi\}$, for any pair of coefficients $a,b>0$. When normalized at $\{0\}$, subset $\mathfrak{K}_1 = \{\phi:\phi(0)=1\} \subset \mathfrak{K}$, becomes a convex compact set in the Banach space $\mathfrak{C}$. Any convex compact K in a Banach space has the set of *extreme points,* $ex(K) = \{\psi:\psi \neq a\phi_1 + (1-a)\phi_2\}$. By the general Krein-Milman Theorem, any point inside K can be represented by a "convex linear combinations" of extereme points. In other words, for any $\phi \in K$, there exists a probability measure $\mathrm{d}\mu_\phi(\psi)$, supported on $ex(K)$, so that ϕ is a baricenter of $\mathrm{d}\mu$,

$$\phi = \int_{ex(K)} \psi \, \mathrm{d}\mu_\phi(\psi). \tag{3.7}$$

Applying this result to positive-definite functions one first shows that $ex(\mathfrak{K}_1)$ consists precisely of all characters $\{\psi_x(\xi) = e^{ix\cdot\xi}\}$ ("extreme directions" in $\mathfrak{K}$ must be translation-invariant!). Hence, by the Krein-Milman, any $\phi(\xi) = \int e^{-ix\cdot\xi} d\mu(x)$, QED.

Remark: The Bochner's Theorem has a far reaching extension to noncommutative groups, known as Gelfand-Naimark-Segal Theorem. The latter characterizes positive-definite functions on G in terms of matrix-entries of unitary representations. The extreme points of $\mathfrak{K}$ are then identified with irreducible representations, and formula (7) gives a decomposition of a unitary representation T^ϕ, associated to ϕ, into the direct sum/integral of irreducibles.

3.5. Mellin transform. Mellin transform of function $f(t)$ on the half-line $t \in [0;\infty)$ is defined by integrating f against a family of *multiplicative characters* $\{\chi_s(t) = t^s; s \in \mathbb{C}\}$ with respect to the multiplicative-invariant measure $d\mu = \frac{dt}{t}$ on $\mathbb{R}_+$;

$d\mu(t\tau) = d\mu(t)$, for all $t, \tau \in \mathbb{R}_+$,

$$\mathcal{M}: f(t) \to \hat{f}(s) = \int_0^\infty f(t) t^{s-1} dt. \tag{3.8}$$

So $\mathcal{M}$ can be viewed as Fourier transform on the multiplicative group $\mathbb{R}_+$. Its inverse $\mathcal{M}^{-1}$ is obtained by integrating $\hat{f}(s)$ over the imaginary line: $\{\Re e s = b\}$ i.e. $\{s = b + i\sigma\}$,

$$\hat{f} \to f(t) = \frac{1}{2\pi i} \int_{b-i\infty}^{b+i\infty} \hat{f}(s) t^{-s} ds. \tag{3.9}$$

A few important examples of the Mellin transform include

1) Γ-function: $\Gamma(s) = \int_0^\infty e^{-t} t^{s-1} dt = \mathcal{M}[e^{-t}]$; and

2) $\lambda^{-s}\Gamma(s) = \int_0^\infty e^{-\lambda t} t^{s-1} dt = \mathcal{M}[e^{-\lambda t}]$

the latter can be also viewed as the Laplace transform, $\mathcal{L}: f(t) = t^{s-1} \to F(\lambda) = \Gamma(s)\lambda^{-s}$;

3) $\Gamma(s)\Gamma(1-s) = \int_0^\infty \frac{t}{t+1} t^{s-1} dt = \mathcal{M}\left[\frac{t}{t+1}\right]$.

4) Bessel function of the 3-rd kind (McDonald):

$$K_\nu(z) = \int_0^\infty \tfrac{1}{2} e^{-\frac{1}{2}(t+1/t)z} t^{\nu-1} dt = \mathcal{M}_{t\to\nu}\left[e^{-\frac{1}{2}(t+1/t)z}\right]. \tag{3.10}$$

5) Riemann Zeta-function:

$$\Gamma(s)Z(s) = \int_0^\infty \frac{e^{-t}}{1-e^{-t}} t^{s-1} dt \Rightarrow Z(s) = \frac{1}{\Gamma(s)} \mathcal{M}\left[\frac{e^{-t/2}}{2\sinh t/2}\right].$$

Formulae (1) and (2) are essentially the definition of Γ; function K_ν of (4) is directly verified to satisfy the *modified Bessel equation* (problem 10); two remaining relations (3), (5), however, would require further explanation. The former (3) is based on the so called *B-function,*

$$B(p;q) = \int_0^1 \tau^{p-1}(1-\tau)^{q-1} d\tau = \int_0^\infty \frac{t^{q-1}}{(t+1)^{p+q}} dt = \frac{\Gamma(p)\Gamma(q)}{\Gamma(p+q)};$$

for special values $p = s$, $q = 1-s$. The latter exploits formula (2) and the definition of Riemann zeta-function,

$$Z(s) = \sum_1^\infty n^{-s} = \frac{1}{\Gamma(s)} \int_0^\infty \left(\sum_1^\infty e^{-tn}\right) t^{s-1} dt = \frac{1}{\Gamma(s)} \int_0^\infty \frac{e^{-t}}{1-e^{-t}} t^{s-1} dt. \tag{3.11}$$

Let us remark that all examples (1)-(5) of Mellin transforms involve integrals of the form

$$I(s)=\int_0^\infty F(t)\,t^{s-1}dt,$$

with certain meromorphic/rational function F. Such integrals can be evaluated by the Residue Theorem, combined with the branch-cut along the half-line $\mathbf{R}_+$,

$$I(s)=\frac{2\pi i}{1-e^{2\pi is}}\sum Res_{z_j}(F)z_j^{s-1};$$

sum over all residues (poles) of F.

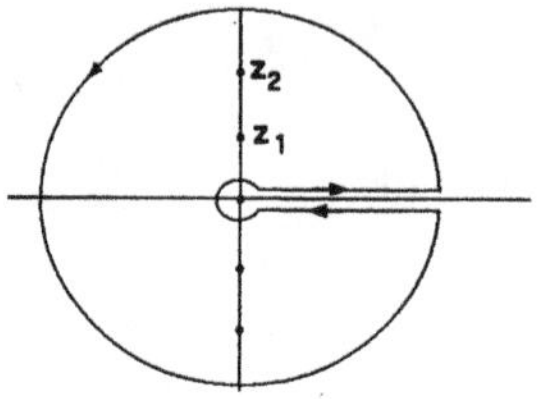

Fig.1: *Branch cut and residues of*

$$F(z)=\frac{e^{-z/2}}{2\sinh z/2}$$

in the Zeta-function transform. Here poles: $z_k=2\pi ik;\ k=\pm 1;\pm 2;\ldots$

Thus we compute the right hand side of (3) and (5), and find

$$\begin{aligned}&(3)\ \mathcal{M}\left[\frac{t}{t+1}\right]=\frac{\pi}{\sin\pi s}\ (\text{for } \Gamma);\\ &(5)\ \mathcal{M}\left[\frac{e^{-t/2}}{2\sinh t/2}\right]=\frac{(2\pi)^s}{\sin(\frac{\pi s}{2})}Z(1-s).\end{aligned}\qquad(3.12)$$

As the result we derive the *functional equations* for Γ and Z-functions, which relate their values at $\{s\}$ to those at $\{-s\}$,

$$\boxed{\Gamma(s)\Gamma(1-s)=\frac{\pi}{\sin\pi s}}\ \text{and}\ \boxed{Z(1-s)=(2\pi)^{-s}\frac{\Gamma(s)Z(s)}{\cos(\frac{\pi s}{2})}}\qquad(3.13)$$

The Mellin transform, in general, and functional equations (3.13), in particular, play extremely important role in the analysis of both functions, Γ and Z, and their numerous applications (see [Tit];[Ven];[Ter];[HST]). We shall see m in the next section, and later in chapter 7 (representations and harmonic analysis of SL_2).

The standard notion of Zeta-function (11) can be extended to *Zeta-functions* of differential operators L (e.g. Laplacian on a manifold),

$$Z_L(s)=tr(L^{-s})=\sum{\lambda_k}^{-s};\ \{\lambda_k\}\text{-eigenvalues of } L.$$

Analytic continuation (regularization) of $Z_L(s)$, particularly, its zeros, poles and residues in C, contain some interesting spectral and geometric information about operator L (§7.6). Among other things it allows to define a seemingly absurd (but highly useful) entity, the "*determinant of L*", $\det(L)=e^{-Z'(0)}$. For the standard Riemann Z one finds from (3.13), $Z(0)=-\frac{1}{2}$; $Z'(0)=-\frac{1}{2}ln(2\pi)$.

Problems and Exercises:

1. Work out an example of the random walk on $\mathbb{Z}$ with uneven probabilities: p (jump to the right) and $q = 1-p$ (jump to the left). What is the most likely position after n (large) steps? What is the probability to find oneself in the range $[(p-q)n + \alpha\sqrt{n};\ (p-q)n + \beta\sqrt{n}]$ after n steps?

2. Apply the finite Fourier transform to the random walk on the n*-cube* $\simeq \underbrace{\mathbb{Z}_2 \times \ldots \times \mathbb{Z}_2}_{n\text{-times}}$ to find $p^{(m)}_{j;\,j+k}$.

3. Verify that function $K = K_\nu(z)$ of (10) satisfies the modified Bessel's equation in variable z,
$$K'' + \tfrac{1}{z}K' + (1+\tfrac{\nu^2}{z^2})K = 0.$$

4. Evaluate integrals (12) and establish functional equations (13) for Γ and Z.

5. Prove Bochner's Theorem in two possible ways.

   i) Using spectral theory of self-adjoint operators (Appendix A). Hint: show that the quadratic form, determined by ϕ,
$$\langle Q_\phi \hat{f} \mid \hat{f}\rangle = \int\int \phi(\xi-\eta)\hat{f}(\xi)\overline{\hat{f}(\eta)}d\xi d\eta > 0,$$
   is positive for all functions $f \in L^1 \bigcap \mathcal{A}$, hence defines an inner product on $\mathcal{A} \bigcap L^2$,
$$\|f\|^2_\phi = \langle Q_\phi \hat{f} \mid \hat{f}\rangle.$$
   The regular representation, $R_a\hat{f} = \hat{f}(\xi-a)$, is unitary in Hilbert space $\mathcal{H}_\phi = \|\ \|_\phi$-closure of $\mathcal{A}\bigcap L^2$. Furthermore, space $\mathcal{H}_\phi$ has *cyclic vectors*: any $\hat{f}$, whose Fourier transform $f(x) \neq 0$, for all x. Observe, that Bochner's Theorem is equivalent to the spectral decomposition (existence of spectral measure $d\mu$) for a family of commuting unitary operators $\{R_a\}$ on $\mathcal{H}_\phi$, or for their infinitesimal generators (Appendix A).

   ii) **Direct proof** of Bochner's Theorem on groups $\mathbf{R}^n$; $\mathbf{Z}^n$, exploits a regularization of distribution $\hat{\phi}(x)$ by an approximate unity (problem 2):
$$\rho_\epsilon(x) = \epsilon^{-n}\rho(\tfrac{x}{\epsilon}),\ \rho(x) \geq 0;\ \int \rho dx = 1.$$
   Set $\hat{\phi}_\epsilon = \rho_\epsilon * \hat{\phi}$, and show that $\{\hat{\phi}_\epsilon\}$ defines a family of positive linear functionals on continuous functions $\mathcal{C}(\mathbf{R}^n)$ (or $\mathbf{T}^n$),
$$|\langle\hat{\phi}_\epsilon \mid f\rangle| \leq \phi(0)\|f\|_\infty;\ \langle\hat{\phi}_\epsilon \mid f\rangle \geq 0, \text{ for } f \geq 0.$$
   Conclude that $\hat{\phi} = \lim_{\epsilon\to 0}\hat{\phi}_\epsilon$ is a finite positive measure $d\mu$ on $\mathbf{R}^n$ of $\|\mu\| = \phi(0)$.

§2.4. Laplacian and related differential equations.

The Fourier transform provides an efficient tool in the study of differential/difference equations on $\mathbf{R}^n$; $\mathbf{T}^n$, due to its diagonalizing properties. Solving a differential equation is then reduced to an algebraic problem, followed by the Fourier inversion. The utility and power of Fourier analysis will be shown by the basic examples of differential equations (heat, wave, Schrödinger), associated to the Laplacian - the most symmetric of all differential operators on $\mathbf{R}^n$. In all cases we shall construct explicit fundamental solutions (Green's functions) and analyze their properties.

After the free-space Green's functions are constructed and analyzed in via the Fourier techniques in part (I), we turn to the boundary-value problems in special regions (part (II)). Our analysis will emphasize group-symmetries of the problem (discrete and continuous). One such techniques exploits the reductions (separation) of variables to bring the problem to exactly solvable ODE. Another one, known as the "method of images", gives solution of the boundary-value problem expressed through the free-space Green's function, subjected to a finite/discrete group of transformations. Our third example, the ball problem, exploits yet another *conformal symmetry* of the Laplacian.

4.1. The role of Fourier transform in differential equations is that $\mathcal{F}$ diagonalizes all constant coefficient differential operators: $L = \sum_{|\alpha| \le m} a_\alpha D^\alpha$, $(D = \frac{1}{i}\partial)$, on $\mathbb{R}^n$; $\mathbb{T}^n$. Namely,

$$\mathcal{F}L[u] = a(\xi)\widehat{u}(\xi), \text{for all } \{u\}, \text{ or } \mathcal{F}L\mathcal{F}^{-1} = a(\xi).$$

So $\mathcal{F}$ takes L into a multiplication with a polynomial function $a(\xi) = \sum a_\alpha \xi^\alpha$, called *symbol of L*. This clearly follows from the *eigenvalue property* of exponentials $\{e^{i\xi \cdot x}\}$ for any such L. In fact, $\{e^{ix \cdot \xi}\}$ form joint eigenfunctions of all such operators[8],

$$L[e^{i\xi \cdot x}] = a(\xi)e^{i\xi \cdot x},$$

with symbol $a(\xi)$ serving as "joint eigenvalue". In many applications operator L is the Laplacian $-\Delta$, or more general *elliptic operator*[9] $L = -\nabla \cdot p\nabla + q$, with a positive definite coefficient-matrix $p = (p_{jk})$. A few basic examples of differential equations associated to operator L include:

I. *Elliptic:* $-(L+\lambda)u = f$; (called Helmholtz, or "reduced wave" equation)

---

[8]On torus $\mathbb{T}^n$ characters $\{e^{im \cdot x}\}$ are honest L^2-eigenfunctions, while on $\mathbb{R}^n$ they represent the generalized eigenfunctions (bounded, but not L^2).

[9]We keep a convention of writing elliptic operators with sign $(-)$, to make L a positive operator: $\langle Lf \mid f\rangle = \int \sum p_{ij}(\partial_i f)(\partial_j f)dx \ge 0$, for all $\{f\}$.

II. *Parabolic (heat-diffusion):* $\partial_t u + L[u] = f$; for $u = u(x;t)$

III. *Elliptic equation in* $n+1$ variables: $(\partial_t^2 - L)u = f$ (called Poisson);

IV. *Hyperbolic (wave):* $u_{tt} + L[u] = f(x;t)$

V. *Schrödinger:* $\partial_t u = iL[u]$.

Each equation must be supplemented by an appropriate *boundary* and *initial conditions.* On regular boundaries (hyper-surfaces $\Sigma \subset \mathbb{R}^n$) one usually takes one of the 3 basic conditions: $u\,|_{\Sigma} = ...$ *(Dirichlet);* $\partial_n u\,|_{\Sigma} = ...$ *(Neumann)*; or more general $(\alpha + \beta\partial_n)u\,|_{\Sigma} = ...$ *(mixed)*. The condition at $\{\infty\}$ ("singular point") typically require u to be bounded, or to have a prescribed asymptotics (incoming/outgoing "radiation condition"). The *initial conditions* depend on the type of equation. One has $u\,|_{t=0} = u_0$ (for elliptic/parabolic problems II-III; V), and $\begin{cases} u\,|_{t=0} = u_0 \\ u_t|_{t=0} = u_1 \end{cases}$; (for hyperbolic IV).

We recall the general definition of the Green's function, for a boundary value problem

$$\begin{cases} M[u] = F; \text{ in } D \\ B[u] = h; \text{ on } \partial D \end{cases}; \tag{4.14}$$

where M could be any differential operator in a space (or space-time) region D (e.g. $M = -\nabla \cdot p\nabla + q$; or $\partial_t^{\,2} - \nabla \cdot p\nabla + q$), and B a suitable boundary/initial conditions on ∂D (e.g. $B = \alpha + \beta\partial_n$). The *Green's function (fundamental solution)* of (4.14) is an integral kernel $K(x;y)$, or $K(x,t;y,s)$ (possibly distributional) on $D \times D$, that satisfies the differential equation M in x-variables with the δ-source at $\{y\}$, (also the formal adjoint equation in y-variables with the δ-source at $\{x\}$), and the zero boundary conditions,

$$\begin{cases} M_x[K] = \delta(x-y) \\ B_x[K] = 0 \end{cases}; \Leftrightarrow \begin{cases} M_y{}^*[K] = \delta(x-y) \\ B_y[K] = 0 \end{cases}. \tag{4.15}$$

For instance, heat-problem (II) has operators $M_{x,t} = \partial_t + L_x$, $M^*{}_{y,s} = -\partial_s + L_y$, and the Green's kernel $K(x,y;t-s)$. Function K of either one of problems (4.15) allows to solve a nonhomogeneous system (4.14) with an arbitrary RHS $\{F;h\}$,

$$u(x) = \int_D K(x;y)F(y)dy + \int_{\partial D} P(x;y)h(y)dS(y); \text{ } dS\text{-surface area element on } \partial D.$$

So the Green's term K gives a contribution of the "continuously distributed sources" $F\,|_D$, while the so called *Poisson term* $P(x;y)$ on $D \times \partial D$, those of the "boundary sources" $h\,|_{\partial D}$. The Poisson kernel $P(x;y)$ is not independent of K, in fact, K is constructed explicitly in terms of K and the boundary operator B. For instance, the Dirichlet condition for $L = -\nabla \cdot p\nabla + q$, yields

$$P(x;y) = p(y)\partial_{n_y} K(x;y) \text{ (normal derivative of } K \text{ in } y \text{ on the boundary),}$$

while the Neumann condition gives

$$P(x;y) = pK(x;y), \text{ for } x \in D;\ y \in \partial D.$$

Similar constructions, based on Green's identities could be given for arbitrary M and boundary operator B.

I. Formal solutions and free-space Green's functions.

We shall construct *fundamental solutions (Green's functions)* for each of problems (I-V) in $\mathbb{R}^n$, by first writing its *formal solution*, then applying the Fourier analysis. Formal solutions of problems (I-V) can be written in the form of certain "functions[10] of L". Thus we get

$$u = (\lambda + L)^{-1}[f], \text{ for the elliptic problem (I);}$$

$$u = e^{tL}[u_0] + \int_0^t e^{(t-s)L}[f(...;s)]\,ds, \text{ for the heat-problem (II),}$$

and similarly,

$$u = e^{itL}[u_0]; \text{ for the Schrödinger problem (V).}$$

Here u_0 represents the initial value of u, and f - its RHS, the "heat-sources". Elliptic "n+1" - problem (III), is solved by

$$u = e^{\pm t\sqrt{L}}[u_0].$$

So in the half-line case: $0 \le t < \infty$ (unbounded range), one takes the negative "forward-stable" exponential $\{e^{-t...}\}$, while finite time-intervals, $a \le t \le b$, with 2-point boundary condition require a suitable combination of $\pm$-exponentials. The wave problem (IV) is solved in either of 2 forms,

$$u = e^{\pm it\sqrt{L}}[...]; \text{ or } \cos t\sqrt{L}[u_0] + \frac{\sin t\sqrt{L}}{\sqrt{L}}[u_1],$$

The choice depends on the type of the "boundary conditions" in t-variable, either "initial" at t=0, or "asymptotic" at $t = \pm\infty$.

In all cases (I-V) fundamental solution $\{\phi(L)\}$ is given by an integral kernel, $K(x;y)$, called the *Green's function* of the problem. For constant-coefficient L these are convolution-integrals $K = K(x-y)$, due to translational invariance of operator L. The Fourier transform yields explicit form of kernels $K(z)$, $z = x-y$. By definition of K we have to find a distribution $K(z)$, z=$x-y$, that satisfies the equation,

[10] "Functions of L" for a self-adjoint operator, could be given precise meaning via diagonalization (spectral resolution) of L. Given a diagonal (multiplication) operator $D{:}\,\phi(\lambda)\to\lambda\phi(\lambda)$, on $L^2(d\mu)$-space of functions/sequences $\{\phi(\lambda)\}$, function $f(D)$ is also a multiplication operator, $f(D){:}\,\phi\to f(\lambda)\phi(\lambda)$. But any self-adjoint L is unitary equivalent to a diagonal operator: $L = UDU^{-1}$, then $f(L) = Uf(D)U^{-1}$ (Appendix A).

$$L[K] = \delta(x-y). \tag{4.16}$$

Applying $\mathfrak{F}$ to both sides of (4.16) we get

$$a_L(\xi)\widehat{K}(\xi) = 1,$$

where $a_L(\xi)$ is the symbol of L. So we immediately get the transformed Green's function

$$\widehat{K}(\xi) = \frac{1}{a_L(\xi)}, \tag{4.17}$$

in other words "symbol of K" is inverse to "symbol of L". Then it remains to Fourier-invert $1/a_L(\xi)$ to recover K. The Fourier-inversion, however, involves the techniques of *complex analysis, distribution theory,* and also *reduction of variables via group-symmetries*, as we shall demonstrate below.

4.2. Laplacian is the most symmetric of all differential on $\mathbb{R}^n$ as it commute all translations and rotations (Euclidian motions). In fact, Δ generates all such operators, in the sense that any operator M on a suitable function-space ($L^2; L^p; C^\infty$), that commutes with $\mathbb{E}_n$ is a convolution with a radially-symmetric function $M(|x|)$,

$$M[u] = \int M(|x-y|)u(y)dy. \tag{4.18}$$

The Fourier-transform of (4.18) gives a multiplication-operator with another radially symmetric function $\widehat{M} = F(|\xi|)$,

$$\widehat{u} \to F(|\xi|)\widehat{u},$$

where $F(\rho)$ is related to $M(r)$ through the reduced Fourier-Bessel (Hankel) transform (§2.1-2.3). Since, ρ^2 is the symbol of $-\Delta$, we can call M a "function of Δ", $M = F(\sqrt{-\Delta})$. The proof of (4.18) is an easy application of the Fourier transform. It contains 2 statements.

Theorem 1: i) *Any translation-invariant operator M on a suitable function-class (e.g. Schwartz) is a convolution: $MR_a u = R_a M u$, for all $a \in \mathbb{R}^n$, and all $u \in \mathcal{S}$,$\Rightarrow$ $M[u] = f*u$, for some distribution f.*

ii) *Rotationally-symmetric convolutions, $MT_g = T_g M$, $g \in SO(n)$, are given by radial functions, $f = f(|x|)$.*

Indeed, any such M is taken by $\mathfrak{F}$ into an an operator $\widehat{M} = \mathfrak{F}M\mathfrak{F}^{-1}$, that commutes with multiplications by characters $\{e^{ib\cdot\xi}\}$,

$$\widehat{M}e^{ib\cdot\xi} = e^{ib\cdot\xi}\widehat{M}, \text{for all } b \in \mathbb{R}^n.$$

Hence $\widehat{M}$ commutes with all multiplications, $\widehat{M}\widehat{h} = \widehat{h}\widehat{M}$, $\widehat{h}$ in the Fourier-algebra $\mathcal{A}$. But

algebra $\mathcal{A}$ contains sufficiently many functions, so any operator, commuting with $\mathcal{A}$ must itself be a multiplication, $\widehat{M} = F(\xi)$, so $Mu = \check{F}*u$. The 2-nd statement is obvious, as $\mathfrak{F}$ takes radial functions into radial.

Our discussion will be limited to the Laplacian, $L = -\Delta$, although many results and techniques extend to more general (2-nd and higher-order) operators. The corresponding Green's functions then become inverse $\mathfrak{F}$-transforms of certain symbols.

i) *resolvent kernel:* $R_\lambda(z) = (\lambda+L)^{-1} = (2\pi)^{-n}\int \frac{e^{i\xi\cdot z}}{|\xi|^2+\lambda}d\xi$,

ii) *heat-semigroup* (Gaussian): $G_t(z) = e^{tL}$ has $\mathfrak{F}[G_t] = \widehat{G}(\xi;t) = e^{-t|\xi|^2}$.

iii) *half-space Poisson kernel:* $P_t(z) = e^{-t\sqrt{L}}$, (a semigroup, generated by $\sqrt{-\Delta}$) has Fourier transform $\widehat{P}(\xi;t) = e^{-t|\xi|}$.

iv) *cos*– and *sin*–type "*wave propagators*": $U_t = \cos t\sqrt{L}$; $V_t = \frac{\sin t\sqrt{L}}{\sqrt{L}}$;

$$U(z;t) = (2\pi)^{-n}\int \cos(t|\xi|)e^{i\xi\cdot z}d\xi.$$

$$V(z;t) = (2\pi)^{-n}\int \frac{\sin t|\xi|}{|\xi|}e^{i\xi\cdot z}d\xi.$$

or the unitary group $\{e^{\pm it\sqrt{L}}\}$, generated by $\sqrt{L}$.

v) *Schrödinger propagator* W_t (a unitary group generated by L),

$$W_t = (2\pi)^{-n}\int e^{i(t|\xi|^2+\xi\cdot z)}d\xi.$$

In some cases the Fourier-inversion is easy and straightforward like for the **Gaussian and Schrödinger propagators** (problem 5 of §2.1),

$$\boxed{G_t(z) = \frac{1}{(4\pi t)^{n/2}}e^{-|z|^2/4t}} \text{ and } \boxed{W_t = e^{it\Delta} = \frac{1}{(4\pi it)^{n/2}}e^{i|z|^2/4t}} \qquad (4.19)$$

The latter is obtained by analytically continuing G_t to imaginary time, $t\to it$.

4.3. The resolvent kernel $R(z;\lambda)$ is expressed through the Bessel functions of the 1-st and 3-rd kind [Leb];[Erd];[WW] of order $\nu = \frac{n}{2} - 1$,

$$\boxed{R(z;\lambda) = \frac{c_n}{|z|^\nu}\begin{cases} K_\nu(\sqrt{\lambda}|z|); \text{ if } \lambda > 0, \text{ i.e. } (-\Delta+\lambda)^{-1} \\ J_\nu(\sqrt{\lambda}|z|); \text{ if } \lambda < 0, \text{ i.e. } (\Delta+\lambda)^{-1} \end{cases}} \qquad (4.20)$$

In the limiting case, eigenvalue $\lambda = 0$, the Green's functions $(-\Delta)^{-1}$ coincides with the classical Newton's potentials:

$$\boxed{K = \frac{1}{\omega_{n-1}}|z|^{2-n} \text{ (for } n \geq 3); \; K = \frac{1}{2\pi}\log|z| \text{ (for } n{=}2)}$$

To derive (4.20) we notice, that all kernels (i-v) are radial functions of z, an easy consequence of the underlying rotational symmetry of Δ. Writing integral $R(z;\lambda)$ in polar coordinates $|z| = \rho$; $|\xi| = r$, we get

$$c_{n-1}\int_0^\infty \frac{r^{n-1}}{r^2+\lambda}dr \int_{S_{n-1}} e^{ir\rho\cos\theta}\sin^{n-2}\theta; \tag{4.21}$$

The inner integral gives the so called Poisson representation of the Bessel function J_ν, $\nu = \frac{n}{2}-1$,

$$J_\nu(z) = c_\nu z^\nu \int_0^\pi e^{iz\cos\theta}\sin^\nu\theta d\theta.$$

Hence we get in (4.21)

$$R(z;\lambda) = c_n\int_0^\infty \frac{r^{n-1}}{r^2+\lambda}(\rho r)^{-(n/2-1)}J_{n/2-1}(r\rho)dr.$$

The latter integral can be evaluated by complex integration in the r-plane of holomorphically extended Bessel function J_ν, whence comes $K_\nu(z) = J_\nu(iz)$. We shall illustrate the foregoing argument in $\mathbb{R}^3$. The inner integral

$$\int_0^\pi e^{ir\rho\cos}\sin\theta d\theta = \frac{2\sin r\rho}{r\rho}; \tag{4.22}$$

since $J_{1/2}(z)$ is an elementary function $\frac{\sin z}{\sqrt{\pi z}}$. We substitute the RHS of (4.22) in the outer integral (4.21) and get

$$R(z;\lambda) = (2\pi)^{-2}\int \frac{2\sin r\rho}{r\rho}\frac{r^2}{r^2+\lambda}dr = \frac{1}{2\pi^2\rho}\int_{-\infty}^{\infty} e^{ir\rho}\frac{r}{r^2+\lambda}dr = \frac{1}{4\pi}\frac{e^{-\sqrt{\lambda}|z|}}{|z|}.$$

4.4. The half-space Poisson kernels. Next we turn to the Poisson kernel $\widehat{P}_t = e^{-t|\xi|}$ in the half-space $\mathbb{R}^{n+1}_+$. The 1-D Poisson kernel is computed directly (problem 5 of §2.1),

$$\boxed{P_t = \mathcal{F}\left(e^{-t|\xi|}\right) = \frac{t}{\pi(z^2+t^2)}}.$$

The multidimensional case is, however, more subtle, and could be derived via the so called subordination principle, which allows to express the Poisson-semigroup, generated by $\sqrt{L}$, in terms of the heat-semigroup, generated by L,

$$e^{-t\sqrt{L}} = \frac{1}{\sqrt{\pi}}\int_0^\infty \frac{1}{\sqrt{s}}e^{-(s+t^2L/4s)}ds. \tag{4.23}$$

The derivation of (4.23) is fairly straightforward, via 1-D Fourier transform (problem 1). In the case of Laplacian $L = -\Delta$ on $\mathbb{R}^n$, we get the Poisson kernel $P(z;t)$ in $\mathbb{R}^{n+1}_+$ in terms of the Gaussian $\{G_t\}$.

By (4.24) and an explicit form of the Gaussian (4.19) we find,

$$P(z;t) = \frac{1}{\sqrt{\pi}}\int_0^\infty \frac{e^{-s}}{\sqrt{s}}\left(\frac{s}{\pi t^2}\right)^{n/2} e^{-|z|^2 s/t^2}ds = \frac{1}{\pi^{(n+1)/2}t^n}\int_0^\infty e^{-s(1+|z|^2/t^2)}s^{(n-1)/2}ds;$$

and the latter integral is reduced to the standard Gamma-function[11].

So we get the n-D Poisson kernel in the half-space $\mathbb{R}_+^{n+1}$,

$$\boxed{P_t = \frac{C_n}{t^n(1+|x|^2/t^2)^{(n+1)/2}} = \frac{C_n t}{(t^2+|x|^2)^{(n+1)/2}}}; \tag{4.25}$$

with constant

$$C_n = \frac{\Gamma(\frac{n+1}{2})}{\pi^{(n+1)/2}}.$$

Solution of the Dirichlet problem for the Laplaces equation in the half-space:

$$\begin{cases} u_{tt} + \Delta u = 0 \\ u\,|_{t=0} = f \end{cases}; \tag{4.26}$$

is then given by the convolution integral: $u(x;t) = P_t * f$, with kernel (4.25). As an application of (4.19), (4.25) one can show

Theorem 2: *Solution $u(x;t)$ of the elliptic Dirichlet problem (4.26), and the initial-value heat-problem: $u_t = \Delta u; u\,|_{t=0} = f$, converges to the boundary value $f(x)$ as $t \to 0$.*

The proof is outlined in problem 2. After the Dirichlet Poisson problem is solved we can proceed to the Neumann problem for the half-space Laplacian,

$$\begin{cases} u_{tt} + \Delta u = 0 \\ \partial_n u = u_t = f\,|_{t=0} \end{cases}; \tag{4.27}$$

The formal solution of (4.27)

$$u = \frac{1}{\sqrt{-\Delta}} e^{-t\sqrt{-\Delta}}[f] = Q_t[f],$$

is given by the *conjugate Poisson* kernel related to P_t via

$$Q(...;t) = \int_t^\infty P(...;s)\,ds, \text{ and } P = \frac{d}{dt}Q. \tag{4.28}$$

[11] Let us remark that Green's functions (i-v) are related one to the other through various transforms. For instance, the resolvent R_λ is obtained by Laplace-transforming the heat-semigroup G_t,

$$(\lambda + L)^{-1} = \int_0^\infty e^{-\lambda t - Lt} dt,$$

while (4.27) gives a similar representation for P_t. Formula (4.27) exemplifies such relations. It represents a special case of the so called Sonin-Mehler integral representation of modified Bessel functions of the 3-rd kind (McDonald functions)

$$K_\nu(r) = \frac{1}{2}\int_0^\infty exp[-(t+\frac{1}{t})\frac{r}{2}]\,t^{-\nu-1}dt.$$

Precisely, (4.28) correspond to order $\nu = \frac{1}{2}$, the only case, when Bessel-function becomes elementary, $K_{1/2}(r) = \sqrt{\frac{\pi}{2r}}e^{-r}$.

as a consequence of an identity, $L^{-1}e^{-tL} = \int_t^\infty e^{-sL}ds$. Substituting (4.28) in (4.25) we find

$$\boxed{Q(z;t) = C_n \int_t^\infty \frac{s}{(s^2 + |z|^2)^{(n+1)/2}} ds = \frac{C_n'}{(t^2 + |z|^2)^{(n-1)/2}}}.$$

So the Neumann problem (4.27) is solved by the convolution kernel

$$u(x;t) = Q(...;t) * f.$$

4.5. Wave-propagators. We shall use Poisson kernels P and Q, found in the previous part to find the *wave-propagators*,

$$U(z;t) = \cos ct\sqrt{-\Delta},\ V = \frac{1}{\sqrt{-\Delta}} \sin ct\sqrt{-\Delta},\ \text{and } \{e^{it\sqrt{-\Delta}}\},$$

i.e. fundamental solutions of the wave-equation: $u_{tt} - c^2\Delta u = 0$, (c = speed of propagation) with 2 different-type initial conditions:

$$\begin{cases} u\,|_{t=0} = \delta \\ u_t|_{t=0} = 0 \end{cases};\ \text{for } U,\ \text{or} \begin{cases} u\,|_{t=0} = 0 \\ u_t\,|_{t=0} = \delta \end{cases};\ \text{for } V.$$

As in the Gaussian case this will be accomplished via analytic continuation of the time parameter through the imaginary axis, $t \to it$. The Fourier transforms of both propagators:

$$\widehat{U}(\xi;t) = \cos ct\,|\,\xi\,|,\ \widehat{V}(\xi;t) = \frac{\sin ct\,|\,\xi\,|}{|\,\xi\,|};$$

as well as Poisson P_t are easily inverted in 1-D. Thus,

$$\sin ct\,|\,\xi\,| = \frac{e^{ict|\xi|} + e^{-ict|\xi|}}{2i\,|\,\xi\,|} \Rightarrow V = \tfrac{1}{2c}H(\,|\,z - ct\,|\,),$$

turns into a *Heaviside jump-function*, truncated at $\{\pm ct\}$. Similarly, the cos-propagator,

$$\cos(ct\,|\,\xi\,|) = \tfrac{1}{2}(e^{ict|\xi|} + e^{-ict|\xi|}) \Rightarrow U = \tfrac{1}{2}\{\delta(x-ct) + \delta(x+ct)\},$$

which is the sum of the right and left-traveling δ-pulses, the classical d'Alambert solution. One has to compute two limiting distributions:

$$U_t(z) = \Re e P(ict) = \tfrac{1}{2}\lim_{\epsilon \to 0} C_n\left(\frac{ct}{(\,|\,z\,|^2 + (\epsilon + ict)^2)^{(n+1)/2}} + \frac{ct}{(\,|\,z\,|^2 + (\epsilon - ict)^2)^{(n+1)/2}}\right);$$

$$V_t(z) = \Im m Q(ict) = \tfrac{1}{2i}\lim_{\epsilon \to 0} C_n'\left(\frac{1}{(\,|\,z\,|^2 + (\epsilon + ict)^2)^{(n-1)/2}} - \frac{1}{(\,|\,z\,|^2 + (\epsilon - ict)^2)^{(n-1)/2}}\right).$$

In dimension $n = 2$ both are given by certain densities,

$$U_t(z) = \frac{t}{4\pi(c^2t^2 - |\,z\,|^2)^{3/2}};\ V_t(z) = \frac{1}{2\pi\sqrt{c^2t^2 - |\,z\,|^2}}.$$

supported in the *light-cone* $\{|x-y| \leq ct\}$. So for each fixed $t>0$ support of U(..;t); V(...;t) belongs to a disk of radius R = ct, centered at the *source*[12] $\{y\}$ (fig.2). In **3-D** space we get

$$V_t = \lim_{\epsilon \to 0} \frac{1}{2i}\left\{\frac{1}{|z|^2+(\epsilon+ict)^2} - \frac{1}{|z|^2+(\epsilon-ict)^2}\right\} = \lim \frac{2\epsilon ct}{(c^2t^2-|z|^2+\epsilon^2)^2+4\epsilon^2c^2t^2}. \quad (4.29)$$

The RHS of (4.29) represents a limiting distribution of the family

$$\left\{\Im\left(\frac{1}{f+i\epsilon}\right) = \frac{\epsilon}{f^2+\epsilon^2}\right\}$$

with a regular function $f = f_\epsilon(z) = c^2t^2 - |z|^2$. From the basic distribution theory of §2.2, we find,

$$\frac{\epsilon}{f^2+\epsilon^2} \to \delta_f(r) = \int_{f=0} \dots \frac{dS}{|\nabla f|}; \text{ or } \sum_{y \in f^{-1}(0)} \frac{1}{|f'(y)|} d(r-y) \text{ (in 1-D); as } \epsilon \to 0,$$

where dS means the natural surface element on the level-set $\{f=0\}$. In other words, the limiting distribution δ_f is made of the "zero-level" δ-function of f, divided by a continuous density $|\nabla f(z)|$. In our case (4.29) we get

$$\boxed{\begin{aligned} V_t(z) &= \frac{1}{2\pi}\delta(ct^2-|z|^2) = \frac{1}{4\pi ct}\delta(|z|-ct) \\ U_t(z) &= \partial_t V = \frac{1}{4\pi(ct)^2}\delta(|z|-ct) + \frac{1}{4\pi ct}\delta'(|z|-ct) \end{aligned}}$$

So the 3-D wave-propagator $\{U;V\}$ can be thought of either as distributions in $(x;t)$-variables supported on the forward light-cone $C = \{(x;t): |x|^2 = ct\} \subset \mathbb{R}^4$, or as a t-parameter family of spherical δ-functions $\{\delta_{S_R}$: of radius $R = ct\}$ in $\mathbb{R}^3$. The corresponding solution of the initial value problem: $u|_0 = f$; $u_t|_0 = g$, in **2-D** and **3-D** are given by convolution kernels,

$$u(x;t) = \frac{1}{4\pi ct}\int_{|y|=ct} g(x-y)dS(y) + \frac{1}{4\pi(ct)^2}\int_{|y|=ct} f(x-y)dS(y);$$

$$u(x;t) = \iint_{|y|\leq ct}\left\{\frac{tf(x-y)}{4\pi(c^2t^2-|y|^2)^{3/2}} + \frac{g(x-y)}{2\pi\sqrt{c^2t^2-|y|^2}}\right\}d^2y. \quad (4.30)$$

Remark: Similar expressions hold for higher-D wave-problems. Namely, the Green's functions $U(z;t)$; $V(z;t)$ are given by distributions supported either inside the light cone C, $\{|z| \leq ct\}$ (even dimensions), or on the surface of the cone $\{|z| = ct\}$, as illustrated in fig.2 below (problem 4). The exacts solution of the n-D wave problem are given in terms of *spherical means* of function f,

$$M_r f(x) = \frac{1}{\omega_{n-1}r^{n-1}}\int_{|y|=r} f(x-y)dS(y),$$

[12] Green functions are often called *point-sources*, as they describe solutions produced by sharp (δ-type) initial impulses.

where ω_{n-1} is the area of the unit sphere in $\mathbf{R}^n$. Then the initial data $u|_0 = 0$; $u_t|_0 = f(x)$, yields [Hel]

$$u(x;t) = \frac{1}{(n-2)!}\partial_t^{n-2}\left[\int_0^t \{M^s f\}(x)\, s\,(t^2-s^2)^{\frac{n-3}{2}} ds\right] \tag{4.31}$$

In special cases n=2;3 one recovers solutions (4.30). Formula (4.31) demonstrates the *Huygens principles* for solutions of the wave equation: any initial disturbance propagates inside the light-cone based on supp$\{f\}$ with a finite speed c. Furthermore, in odd dimensions disturbance of any finite extent vanishes after a signal passes (*sharp Huygens principle),* whereas in odd dimensions once a particular space-point is reached by the disturbance will remain forever, dying out exponentially fast with time. Indeed, derivatives of integral (4.31) in odd dimensions take the form

$$\sum P_k(t)\partial_t^k[M^t f],$$

with some polynomial coefficients $\{P_k(t)\}$.

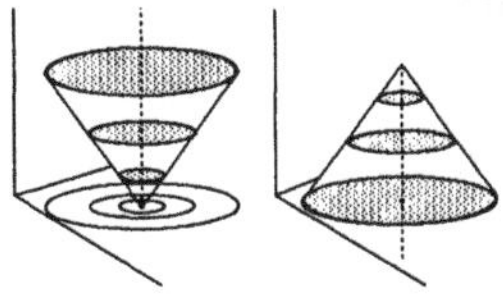

Fig. 2. *The forward and backward light-cones for the wave equation. The left figure (forward) shows a point-source solution (Greens function) at 3 different time cross-sections. At each t the disturbance is localized either on the surface (sphere $S_R(x)$ of radius $R = ct$, centered at the source $\{x\}$), or inside the ball $B_R(x) = \{|y-x| \le ct\}$. The right figure shows the domain of dependence: any solution $u(x;t)$ depends on the initial data inside the ball $\{|y-x| \le ct\}$.*

There are different way to derive (4.31), all exploiting to some degree the underlying group symmetry of the problem. We shall mention 2 of them: [Hel2] (chapter 1) is based on the Radon transform on $\mathbb{R}^n$ (and rotational symmetries), whole R. Howe [How] presents an elegant derivation based on the representation theory of the Heisenberg group (cf. §6.2).

We have summarized the main results of the section in the table at the end of the section.

II. Symmetries and differential equations in regions of $\mathbb{R}^n$.

After we obtained the free-space fundamental solutions in $\mathbb{R}^n$, let us turn to boundary-value problems. The analysis below will emphasize group-symmetries (discrete and continuous), and whenever possible provide explicit formulae and constructions of Green's functions. Of course, our discussion will be limited to special classes of symmetric regions and differential equations.

4.6. Symmetry reduction and the source condition. Green's function $K=f(L)$ clearly inherits all original symmetries of the problem[13] (operator L; region Ω), which allows to reduce the number of variables in the Green's function K. For instance, hermitian symmetry, $L=L^*$, implies symmetry of integral kernel, $K(x,y)=K^*(y,x)$ - complex conjugate. Time-independence of L means that $K=K(t-s;x,y)$ are time-convolutions kernels. Similarly, constant-coefficient, i.e. translational-invariance of L in $\mathbb{R}^n$ yield translational-invariant (convolution) kernels $K=K(x-y)$.

Larger symmetry groups play the role for special operators L, like orthogonal rotations $SO(n)$ for the Laplacian/Helmholtz operators: Δ, $\Delta+m^2$ on $\mathbb{R}^n$, and hyperbolic rotations $SO(1;n)$ for the *d'Alambertian* (wave operator) $\square=\partial_t^2-\Delta$, or the *Klein-Gordon (KG)* operator: $\partial_t^2-\Delta+m^2$. The corresponding Green's functions, $K(z)$ or $K(t;z)$, $z=x-y$, retain the symmetries of L, which allows to reduce K to a single variable function $K(r)$, of the Euclidian or hyperbolic radius: $r=|x|$, or $r=\sqrt{t^2-|x|^2}$. In both cases the radial function K solves the reduced ODE,

$$\boxed{K_{rr}+\tfrac{n-1}{r}K_r\pm m^2K=0,\ \text{for } r>0} \tag{4.32}$$

which is singular at the end-points $r=0;\infty$, hence requires an appropriate boundary condition at both. The "source condition" refers to a proper choice of the boundary condition at a "singular (source) point". We shall demonstrate with 2 familiar examples.

4.7. Elliptic Case. Laplacian/Helmholtz-type operator: $L=-\Delta+q$. The condition at $\{0\}$ represents the so called "***point-source***",

$$r^{n-1}K'(r)\,|_{r=0}=\frac{1}{C_{n-1}}, \tag{4.33}$$

where $C_{n-1}=\frac{2\pi^{n/2}}{\Gamma(n/2)}$ - surface area of the unit sphere in $\mathbb{R}^n$. To derive (4.33) we take any function $u(y)$ on $\mathbb{R}^n$, write $L[u]=F(y)$, apply kernel $K(x,y)$ to the equation, and integrate by parts (Green identity) over the complement $\Omega=\mathbb{R}^n\backslash B_\epsilon(x)$, of a small ball $B_\epsilon(x)$ centered at $\{x\}$. On one hand,

$$u(x)=\lim_{\epsilon\to 0}\int_\Omega K(x;y)\,L[u]\,dy,$$

[13]We shall not pursue any systematic study of symmetries in differential equations here, but refer to the recent book [Olv], which explores the subject in great depth (see also [Ovs];[Mil]). Our goals will be limited to exploiting some apparent and well know (geometric) symmetries, rather than investigate all of them.

but

$$\int_\Omega KL[u]dy = \int_\Omega L_y[K]udy + \oint_\Gamma (\partial_n K u - K\partial_n u)dS.$$

First integral in the RHS vanishes, since $L_y[K]=0$ outside of the source. Remembering that $K=K(|x-y|)$, we get the second integral

$$\oint_{|x-y|=\epsilon} \ldots = C_{n-1}\epsilon^{n-1}K'(\epsilon)\,(u(x)+o(\epsilon)),$$

while the third $\oint_{|x-y|=\epsilon} \ldots \approx \epsilon^n K(\epsilon)\,\Delta u(x) \to 0$, as $\epsilon\to 0$.

Thus

$$u(x) = \lim_{\epsilon\to 0}\{C_{n-1}\epsilon^{n-1}K'(\epsilon)\}\,u(x) + \{\epsilon^n K(\epsilon)\}\,\Delta u(x),$$

for any function u, which yields the source condition (4.33). Equation (4.32) for $q=0$ takes the form

$$K_{rr} + \tfrac{n-1}{r}K_r = 0,$$

it has 2 solutions (regular and singular): $\{1; \log r\}$ in $n=2$, and $\{1; r^{2-n}\}$ for $n\geq 3$. So

$$K = c_1 + c_2 r^{2-n} \text{ (or } \log r).$$

The boundary condition for K at $\{\infty\}$ eliminates the constant term, while the source condition (4.33) yields c_2, whence the familiar form of the Newton/Gauss potential, the Green's function of the Laplacian,

$$\boxed{K = \frac{1}{C_{n-1}|x-y|^{n-2}};\ \text{for } n\geq 3;\ K = -\frac{1}{2\pi}\log|x-y| \text{ in } \mathbb{R}^2}$$

A similar treatment applies to the case $q\neq 0$. But this time (4.32) becomes the equation of *Bessel-type*,

$$y'' + \tfrac{n-1}{r}y' \pm \lambda y = 0, \text{ with } \lambda = q.$$

It is easily reduced to the standard (1-st or 3-nd kind) Bessel equation

$$Y'' + \tfrac{1}{r}Y' \pm \left(1 - \tfrac{\nu^2}{r^2}\right)Y = 0, \text{ of order } \nu = \tfrac{n-2}{2}, \tag{4.34}$$

by the change,

$$y(r) = r^{-\nu}Y_\nu(\sqrt{\lambda}\,r).$$

Thus solution of (4.32) with positive m^2 becomes a combination of the 1-st and 2-nd kind Bessel functions: $K = r^{-\nu}\{c_1 J_\nu(mr) + c_2 Y_\nu(mr)\}$, of order $\nu = \frac{n-2}{2}$. The source-condition at $\{0\}$ is furnished by a non-vanishing coefficient c_2. The condition at $\{\infty\}$ should be chosen either as *incoming*, or *outgoing "radiation"* condition

$$K \sim r^{(\frac{n-1}{2})}e^{\pm imr}.$$

The latter with a proper normalization at $\{0\}$ uniquely determines K as the *MacDonald function*, $M_\nu = Const\,(J_\nu \pm iY_\nu)$, so

$$K(r) = (\Delta + m^2)^{-1} = \tfrac{c_n}{m} M_\nu(mr) \sim \begin{cases} \frac{1}{\omega_{n-1}} r^{2-n} \\ r^{\frac{1-n}{2}} e^{imr} \end{cases}; \ r = |x-y|\cdot$$

Similarly, for $\Delta - m^2$, the reduced equation changes into the modified Bessel,

$$K_{rr} + \tfrac{n-1}{r} K_r - m^2 K = 0,$$

whose solutions are

$$I_\nu \sim r^{\frac{1-n}{2}} e^{mr}, \text{ and } K_\nu \sim r^{\frac{1-n}{2}} e^{-mr}.$$

We obviously choose the exponentially decaying Kelvin function K_ν, which could be normalized to satisfy the source condition at $\{0\}$.

4.8. Hyperbolic Case. The wave and *KG*-equations:

$$\Box u = u_{tt} - c^2 \Delta u = \delta; \text{ and } (\Box \pm m^2)u = \delta,$$

have all the symmetries of $\mathbf{R}^{n+1}$ space-time, translations as well as hyperbolic rotations $SO(1;n)$. The corresponding Green's functions $K = K(t,x)$ are reduced as above to a single variable, the hyperbolic radius $r = \sqrt{(ct)^2 - |x|^2}$. They satisfy the Bessel-type equation

$$K_{rr} + \tfrac{n}{r} K_r \pm m^2 K = 0. \tag{4.35}$$

But unlike the elliptic case it is not easy to write down the proper "source condition" at $r = 0$. In the last section we used the Fourier analysis to derive the wave-propagators, which yields for dimensions $n = 1, 2, 3$,

$$K = \tfrac{1}{2c} H(ct - |x|); \ K = \frac{1}{2\pi\sqrt{(ct)^2 - |x|^2}}; \ K = \tfrac{1}{4\pi ct}\delta(ct - |x|). \tag{4.36}$$

In higher dimensions K becomes yet more singular distribution, supported in the light cone $\{|x| \le ct\}$ or its surface $\{|x| = ct\}$ (4.31), i.e. $K = \phi(r)$ - a distribution in $r = ct - |x|$, or hyperbolic $r = \sqrt{(ct)^2 - |x|^2}$. The *KG*-Green's functions satisfy the Bessel-type (4.35) equation in hyperbolic r, and we can use solutions (4.35), as proper "source-conditions". This yields

$$\boxed{K = J_0(mr); \ K = \frac{C}{\sqrt{r}} Y_{\frac{1}{2}}(mr); \text{ and } K = \tfrac{C}{r} Y_1(mr)},$$

in spatial dimensions $n = 1, 2, 3$.

4.9. Discrete symmetries and the method of images. The method of images (reflected sources) applies to differential equations in regions of $\mathbb{R}^n$, and other symmetric spaces (see §7.6 of chapter 7), obtained by "discrete symmetries" (reflections, translations, conformal mappings), like the half-space, quadrants, strips/ slabs, rectangles, spheres, etc. The Green's function of the corresponding problem is obtained from the free-space Green's function K_0 by subjecting K_0 to a (discrete) set of symmetry transformations, or by subtracting from K_0 a suitable regular solution.

amely, K_0 satisfies the source (δ − function) condition, $L_x[K_0] = \delta(x-y)$ in Ω, but it ils to vanish on the boundary, $B[K_0]|_{\Sigma} \neq 0$. So we want to find a family of regular nctions $\{u(x;y)\}$, depending on the position of the source $\{y\}$, so that

$$L_x[u(...;y)] = 0 \text{ in } \Omega; \text{ and } B_x[u]\,|_{\Gamma} = K_0(x;y). \tag{4.37}$$

4.10. Reflections and half-space problems. We consider the half-space $\mathbb{R}^{n+1}_{+} = \{(x,t): x \in \mathbb{R}^n;\ t > 0\}$, and assume operator L to be reflection-invariant: $(x;t) \to (x;-t)$, for instance $L = \partial_t^2 + \Delta$ - the half-space Laplacian. The corresponding Green's function depends on the type of boundary condition. For *Dirichlet Problems* the Green's function

$$K^D = K(z;t,s) = K_0(z;t-s) - K_0(z;t+s),$$

represents the difference between the "source-solution" $K_0(y;s)$ (e.g. Newton's potential) and the "*reflected source*" $K_0(y;-s)$ (fig.3). The reflected source gives the requisite correction u in (4.37). So the half-space Dirichlet Green's function represents a convolution kernel in x-variable, $K(x-y;\,t,s)$ of the form

$$K = \frac{1}{\omega_n}\left\{(\,|\,z\,|^2 + (t-s)^2)^{-\frac{n-1}{2}} - (\,|\,z\,|^2 + (t+s)^2)^{-\frac{n-1}{2}}\right\}.$$

This yields, in particular, the half-space Poisson kernel $P(x-y;t)$, derived earlier in by the Fourier-transform methods

$$\boxed{P(z;t) = e^{-t\sqrt{L}} = \partial_s K\,|_{s=0} = \frac{C_n t}{(\,|\,z\,|^2 + t^2)^{(n+1)/2}}} \tag{4.38}$$

Let us observe, that solution of the Dirichlet problem: $\Delta u = F;\ u\,|_{\Sigma} = f$; in any region Ω with boundary $\Sigma = \partial\Omega$, is represented by the *Green's identity*, as

$$u(x) = \int_{\Omega} K(x;y)F(y)dy + \int_{\Sigma} \partial_n K(x;y)f(y)dS(y),$$

where ∂_n denotes the normal derivative of the Dirichlet Green's function in $y \in \Sigma$. So the Green's function K picks up the "continuously distributed sources" F over Ω, while its normal derivative $P = \partial_n K$ - the *Poisson kernel* gives the contribution of the "boundary sources".

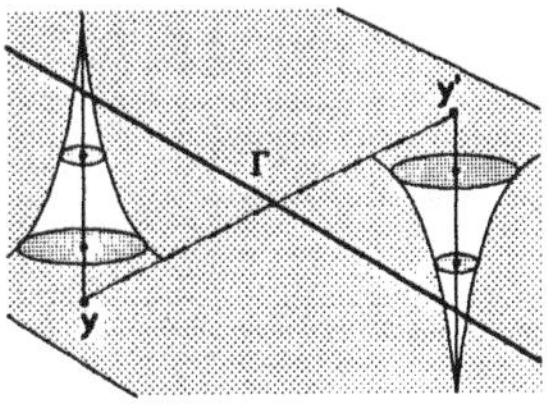

Fig.3: *Free-space Green's functions at a source $\{y\}$ and the reflected source $\{y'\}$ cancel each other on the boundary Γ, separating two half-spaces (Dirichlet problem). In the Neumann problem 2 "sources" add up, so the normal derivative vanishes on the boundary.*

For *Neumann Problem* the difference of two sources is replaced with the sum

$$K^N = K_0(z;t-s) + K_0(z;t+s),$$

(so that normal derivatives would cancel each other). As application we obtain the *conjugate Poisson kernel* Q in the half-plane[14],

$$Q(z;t) = \frac{e^{-t\sqrt{L}}}{\sqrt{L}} = K\,|_{s=0} = \frac{C'_n}{(|z|^2+t^2)^{(n-1)/2}};$$

found earlier by the Fourier transform methods. In a similar vein one constructs Green's functions for other problems, e.g. the half-space Dirichlet/Neumann "heat problem" has

$$K(x,y;t) = G(x-y;t) \mp G(x-y';t)$$

- the difference/sum of two Gaussians with "sources" at $\{y=(y_1;...y_n)\}$ and the reflected point: $\{y'=(-y_1;...y_n)\}$. Quadrants and other product-type regions are treated in a similar fashion. Thus the quadrant $\{(x_1;x_2)>0\}$ Green's function consists of 4 reflected sources:

$$K(x,y...) = K_0(x-y) - K_0(x-y') - K_0(x-y'') + K_0(x+y);$$

where y'; y'' represent reflections of $\{y\}$ relative to the 1-st, respectively 2-nd coordinate axis.

4.11. Green's functions in finite regions. Green's functions of elliptic differential operators in bounded regions Ω (compact manifolds) can always be expanded in terms of eigenfunctions $\{\psi_k\}$ of L, subject to a proper boundary condition. We leave to the reader as as exercise to write expansions for the inverse operator L^{-1}, the heat semigroup e^{-tL}; Poisson kernel $\exp(-t\sqrt{L})$; wave-propagators, etc. Let us remark that the Green's function of any regular S-L problem on interval [0;T] can be represented in two different ways: via the eigenfunction expansion[15]

$$K(x;y) = \sum \frac{1}{\lambda_k}\psi_k(x)\bar{\psi}_k(y), \tag{4.39}$$

or in terms of a *fundamental pair* of solutions: $L[u_{1,2}]=0$, one of which satisfies the zero boundary condition on the left ($u_1\,|_{t=0}=0$), and the other on the right ($u_2\,|_T=0$). Given a fundamental pair $\{u_1;u_2\}$, the Green's function is constructed as,

$$K(t;s) = \frac{1}{W(s)}\begin{cases} u_1(t)\,u_2(s);\ t\le s \\ u_2(t)\,u_1(s);\ t\ge s\end{cases};\ W\text{=Wronskian of } \{u_1;u_2\}. \tag{4.40}$$

For constant coefficient operators $L=-\partial^2+q$, two representations are related by another version of the *Method of images.* Namely, we take source $y\in[0;T]$, and "reflect it" infinitely

---

[14] Indeed, solution of the Neumann problem: $\Delta u = F$; $\partial_n u\,|_\Sigma = f$; is represented via Green's identity, as

$$u(x) = \int_\Omega K(x;y)F(y)dy + \int_\Sigma K(x;y)f(y)dS(y),$$

where $K=K^N$ is the Neumann Green's function, so $Q(x;y)=K(x;y)$, for $x\in\Omega, y\in\Sigma$.

[15] If one of eigenvalues $\lambda_0=0$, then kernel K (4.5) (sum over nonzeros λ's defines the so called *modified Green's function,* i.e. inverse of L on the subspace spanned by nonzeros eigenfunctions $\{\psi_k\}$, so the product of operators $LK=\delta(x-y)-\psi_0(x)\bar{\psi}_0(y)$.

many times by the lattice $\mathbf{Z} = \{mT\}$. So the free-space Green's function of L on $\mathbf{R}$.

$$K_0(|x-y|) = \frac{1}{2\sqrt{q}} e^{-\sqrt{q}|x-y|};$$

is shifted and summed over the lattice of period T. Of course, the corresponding Green's function K on $[0;T]$ has different representations depending on the type of boundary condition. Namely, the *periodic boundary condition (torus)*, gives

$$K(x,y) = \sum_m K_0(x-y+mT) = \frac{1}{T}\sum_k \frac{1}{\lambda_k} exp(i\frac{2\pi k}{T}(x-y)); \tag{4.41}$$

with eigenvalues: $\{\lambda_k = (\pi k/T)^2 + q: k \in \mathbf{Z}\}$, while for the *Dirichlet/Neumann* condition the source at $\{y\}$ is reflected to $\{-y\}$, and then the process is repeated periodically with period $2T$. So one has

$$K(x,y) = \sum_m K_0(x-y+2mT) \mp K_0(x+y+2mT) = \frac{2}{T}\sum_k \frac{1}{\lambda_k} sin\frac{\pi kx}{T} sin\frac{\pi ky}{T}. \tag{4.42}$$

The reader should recognize relations (4.41) and (4.42), as special cases of the Poisson summation for function $f(t) = K_0(|x-y|+t)$ on the lattice $\mathbf{Z} = \{mT\}$. Similar considerations apply to the heat, wave, *KG* and other equations, and to equations on multi-dimensional tori. Two such examples are

i) $R(z;\lambda) = \sum_m \frac{e^{im \cdot z}}{m^2+\lambda} = (2\pi)^{-1}\sum_\nu \frac{1}{\sqrt{\lambda}} e^{-\sqrt{\lambda}|z-2\pi\nu|}$ (in 1-D);

ii) $G(z;t) = \sum_m exp(-\frac{1}{2}tm^2 + im \cdot z) = (2\pi t)^{-n}\sum_\nu exp(-\frac{|z-\nu|^2}{2t})$ (Jacobi *theta-function*).

4.12. Dirichlet Laplacians on triangles.

• **Right-isosceles triangle** makes up $\frac{1}{2}$ of the square. The $(\pi \times \pi)$-square eigenmodes are made of products of the even and odd Fourier modes $\{c_k = \cos(k+\frac{1}{2})x\}$, $\{s_m = \sin 2mx\}$ (in variables x and y). So $\{\psi_{km}\}$ consist of products $\{c_k(x)\,c_m(y)\}, \{s_k(x)\,s_m(y)\}$ and $\{c_k(x)\,s_m(y); s_k(x)\,c_m(y)\}$. The former two ($cc-$ and $ss-$types) correspond to different families of eigenvalues $\{(k+\frac{1}{2})^2+(m+\frac{1}{2})^2\}$, and $\{4(k^2+m^2)\}$, so they never mix. The latter ($cs-$ and $sc-$ types) produce eigenmodes ψ_{km} of the form $\frac{c_k(x)s_m(y) \pm s_m(x)c_k(y)}{2}$, that vanish on diagonals $x = \pm y$, hence yield the triangle Dirichlet eigenfunctions.

• **Equilateral triangle** T makes $\frac{1}{2}$ of the diamond, the fundamental region of the hexagonal lattice $\Gamma \subset \mathbf{C}$, spanned by $\{1;\alpha = e^{i\pi/3}\}$ (fig.4). Its Dirichlet eigenfunctions fall into 3 families, according to the natural action of dihedral symmetry group $\mathbf{D}_3 = \mathbf{Z}_3 \rhd \mathbf{Z}_2$ on T. The latter has 3 irreducible representations: trivial $\chi_0 = 1$, $\chi_1 = \pm 1$ (trivial on rotations $\{1;\omega;\bar{\omega}:\omega = e^{i2\pi/3}\}$ and equal -1 on any reflection), and a 2-D representation σ, $\sigma_\omega:(\xi;\eta) \to (\omega\xi;\bar{\omega}\eta)$ (for rotations $\{1;\omega;\bar{\omega}\}$), and $\sigma_r(\xi;\eta) = (\eta,\xi)$, for reflection r about the z-axis (see §3.3). So family (I) consists of eigenfunctions[16], symmetric relative to the $\mathbf{D}_3$-action , family (II) are rotation-invariant, but

[16]Dirichlet eigenfunctions are uniquely determined by the values of their normal derivatives on the boundary, so the description here refers to the values of $\partial_n\psi$ on all 3 sides of T.

mapped to $-\psi$ by reflections, finally (III) gives a pair of conjugate values $\{\alpha;\bar{\alpha}\}$, $\{-1;-1\}$,$\{\bar{\alpha};\alpha\}$ on each of 3 sides of T (see fig.5c).

Our goal is to link eigenvalues of Δ_T to the well known eigenvalues of the Γ-periodic Δ on torus $\mathbf{C}/\Gamma$,

$$\lambda_{km} = |ik + \alpha m|^2 = k^2 + m^2 - km.$$

Figure 4 demonstrates how each eigenfunction of family (I), (II) and (III) extends to a Γ-periodic eigenfunction of Δ: for (I)-(II) is equal to the size of the triangle (the plane is cut into the union of copies of T, and ψ is taken into all images $\{T'\}$ of T with signs $\pm$, depending on the orientation of T'. That same reflection procedure applied to a III-type eigenfunction requires to triple the period. Since each periodic eigenfunction is a combination of Fourier modes $\{\exp[i(kx+m\Im(\bar{\alpha}\, y)]: k,m\in\mathbf{Z}\}$, we obtain the eigenvalue-spectrum of Δ_T, to be made of 2 sequences:

$$\boxed{\lambda_{km} = k^2+m^2-km \text{ (for I-II) } \lambda'_{km} = \tfrac{1}{9}(k^2+m^2-km) \text{ (for III)}}$$

Fig.6 below demonstrates a II-type eigenfunction as a combination of rotated and reflected exponents, $\psi = \chi_1 - \chi_2 + \chi_3 - \chi_4 + \chi_5 - \chi_6$.

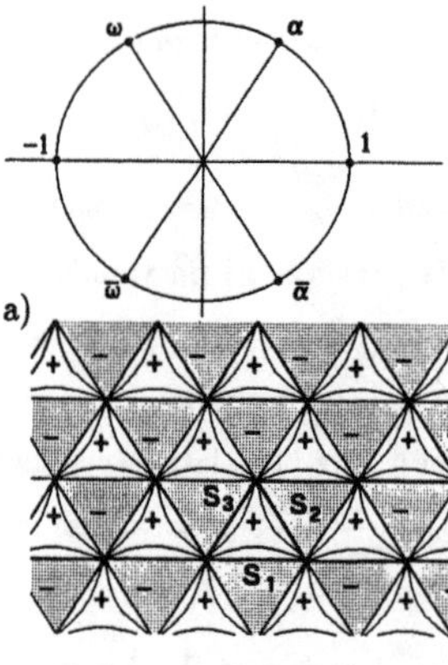

Fig.4: shows cubic roots $\{\omega,\bar{\omega} = \omega^2\}$ of 1, along with their negatives

$$\{\alpha = -\bar{\omega};\bar{\alpha} = -\omega\}.$$

Numbers $\{1;\omega;\bar{\omega}\}$ give 3 irreducible characters of subgroup $\mathbf{Z}_3 \subset \mathbf{D}_3$, while reflection (complex conjugation) defines a 2-D representation of $\mathbf{D}_3$.

a)

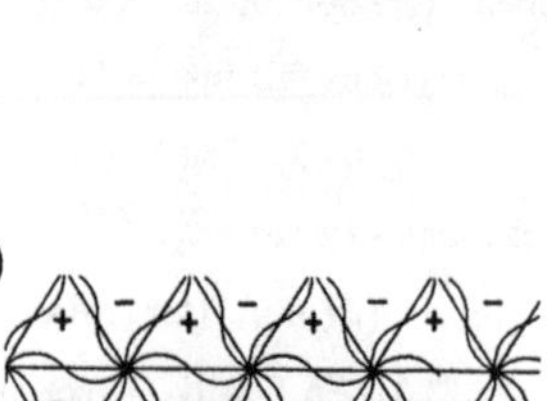

b)

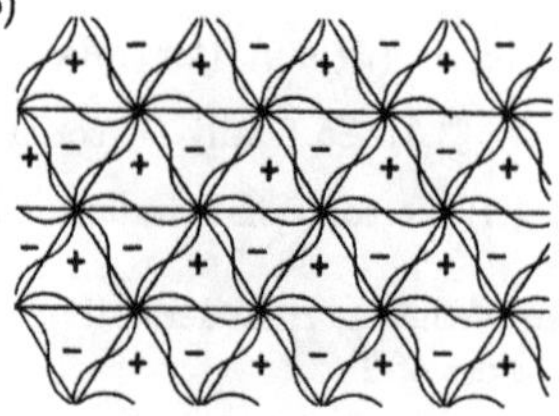

Fig.5 illustrates a periodic extension of eigenfunctions of types I, II, and III. A Dirichlet eigenfunction ψ in T is determined by its normal derivatives $\{f_i = \partial_n \psi$, on the i-th side S_i, $i = 1,2,3\}$.

We schematically represent each $\{f_i\}$ by a curve along S_i; the portion of the curve inside the triangle corresponds to a positive derivative $\partial_n f$, while the outside portion to the negative derivative. For type-I all functions $\{f_i\}$ are equal and symmetric with respect to the center of the side.

So given a type-I ψ in a "+"-triangle, we reflect it to a neighboring "-"triangle, changing the sign $\psi \to -\psi$, and repeat the process. The resulting Γ-periodic function on $\mathbf{C}$ is clearly an eigenfunction of Δ.

For type-II ψ (b) boundary function f is odd on

c)

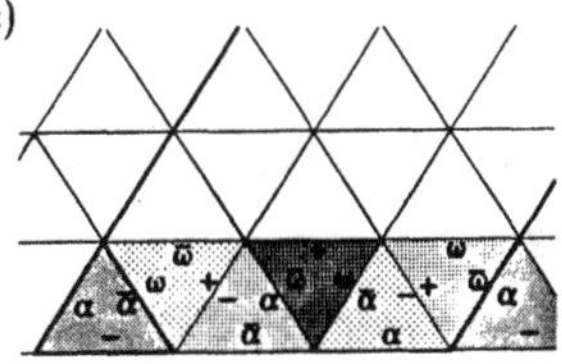

each side S_i and repeats itself under rotational symmetries with $\{1;\omega;\overline{\omega}\}$. Once again reflections $\psi \to -\psi$, from T_+ to T_- across $\{S_1;S_2;S_3\}$ produces a Γ-periodic eigenfunction of Δ in $\mathbb{C}$.

The type-III (c) is somewhat more complicated. Here we pick a pair of complex conjugate functions $\{\psi;\overline{\psi}\}$. Then one can easily verify that the boundary functions $\{f_i;\overline{f}_i\}$ have absolute values $|f_1| = |f_2| = |f_3| = \rho$ (an even positive function), and differ one from the other by a factor ω, $f_2 = \omega f_1; f_3 = \overline{\omega} f_1$ (and similarly for $\overline{f}_i$). If we choose f_1 (left side) to have phase $\alpha = \sqrt{\omega}$, then f_2 (horizontal) have phase -1, while $f_3 = \overline{\alpha}\,\rho$. The reflection $\psi \to -\psi$ about side S_3 takes triple $\{\alpha;-1;\overline{\alpha}\}$ into $\{\omega = -\overline{\alpha};+1;\overline{\omega} = -\alpha\}$, and the process has to continued 3-times along the (horizontal) line to get the original triple $\{\alpha;-;\overline{\alpha}\}$. So type-III ψ yields a 3Γ-periodic eigenfunction of the Laplacian in $\mathbb{C}$.

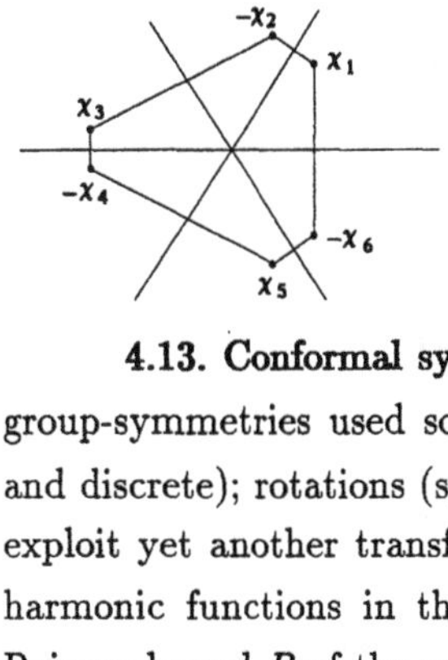

Fig.6 shows a type-III eigenfunction constructed as a combination of 6 exponentials: $\psi = \chi_1 - \chi_2 + \chi_3 - \chi_4 + \chi_5 - \chi_6$.

4.13. Conformal symmetry: Green's function and Poisson kernel in the ball. The group-symmetries used so far in our discussion involved space-translations (continuous and discrete); rotations (spherical and hyperbolic) and reflections. Our last example will exploit yet another transformation, the *conformal symmetry* of the Laplacian, to study harmonic functions in the balls $B_R = \{|x| \le R\} \subset \mathbb{R}^n$. The Green's function K and Poisson kernel P of the unit disk $\mathbb{D} \subset \mathbb{R}^2 = \mathbb{C}$ are well known from the complex analysis, and can be derived in many different ways. We just state the result, referring to problem 6 for details,

$$K(z;w) = \frac{1}{2\pi} ln\left|\frac{z-w}{z-1/\overline{w}}\right|;\ P(r;t-\theta) = \frac{1-r^2}{\pi(1-2r\cos(t-\theta)+r^2)}; \tag{4.43}$$

where $z = re^{it}$; $w = \rho e^{i\theta}$ are complex points in $\mathbb{D}$ ($\rho = 1$ for P). Our goal is to extend (4.43) to higher dimensions. Namely, the Poisson kernel in the n-ball is given by

$$\boxed{P(x;y) = \frac{1}{\omega_{n-1}}\frac{1-|x|^2}{|x-y|^n} = \frac{1-r^2}{\omega_{n-1}(1+r^2-2r\cos\theta)^{n/2}}} \tag{4.44}$$

where $|x| = r$; $|y| = 1$; $\theta =$ *angle between* x *and* y, *and* $\omega_{n-1} = \frac{2\pi^{n/2}}{\Gamma(n/2)}$ - *volume of the unit sphere in* $\mathbb{R}^n$.

The derivation of (4.44) will utilizes a particular *conformal transformation* in $\mathbf{R}^n$, the *inversion*,

$$\sigma: x \to x^* = \frac{x}{|x|^2}; \tag{4.45}$$

We recall that name *conformal* refers to diffeomorphisms $\{\phi\}$ of $\mathbf{R}^n$ (or any Riemannian manifold) that preserve angles between vectors, but not necessarily norms. The Jacobian of such map ϕ, $A = \phi'$, is a *conformal matrix*, product "scalar × orthogonal", i.e. ${}^tAA = \lambda I$ (problem 8). Conformal map σ of (4.45) takes interior of the unit ball in $\mathbf{R}^n$ into the exterior, and vice versa, and acts identically on the unit sphere $\{|x| = 1\}$. To construct the Green's function of Δ on B we take the free-space Green's function (Newton's potential) $K_0 = C_n|x-y|^{2-n}$ (for $n > 2$), and $\frac{1}{2\pi}ln|x-y|$ in $\mathbf{R}^2$. It satisfies the differential equation: $\Delta_x K = \delta(x-y)$, but not the boundary condition, $K_0(x-y)|_{\partial\Omega} = h_y(x) > 0$, so we need to correct K_0 by a regular harmonic function $u_y(x)$ in Ω, that satisfies $\Delta_x[u_y] = 0$; $u_y(x)|_{x \in \partial\Omega} = h_y(x)$,

$$K(x;y) = K_0(x-y) - u_y(\mathrm{x}).$$

Such u_y can be constructed by inversion of K_0. We shall use the following identity valid for any function u,

$$\Delta\left(|x|^{2-n}u\left(\frac{x}{|x|^2}\right)\right) = |x|^{-n}(\Delta u)\left(\frac{x}{|x|^2}\right), \tag{4.46}$$

The proof of invariance-relation[17] (4.46) is outlined in problem 8. Its immediate corollary is

Proposition: *Function* u *is harmonic iff* $v(x) = |x|^{n-2}u(x/|x|^2)$ *is harmonic.*

Applying Proposition to the Newton potential $K_0 = C|x-y|^{2-n}$ we get the requisite harmonic correction:

$$u_y(x) = \frac{C_n}{|(x/|x| - |x|y)|^{n-2}}.$$

Obviously, u_y is harmonic and coincides with $K_0(x-y)$ on the unit sphere. Thus the Green's function of the unit ball

$$K(x;y) = C\left(\frac{1}{|x-y|^{n-2}} - \frac{1}{|(x' - |x|y)|^n}\right); \text{ where } x' = x/|x|. \tag{4.47}$$

Finally, to compute the Poisson kernel we take the normal derivative $\partial_n K = n_y \cdot \nabla_y K$, and observe that for y on the boundary ($|y| = 1$), $n_y = y$. Hence,

$$P(x;y) = \frac{(y-x)\cdot y}{|y-x|^n} - \frac{(y-x^*)\cdot y}{|x|^{n-2}|y-x^*|^n} = = \frac{1-x\cdot y}{|y-x|^n} - \frac{|x|^2 - x\cdot y}{\|\,|x|y - x/|x|\,\|^n}. \tag{4.48}$$

It remains to note that both denominators in (4.48) are equal (see fig.7), so we get Poisson

---

[17]The $\mathbf{R}^n$-Laplacian is not invariant under the conformal map σ, but comes "very near" to it. Indeed, (4.30) means that the unitary operator $U_\sigma: f(x) \to |x|^{-n}f(x^*)$; ($|x|^{-n} = \sqrt{\det|\sigma'|}$!), intertwines operators $\Delta|x|^2$ and Δ, $\quad (\Delta|x|^2)U_\sigma = U_\sigma\Delta$.

kernel (4.44),

$$P = \frac{1-|x|^2}{C_{n-1}|x-y|^n}.$$

From the Poisson kernel on the unit ball we can easily derive P for any radius ρ,

$$P(x;y) = \frac{1}{C_{n-1}} \frac{1-(r/\rho)^2}{[1+(r/\rho)^2 - 2\,r/\rho\cos\theta]^{n/2}};\ r = |x|, \rho = |y|. \tag{4.49}$$

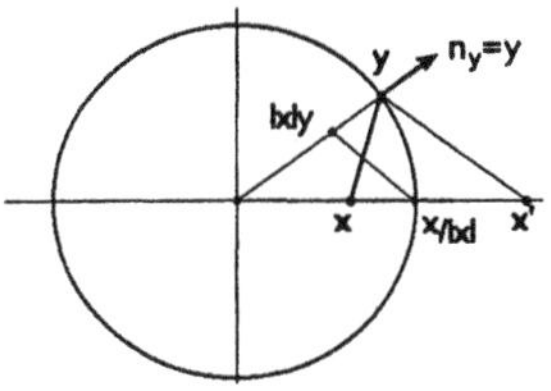

Fig.7 *demonstrates equality of two denominators of (4.49), that measure the distances between pairs* $\{x;y\}$ *and* $\{x/|x|;|x|y\}$.

Formula (4.49) has many applications in the theory of harmonic functions. It follows from (4.49) that any harmonic $u(x)$ is real analytic, since $P(x;...)$ is real-analytic in x for each y on the boundary. In fact, one could estimate the radius of convergence of the Taylor series $\sum\frac{1}{\alpha!}u^{(\alpha)}(x-x_0)^\alpha$ of harmonic function $u(x)$ at each interior point $\{x_0\}$ in terms of the distance from $\{x_0\}$ to the boundary Σ. Furthermore, (4.49) gives the minimal and maximal value of the Poisson kernel for each $0<r<\rho$, they correspond to $\cos\theta = \pm 1$,

$$\boxed{c_n\frac{1-r/\rho}{(1+r/\rho)^{n-1}} \le P(r/\rho;\theta) \le C_n\frac{1+r/\rho}{(1-r/\rho)^n}} \tag{4.50}$$

As a consequence of (4.50) we get an important result in the theory of harmonic functions.

Harnack inequality: *The ratio between the maximal and minimal values of a harmonic positive function* $u(x)\ge 0$ *in the ball of radius* r *is estimated by*

$$\boxed{\frac{max\,\{u(x): |x| = r\}}{min\,\{u(x): |x| = r\}} \le \left(\frac{\rho+r}{\rho-r}\right)^n} \tag{4.51}$$

4.13. Green's functions and Poisson kernels in arbitrary domains D are constructed, as in solid sphere case, from the free-space Green's function (Newton/Bessel potential) K_0, and an auxiliary (harmonic) function $u(x;y)$ $(x,y\in D)$, that satisfies the homogeneous equation: $\Delta_x u = 0$; and the boundary condition: $u(...;y)|_{x\in\partial D} = -K_0(x-y)$ (Dirichlet); or $\partial_n u|_{\partial D} = -\partial_n K_0(x-y)$ (Neumann). Then, the Green's function of D,

$$K(x;y) = K_0(x-y) - u(x;y);$$

while the Poisson kernel, $P(x;y) = -\partial_n K(x;y)$(Dirichlet); and $P(x;y) = -\partial_n K(x;y)$ (Neumann). Solution of any inhomogeneous problem: $\Delta u = F$; $u\mid_D = f$; is given by

$$u(x) = \int_D K(x;y)F(y)dy + \int_{\partial D} P(x;y)f(y)dS(y).$$

Another construction of $P(x;y)$ in terms of the solid-angle form is outlined in problem 12.

Additional comments and results.

Our discussion of differential operators was limited to specific examples and methods. Here we give a cursory introduction to the general elliptic theory, based on symbolic calculus. For detailed exposition see [Hör];[Ta1].

1. Symbolic calculus. We write differential operators on $\mathbf{R}^n$, as $L = \sum a_\alpha(x)D^\alpha$, where $\alpha = (\alpha_1;...\alpha_n)$ means a multi-index, D^α- the corresponding partial derivative ($D = \frac{1}{i}\partial$ - partial divided by i). We define a *symbol* $\sigma = \sigma_L(x;\xi)$ of L, as a polynomial $\sum a_\alpha(x)\xi^\alpha$. Clearly, symbol is obtained by conjugating L with exponentials, $\sigma_L = e^{-ix\cdot\xi}L[e^{ix\cdot\xi}]$, and the Fourier transform links symbols to operators,

$$L[f](x) = \frac{1}{(2\pi)^n}\int \sigma(x;\xi)e^{i(x-y)\cdot\xi}\hat{f}(\xi)\,d\xi. \tag{4.52}$$

Another way to write (4.52) is

$$L = \mathfrak{F}\circ\sigma_L\circ\mathfrak{F}^{-1},$$

here σ_L acts on $\hat{f}$ by a multiplication in ξ. Formula (4.52) could be extended from polynomial symbols of of pdo's (partial differential operators), to more general classes of functions $\{a(x;\xi)\}$ with "nice" asymptotic behavior at large ξ. This gives rise to a wider class of operators $A = a(x;D)$, called *pseudo-differential.* A ψdo A (4.52) is given by an integral (possibly distributional) kernel,

$$K(x;x-y) = \frac{1}{(2\pi)^n}\int e^{i(x-y)\cdot\xi}a(x;\xi)d\xi. \tag{4.53}$$

Pseudo-differential operators (ψdo's) retain many basic properties of differential operators (see §2.2). For instance, one can introduce a notion of order: *symbol* $a(x;\xi)$ is said to be of order $m \in \mathbf{R}$ (positive, negative, fractional), if $a(x;\xi) = O(\mid\xi\mid^m)$, and each differentiation in ξ lowers the order by 1: $\partial_\xi^\alpha a(x;\xi) = O(\mid\xi\mid^{m-\alpha})\}$. The reflects the natural properties of the corresponding ψdo's, $A = a(x;D)$. Namely,

Proposition: i) *A ψdo's A of order 0 is bounded in L^2 (and all Sobolev $\mathcal{H}_s$) spaces.*

ii) Furthermore, a ψdo A of order m maps properly the Sobolev scale,

$$A: \mathcal{H}_s \to \mathcal{H}_{s-m}, \text{ for all } s. \tag{4.54}$$

So positive-order ψdo's deregularize smooth functions, while negative-order A smooth them out. The proof of (4.54) exploits the Fourier analysis of §§2.1-2 (interpolation and Sobolev estimates).

ψdo's could be multiplied $A\cdot B$ (product rule), inverted (in some cases) and conjugated, $A\to A^*$, maintaining the proper order, $m(AB)=m(A)+m(B)$, $m(A^{-1})=\frac{1}{m(A)}$ (for elliptic A). Both operations have natural characterization in terms of symbols. To write both rules we shall use the convention: $a^{(\alpha)}$ - for partial derivatives $D^\alpha a$ in ξ, and $a_{(\alpha)}$ - for partial ∂^α in x. Then

$$\begin{aligned} \sigma_{AB} &\sim \sum \tfrac{1}{\alpha!} a^{(\alpha)} b_{(\alpha)} \text{ (product);} \\ \sigma_{A^*} &\sim \sum \tfrac{1}{\alpha!} D^\alpha \partial^\alpha [\bar{a}\,] \text{ (adjoint)} \end{aligned} \tag{4.55}$$

the expansions are taken in all multi-indices $\alpha=(\alpha_1;...\alpha_n)$. Asymptotic sign $\sim$ reflects our emphasis on *local* (in x-variable) properties of ψdo's $\{A\}$, determined (via $\mathcal{F}$) by the behavior of symbols at $\xi\to\infty$. So "large α-terms" in (4.55) define "locally small (smoothing)" ψdo's $A_\alpha=a^{(\alpha)}(x;D)$ (each differentiation in ξ lowers the order of a by 1!). Moreover, "smallness" of $\{A_\alpha\}$ could be established not only in terms of their smoothing (order), but also operator norms of $\{A_\alpha\}$ could be estimated, and shown to decay to 0 sufficiently fast. So the resulting expansions (4.55) can be interpreted as exact relations with "small remainders" in either sense.

Let us remark, that both rules correspond to the standard convention of writing differential operators: derivatives $\{\partial^\alpha\}$ (on the right) followed by multiplications with $\{a_\alpha(x)\}$. Other conventions (multiplication followed by differentiation, i.e. $A=\sum\partial^\alpha[a_\alpha(x)...])$, or *Weyl-convention* (where ∂ and x are symmetrized, chapter 6) give somewhat different rules. Thus ψdo's $\{a(x;D)\}$ form an algebra, that extends the algebra of all (variable-coefficient) differential operators. Product of ψdo's is easily seen to maintain the order,

$$m(A\cdot B)=m(A)+m(B);$$

0-order ψdo's are bounded operators, and negative-order A's (*order* $-m$) have "nice integral kernels", estimated by

$$|K(x;y)| \leq Const\,|x-y|^{m-n}. \tag{4.56}$$

Constant in (4.56) is bounded by certain seminorms, that involve symbol $\{a(x;\xi)\}$ and its derivatives in $(x;\xi)$. The goal of symbolic calculus is to utilize symbols and estimates, like (4.56), to (approximately) construct inverses $\{A^{-1}\}$, powers $\{A^s\}$, or more general "functions of A" $\{f(A)\}$; to find the kernels of $\{f(A)\}$, and to analyze their properties.

The crucial role here is played by the Fourier analysis.

2. Elliptic theory. Differential operators (pdo's) or ψdo's with nonvanishing (at $\{\infty\}$!) symbol $a(x;\xi)$ are called *elliptic*. For 2-nd order pdo's, $A = \sum a_{ij}(x)\partial^2_{ij}$, this means that their symbol, quadratic form $a(x;\xi) = \sum a_{ij}\xi_i\xi_j$ is definite (positive or negative),

$$a(x;\xi) > 0 \text{ or } a(x;\xi) < 0, \text{ for all } \xi \neq 0.$$

Hence,

$$C_1 \mid \xi \mid^2 \leq a(x;\xi) \leq C_2 \mid \xi \mid^2; \text{ for all } \xi, \tag{4.57}$$

the standard notion of ellipticity. Estimate (4.57) plays the crucial role in the approximate inversion of elliptic operators. Indeed, inverse ψdo A^{-1} (if it exists) has the principal symbol given by the product-rule,

$$\sigma_{A^{-1}} \sim \frac{1}{a(x;\xi)} + \dots \text{ (l.o.t)},$$

and $\frac{1}{a(x;\xi)}$ is also an elliptic ψdo or order $-m$. As a consequence, one can construct an approximate inverse $B \simeq A^{-1}$ for any elliptic ψdo A. Namely, one defines a ψdo B by (4.37) with symbol $\frac{1}{a(x;\xi)}$, then shows via the product-rule (4.39) and norm estimates (4.40), that

$$A \cdot B = I + R \text{ (remainder)},$$

where remainder R has negative order $-\epsilon$, hence operator R is smoothing in the Sobolev scale (4.38),

$$R: \mathcal{H}_s \to \mathcal{H}_{s+\epsilon}.$$

When restricted on a compact domain $\Omega \subset \mathbb{R}^n$, any such R becomes a compact operator, due to compactness of the Sobolev embedding: $\mathcal{H}_s(\Omega) \subset \mathcal{H}_t(\Omega)$, for any pair $s > t$ (Theorem 1 of §2.2). In other words approximate inverse $B = \frac{1}{a}(x;D)$ becomes the *Fredholm inverse* of A (inverse modulo compact operators). This result along with the basic spectral theory of compact operators (Appendix A) has important consequences for the theory of elliptic equations. Before we state the general results, let us mention that the notions of symbol, ψdo, Sobolev space, etc., could be extended from $\mathbb{R}^n$ to manifolds and domains[18], i.e. boundary value problems (see [Hör];[Tal]).

Theorem: *Let A be an elliptic operator on a compact manifold, or domain (with proper boundary conditions). Then* (i) *spectrum of A consists of a discrete set of eigenvalues $\{\lambda_k \to \infty\}$; each eigenspace E_k (or root subspace for non-self-adjoint A) being finite-dimensional;*

[18]To construct ψdo's on manifolds $\{\mathcal{M}\}$ one exploits local formula (4.36), then patches together "local pieces". Such process, however, defines a ψdo A only approximately to the leading order. So $\{\sigma_L(x;\xi)\}$ on manifolds could be understood only as *principle symbols*, a consistent choice of other lower-order terms, requires additional structures. In some cases, group-structure or geometry provide necessary tools to build a "complete symbolic calculus on manifolds" ([Be];[Un];[Ur];[Wi]).

(ii) *for any* $\lambda \notin \mathrm{spec}(A)$, *operator* $(A-\lambda)$ *is invertible and* $(A-\lambda)^{-1}: \mathcal{H}_s \to \mathcal{H}_{s+m}$; $(m = \mathrm{order}(A))$. *Hence, differential equation:* $(A-\lambda)u = f$, *with any* $f \in \mathcal{H}_s$, *has solution* $u \in \mathcal{H}_{s+m}$. *In particular,* $f \in C^\infty$, *yields* ∞*-smooth solutions. Also*

(iii) *for all* $\lambda \in \mathbb{C}$ *operator* $(A-\lambda)$ *is Fredholm-invertible. So equation* $(A-\lambda)u = f$, *has solutions for all f, orthogonal to the null-space* $\{\psi\}$, *of the adjoint operator:* $(A^* - \bar{\lambda})\psi = 0$ *(Fredholm alternative);*

(iv) *eigenfunctions:* $A\psi = \lambda_k \psi$, *are* ∞*-smooth.*

In case of (real) analytic A "C^∞" could be replaced by (real) analyticity. Further analysis of operators $\{A\}$, via ψdo techniques, reveals much more. One can find for instance, asymptotic distribution of eigenvalues $\{\lambda_k\}$, given by the celebrated *Weyl (volume-counting) principle:*

$$N(\lambda; A) = \#\{\lambda_k(A) \le \lambda\} \sim \frac{1}{(2\pi)^n} \mathrm{vol}\{(x;\xi): a(x;\xi) \le \lambda\}, \tag{4.58}$$

here the volume is taken in the the *cotangent bundle* $T^*(\mathcal{M}) = \{(x;\xi): \xi \in T_x^*\}$. The key idea in derivation of (4.58) is to compute the trace of a suitable function $f(A)$, e.g. $\mathrm{tr}(e^{-tA})$, or $\mathrm{tr}[(\zeta - A)^{-s}]$ (for sufficiently large s). On the one hand such traces represent certain transforms (Laplace, Cauchy-Stieltjes) of the counting function $N(\lambda; A)$, e.g.

$$\mathrm{tr}(e^{-tA}) = \sum e^{-t\lambda_k} = \int_0^\infty e^{-\lambda t} dN(\lambda).$$

On the other hand symbolic calculus yields kernels and symbols of ψdo's $\{B = f(A)\}$, $\sigma_{f(A)} \sim f \circ \sigma_A$. The connection between tr(B), and its symbol σ_B results from the basic integral representation (4.37),

$$\mathrm{tr}(B) = \int\int b(x;\xi) dx d\xi.$$

Hence, $\mathrm{tr}(e^{-tA}) \sim \int\int e^{-ta(x;\xi)} dx d\xi$, where the RHS gives the Laplace transform of the volume-function $V(\lambda) = vol\{(x;\xi): a \le \lambda\}$. Once Laplace-transformed "$N(\lambda)$ and $V(\lambda)$" are found asymptotically equivalent (at small t), one can go back to the original "N and V", via *Tauberian Theorems.*

Remark: In quantum mechanics, $T^*(\mathcal{M})$ gives the *phase space* of the classical-mechanical system (with coordinate x being the position-variable, and ξ - the momentum variable); symbol $a(x;\xi)$ means the *classical hamiltonian* (e.g. energy-function), while differential/pseudo-differential $A = a(x; D)$ - the corresponding *quantum hamiltonian.* We shall discuss the *quantization procedure* in chapters 6 and 8 (see [How]).

Problems and Exercises:

1. Check the subordination principle by Fourier-transforming both sides of (4.47) in t.

2. **Regularity and convergence** of solutions of the heat- and Laplaces equations. Show that for any $f \in L^p$ solutions of the heat and Laplaces problems: $u_t = \Delta u$, *and* $u_{tt} = \Delta u$, with initial value f satisfy

   i) $u(x;t)$ is C^∞ for all $t > 0$;

   ii) $\|u(...;t)\|_{L^p(dx)} \leq \|f\|_{L^p}$

   iii) $u(x;t) \to f(x)$, as $t \to 0$, in L^p-norm

   Use the convolution representation of solutions, $u = \Phi_t * f$, with the Gaussian or Poisson kernel, $\Phi_t = t^{-n}\Phi(|x|/t)$, and show that both kernels form an approximate unity as defined in problem 2 of §2.1. In fact, one could show that $u(x;t) \to f(x)$, pointwise almost everywhere inside any cone: $|x - x_0| < ct$, but this would require more powerful tools of Fourier analysis [SW].

3. Use the Fourier transform to solve Dirichlet problem for the Laplaces equation in the strip: $0 < y < b$, $\Delta u = 0$; $u|_{y=0} = f(x)$; $u|_{y=b} = 0$.

4. Calculate wave propagators $U; V; W$ in higher dimensions, using the relation:

$$\Im \frac{1}{(f+i\epsilon)^m} = \frac{1}{(m-1)!}(-\partial)^{m-1}\Im\frac{1}{f+i\epsilon} \to \frac{(-1)^{m-1}}{(m-1)!}\,\delta^{(m-1)}(f).$$

5. Obtain the fundamental solution of the **heat-problem:**

$$\begin{cases} K_t - \Delta K = 0 \\ K|_{t=0} = \delta(x) \end{cases} \Rightarrow K(x;t) = (4\pi t)^{-n/2} exp(-\tfrac{x^2}{4t}) \text{ - Gaussian,}$$

   by *symmetry reduction*, using non-isotropic dilations in the $(x;t)$-space:

$$D_\alpha:(x,t) \to (\alpha x; \alpha^2 t),\ \alpha \geq 0.$$

   (i) Show, if $u(t,x)$ solves the homogeneous heat-equation, then $u^* = u(\alpha^2 t; \alpha x)$ does so.

   (ii) Write the fundamental solution $K^* = K(\alpha^2 t; \alpha x)$ as a multiple of $K(t;x)$, $K^* = c(\alpha)K$, compute coefficient $c(\alpha)$ from the "δ source" condition at $t = 0$, and prove $K(t;x) = \alpha^n K(\alpha^2 t; \alpha x)$. Thus K is reduced to a single variable function $K(r)$,

$$K = t^{-n/2}K(rt^{-1/2}).$$

   (iii) Verify that $K(r)$ solves an ODE,

$$K'' + (\tfrac{n-1}{r} + \tfrac{r}{2})K' + \tfrac{n}{2}K = 0;$$

   Change variable: $r \to \frac{r^2}{2} = z$, and show that in the new variable K satisfies an ODE:

$$2z(K' + \tfrac{1}{2}K)' + n(K' + \tfrac{1}{2}K) = 0;$$

   (iv) Obtain the general solution of the latter,

$$K = e^{-z/2}(C_1 + C_2 \textstyle\int e^{z/2} z^{-n/2} dz),$$

   show that the second (singular term) must vanish. Hence, $K = e^{-z/2} = e^{-r^2/4}$, QED.

6. (i) Derive the Poisson kernel of the Laplacian in the unit disk $\mathbf{D}$ as Fourier series expansion

$$P(r;\theta - t) = \sum_{-\infty}^{\infty} r^{|k|} e^{ik(\theta - t)}.$$

   (ii) Sum the resulting series to get

$$P(r;\theta - t) = \frac{1}{2\pi}\left(\frac{1 - r^2}{1 + r^2 - 2r\cos(\theta - t)}\right) = \frac{1 - |z|^2}{2\pi\,|z - expi(t-\theta)|^2},$$

in complex variables $z = re^{i\theta}$.

(iii) Show that $P \to \delta(\theta - t)$, as $r \to 1$, and verify the equation $\Delta P = 0$ (use polar coordinates).

(iv) Expand the Green's function $K(z;w)$ ($z = re^{i\theta}$; $w = \rho e^{i\phi}$) into the Fourier series

$$\sum K_m(r;\rho)e^{im(\theta - \phi)}, \tag{4.59}$$

and demonstrate that coefficients $\{K_m\}$ are Green's functions of the S – L Problems:

$$K'' + \tfrac{1}{r}K' + \tfrac{m^2}{r^2}K = \delta(r - \rho), \tag{4.60}$$

(v) Find a fundamental pair $\{u_1(r); u_2(r)\}$ of each S-L problem (4.60) and compute functions $\{K_m\}$ using (4.40).

(vi) Sum series (4.59) to get

$$K(z;w) = \frac{1}{2\pi} ln\left(\frac{|z-w|}{|z-1/\bar{w}|}\right).$$

Interpret this form of K in terms of reflected sources and P is equal to the normal (radial) derivative of K on the boundary $\{|w| = 1\}$.

7. Verify the Laplace equation $\Delta_x P = 0$, and the boundary condition, $P(x;y) = \delta(x - y)$, on ∂B - the unit sphere, for the Poisson kernel (4.44).

8. Prove the identity (4.46) for any function u on $\mathbb{R}^n$. Steps:

(i) Use a general coordinate-change formula for the Laplacian Δ: if map $\phi: x \to y = (\ldots\phi_i(x)\ldots)$, has Jacobian matrix $A = \phi'$, with determinant $J = det A$, then

$$\Delta_y \to \tfrac{1}{J}\nabla \cdot J({}^{\mathrm{T}}AA)^{-1}\nabla.$$

Written explicitly this yields,

$$\Delta(u \circ \phi) = \left\{\sum_{jk}\left(\sum_i \partial_j\phi_i\, \partial_k\phi_i\right)\partial^2_{jk}u + \sum_i \Delta\phi_i\, \partial_i u\right\} \circ \phi$$

(ii) Check that inversion $\sigma: x \to x/|x|^2$, gives a conformal map, so Jacobian $\sigma' = (\partial_1\sigma; \ldots \partial_n\sigma)$ satisfies ${}^t\sigma' \cdot \sigma' = \rho(x)\, I$, with factor $\rho = |x|^{-2}$. Hence,

$$\Delta\left[u\left(\frac{x}{|x|^2}\right)\right] = |x|^{-2}\left\{\Delta u - 2(n-2)\frac{x}{|x|^2} \cdot \nabla u\right\}\left(\frac{x}{|x|^2}\right). \tag{4.61}$$

(iii) Combine (4.61) with the product rule: $\Delta(fg) = \Delta(f)g + 2\nabla f \cdot \nabla g + f\Delta g$, applied to $f = |x|^{\alpha}$ and $g = u(x/|x|^2)$, to get

$$\Delta(|x|^{\alpha} u \circ \sigma) = |x|^{\alpha - 2}\left\{\alpha(\alpha + n - 2)u - 2(\alpha + n - 2)\frac{x}{|x|^2} \cdot \nabla u + \Delta u\right\} \circ \sigma,$$

whence follows (4.46) for $\alpha = 2 - n$.

9. **Mean-value property for harmonic functions:** Show that any harmonic function $u(x)$ on $\mathbb{R}^n$ satisfies

$$u(x) = \frac{1}{r^{n-1}\omega_{n-1}} \int_{|y|=1} u(x - ry)dS_y \text{ - mean over the sphere of radius } r; \Leftrightarrow$$

$$u(x) = \frac{1}{vol B(r)} \int_{B(r)} u(x - y)\, dy \text{ — mean over the ball } B(r) \text{ of radius } r.$$

Hint: Apply the Green's identity: $\int_D v\Delta u - u\Delta v = \int_{\partial D} \left(v\frac{\partial u}{\partial n} - u\frac{\partial v}{\partial n}\right)dS$;

in region $D \subset \mathbb{R}^n$ ($\frac{\partial}{\partial n}$ - normal derivative on ∂D), to the pair: $\{u(x);\ v = K_0(x - y)$- Newton's potential$\}$, integrated over the ball $D = \{y: |x - y| \le r\}$.

10. Apply (4.51) to show that there are no non-constant positive harmonic functions on

$\mathbb{R}^n$. More elementary fact: *there are no bounded harmonic functions on* $\mathbb{R}^n$, follows from the mean-value property (problem 9).

11. Use the mean – value property (problem 9) to prove that a nontrivial harmonic function $u(x)$ on $\mathbb{R}^n$ is unbounded.

Hint: assume $|u(x)| \le C$; pick a pair of points x, y; write $u(x)$, $u(y)$ by their mean values in large balls $\{B(r)\}$; estimate $|u(x) - u(y)| \le ...(r)$, and let $r \to \infty$.

12. The Poisson kernel $P(x;y)$, $x \in D, y \in \Sigma = \partial D$, in any (convex) region D, can be constructed in terms of the *solid-angle kernel*,

$$T(x,y) = \frac{1}{C_{n-1}} \frac{\partial \theta}{\partial S} = \frac{\cos\alpha}{C_{n-1} r^{n-1}};$$

where α is the angle between normal n_y and vector $(x-y)$, (see fig.8). Geometrically kernel $T(x;y)$ represents the density of the solid angle with vertex at $\{x\}$, subtended by a surface area element at $\{y\}$. It can be obtained by restricting the $(n-1)$-differential form (solid-angle form),

$$\Theta = \frac{1}{r^n}(x_1 dx_2 \wedge ... \wedge dx_n - x_2 dx_1 \wedge dx_3 \wedge ... \wedge dx_n + ...),$$

on the boundary surface $\Sigma = \partial D$.

i) Show that $T = \Theta | \Sigma$ (parametrize Σ by map $\Phi: \mathbb{R}^{n-1} \to \mathbb{R}^n$; $x = \Phi(y)$), and show that vector surface-area element

$$n dS = \Phi_{y_1} \wedge ... \wedge \Phi_{y_{n-1}} d^{n-1} y,$$

where $\{\Phi_y\}$ are partial derivatives of Φ in variables $\{y\}$, n-unit normal to Σ.

ii) Show $u(x) = \int T(x;y) f(y) dS(y)$, solves the Laplaces equation, $\Delta u = 0$; but not the boundary condition,

$$u | \Sigma = \tfrac{1}{2} f + Q[f],$$

where operator Q is obtained by restricting kernel $T(x;y)$ on Σ.

iii) Integral operator Q is compact on $L^2(\Sigma)$ (see Appendix B), since Q has integrable singularity, $|Q(x;y)| \le C |x-y|^{2-n}$. So , operator $(\frac{1}{2} I + Q)$ is Fredholm-invertible (modulo possible finite-dimensional eigensubspace $\lambda = -\frac{1}{2}$), and the Poisson kernel,

$$P = T(\tfrac{1}{2} I + Q)^{-1} = 2(T - 2TQ + 4TQ^2 - ...).$$

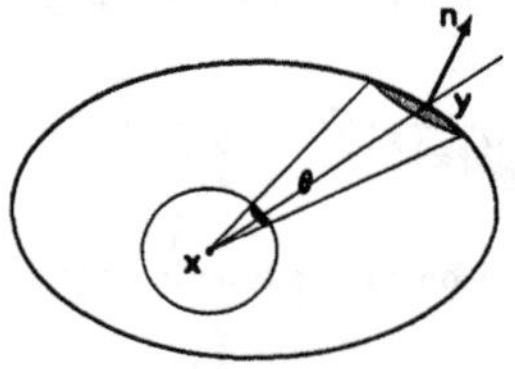

Fig.8: *Geometric view of the solid angle form, that gives a Poisson kernel in a convex region:*

$$\Theta(x;y) = \frac{1}{C_{n-1}} \left| \frac{\partial \theta}{\partial s} \right|,$$

where $\theta = \theta(x;y)$ *- solid angle, centered at* $\{x\}$, *as a function* $y \in \partial D$; $S = S(y)$ *- surface area element.*

Table of basic differential equations and their Greens functions in $\mathbf{R}^n$.

Equation		*Formal solution*	*Greens function*
Laplace:	$-\Delta u = \delta$	$K = \Delta^{-1}\delta$	$K(z) = \frac{1}{\omega_{n-1}} \lvert z \rvert^{2-n}$ $(n \geq 3)$; $\frac{1}{2\pi} \log \lvert z \rvert$ $(n{=}2)$
Resolvent:	$(\lambda - \Delta)u = \delta$	$R_\lambda = (\lambda - \Delta)^{-1}$	$R(z;\lambda) = \frac{c_n}{\lvert z \rvert^\nu} \begin{cases} K_\nu(\sqrt{\lambda} \lvert z \rvert); \text{ if } \lambda > 0, \\ J_\nu(\sqrt{\lambda} \lvert z \rvert); \text{ if } \lambda < 0, \end{cases} \nu = \frac{n}{2} - 1$ $R_\lambda(z) = \frac{1}{4\pi} \frac{e^{-\sqrt{\lambda} \lvert z \rvert}}{\lvert z \rvert}$; for $n = 3$
Heat:	$u_t - \Delta u = 0$ $u\rvert_{t=0} = \delta$	$G_t = e^{t\Delta}$	$G_t(z) = \frac{1}{(4\pi t)^{n/2}} e^{-\lvert z \rvert^2/4t}$
Schrödinger:	$u_t - i\Delta u = 0$ $u\rvert_{t=0} = \delta$	$W_t = e^{it\Delta}$	$W_t = \frac{1}{(4\pi i t)^{n/2}} e^{i \lvert z \rvert^2/4t}$
Poisson:	$u_{tt} + \Delta u = 0$ $u\rvert_{t=0} = \delta$	$P_t = e^{-t\sqrt{-\Delta}}$	$P_t = \frac{C_n t}{(t^2 + \lvert z \rvert^2)^{(n+1)/2}}$
Conj. Poisson:	$u_{tt} + \Delta u = 0$ $u_t\rvert_{t=0} = \delta$	$Q_t = \frac{1}{\sqrt{-\Delta}} e^{-t\sqrt{-\Delta}}$	$Q(z;t) = \frac{C'_n}{(t^2 + \lvert z \rvert^2)^{(n-1)/2}}$
Wave (initial velocity): $u_{tt} - c^2\Delta u = 0$ $u\rvert_0 = 0; u_t\rvert_{t=0} = \delta$		$V_t = \frac{\sin t\sqrt{L}}{\sqrt{L}}$	$V(z;t) = \begin{cases} \frac{1}{4\pi(ct)^2} \delta(\lvert z \rvert - ct); \text{ (3-D)} \\ \frac{1}{2\pi\sqrt{(ct)^2 - \lvert z \rvert^2}}; \text{ (2-D)} \end{cases}$
Wave (initial disturbance): $u_{tt} - c^2\Delta u = 0$ $u\rvert_0 = \delta;\ u_t\rvert_{t=0} = 0$		$U_t = \cos t\sqrt{L}$	$U_t(z) = \begin{cases} \frac{1}{4\pi(ct)^2} \delta(\lvert z \rvert - ct) + \frac{1}{4\pi ct} \delta'(\lvert z \rvert - ct) \\ \frac{t}{4\pi(c^2t^2 - \lvert z \rvert^2)^{3/2}} \end{cases}$

§2.5. The Radon transform.

The Radon transform appears in many guises and different geometric settings. The standard $\mathbf{R}^2$-transform takes a function f in $\mathbf{R}^2$, and integrates it along all lines $\{\ell\}$, $\mathfrak{R}: f \to \hat{f}(\ell)$. Similarly, the $\mathbf{R}^n$-functions can integrated along lines (X-ray transform), hyperplanes (Radon), or all intermediate p-planes $(1 \le p \le n-1)$. One can similarly define the Radon transform on n-spheres, integrating either over great circles (geodesics) or hyperspheres, or intermediate p-spheres. The basic problem is to invert $\mathfrak{R}$, i.e. recover function f from its integrals over a suitable family of lines, surfaces, etc. It turns out that the Radon transform is closely connected to the Fourier analysis in the Euclidian setup, and to harmonic analysis of relevant groups (orthogonal, Lorentz, etc.) in other cases.

In this section we shall give a brief introduction to the Radon transforms on Euclidian spaces, and its connections to the Fourier transform, then mention how these results extend to other rank-one symmetric spaces.

5.1. General framework for the Radon transform is a *dual pair* of manifolds: $\mathfrak{X} \leftrightarrow \Xi$, where each point $x \in \mathfrak{X}$ is assigned a subset $[x] = \{\xi\} \subset \Xi$ with a natural measure $d_x(\xi)$. Similarly, each point $\xi \in \Xi$ corresponds to a subset $[\xi] = \{x\} \subset \mathfrak{X}$, with measure $d_\xi(x)$. The generalized *Radon transform* $\hat{\mathfrak{R}}$ takes $\mathfrak{X}$-functions into Ξ-functions,

$$\hat{\mathfrak{R}}: f(x) \to \hat{f}(\xi) = \int_{[\xi]} f(x) d_\xi(x);$$

and the *dual transforms,*

$$\check{\mathfrak{R}}: f(\xi) \to \check{f}(x) = \int_{[x]} f(\xi) d_x(\xi).$$

In general, little can be said about the product and inversion of $\mathfrak{R}$-transforms,

$$\hat{\mathfrak{R}}\check{\mathfrak{R}} \text{ - ? } \check{\mathfrak{R}}\hat{\mathfrak{R}} \text{ - ? } \hat{\mathfrak{R}}^{-1}\text{-?} \tag{5.1}$$

The situation becomes more manageable when both manifolds are homogeneous spaces of some (Lie) group G, and G-actions on $\mathfrak{X}$ and Ξ are consistent:

$$[\xi^g] = \{x^g : x \in [\xi]\}, \text{ and } [x^g] = \{\xi^g : \xi \in [x]\}, \text{ for all } x; \xi.$$

We also assume that measures $\{d_x(\xi)\}$ on leaves $[x] \subset \Xi$, and $\{d_\xi(x)\}$ on leaves $[\xi] \subset \mathfrak{X}$, are transformed accordingly,

$$g: d_\xi(x) \to d_{\xi^g}(x); \text{ and } d_x(\xi) \to d_{x^g}(\xi).$$

Group G obviously commutes with both transforms, as well as their products. Therefore, product-operators: $\hat{\mathfrak{R}}\check{\mathfrak{R}}$ on $L^2(\Xi)$, and $\check{\mathfrak{R}}\hat{\mathfrak{R}}$ on $L^2(\mathfrak{X})$ commute with the regular representations R of G on both L^2-spaces. The analysis of the resulting intertwining operators can often be reduced to special classes of functions (*spherical* $\{\phi\}$), where calculation of (5.1) can be performed explicitly. The foremost is the case of the so called *rank-one symmetric space* $\mathfrak{S}$ (see examples below and §5.7 of chapter 5). Here the algebra of intertwining operators is generated by a single element, the Laplacian Δ. So the product of two transforms

$$\boxed{\check{\mathfrak{R}}\hat{\mathfrak{R}} = F(\Delta)} \tag{5.2}$$

and one only needs to compute the proper (1-variable!) function $F(\lambda)$. Formula (5.2) immediately yields the Radon inversion:

$$\boxed{\hat{\mathfrak{R}}^{-1} = 1/F(\Delta)\,\check{\mathfrak{R}}}$$

The following 3 examples will illustrate (5.2).

5.2. Euclidian space: The Radon transform on $\mathbb{R}^n$ is defined by integrating function $f(x)$ over all hyperplanes $\{\xi \subset \mathbb{R}^n\}$, that make up the dual space Ξ. The latter can be parametrized by all pairs $\{(r;\omega): r \geq 0; \omega\text{-unit vector}\}$, $\xi = \xi_{r,\omega} = \{x: x\cdot\omega = r\}$. Each point $x \in \mathbb{R}^n$ defines a subset $[x]$ of all hyperplanes passing through it, which can be identified with the *projective space* $\mathbb{P}^{n-1} \simeq S^{n-1}/\mathbb{Z}_2 = \{\text{lines in } \mathbb{R}^n \text{ through } 0\}$, equipped with the natural (rotation-invariant) measure $d_x\xi = d\xi$. So the entire space Ξ factors into the product: $\mathbb{R}^+ \times \mathbb{P}^{n-1}$. Then the Radon transform,

$$\hat{\mathfrak{R}}: f(x) \to \widehat{f}(r;\omega) = \int_{\xi_{r,\omega}} f dS = \int_{\omega\cdot x = r} f(x) d^{n-1}x,$$

while the dual transform

$$\check{\mathfrak{R}}: \phi(r;\omega) \to \check{\phi}(x) = \int_{[x]} \phi d\xi = \int_{\omega\in S^{n-1}} \phi(\omega\cdot x;\omega) dS(\omega).$$

The Euclidean motion group $\mathbb{E}_n = \mathbb{R}^n \rhd SO(n)$ acts naturally on both spaces,

$$g = (a;u): x \to x^u + a \ (x \in \mathbb{R}^n);\ \xi_{r,\omega} \to \xi_{r+a\cdot\omega;\,\omega^u} \ (\xi \in \Xi),$$

and transforms the corresponding function-spaces, $f \to R_g f$. Operators $\hat{\mathfrak{R}}, \check{\mathfrak{R}}$ clearly intertwine the resulting regular representations, $\hat{\mathfrak{R}}(R_g f) = R_g \widehat{f}$. Space $\mathbb{R}^n$ has *rank 1* relative to $\mathbb{E}_n$, in the sense that all $\mathbb{E}_n$-invariant polynomials, respectively constant-coefficient differential operators, are functions of the Laplacian Δ (see Theorem 1 of §2.4 and §§4.4-5 and §5.7). So (5.2) follows. Function F can be computed explicitly [Hel], and is found to be,

$$F(\lambda) = C\lambda^{-\frac{n-1}{2}}, \text{ with constant } C = C_n = (4\pi)^{\frac{n-1}{2}} \frac{\Gamma(n/2)}{\Gamma(\frac{1}{2})}.$$

In other words the $\mathbb{R}^n$-Radon transforms satisfies

$$\boxed{\check{\mathfrak{R}}\hat{\mathfrak{R}} = C(-\Delta)^{-\frac{n-1}{2}};\ \text{with } C = (4\pi)^{\frac{n-1}{2}}\frac{\Gamma(n/2)}{\Gamma(\frac{1}{2})}} \tag{5.3}$$

To establish (5.3) we observe that set $[x]$ of hyperplanes $\{\xi\}$ passing through x, is obtained from a single (coordinate) hyperplane ξ_0, rotated by all elements $u \in K = SO(n+1)$. So $[x] = \{x+\xi_0^u : u \in K\}$, and we can write (fig.9),

$$(\hat{f})\check{}(x) = \int_K \hat{f}(x+\xi_0^u)du = \int_K\int_{\xi_0} f(x+y^u)d_{\xi_0}(y)du = \int_{\xi_0}(M_{|y|}f)(x)d^{n-1}y = C_{n-2}\int_0^\infty (M_r f)(x)r^{n-2}dr. \tag{5.4}$$

Here du denotes the normalized Haar measure on K; $d_\xi(y) = d^{n-1}y$ – the natural (Lebesgue) measure on ξ_0, and $M_r f$ - the spherical mean of function f over the sphere $S_r(x) = \{y: |y-x| = r\}$ in $\mathbb{R}^n$,

$$(M_r f)(x) = \frac{1}{C_{n-1}r^{n-1}}\int_{|y|=r} f(x+y)dS(y) = \int_K f(x+y^u)du,$$

Constant C_k measures the volume of the unit k-sphere $S^k \subset \mathbb{R}^{k+1}$. Now it remains to rewrite integral (5.4), as

$$C_{n-2}\int_0^\infty r^{n-2}dr\frac{1}{C_{n-1}}\int_{S^{n-1}} f(x+ry)dS(y) = \frac{C_{n-2}}{C_{n-1}}\int_{\mathbb{R}^n} |x-y|^{-1}f(y)dy. \tag{5.5}$$

Here we used the volume element in spherical-coordinate, $dy = r^{n-1}drdS$, and noticed that factor $r^{n-2}drdS$, in the LHS of (5.5) turns into $\frac{1}{r}dy$. So integral (5.5) becomes a convolution, $f*\frac{1}{r}$. But the convolution-kernel $\frac{1}{r} = \frac{1}{|x-y|}$ gives (up to a constant factor) a fractional power of the Laplacian, Riesz potential: $I_s = (-\Delta)^{-s}$, of order $s = \frac{n-1}{2}$ (see §2.2), which completes the proof.

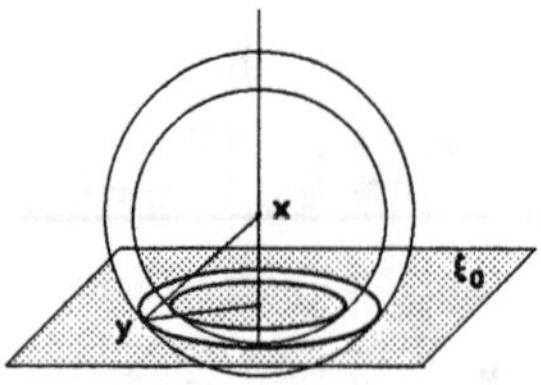

Fig.9: The double Radon transform: $\check{\mathfrak{R}}\hat{\mathfrak{R}}[f]$ *at point* x *is reduced in* (5.4) *to taking the* $\mathbb{R}^n$-spherical means $\{M_{|y|}f(x)\}$ *through all points* $\{y\}$ *in the hyperplane* ξ_0, *and integrating them over* ξ_0.

By analogy with hyperplane transforms one can introduce the family of d-plane Radon transforms for all intermediate dimensions $1 \le d \le n-1$. Here we integrate function f on $\mathbb{R}^n$ over all d-planes $\xi \subset \mathbb{R}^n$. The standard $\mathfrak{R}$ corresponds to $d=n-1$, while another extreme case $d=1$ gives the famous X-ray transform[19].

The space of all d-planes is parametrized by the points $\{\tau\}$ of the *Grassmanian* $Gr_n(n-d)$, and all points $\xi \in \tau$. The Grassmanian $Gr_n(d)$ can be identified with the double quotient space $SO(d)\backslash SO(n)/SO(n-d)$, of dimension,

$$\dim Gr_n(d) = d(n-d).$$

The Radon space Ξ_d can be thought of as a natural vector $(n-d)$-bundle over $Gr_n(n-d)$, so its dimension

$$\dim(\Xi_d) = d(n-d) + (n-d) = (n-d)(d+1).$$

Once again we have a correspondence between points $\{x \in \mathbb{R}^n\}$ and subsets $[x] \subset \Xi_d$ (all d-planes through x) on the one hand, and all points (d-planes) $\xi \in \Xi_d$ and subsets $\xi = \{x \in \xi\}$ in $\mathbb{R}^n$. So can define a pair of transforms $\hat{\mathfrak{R}}$, $\check{\mathfrak{R}}$, and those turn out to be related by an analog of (5.5)

$$\boxed{\check{\mathfrak{R}}_d \hat{\mathfrak{R}}_d = C(-\Delta)^{-\frac{d}{2}};\ \text{with constant } C = C(d) = (4\pi)^{\frac{d}{2}} \frac{\Gamma(n/2)}{\Gamma(\frac{n-d}{2})}} \tag{5.6}$$

5.3. The spherical and hyperbolic spaces. The standard Radon transform on S^n is given by integrating f over either totally geodesic (equatorial) hyperspheres $\sigma \simeq S^{n-1} \subset S^n$ (intersections of S^n with hyperplanes), or geodesics/great circles (spherical X-ray transform), or intermediate totally geodesic d-spheres. This time space Ξ is identified with the Grassmanian of 2-planes (X-ray transform), 3-planes, etc., in $\mathbb{R}^{n+1}$. The symmetry group here is still $K = SO(n+1)$, and space S^n has rank one. As a consequence of Frobenius reciprocity (§5.7), the regular representation R on S^n has multiplicity-free spectrum. Hence formula (5.2) holds with a function $F(\lambda)$ depending on d. It turns out that for all even d, F_d is polynomial of degree $\frac{d}{2}$ (like in $\mathbb{R}^n$). Precisely [Hel3],

$$P_d(\lambda) = \underbrace{[\lambda + (d-1)(n-d)]...[\lambda + 1(n-2)]}_{d/2-factors}. \tag{5.7}$$

So we get

---

[19] The energy (intensity) of an X-ray passing through a medium of variable density $\rho(x)$ is attenuated at a rate proportional to the integral along its (straight-line) path l:

$$ln\frac{E(\text{incoming})}{E(\text{outgoing})} = \int_l \rho ds.$$

So X-ray measurements produce density integrals of ρ along all lines, passing through the body (of a patient), and one would like to recover the density variations (e.g. tumor).

$$\check{\mathfrak{R}}_d\hat{\mathfrak{R}}_d = C\,/P_d(-\Delta);\ \text{with constant } C = C(d) = (4\pi)^{\frac{d}{2}}\frac{\Gamma(n/2)}{\Gamma(\frac{n-d}{2})} \tag{5.8}$$

Similar result holds for the rank-one hyperbolic spaces: $\mathbb{H}^n = SO(1;n)/SO(n)$. The role of circles and planes is played here by the closed geodesics or totally geodesic d-submanifolds. Once again relation (5.8) holds with the same constant $C = C(d)$, but polynomial F_d (5.7) has all "+" signs in factor $[\lambda + ...]$ changed to "-", so the hyperbolic

$$P_d(\lambda) = \underbrace{[\lambda - (d-1)(n-d)]...[\lambda - 1(n-2)]}_{d/2-factors}. \tag{5.9}$$

Remark: Another version of the Radon transform is defined by integrating function u on $\mathbb{R}^n$ over all spheres, passing through the origin. Any such sphere is parametrized by the radius r, and the direction-vector ω, $S_{r,\omega} = \{x: |x - r\omega| = r\}$,

$$\mathfrak{R}: u(x) \to \hat{u}(r;\omega) = \int_{|x-r\omega|=r} u(x)dS(x),$$

where dS is the natural surface area element on $S_{r,\omega}$. We refer to [CQ] for further details. Let us only remark that geometrically the space of "spheres through the origin" is identical to the space of "hyperplanes", via the inversion map of §2.4, $y \to y^* = \frac{y}{|y|^2}$. The difference is in the family of measures $\{d_\xi(x)\}$ on surfaces $\{\xi_{r,\omega} \sim S_{r,\omega}\}$. The standard Euclidean dy on hyperplanes $\{\xi\}$ is repalced by the spherical measure $d\mu_r(y) = \frac{4}{r^2 + |y|^2}dy$ (see fig.10). The Radon transform on spheres finds an interesting applications to the Darboux problem for the wave equation, $u_{tt} = \Delta u$; with characteristic boundary conditions, $u(x, |x|) = f(x)$, on the light-cone $\{t = |x|\}$ (see[CQ]).

For complete proofs and further details of Radon transforms we refer to a superb exposition of the subject in the Helgason's books [Hel] (see also [Ter]; [Wal]; [Kar]).

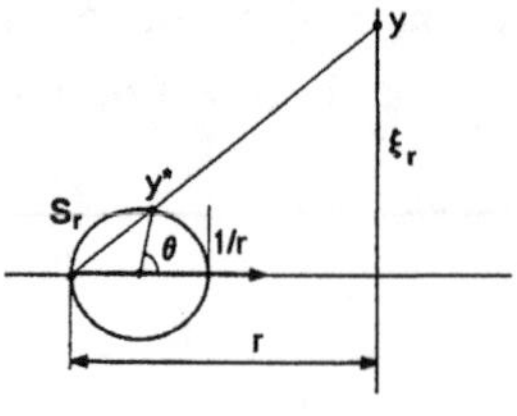

Fig.10. The inversion $y \to y/|y|^2$ takes a hyperplane $\xi_r = \{y \cdot \omega = r\}$ into the sphere $S_{2/r,\omega}$ (of radius $2/r$), passing through the origin, and transforms the Euclidean measure dy on ξ into the measure,

$$\frac{2}{r}d\theta = \frac{4\cos^2\theta}{r^2}dy = \frac{4}{r^2+y^2}dy.$$

Chapter 3. Representations of compact and finite groups.

In the first section of chapter 3 we develop the general representation theory of compact groups G, known as *Peter-Weyl theory*. We show among other results the orthogonality and completeness of irreducible matrix entries and characters of G, and complete reducibility for any representation T of G.

Then (§3.2) we shall study irreducibility and decomposition of regular and induced representations of compact and finite groups, based on Mackey's imprimitivity systems and Frobenius reciprocity. Several examples illustrates the general concepts and methods of §3.2: symmetric and alternating groups, symmetry-groups of regular polyhedra. We apply these results to "Laplacians on Platonic solids".

In the last section §3.3 we carry out a detailed study of semidirect products. Our presentation here will be motivated and illustrated by specific examples of finite group: symmetric and alternating groups; unimodular group SL_2 over finite fields, finite Heisenberg groups $\mathbb{H}_n$. The latter are closely to Clifford groups and algebras, and find an interesting application to a topological problem of determining the number of linearly independent vector fields over spheres.

§3.1. The Peter-Weyl Theory.

1.1. Matrix entries and characters. For any representation π of group G in a finite-dimensional space $\mathcal{V} \simeq \mathbb{C}^n$, we choose a basis $\{e_1;...;e_n\}$ in $\mathcal{V}$, and define matrix entries

$$\{\pi_{jk}(u)=\langle\pi_u e_j \mid e_k\rangle\}_{jk}. \tag{1.1}$$

More generally, we pick any pair of vectors $\xi;\eta \in \mathcal{V}$, and set

$$\pi_{\xi\eta}(u)=\langle\pi_u\xi \mid \eta\rangle. \tag{1.2}$$

The character of a representation π is obtained by taking the trace of π_u,

$$\chi_\pi(u)\overset{def}{=} tr(\pi_u)=\sum_{j=1}^{n}\pi_{jj}(u). \tag{1.3}$$

Definition (1.3) strictly speaking applies only to finite-D representations $\{\pi\}$. However, we can define character χ_T of infinite-D (unitary) representations, as a distribution on G,

$$\langle\chi_T \mid f\rangle=tr(T_f);$$

for a suitable class of test-functions $\mathfrak{C}=\{f(u)\}$ on G, provided operators $\{T_f=\int f(u)T_u du: f \in \mathfrak{C}\}$ belong to the trace-class.

The latter holds for regular representations $R^G; R^X$ on G, or its quotients $X = H\backslash G$, for all reasonable function-spaces $\mathcal{C}$; L^2; L^p. Indeed, operator $\{R_f: f \in \mathcal{C}\}$ are given by continuous integral kernels, e.g. $K(x;y) = f(xy^{-1})$, on the compact space G or X. Any such K is well known to be of trace-class (see Appendix B), with

$$trK = \int K(x;x)dx.$$

In particular, character of the regular representation R on G,

$$\langle \chi_R \mid f\rangle = trR_f = \int f(e)dx = vol(G)f(e), \text{ for all } f \in \mathcal{C}(G),$$

i.e. (problem 3),

$$\chi_R = |G|\delta_e. \tag{1.4}$$

We already know (1.4) for commutative groups $\mathbb{R}^n$; $\mathbb{T}^n$, where it can be interpreted as orthogonality relations for exponentials/characters $\{\chi_\xi(x) = e^{ix\cdot\xi}\}$, and was shown to be equivalent to the Inversion/Plancherel formula. In this section we shall establish a (noncommutative) compact version of these results.

Matrix entries (1.1)-(1.2) and characters are continuous functions on G, hence belong to L^2-space with respect to the *Haar measure*[1] du. The characters are easily seen to be *conjugate-invariant* functions on G,

$$\chi(uxu^{-1}) = \chi(x), \text{ for all } x, u \in G,$$

so they make up the center of the group algebra $\mathfrak{Z}(G)$ (problem 1).

The following Theorem due to Peter-Weyl summarizes the main results of the representation theory of compact groups.

1.2. Peter-Weyl Theorem: (i) *Any irreducible representation T of a compact group G is finite dimensional.*

(ii) *Matrix entries and characters of irreducible representations $\{\pi\}$ satisfy the following orthogonality relations*

$$\langle \pi_{\xi\eta} \mid \pi_{\xi'\eta'}\rangle = \frac{\langle \xi \mid \xi'\rangle\langle \eta' \mid \eta\rangle}{d(\pi)};\ \langle \pi_{jk} \mid \pi_{j'k'}\rangle = \frac{\delta_{jj'}\delta_{kk'}}{d(\pi)};\ d(\pi) = deg\pi;$$
$$\langle \pi_{jk} \mid \sigma_{mn}\rangle = 0, \textit{ for all pairs } jk,\ mn, \textit{ if } \pi \neq \sigma. \tag{1.5}$$

(iii) *The orthogonality relations for characters*

$$\langle \chi_\pi \mid \chi_\sigma\rangle = \begin{cases} 0; & \pi \neq \sigma \\ 1; & \pi \sim \sigma \end{cases}. \tag{1.6}$$

(iv) *Completeness: matrix entries $\{\pi_{jk}: 1 \le j,k \le d(\pi); \pi \in \hat{G}\}$ form a complete orthogonal system in $L^2(G)$; characters $\{\chi_\pi: \pi \in \hat{G}\}$ are complete and orthogonal in the subspace of conjugate-invariant functions*

[1] We have shown in §1.2 that the right/left invariant Haar measure exists on any compact (even locally compact) group. It defines the scale of L^p-spaces ($1 \le p \le \infty$), and the convolution structure on G.

$$\mathfrak{Z}(G) = \{f \in L^2 : f(u^{-1}xu) = f(x); x, u \in G\},$$
center of the group algebra[2] $L^2(G)$.

Proof involves two steps: first we show that any irreducible π can be embedded into the regular representation on $L^2(G)$, then we apply the *averaging procedure* to certain rank-1 operators in $L^2(G)$, and utilize Schur's Lemma.

Step 1: Embedding. We pick a vector η in the representation space $\mathcal{V} = \mathcal{V}(\pi)$ and define a map

$$W = W_\eta : \xi \rightarrow \langle \pi_x \xi \mid \eta \rangle = f(x),$$

from $\mathcal{V}$ into $L^2(G)$. Map W is bounded:

$$\| f \|^2 = \int |\langle \pi_u \xi \mid \eta \rangle|^2 du \leq \| \eta \|^2 \| \xi \|^2,$$

so $\| W \| \leq \| \eta \|$, and it intertwines representations "$\pi \mid \mathcal{V}$" and "$R \mid range\ W \subset L^2(G)$". From the unitary version of Schur's Lemma we easily deduce that W is an isometry, modulo scalar factor. Indeed, W^*W commutes with π, hence $W^*W = cI$ - scalar. Thus an irreducible π can be identified with a subrepresentation of R on a subspace $\mathcal{V} \subset L^2(G)$.

Step 2: Averaging. If a compact group G acts on a vector space $\mathcal{V}$ by linear transformations $\{T_x\}$, all associated objects (vectors, operators, tensors) could be *averaged over* G to yield G-invariant objects (problem 7). In particular, *average of a linear operator* Q on $\mathcal{V}$ is defined via

$$\bar{Q} \overset{def}{=} \int_G T_x^{-1} Q T_x dx, \tag{1.7}$$

and is easily seen to commute with all $\{T_x\}$.

Given an irreducible representation π on $\mathcal{V} \subset L^2(G)$, we pick a pair of functions φ, ψ in $\mathcal{V}$ and define a rank-one operator

$$Q(f) = Q_{\varphi\psi}(f) = \langle f \mid \varphi \rangle \psi; \quad Q: L^2(G) \rightarrow Span\{\psi\}.$$

Its average $\bar{Q}$ has the form

$$\bar{Q}(f) = \int R_u Q R_u^{-1}(f) du = \int \langle R_u^{-1} f \mid \varphi \rangle R_u \psi du = \langle f \mid \int R_u \varphi \rangle\, R_u \psi\, du.$$

Notice that $\bar{Q}$ maps $L^2(G)$ onto the *cyclic subspace* of ψ,

$$\mathcal{V}(\psi) = Span\{R_u \psi : u \in G\},$$

and $\bar{Q}$ is 0 on the orthogonal complement of $\mathcal{V}(\phi) = Span\{R_u \phi; u \in G\}$. In the case of irreducible π, both cyclic spaces coincide $\mathcal{V}(\psi) = \mathcal{V}(\phi) = \mathcal{V}$, hence $\bar{Q} = \lambda I$ is scalar on $\mathcal{V}$. Constant λ can be computed from the bilinear form

$$\langle \bar{Q}\phi \mid \psi \rangle = \lambda \langle \phi \mid \psi \rangle = \int \langle R_u \phi \mid \phi \rangle \overline{\langle R_u \psi \mid \psi \rangle} du,$$

---

[2] We remind the reader that L^2-functions on compact G form a convolution algebra, as $\| f * h \|_2 \leq \| f \|_2 \| h \|_2$.

whence

$$\lambda = \frac{\langle r_{\phi\phi} \mid r_{\psi\psi}\rangle}{\langle \phi \mid \psi\rangle}.$$

Here $\{r_{\phi\psi}\}$ denote matrix entries of the representation R. This shows that the average operator $\bar{Q}$ is indeed a λ-multiple of the orthogonal projection $P: L^2 \to \mathcal{V}$. Let us observe that both Q and $\bar{Q}$ are given by *integral kernels:*

$$Q(f) = \int K(x;y)f(y)dy; \text{ where } K(x,y) = \psi(x)\bar{\phi}(y),$$

$$\bar{Q}(f) = \int \bar{K}(x,y)f(y)dy, \text{ where } \bar{K}(x,y) = \int \psi(xg)\bar{\varphi}(xg)\,dg.$$

Obviously, kernel $\bar{K}(x;y) \in L^2(G\times G)$, in fact $\bar{K}(x,y)$ is continuous on $G\times G$. Therefore, *integral operator* $\bar{Q}: L^2(G) \to L^2(G)$ is *compact (even Hilbert-Schmidt)*. But a compact projection has always finite rank, hence $dim\mathcal{V} < \infty$. This proves the first statement of the Theorem.

Step 3: Orthogonality. Given an irreducible representation π we once again pick a rank-one operator $Q = \langle ... \mid \xi\rangle\eta$, and average it over G,

$$\bar{Q} = \int_G \pi_u^{-1} Q\, \pi_u du.$$

Obviously, $\bar{Q}$ commutes with π, and irreducibility of π, implies by Schur's Lemma, that $\bar{Q} = \lambda I$ is scalar. But $trQ = tr\,\bar{Q} = \lambda d(\pi)$, whence we find

$$\lambda \; \frac{trQ}{d(\pi)} = \frac{\langle \eta \mid \xi\rangle}{d(\pi)}.$$

Now we evaluate the bilinear form $\langle \bar{Q}\xi' \mid \eta'\rangle$, for a pair of vectors $\xi';\eta'$. On the one hand

$$\int_G \langle Q\pi_u(\xi') \mid \pi_u(\eta')\rangle du = \int_G \pi_{\xi'\xi}(u)\bar{\pi}_{\eta'\eta}(u)du = \langle \pi_{\xi'\xi} \mid \pi_{\eta'\eta}\rangle.$$

On the other hand

$$\langle Q\xi' \mid \eta'\rangle = \lambda\langle \xi' \mid \eta'\rangle = \frac{\langle \eta \mid \xi\rangle\langle \xi' \mid \eta'\rangle}{d(\pi)},$$

which proves orthogonality relation (1.5) for matrix entries of the same representation π. If entries π_{jk} and σ_{nm} come from two different (nonequivalent) representations, we take a rank-one operator Q from $\mathcal{V}(\pi)$ into $\mathcal{V}(\sigma)$, and average it with respect to the pair (π,σ),

$$\bar{Q} = \int \sigma_u^{-1} Q \pi_u du.$$

Such $\bar{Q}$ is easily seen to intertwine π and σ, hence by Schur's Lemma $\bar{Q} = 0$, whence follows the orthogonality of π_{jk} and σ_{nm}. The orthogonality of characters (1.6) is a direct consequence of (1.5)

Step 4: Completeness of matrix entries and characters in $L^2(G)$. Careful analysis of the argument in step 1, shows that any *cyclic representation* T (a representation with a cyclic vector ξ_0), can be embedded in $L^2(G)$, also any cyclic subrepresentation T *of* R contains a finite-dimensional (hence an irreducible) subrepresentation π. So if the span of all matrix entries $\{\pi_{jk}: j, k; \pi \in \widehat{G}\}$ were incomplete (had an orthogonal complement $\Upsilon \subset L^2$), then the restriction $T = R \mid \Upsilon$ would contain an irreducible π, a contradiction! Completeness of characters results from completeness of matrix entries $\{\pi_{jk}\}$ via the orthogonal projection (averaging),

$$P: f \to \overline{f}(x) = \int_G f(u^{-1}xu)du,$$

that maps $L^2(G)$ onto the subspace of conjugate-invariant functions $\mathfrak{Z}(G) =$ "center $L^2(G)$", QED.

Corollaries: *(i) Matrix entries of irreducible representations*

$$\{\pi_{jk}: 1 \leq j; k \leq d(\pi); \pi \in \hat{G}\}$$

form a complete, orthogonal system in $L^2(G)$, a noncommutative analog of characters $\{\chi(u) = e^{im \cdot u}\}$ on torus. So any $f \in L^2$ can be decomposed into a generalized "Fourier series" in matrix entries,

$$f(x) = \sum a_{\pi, jk}\, \pi_{jk}(x),$$

with coefficients $a_{\pi, jk} = \langle f \mid \pi_{jk} \rangle d(\pi)$.

The series converges in the L^2-norm, as in the torus-case, and one can also show convergence in all L^p-norms $(1 < p < \infty)$, via interpolation.

(ii) *Convolution formulae for matrix entries and characters,*

$$\boxed{\pi_{jk} * \pi_{im} = \frac{1}{d(\pi)} \begin{cases} 0; & \text{if } k \neq i \\ \pi_{jm}; & \text{if } k = i \end{cases}} \tag{1.8}$$

and

$$\boxed{\chi_\pi * \chi_\sigma \quad \begin{cases} \chi_\pi; & \text{if } \pi = \sigma \\ 0; & \text{if } \pi \neq \sigma \end{cases}} \tag{1.9}$$

Those follow by expanding each entry into the product,

$$\pi_{jk}(xy) = \sum_i \pi_{ji}(x)\pi_{ik}(y),$$

using unitarity, $\pi_{jk}(x^{-1}) = \overline{\pi_{kj}(x)}$, and orthogonality relations (1.5)-(1.6).

1.3. Decomposition. Formula (1.9) implies that characters $\{\chi_\pi\}$ form a system of mutually disjoint *central projections* in the convolution group algebra $\mathcal{C}(G)$. This property of characters will be applied now to construct the *canonical decomposition* of representations. In section 1.3 (proposition 3.2) we have shown that any unitary finite-D representation T is uniquely decomposed into the direct sum of primary components (multiples of irreducibles). For compact groups this result can be extended to all (∞-D) representations.

Decomposition Theorem 2: *Any unitary representation T of a compact group G is uniquely decomposed into the direct sum of primary components, i.e. $\mathcal{H} = \bigoplus_{\pi} \mathcal{H}_{\pi}$ - direct sum of invariant subspaces, and the restrictions $T \mid \mathcal{H}_{\pi} \simeq \pi \otimes m$ - are multiples of $\pi \in \widehat{G}$, with finite or infinite multiplicities $m = \nu(\pi;T)$.*

The result is established by passing from the group representation to the convolution algebra $\mathcal{C}(G)$, or $L^1(G)$, $\quad f \to T_f = \int_G f(u) T_u du.$

Irreducible characters $\{\chi\}$ yield a family of mutually disjoint (orthogonal) projections $\{P_{\pi} = T_{\chi_{\pi}} : \pi \in \hat{G}\}$ in $\mathcal{H}$. The fact that images, $\{\mathcal{H}_{\pi} = P_{\pi}(\mathcal{H})\}$ are primary subspaces follows from the characterization of primary subrepresentations: $S \simeq \pi \otimes m$ *is primary, iff all matrix entries of* S $\{s_{jk}(u) = \langle S_u e_j \mid e_k \rangle\}$ *are linear combinations of the matrix entries of* π.

Remarks: 1) Irreducible characters, as well as characters of more general *finite-multiplicity* representations, $T \simeq \bigoplus_{\pi} \pi \otimes m(\pi)$; $m(\pi) < \infty$, provide a convenient way to label $\widehat{G}$, also to analyze irreducibility and decomposition of T. Indeed, by general (functorial) properties of characters (problem 6),

$$\boxed{\chi_{T \oplus S} = \chi_T + \chi_S;\ \chi_{T \otimes S} = \chi_T \chi_S}; \tag{1.10}$$

it follows that the correspondence $T \to \chi_T$ is 1-1, and $\chi_T = \sum m(\pi) \chi_{\pi}$.

As a consequence, one can show

i) *equivalence:* $T \sim S$ *iff* $\chi_T = \chi_S$;

ii) $deg(T) = \langle \chi_T \mid \chi_T \rangle$; hence T is irreducible iff $\|\chi_T\|^2 = 1$ (1.11)

iii) *intertwining spaces:* $dim\ Int(T;S) = \langle \chi_T \mid \chi_S \rangle$.

Formulae (1.10) along with orthogonality relations for $\{\chi_{\pi}\}$ yield a decomposition of tensor (Kronecker) products of irreducible representations,

$$\boxed{\pi \otimes \sigma = \bigoplus_{\tau \in \widehat{G}} \tau \otimes C(\pi;\sigma \mid \tau)} \tag{1.12}$$

The multiplicities $\{C(\pi;\sigma \mid \tau)\}$ called *Clebsch-Gordan coefficients* can computed from (1.12)-(1.10)

$$\boxed{C(\pi;\sigma \mid \tau) = \int_G \chi_{\pi}(u) \chi_{\sigma}(u) \overline{\chi}_{\tau}(u)\, du}. \tag{1.13}$$

2) The decomposition Theorem remains valid for Banach-space representations, in

particular, the regular representation R in L^p-spaces or $\mathcal{C}(G)$. In each case we get a complete system of mutually disjoint central projections $\{P_\pi\}$. So any vector ξ in $\mathcal{V}$ has a generalized Fourier series expansion, $\xi = \sum_\pi \xi_\pi$; $\xi_\pi = P_\pi(\xi)$. However, convergence of such series is usually more subtle issue, it holds in all spaces $\{L^p(G)\colon 1 < p < \infty\}$, but not in $\mathcal{C}(G)$.

1.4. Noncommutative Fourier transform. Given a function f on a compact group G we define its *operator-Fourier coefficients*,

$$\hat{f}(\pi) = \int_G f(x)\pi_x dx; \text{ so } \mathfrak{F}: f \to \{\hat{f}(\pi)\}.$$

Map $\mathfrak{F}$ takes convolution of functions into the product of Fourier coefficients, $(f*g)\hat{}(\pi) = \hat{f}(\pi)\hat{g}(\pi)$. So any convolution algebra on G ($\mathcal{C}$, L^1, etc.) is taken into the direct sum of matrix algebras: $\oplus_\pi Mat_{d(\pi)}$. Map $\mathfrak{F}$ share many properties of its commutative counterpart,

Inversion and Plancherel Theorem: *for any function* $f \in \mathcal{C}(G)$, or $L^1 \cap L^2(G)$,

$$\boxed{f(e) = \sum_{\pi \in \hat{G}} d(\pi)\, tr\, \hat{f}(\pi)} \text{ and } \boxed{f(g) = \sum_{\pi \in \hat{G}} d(\pi)\, tr[\hat{f}(\pi)\pi_g^{-1}]} \tag{1.14}$$

for any $f \in L^2(G)$,

$$\boxed{\int_G |f|^2 dx = \sum_{\pi \in \hat{G}} \|\hat{f}(\pi)\|_{HS}^2 d(\pi)} \tag{1.15}$$

Here $\|A\|_{HS}$ denotes the *Hilbert-Schmidt norm* of the matrix/operator A $\hat{f}$, $\|A\|_{HS}^2 = tr(AA^*)$. Inversion easily follows from the orthogonality relations for matrix entries and characters, while Plancherel (1.15) is obtained from (1.14), specified to functions $g = f*f^*$. Sequence of degrees $\{d(\pi)\colon \pi \in \hat{G}\}$ defines the so called *Plancherel measure* on $\hat{G}$. Later (chapters 4-5) we shall compute them explicitly for the classical compact Lie groups.

Finally, we shall specify the decomposition Theorem to the regular representation $R_u f = f(xu)$, and describe the structure of the group algebra $L^2(G)$.

Theorem 3: i) *regular representation* $R \mid L^2(G)$ *is decomposed into the direct sum of irreducibles,*

$$R \simeq \bigoplus_{\pi \in \hat{G}} \pi \otimes d(\pi),$$

each π *appearing with multiplicity equal to its degree* $d(\pi)$. *Furthermore,* $L^2(G) = \oplus_\pi \mathcal{L}(\pi)$ - *direct orthogonal sum of subspaces,*

$$\mathcal{L}(\pi) = Span\{\pi_{jk}: 1 \le j,k \le d(\pi)\},$$

obtain by central projections, χ_π: $L^2 \to \mathcal{L}(\pi)$,

$$\chi_\pi(f) = \chi_\pi * f = f * \chi_\pi.$$

ii) *Each subspace* $\mathcal{L}(\pi)$ *is closed under the convolution:* $\mathcal{L}(\pi)*\mathcal{L}(\pi) \subset \mathcal{L}(\pi)$, *so* $\mathcal{L}(\pi)$ *forms a subalgebra of* $L^2(G)$, *and* $\mathcal{L}(\pi)*\mathcal{L}(\sigma) = 0$ *for* $\pi \neq \sigma$.

iii) Each subalgebra $\mathcal{L}(\pi) \simeq Mat_{d(\pi)}$ *- the full matrix algebra of degree* $d(\pi)$.

Indeed, convolution formulae (1.8) identify basic matrix entries $\{\pi_{jk}\}$ in $\mathcal{L}(\pi)$ with the Kronecker $\{\delta_{jk}\}$-basis in Mat_d. So the group algebra $L^2(G)$ on group compact G is decomposed into the ℓ^2-direct sum of matrix algebras,

$$\boxed{L^2(G) \simeq \bigoplus_{\pi \in \widehat{G}} Mat_{d(\pi)}}$$

1.5. Finite groups. We shall illustrate the foregoing by a few examples of finite groups. All the above results apply here, but some can be further refined. Namely,

Theorem 4: (i) *There are finitely many irreducible representations* $\{\pi\}$ *of* G, *their number* $|\hat{G}| = \#\{$*conjugacy classes of* $G\}$.

(ii) *The group algebra* $\mathcal{L}(G) \simeq \bigoplus_{\pi} Mat_{d(\pi)}$, *hence degrees of irreducible representations satisfy:*

$$\boxed{\sum_{\pi \in \widehat{G}} d(\pi)^2 = |G|} \text{ - the order of } G. \tag{1.16}$$

(iii) *Degrees of irreducible representations* $\{d(\pi)\}$ *divide the order of* G.

The first two statements readily follow by labeling irreducible representations by their characters $\{\chi_\pi\}$, then identifying $\{\chi_\pi\}$ with a basis of the center $\mathfrak{Z}(G)$. The last statement is less obvious, it exploits some properties of algebraic integers (see problem 5 and).

Example 5: Symmetric group $\mathbb{W}_n$ has conjugacy classes labeled by all ordered p-tuples of integers: $m_1 \ge m_2 \ge ... \ge m_p$; $\sum m_j = n$. A representative of conjugacy class $\alpha = (m_1;...m_k)$, can be chosen as a product of cyclic permutations: $\sigma_1...\sigma_k$; $\sigma_1 = (12...m_1)$, $\sigma_2 = ([m_1+1]...[m_1+m_2])$, ... (i.e. cyclic reshuffling σ_1 of the first m_1 numbers, then m_2 subsequent numbers, etc.).

Two obvious examples of irreducible representations of $\mathbb{W}_n$ are the trivial representation $\pi_g^0 = 1$; and $\pi_g^1 = sgn\ g$. Another one arises from the natural action of $\mathbb{W}_n$

by permutations of $\Sigma = \{1,2,...n\}$, $(R_g f)(k) = f(k^g)$, for $k \in \Sigma$. The regular representation R on $\mathcal{L}(\Sigma) \simeq \mathbb{C}^n$ has an invariant subspace $\mathcal{V}$ $\{f: \sum f(k) = 0\} \simeq \mathbb{C}^{n-1}$, and one can show that $\pi^{n-1} = R \mid \mathcal{V}$ is irreducible (problem 4, of §1.3), so R is decomposed into the sum, $R \sim \pi^0 \oplus \pi^{n-1}$. The characters of all three are easy to compute,

$$\chi_{\pi^0}(g) = 1;$$
$$\chi_R(g) = \#\{k \in (1,2,...n), \text{ fixed by } g\}$$
$$\chi_{\pi^{n-1}(g)} = \chi_R - \chi_{\pi^0} = \#\{\text{fixed } k\text{'s}\} - 1.$$

Specifically, group $\mathbb{W}_3$ of order 6 has 3 conjugacy classes (by the number of partitions α: $m_1 + m_2 + ... = 3$) labeled by

$(a) = \{e\}$;
(ab) - made of transpositions $\{(12); (13); (23)\}$ (2-cycles);
(abc) - 3-cycles $\{(123);(213)\}$.

So $\mathbb{W}_3$ has 3 irreducible representations, whose degrees can be found from (1.16) $|G| = 6 = 1^2 + 1^2 + 2^2$ (the only way to decompose 6 into the sum of 3 squares!). Thus we get a complete table of the representations and characters of $\mathbb{W}_3$.

	(a)	(ab)	(abc)
χ_0	1	1	1
χ_1	1	-1	1
χ_2	2	0	-1

The relative sizes of conjugacy classes $\{w_j = C_j / |G|\}$ for $\mathbb{W}_3$ are $\left\{\frac{1}{6};\frac{1}{2};\frac{1}{3}\right\}$ (for classes (a), (ab), (abc) respectively), and one can check directly the orthogonality of characters $\{\chi_0; \chi_1; \chi_2\}$ with weights $\{w_j\}$.

Example 6: An *alternating group* $G = \mathbb{A}_4$, consists of all even permutations of 4 objects $\{1;2;3;4\}$. It has order 12 and 4 conjugacy classes:

$(a) = \{e\}$, $|C_1| = 1$;
$(ab)(cd)$ - products of 2-cycles, $|C_2| = 3$
(abc) - the class of a 3-cycle (123), $|C_3| = 4$,
$(abc)^2$ - the class of $(123)^2 = $ class of (132), $|C_4| = 4$.

Group $\mathbb{A}_4$ is split into a semidirect product of an abelian normal group H $\mathbb{Z}_2 \times \mathbb{Z}_2 = \{e; (12)(34); (13)(24); (14)(23)\}$, and a subgroup $\mathbb{Z}_3$, generated by a 3-cycle, $\mathbb{A}_4 = H \triangleleft \mathbb{Z}_3$. In the next section we shall develop the representation theory of semidirect products, but here we shall use more elementary tools. We already know two of $\mathbb{A}_4$-representations: the trivial π^0, with character $\chi_0 = 1$, and a 3-D representation π^3 on

$\mathcal{V}_0 \subset \mathbb{C}^4$. Two other are easy to find from the decomposition (1.16): $12 = 1^2+1^2+1^2+3^2$. Thus $\pi^1 = \chi_1$ and $\pi^2 = \chi_2$ are both one dimensional. In fact both are nontrivial characters of the Abelian factor-group $\mathbb{A}_4/\mathbb{Z}_2\times\mathbb{Z}_2 \simeq \mathbb{Z}_3$. So the character table of $\mathbb{A}_4$ takes the form:

	(a)	$(ab)(cd)$	(abc)	(acb)
χ_0	1	1	1	1
χ_1	1	1	$(-1+i\sqrt{3})/2$	$(-1-i\sqrt{3})/2$
χ_2	1	1	$(-1-i\sqrt{3})/2$	$(-1+i\sqrt{3})/2$
χ_3	3	-1	0	0

As above one can easily check the orthogonality relations for $\{\chi_j\}$ with weights: $\left\{\frac{1}{12};\frac{1}{4};\frac{1}{3};\frac{1}{3}\right\}$.

Example 7: *The symmetric group* $\mathbb{W}_4$ has order 24 and 5 conjugacy classes, (according to partitions α: $m_1+m_2+\ldots=4$), labeled by $(a)=\{e\}$-trivial; (ab) - a 2-cycle; $(ab)(cd)$ - a pair of 2-cycles; (abc) - a 3-cycle; $(abcd)$ - a 4-cycle.

Hence there are 5 irreducible representations. We already know 3 of them: $\chi^+=1$; $\chi^-=sgn$; and a 3-D representation π^3, realized in functions on a 4-element set (a homogeneous space of $\mathbb{W}_4$!). Counting their degrees $1^2+1^2+3^2=11$, we find that the remaining two representations have degrees: $a^2+b^2=13$, so $a=2; b=3$ (the only solution!). We shall construct them explicitly in §3.3 (see also chapter 5, §5.5).

A similar analysis applies to $\mathbb{W}_5$ of order $5!=120$. It has 7 conjugacy classes, labeled by $\{5\}$; $\{4;1\}$; $\{3;2\}$; $\{3;1;1\}$; $\{2;2;1\}$; $\{2;1;1;1\}$ and $\{1;1;1;1;1\}=\{e\}$ (by the number of partitions). We already know 3 of its representations: $\chi^+=1$ (trivial); $\chi^-=sgn$; and π^4 (of degree 4). The remaining 4 could be shown (problem 8) to have degrees: $d=6;5;5;4$. So the entire dual object of $\mathbb{W}_5$ consists of 3 pairs: $\{\chi^\pm\};\{\pi^{4\pm}\};\{\pi^{5\pm}\}$, and $\{\pi^6\}$. In the next sections we shall explain the "breaking into pairs", and the relation between representations of $\mathbb{W}_5$ and its subgroup $\mathbb{A}_5$.

Let us also mention that group $\mathbb{A}_5$ has 5 conjugacy classes, hence 5 irreducible representations of degrees: $d=1;3;3;4;5$ (problem 9).

Problems and Exercises:

1. Show that center $\mathfrak{Z}$ of any group algebra $\mathcal{L}(G)$ consists of class-functions $f(uxu^{-1}) = f(x)$ (Elements of $\mathfrak{Z}$ commute with the regular representation!).

2. Show that an irreducible character χ_π defines a homomorphism, $f \to \frac{1}{d(\pi)}(\chi * f)(e)$, from $\mathfrak{Z}$ into C.

3. Find the character of the regular representation R on G, and show that $\chi_R = \delta(u) = \sum d(\pi)\chi_\pi$.

4. Calculate the Clebsch-Gordan coefficients for $\pi_3 \otimes \pi_3$ and $\pi_4 \otimes \pi_4$ in Examples 5 and 6: groups $G = S_3$ and $G = \mathbf{A}_4$.

5. **Theorem** [Ser3]: *For any finite group order* $|G|$ *divides the degree* $d(\pi)$ *of any irreducible representation* $\pi \in \widehat{G}$.

   Follow steps:
   i) An *algebraic integer* in the center $\mathfrak{Z}$ of the group algebra, is an element u that solves an algebraic equation:
   $$u^n + \sum_1^n a_j u^{n-j} = 0,$$
   with integer coefficients $\{a_j\} \in \mathbf{Z}$.

   ii) All integers in $\mathfrak{Z}$ form a ring, closed under multiplication with complex algebraic integers (numbers), i.e. any combination $\sum b_j u_j$ of $\mathfrak{Z}$-integers $\{u_j\}$ with algebraic integer coefficients $\{b_j\}$ in C, is itself a $\mathfrak{Z}$-integer.

   iii) Take the natural basis in the space $\mathfrak{Z}$, which consists of class-functions
   $$e_j = \begin{cases} 1; \text{ on } C_j \text{ the } j\text{-th conjugacy class} \\ 0; \text{ elsewhere} \end{cases}$$
   Show that functions $\{e_j\}_j$ are $\mathfrak{Z}$-integers ($e_i * e_j = \sum m_{ij}^k e_k$; with integer m_{ij}^k).

   So any class-function whose values $\{u(g): g \in G\}$ are algebraic integers is itself an integer in $\mathfrak{Z}$, therefore any character χ is an integer.

   iv) Any irreducible character χ defines an algebra homomorphism (problem 2),
   $$\phi_\chi: \mathfrak{Z} \to \mathrm{C},\ \phi_\chi(u) = \frac{1}{d}\sum u(g)\overline{\chi}(g);\ d = \text{degree of } \pi,$$
   that maps algebraic $\mathfrak{Z}$-integers into algebraic complex integers.

   v) Apply ϕ_χ to χ to show:
   $$\phi_\chi(\chi) = \frac{|G|}{d}\langle\chi \mid \chi\rangle = \frac{|G|}{d} \text{ - algebraic integer.}$$
   But any rational algebraic integer in C is an integer, QED.

6. *Characters:* (i) check the direct sum, tensor product and conjugation formulae for characters:
   $$\chi_{T \oplus S} = \chi_T + \chi_S;\ \chi_{T \otimes S} = \chi_T \chi_S;\ \chi_{\overline{T}} = \overline{\chi}_T.$$
   (ii) Use these formulae and orthogonality relations for characters to establish (1.10)-(1.13).

7. *Averaging*: Show that any representation T of a compact group G in an inner product space $\mathcal{H}$ is equivalent to a unitary representation (Take the product $\langle\xi \mid \eta\rangle$ in $\mathcal{H}$ and average it over G, $\langle\xi \mid \eta\rangle_0 = \int \langle T_u\xi \mid T_u\eta\rangle du$!).

8. Analyze 4 remaining irreducible representations of $\mathbf{W}_5$ (example 7), and show their degrees to be 6; 5; 5; 4. This is the only solution of the equation $a^2+b^2+c^2+d^2 = 102$, subject to constraint of problem 5: a, b, c, d divide 120, and another constraint: all four are greater than 1 (as any $\mathbf{W}_n$ has only two 1-D representations, those of the quotient

$\mathbf{W}_n/\mathbf{A}_n!$).

9. i) Show that alternating group $\mathbf{A}_5$ has 5 conjugacy classes, represented by the following elements (cycles and products of cycles): (12345); (21345); (123); (12)(34); and $\{e\}$.

Hint: take "even" conjugacy classes in $\mathbf{W}_5$, and show that all of them but one, (12345), coincide with the "alternating conjugacy classes"; the W-class of (12345) though splits into two A-classes. Splitting or unsplitting of the W-conjugacy class of $g \in \mathbf{A}$, depends on whether the commutant (centralizer) of g in $\mathbf{W}$, $Z(g) = \{h \in \mathbf{W}: h^{-1}gh = g\}$, "has" or "has no" an odd permutation (check)!

ii) Use Theorem 4 (problem 5) to derive the degrees of irreducible representations of $\mathbf{A}_5$: $d = 1; 3; 3; 4; 5$.

§3.2. Induced representations and Frobenius reciprocity.

Induction of group representations was introduced in section 1.2 (chapter 1). This procedures allows one to construct in a canonical way a representation $\{T\}$ of group G, starting from a representation of its subgroup H, and provides one of the basic constructions in the representation theory. Here we shall study induced representations of compact groups, and address both problems: irreducibility and decomposition. The former is given by Mackey's test, the latter by the Frobenius Reciprocity Theorem. Several examples will illustrate the general results, but many more will come in subsequent sections and chapters.

2.1. Induction. The induction procedure arises in the context of group G acting on a homogeneous space $X \simeq H\backslash G$, $g{:}\,x \to x^g$, where H denotes the stabilizer of a fixed point $x_0 \in X$. We pick a system of *coset representatives* $\{\gamma_x\}_{x \in X}$ in each class $\{x = Hg\}$. Then any element $g \in G$ is uniquely decomposed into the product $g = h\gamma_x$, where $x = Hg$ is the class of g $(mod\ H)$, and $h \in H$ depends on g, and on the choice of coset representatives.

Given a pair $x \in X$, $g \in G$ we take the product $\gamma_x g$ and decompose it as

$$\gamma_x g = h(x,g)\,\gamma_{x^g}.$$

The resulting function $h{:}\,X \times G \to H$ satisfies the cocycle condition

$$h(x;gu) = h(x;g)h(x^g;u), \text{ for all } x \in X \text{ and } g,u \in G. \tag{2.1}$$

The induced representation $T = ind(S \mid H;G)$ of group G is constructed from a representation S of H, acting in space $\mathcal{V}$, and could be realized in many different ways.

Construction 1: We take a suitable space of $\mathcal{V}$-valued functions on X, $\mathcal{C} = \mathcal{C}(X;\mathcal{V})$, or L^2; *etc.*, and define operators

$$\boxed{T_g f(x) = S_{h(x,g)}[f(x^g)];\ g \in G,\ f \in \mathcal{C}} \tag{2.2}$$

It is an easy exercise to check, using (2.1), that T is a group-representation, independent of a particular choice of coset representatives $\{\gamma_x\}$. Furthermore, equivalent representations S^1; S^2 of H induce equivalent T^1, T^2 of G.

The natural examples of induction are furnished by regular representations on G and X, both induced by the trivial representation $\mathbf{1}$,

$$R^G = ind(\mathbf{1} \mid \{e\};G);\ R^X = ind(\mathbf{1} \mid H;G).$$

Construction 2. We take a subspace $\mathcal{C}(G;S) \subset \mathcal{C}(G;\mathcal{V})$ of all $\mathcal{V}$-valued functions on G that *transform according to* S under the left translates with H,

$$\mathfrak{C}(G;S) = \{f(g): f(hg) = S_h[f(g)], \text{ all } h \in H, \ g \in G\}.$$

Obviously, $\mathfrak{C}(G;S)$ is invariant under right translates with $g \in G$, so we get a representation

$$\boxed{T_g f(x) = f(xg); \ all \ x, g \in G}. \tag{2.3}$$

Here T is realized as a subrepresentation of the $d(S)$-multiple of the regular representation on G, $T \subset R \otimes I_{d(S)}$.

One can easily check that both definitions (2.2) and (2.3) of T are equivalent, by writing an intertwining operator W: $\mathfrak{C}(G;S) \to \mathfrak{C}(X;\mathcal{V})$. Indeed, each function $F(g)$ on G, that transforms according to S, is uniquely determined by its values at coset representatives $\{\gamma_x : x \in X\}$. So we get $W: F(g) \to f(x) = F(\gamma_x)$. The second construction is clearly independent of a particular choice of $\{\gamma_x\}$!

Construction 3. A subgroup $H \subset G$ and a representation $S \mid H$ in $\mathcal{V}$ give rise in the standard way to an associated *vector bundle* $\mathcal{W} = \bigcup_{x \in X} \mathcal{V}_x$, over $X = H \backslash G$, with fibers $\mathcal{V}_x \simeq \mathcal{V}$.

Group G acts on the bundle $\mathcal{W}$ and the base-space X, and two actions commute with the natural projection $\mathcal{P}: \mathcal{W} \to X$. Moreover on each fiber-space map $g: \mathcal{V}_x \to \mathcal{V}_{x^g}$ is linear. We call it $\sigma(x,g)$. Function $\sigma(x,g)$ is easily seen to satisfy a cocycle condition (2.1), with group multiplication turning into the product of operators.

The G-action on the vector-bundle results in the *induced action* by linear transformations on vector-spaces of *cross-sections* $\Gamma(X;\mathcal{W}) = \{f(x) \in \mathcal{V}_x\}$, equipped with usual ($L^2, L^\infty$, etc.) norms[3],

$$\boxed{(T_g f)(x) = \sigma(x,g)^{-1}[f(x^g)]; \ f \in \Gamma(X;\mathcal{W})}. \tag{2.4}$$

As above one can easily construct an intertwining map $W: \Gamma(X;\mathcal{W}) \to \mathfrak{C}(X;\mathcal{V})$; or to $\mathfrak{C}(G;S)$, and to show that all three definitions (2.2); (2.3); (2.4) are equivalent.

2.2. Commutator algebra and the irreducibility test for Ind. Let $X = H \backslash G$ and σ be a representation of H in $\mathcal{V}$. We choose a particular set of coset representatives $\{\gamma_x\}$, and construct $T = Int(\sigma \mid H; G)$ in space $\mathfrak{C}(X;\mathcal{V})$. Let $H_x = \gamma_x^{-1} H \gamma_x$ denote a stabilizer of $x \in X$. Notice that the cocycle $\sigma_{h(x;g)}$ (for each fixed x) defines a representation of H_x in $\mathcal{V}$. We call it σ^x, the "*pull back of* $\sigma \mid H$" by the adjoint (inner) automorphism ϕ_x: $H \to H_x = \gamma_x H \gamma_x^{-1}$;

[3]We assume that all fiber-spaces are furnished with certain norms (inner products), and X has a suitable (G-invariant) volume element.

$$\sigma^x(g) = \sigma(x;g) = \sigma(\gamma_x^{-1} g \gamma_x). \tag{2.5}$$

Our goal is to characterize the commutator algebra $Com(T)$, in particular to find conditions for $Com(T)$ to be scalar, i.e. T - irreducible. To this end we use a simple analytical result: *any linear operator W on the function-space $\mathfrak{C}(X;\Upsilon)$ is given by an operator-valued kernel (in the "finite case" a matrix, with entries labeled by $x,y \in X$);* $A(x,y): X \times X \to \mathfrak{B}(\Upsilon)$ - the algebra of linear operators on Υ

$$Wf(x) = \int_X A(x;y)f(y)dy; \text{ or } \sum_{y \in X} A(x;y)f(y).$$

If W commutes with the induced representation, $WT_g = T_g W$, it follows that

$$A(x;y)\sigma_{h(y;g)}[f(y^g)] = \sigma_{h(x;g)} A(x^g;\, y)[f(y)]; \tag{2.6}$$

for any f in $\mathfrak{C}(X;\Upsilon)$. Changing the variable $y \to y^g$ in (2.6) we get:

$$A(x;y)\sigma_{h(y;g)} = \sigma_{h(x,g)} A(x^g;y^g); \text{ for all } x,y \in X;\ g \in G$$

and the latter can be rewritten as

$$\boxed{A(x^g;y^g) = \sigma^{-1}_{h(x;g)}\, A(x;y)\sigma_{h(y;g)}} \tag{2.7}$$

The G-action on the product-space $X \times X = \{(x,y)\}$ splits it into the union of G-orbits: $\bigcup_1^n \omega_j$; each $\omega_j = \{(x,y)^g = (x^g;y^g): g \in G\}$. One of the orbits is the diagonal $\omega_0 = \{(x;x)\}$. Obviously, the value of kernel A at each diagonal point (x,x) is determined by its value at a single point $(x_0;x_o) \in \omega_0$. Moreover, if H denotes the stabilizer of $\{x_0\}$ *in* G, then the value $A(x_0;x_0) = A_0 \in \mathfrak{B}(\Upsilon)$, commutes with the representation σ of H, $A_0 \in Com(\sigma \mid H)$. Conversely, any $A_0 \in Com(\sigma \mid H)$ can be extended to a matrix/operator-valued kernel $A(x;x)$ on ω_0, satisfying (2.7).

Next we shall analyze orbits spanned by non-diagonal pairs $\{p = (x;y)\}$, and their stabilizers $H_p = H_x \cap H_y$. Let us observe that any cocycle $\sigma(x,g)$ defines a family of representations $\{\sigma^x\}_{x \in X}$ of stabilizers $\{H_x\}$ in space Υ. The RHS of (2.7) can be written in terms of the family of representations $\{\sigma^x\}$ of stabilizers $\{H_x\}$,

$$(\sigma_g^x)^{-1} A(x;y)(\sigma_g^y) = A(x,y), \text{ for all } g \in H_{(x;y)} = H_x \cap H_y.$$

This implies in particular that each operator $A_0 = A(x;y)$ on space Υ intertwines representations: $\sigma^y \mid H_x \cap H_y$ and $\sigma^x \mid H_x \cap H_y$. Conversely, any A_o in the intertwining space of two representations σ^x and σ^y, restricted on $H_x \cap H_y$, can be extended to an operator-valued kernel $A(x;y)$, satisfying (2.7) on the entire orbit ω of $(x;y)$.

Combining both results of our analysis (diagonal and non-diagonal parts of $X \times X$), we get a complete characterization of the commutator algebra for induced representations.

Mackey's Theorem 1: *The commutator algebra of the induced representation* $T = ind(\sigma \mid H;G)$ *consists of all operator-valued kernels* $A(x,y)$, *satisfying* (2.7). *Hence,*

$$ComT \simeq Com(\sigma \mid H) \bigoplus_{j=1}^{n} Int(\sigma^{y_j} \mid H_{x_j} \cap H_{y_j}; \sigma^{x_j} \mid H_{x_j} \cap H_{y_j}) \tag{2.8}$$

- *sum over all orbits* $\{\omega_j\}$ *in* $X \times X$, *where* $(x_j;y_j)$ *is a point in* ω_j.

Since any G-orbit in $X \times X$ has a point of the form $(x_0;y)$, where x_0 is a fixed (distinguished) point in X, we can state (2.8) as

$$\boxed{ComT = Com(\sigma \mid H) \bigoplus_{\omega} Int(\sigma \mid H \cap H_\omega; \sigma^\omega \mid H \cap H_\omega)} \tag{2.9}$$

sum over all orbits $\omega \subset X \times X$. Here we picked a point $y = y(\omega)$ in each orbit ω, and denoted by H_ω its stabilizer, and by σ^ω the pull-back σ^y (2.5) of σ on $H_{y(\omega)}$. As a corollary we obtain the folowing Irreducibility Test for induced representations.

Theorem 2: $T = ind(\sigma \mid H;G)$ *is irreducible iff* σ *is irreducible, and all intertwining spaces:*

$$Int(\sigma^x \mid H_x \cap H_y; \sigma^y \mid H_x \cap H_y) = 0, \text{ for any pair } x \neq y. \tag{2.10}$$

Remarks: 1) For the sake of presentation we stated the above results for finite/discrete spaces X. Both of them extend with some technical modifications to infinite/continuous spaces, the direct sum (2.9) in Mackey's Theorem being replaced by the direct integral over the orbit-space $\Omega = (X \times X)/G$, while the intertwining condition of Irreducibility Test required for *almost any* pair $(x;y)$, or almost any orbit $\omega \in \Omega$.

2) In many cases, like semidirect products to be discussed in subsequent chapters, the "off-diagonal" part of $A(x,y)$ vanishes for obvious reasons, so irreducibility of T reduces to irreducibility of σ!

A special case of (2.9) arises when H is a normal subgroup of G. Then space $X = H \backslash G$ is the factor-group, and all conjugate stabilizers $\{H_x\}_{x \in X}$ coincide with H. The action of G on H, $h \to g^{-1}hg$, induces the "dual" action[4] on "representations of H",

$$g: \sigma \to \sigma_h^g = \sigma(g^{-1}hg).$$

Note that elements $h \in H$ transform σ into an equivalent representation: $\sigma^h \sim \sigma$. So σ^g depends only on the class of g in $H \backslash G$, i.e. $\sigma^g = \sigma^x$ $(x = Hg \in X)$.

[4]The same holds for any automorphism $\alpha \in Aut(H)$, $\alpha: \sigma \to \sigma_h^\alpha = \sigma(h^\alpha)$!

Now part relation (2.10) of Theorem 2 takes the form

$$\boxed{Com(T) = Com(\sigma \mid H) \bigoplus_1^n Com(\sigma^{x_j} \mid H; \sigma^{y_j} \mid H)}$$

If σ is irreducible and all $\{\sigma^x : x \neq x_0\}$ belong to different equivalence-classes in $\widehat{H}$, then T is irreducible! This observation will be essential in the study of semidirect products in the next section and in chapter 6. But the importance of induction extends far beyond the class of semidirect products and group extensions.

Here we shall briefly describe two such examples.

Example 1. Let $G = SL(2;\mathbb{F})$ be the group of 2×2 matrices of determinant 1 over a finite (or infinite) field $\mathbb{F}$, and H denote a subgroup of upper triangular matrices $\left\{\begin{pmatrix} a & b \\ 0 & a^{-1}\end{pmatrix}\right\}$, called the *Borel subgroup* of G.

We shall pick a character (1-D representation) $\chi(h)$ of H, and study $T = ind(G;\chi \mid H)$. To compute operators $\{T_g\}$ explicitly we choose coset representatives of the form $\gamma_x = \begin{pmatrix} 1 & 0 \\ x & 1\end{pmatrix}$, and write the product:

$$\gamma_x g = \begin{pmatrix} 1 & 0 \\ x & 1\end{pmatrix}\begin{pmatrix} a & b \\ c & d\end{pmatrix} = \begin{pmatrix} a & b \\ ax+c & bx+d\end{pmatrix} = \begin{pmatrix} \frac{1}{bx+d} & b \\ 0 & bx+d\end{pmatrix}\begin{pmatrix} 1 & 0 \\ \frac{ax+c}{bx+d} & 1\end{pmatrix}.$$

This factorization shows that the homogeneous space $X = H\backslash G$ can be identified with the projective space: $\mathbb{P}_1(\mathbb{F}) \simeq \mathbb{F}\bigcup\{\infty\}$, group G acts on X by fractional-linear transformations $g: x \to \frac{ax+b}{cx+d}$, and the representation T transforms functions $f(x)$ by

$$T_g^\chi f(x) = \chi(bx+d) f\left(\tfrac{ax+c}{bx+d}\right);\ g = \begin{pmatrix} a & b \\ c & d\end{pmatrix}. \tag{2.11}$$

Using Mackey's test one can check that T^χ is irreducible iff $\chi^2 \neq 1$ (Problem 3). Formula (2.11) defines the so called *principal series irreducible representations* of SL_2.

Example 2: *Alternating group* $\mathbb{A}_5$. We already know the degrees of irreducible representations of $\mathbb{A}_5$ (problem 9 of §3.1): 1; $\pi^{3\pm}$; π^4; and π^5 (superscript indicates the degree of π). Two of the list are known to us: 1, and a 4-D representation in "functions on a 5-point homogeneous space",

$$\Sigma_5 = \{1,2,3,4,5\} \simeq \mathbb{A}_5/\mathbb{A}_4.$$

In other words the regular representation $R = ind(1 \mid \mathbb{A}_4;\mathbb{A}_5)$ of $\mathbb{A}_5$ on Σ_5 is decomposed into the direct sum: $R \simeq 1 \oplus \pi^4$. It turns out π^5 can also be realized, as an induced representation on Σ_5. But this time the inducing character is nontrivial: $\chi = \chi^\pm$, either one of (nontrivial) 1-D representations of $\mathbb{A}_4$ (rather $\mathbb{A}_4/[\text{commutator}] \simeq \mathbb{Z}_3$), described in example 6 of §3.1. So

$$\pi = ind(\chi \mid \mathbb{A}_4;\mathbb{A}_5). \tag{2.12}$$

Mackey's test applies to show that π is irreducible, and 2 representations $\pi^{\pm}$, induced by $\chi^{\pm}$ are equivalent (problem 6). To complete the analysis of $\mathbb{A}_5$ we need to construct a pair of 3-D irreducible representations $\{\pi^{3\pm}\}$. Let us remember that $\mathbb{A}_5$ acts naturally by orthogonal matrices (dodecahedral symmetries) on $\mathbb{R}^3$ (§1.1). This action is easily seen to be irreducible (leaves no invariant lines or planes!), and to be of real type (§3.1), i.e. scalar commutator. Another 3-D representation of $\mathbb{A}_5$ is obtain by conjugating π^3 with an outer automorphism $\sigma = (12) \in \mathbb{W}_5$, so $\pi^{3-}(g) = \pi^3(\sigma g \sigma)$.

2.3. Characters of induced representations. Since the equivalence class of $T = ind(G;\sigma \mid H)$ is uniquely determined by σ we should be able to express its character χ_T in terms of χ_σ. Indeed, from (2.2),

$$T_g f(x) = \sigma(h(x;g))[f(x^g)],$$

it follows that the only nonzeros contribution to $tr(T_g)$ comes from fixed points $\{x\}$ of transformation g, i.e. $\{x: \gamma_x g \gamma_x^{-1} \in H\}$. Then

$$\chi_T(g) = \sum tr\,\sigma(\gamma_x g \gamma_x^{-1}),$$

summation over all $\{\gamma_x\}$, that conjugate g into H. We can rewrite it in the form,

$$\boxed{\chi_T(g) = \frac{1}{|H|} \sum \chi_\sigma[(\gamma_x g \gamma_x^{-1})]},$$

where summation extends over all fixed points $\{x\}$ of element g. In subsequent chapters we shall derive similar expressions for induced representations of Lie groups.

After the nature of *irreducible induced* representations was analyzed we turn to the decomposition problem for induced T.

Frobenius Reciprocity Theorem: *Given an induced representation* $T = ind(\sigma \mid H;G)$, *and a representation* S of G, *the intertwining spaces* $Int(S;T)$ *and* $Int(S \mid H;\sigma \mid H)$ *are naturally isomorphic.*

In particular multiplicity of an irreducible $\pi \in \hat{G}$ *in* T *is equal to the intertwining number* $m(\sigma;\pi|H) =$ *"multiplicity of* σ *in* $\pi \mid H$*".*

Proof: We denote by $\mathcal{V}(S)$; $\mathcal{V}(\sigma)$; $\mathcal{V}(T)$ the representation spaces of S, σ and T. Each linear map $W: \mathcal{V}(S) \to \mathcal{V}(T)$ is given by an operator-valued function $F(x): \mathcal{V}(S) \to \mathcal{V}(\sigma)$, defined via

$$F(x)u = (Wu)(x), \text{ for all } u \in \mathcal{V}(S).$$

For operator W to intertwine S and T function $F(x)$ must obey the relation:

$$F(x)S_g u = \sigma \circ h(x,g)F(x^g)u, \text{ for all } x \in X, \text{ and } g \in G.$$

So $F(x)$ is uniquely determined by its value at a single point, $x_0 = \{H\} \in H\backslash G$, and

$$F(x_0) = \sigma^{-1}(h)F(x_0)S_h, \text{ for any } h \in H,$$

so $F(x_0)$ intertwines representations S and σ of H. Conversely, each intertwining linear map $F_0 = F(x_0) \in Int(S;\sigma)$, extends to an "intertwining function"

$$F(x) = \sigma^{-1} \circ h(x_0; g_x)F_0 S(g_x),$$

which proves the requisite isomorphism of the Int-spaces, QED.

As a corollary we recover the multiplicities of irreducible $\{\pi\}$ in the canonical decomposition of the regular representation $R^G = ind(1 \mid \{e\};G)$, established in the previous section,

$$R^G = \bigoplus_{\pi \in \widehat{G}} \pi \otimes d(\pi),$$

so the multiplicity of each irreducible π is equal to its degree. We also obtain a decomposition of the regular representation $R^X = ind(1 \mid H;G)$ on quotients $X = H\backslash G$. By Frobenius Reciprocity R^X is made up of $\{\pi \in \hat{G}\}$, whose restriction on H contains a trivial subrepresentation 1 of H, and the multiplicities, $m(\pi; R^X) = \#\{H\text{-invariant vectors in } \mathcal{V}(\pi)\}$.

A specific example in case is the regular representation of $G = SO(3)$ on the 2-sphere, $R_u f(x) = f(x^u)$, $f \in L^2(S^2), u \in G$, induced by a subgroup $H = SO(2)$. In the next chapter 4, we shall obtain a complete decomposition of L^2 in the sum of irreducible components $\bigoplus_0^\infty \mathcal{H}_m$ $(dim\mathcal{H}_m = 2m+1)$, each appearing with multiplicity 1, and each subspace containing a unique H - invariant vector (function), $Y_m(\theta)$, called zonal spherical harmonics on S^2. The same holds for higher dimensional spheres $S^n \simeq SO(n+1)/SO(n)$. Once again the regular representation R of group $SO(n+1)$ on $L^2(S^n)$ has simple spectrum (direct sum of multiplicity-free irreducible components), as each irreducible subspace contains a unique $SO(n)$-invariant (see §5.7).

2.4. Laplacians on Platonic solids. We retell a horror story from a popular thriller [Kirl]: A student of mathematics having once spent a long but disappointing night playing dice, decided to make the dice "fair". He relabeled its faces from the standard (integer) set $\{1;2;3;4;5;6\}$, *to another sextet, by averaging each face over its* 4 *neighbors. The new fractional dice played somewhat better, but not quite to his satisfaction, so the student kept on relabeling the faces again and again, maintaining the utmost accuracy he could. How close did he come to the fair mean* 3.5 *after* 30 *laborious steps*

The reader has probably recognized lurking behind this not "very serious" problem, the cubical Laplacian,

$$Lf(\alpha) = \tfrac{1}{4}\sum f(\beta),$$

sum over 4 neighboring faces, acting on "faces of the cube" $\mathcal{F}$, better to say function-space $\mathcal{V} = \mathbb{C}(\mathcal{F})$. Clearly, Laplacian L commutes with the action (regular representation) of symmetry group $G = G_{cube}$ on $\mathcal{V}$. To diagonalize L we need to decompose $\mathcal{V}$ into the sum of irreducible components $\oplus \mathcal{V}_j$, and compute eigenvalues $\{\lambda_j = L\,|\,\mathcal{V}_j\}$. The number of components equal to the number of G-orbits: $\omega \subset \mathcal{F}\times\mathcal{F}$, and the latter has 3 terms: diagonal $\omega_0 = \{(\alpha\alpha)\}$; neighboring pairs $\omega_1 = \{(\alpha\beta)\}$; and opposite pairs, $\omega_2 = \{(\alpha\alpha')\}$ (check that G acts transitively on all 3). So there are 3 irreducible components in $\mathcal{V}$, and these are easily identified:
$\mathcal{V}_0 = \{\text{constants}\}$ (deg $= 1$);
$\mathcal{V}_1 = \{$even functions of total sum 0: $\sum_{\alpha\in\mathcal{F}} f(\alpha) = 0$ (deg $= 2$);
$\mathcal{V}_2 = \{\text{odd functions}\}$ (deg $= 3$).

The value of L on $\mathcal{V}_0$ is $\lambda_0 = 1$. To find λ_1 we pick a particular $f_1 = 1$ on top and bottom, and $-\frac{1}{2}$ on walls of the cube, hence $\lambda_2 = -\frac{1}{2}$. Finally, subspace $\mathcal{V}_2$ is represented by $f_2 = 1$ (top); -1(bottom); 0 (walls). Hence, $\lambda_2 = 0$. Triple $\{1;-\frac{1}{2};0\}$ completely determines L and all its iterates. So we find $L^{30}[f]$ $(f = (1;2;3;4;5;6))$ to be within $\epsilon = 2^{-30}\|f - \bar{f}\|$ proximity of the mean $\bar{f} = (3.5;...;3.5)$.

A similar argument applies to other regular dice,
tetrahedron: $\lambda = 1; -\frac{1}{3}$ (deg $= 3$);
octahedron: $\lambda = 1;-1;\,0$;
dodecahedron: $\lambda = 1;\ -\frac{1}{5}$ (deg $= 5$); $1/\sqrt{5}$ (deg $= 3$); $-1/\sqrt{5}$ (deg $= 3$).

We shall briefly sketch the last case. The dodecahedron has 12 faces and 4 orbits in $\mathcal{F}\times\mathcal{F}$: diagonal $(\alpha\alpha)$; adjacent pairs $(\alpha\beta)$; pairs one-face-apart $(\alpha\gamma)$; and opposite pairs $(\alpha\alpha')$. To identify irreducible components of $\mathcal{V}(\mathcal{F}) \simeq \mathbb{C}^{12}$, we break it into the even and add parts, write

$$\mathcal{V}_{ev} = \mathcal{V}_0\{\text{const}\} \oplus \mathcal{V}_5 \text{ (all even } f \text{ of sum } \sum_\alpha f(\alpha) = 0),$$

while

$$\mathcal{V}_{odd} = \mathcal{V}_{3+} \oplus \mathcal{V}_{3-}.$$

The former $(\mathcal{V}_{3+})$ consists of linear functions (vectors) $\xi \in \mathbb{R}^3$ restricted on faces $\{\alpha \in \mathcal{F}\}$. We associate unit ort e_α with each face α, and set

$$\xi \to f_\xi(\alpha) = \xi \cdot e_\alpha.$$

The latter (Υ_{3-}) is made of bi-vectors (tensors) $\xi \wedge \eta$ by a similar rule

$$\xi \wedge \eta \to f_{\xi \wedge \eta}(\alpha) = \xi \wedge \eta \cdot e_\alpha = \det[\xi;\eta;e_\alpha].$$

The eigenvalues of the even part: $\lambda_0 = 1$; $\lambda_5 = -\frac{1}{5}$. For Υ_5 we pick $f = 1$ on top face, and $f = -\frac{1}{5}$ on 5 adjacent. For Υ_{3+} we pick vertical ort ξ and compute the ξ-components of 5 adjacent $\{e_\alpha\} = 1/\sqrt{5}$, which yields $\lambda_{3+} = 1/\sqrt{5}$. A similar argument applies to $\lambda_{3-} = -1/\sqrt{5}$. But we can also observe that total trace of $L = 0$, hence

$$\lambda_{3-} = -\lambda_{3+}.$$

Based on our results one can easily analyze more general symmetric random walks of regular polyhedra of $(p;q)$-type, i.e. probability p to stay at a given vertex/face and probability q to jump to an adjacent one, where $p + 3q = 1$ (or $p + 4q = 1$; $p + 5q = 1$), depending on polyhedron. The corresponding operators (stochastic matrices) are combinations: $qL + pI$.

Problems and Exercises:

1. Show that the equivalence class of $T = ind(S \mid H; G)$ is unequally determined by the class of S, i.e. given $W_0 \in Int(S_1; S_2)$ construct $W \in Int(T_1; T_2)$.

2. Calculate $F(\gamma_x g)$ and show that W intertwines both representations (2.2) and (2.3).

3. Show that the representation T of $SL(2; \mathbb{F})$ given by (2.7) is irreducible iff $\chi^2 \neq 1$. (Use Mackey's test).

4. Calculate the character of T in problem 3.

5. Find the character of the regular representation R^X on the quotient-space $X = H \backslash G$ and expand it in the sum of irreducible characters.

6. Use Mackey's test to show that induced representation (12) of $\mathbb{A}_5$ is irreducible. Show that representations $\pi^{\pm}$ (2.12), induced by characters $\chi^{\pm}$ of $\mathbb{A}_4$, are equivalent.

7. i) Compute irreducible characters of $\mathbb{A}_5$ on all 5 conjugacy classes, represented by elements: $a = (12345)$; $a^2 = (13524)$; $b = (12)(34)$; $c = (123)$, and $\{e\}$.

   ii) decompose the regular representation R of $\mathbb{A}_5$ on space $\mathbb{C}(X_f) \simeq \mathbb{C}^{12}$ of dodecahedral faces ($X = \mathbb{A}_5/\mathbb{Z}_5$), into the sum of 4 irreducible representations: $1 \oplus \pi^5$ (even part of $\mathbb{C}$); $\pi^{3+} \oplus \pi^{3-}$ (odd part).

§3.3. Semidirect products.

In this section we shall examine the structure of irreducible representations of semidirect products: $G = H \triangleright K$. Although the answer becomes more complicated, compared to the direct-product case, there is a systematic procedure to describe and construct them completely, in the language of Mackey's *group extension theory* and *induced representations*. Our presentation here will emphasize the finite/compact examples, but later (chapter 6) many of the results will be extended to continuous (Lie) groups.

3.1. Definition and examples. Let us remind the definition of a semidirect product. Given a pair of groups H and K, K acting by automorphisms of H, $u: x \to x^u$, $(u \in K; x \in H)$, a *semidirect product* $H \triangleright K$ consists of all pairs $g = (x, u)$ with the multiplication rule

$$(x,u) \cdot (y,v) = (x \cdot y^{u^{-1}}; uv).$$

It can be easily seen that $H = \{(x;e)\}$ forms a normal subgroup of G, while $K = \{(e;u)\}$ a subgroup, so that $H \cap K = \{e\}$, and the whole group $G = H \cdot K$.

Natural examples of semidirect products include:

i) *Dihedral group:* $\mathbb{D}_n = \mathbb{Z}_n \triangleright \mathbb{Z}_2$; $K = \mathbb{Z}_2$ acting on $\mathbb{Z}_n$ by inversion, $u: x \to (-x)$.

ii) *Alternating group:* $\mathbb{A}_4 = (\mathbb{Z}_2 \times \mathbb{Z}_2) \triangleright \mathbb{Z}_3$. Two subgroups $\mathbb{Z}_2$ are generated by double cyclic elements: (12)(34) and (13)(24), while $\mathbb{Z}_3$ is generated by a triple cyclic permutation (123).

iii) *Symmetric group:* $\mathbb{W}_n = \mathbb{A}_n \triangleright \mathbb{Z}_2$. Any transposition can be chosen to represent $K = \mathbb{Z}_2$ in $\mathbb{W}_n$.

iv) *Affine transformations:* $(a,b): x \to ax + b$, over a finite (or infinite) field $\mathbb{F}$. Here translations, $b: x \to x+b$, form a normal subgroup, isomorphic to the additive group $\mathbb{F}$, while multiplications, $x \to ax$, make up a multiplicative subgroup $\mathbb{F}^*$ of $\mathbb{F}$. So

$$Aff(\mathbb{F}) \simeq \mathbb{F} \triangleright \mathbb{F}^*.$$

The affine group can also be realized by the upper-triangular matrices $\left\{\begin{bmatrix} a & b \\ & 1 \end{bmatrix}\right\}$.

There are many examples of (infinite) Lie-group semidirect products. Here we mention just two of them (the details of both will appear in chapter 6).

v) *Euclidian motion group:* $\mathbb{E}_n = \mathbb{R}^n \triangleright SO(n)$. Here normal subgroup $\mathbb{R}^n$ consists of all translations, while subgroup $SO(n)$ acts on $\mathbb{R}^n$ by rotations.

vi) Another well-known example is the *Poincare group* of Special relativity:

$\mathbb{P}_4 = \mathsf{M}^4 \rhd SO(1,3)$, here M^4 denotes the Minkowski 4-space with indefinite metric $(+---)$, and $SO(1,3)$ the Lorentz group of all linear transformations that preserve it.

Let $\widehat{H}$ denote the dual object of H, the set of all equivalence classes of irreducible representations of H. Group K (respectively G) acts on H by automorphisms and this action is transferred to $\widehat{H}$,

$$g: \chi \to \chi^g(h) = \chi(h^{g^{-1}}).$$

Thus the dual object $\widehat{H}$ splits into the union of K-orbits

$$\omega = \{\kappa : \kappa = \chi^g;\ g \in K\} \in \widehat{H} \backslash K = \Omega\text{-orbit space}.$$

For each χ in $\hat{H}$ we denote by K_χ its stabilizer in K, and $G_\chi = H \rhd K_\chi$ its stabilizer in G.

3.2. Semidirect products with commutative normal subgroups. We shall first study this special case of semidirect products, as it makes the result somewhat easier to formulate and to prove.

Theorem 1: *Irreducible representations T of a semidirect product, $G = H \rhd K$, with commutative normal subgroup H, are parametrized by pairs $\{(\omega;\sigma)\}$, $\omega \in \Omega$ - an orbit of K in $\hat{H}$, and $\sigma \in \hat{K}_\chi$ - an irreducible representation of stabilizer K_χ of point $\chi \in \omega$. Furthermore, T is equivalent to induced representation,*

$$T \sim ind(\chi \otimes \sigma \mid H \rhd K_\chi; G).$$

Proof: Take a representation T of G, and decompose its restriction, $T \mid H$, into a sum of primary components:

$$T \mid H = \bigoplus_{\chi \in \widehat{H}} \chi \otimes m(\chi). \tag{3.1}$$

Let us observe that the set of characters $\{\chi \in \widehat{H}\}$ in decomposition (3.1), splits into the union of G-orbits $\{\omega\}$, under the natural action of G on $\hat{H}$. Note also that any two χ's in the same orbit, $\chi_1 = \chi_2^g$, have equal multiplicities, the corresponding subspaces being transformed one into the other by

$$T_g: \mathcal{V}(\chi_1) \to \mathcal{V}(\chi_2).$$

Furthermore, direct sum of primary subspaces over ω,

$$\mathcal{V}(\omega) = \bigoplus_{\chi \in \omega} \mathcal{V}(\chi),$$

is invariant under the entire group G. We denote by $S = S^\chi$ a subrepresentation, $T_g \mid \mathcal{V}(\chi)$, of stabilizer $G_\chi = H \rhd K_\chi$ on the χ-th primary subspace. Then irreducibility of $T \mid G$ implies that the primary decomposition of $T \mid H$ is made of a single orbit ω. Picking a point χ_0 in ω with stabilizer, $G_0 = H \rhd K_0$, one can easily show that subspace

$\Upsilon(\chi_0)$ is irreducible under representation $S \mid G_0$, and $T = ind(G; S \mid G_0)$. Indeed, G acts on the homogeneous space $\omega = G_0 \backslash G$, and the associated family of subspaces $\{\Upsilon(\chi)\}_{\chi \in \omega}$ forms a G-invariant vector bundle over ω. The resulting G-action on "sections" of the bundle is obviously $ind(G; S \mid G_0)$. Hence, the Mackey test applies to show that T is irreducible, iff G_{χ_0} acts irreducibly on $\Upsilon(\chi_0)$. The verification of the Mackey's criteria for

$$T = ind(G; \chi \otimes \sigma \mid H \triangleright K_0),$$

is fairly straightforward. All pair-intersections: $G_\chi \bigcap G_{\chi_0} \supseteq H$ - normal subgroup of G, and the fiber-actions of H at two different points, $\chi \neq \chi_0$ are clearly disjoint,

$$S^{\chi_0} \mid H \simeq \chi_0 \otimes I \neq S^{\chi} \mid H \simeq \chi \otimes I.$$

Hence, $S^{\chi_0} \mid G_\chi \bigcap G_0$ and $S^{\chi} \mid G_\chi \bigcap G_0$ are also disjoint!

So the irreducibility problem is reduced from representation T of G to a subrepresentation $S = S^{\chi}$ of a stabilizer $G_\chi = H \triangleright K_\chi$ on a single fiber $\Upsilon(\chi)$. But the normal subgroup $H \subset G_\chi$ acts on $\Upsilon(\chi)$ by scalars, $\{S_h = \chi(h) I\}$, so irreducibility of $S^{\chi} \mid G_\chi$ is reduced to a representation $\sigma(u) = S^{\chi}(u)$ of stabilizer subgroup $K_\chi \subset K$. This completes the proof.

Remark: Theorem is a special (finite) case of Mackey's *imprimitivity system* [Mac]. In general such systems are made of a group G, acting on manifold X, and a pair of representations: $\{T\}$ of G, and $\{L_f\}$ of the function-algebra $\mathbb{C}(X)$ in space $\mathcal{H}$, that obey the natural consistency condition, $T_g{}^{-1} L_f T_g = L_{f^g}$, for all f, g, where $f^g(x) = f(x^g)$. G. Mackey shows that any irreducible imprimitivity system must be realized in L^2 vector-valued functions $\{\psi(x)\}$ on the orbit $\omega \subset X$ (or a closure of the orbit), by operators, $T_g \psi = \psi(x^g)$; and $L_f \psi = f(x) \psi(x)$.

We shall illustrate Theorem 1 with several examples:

Example 2: *The dihedral group,* $\mathbb{D}_n = \mathbb{Z}_n \triangleright \mathbb{Z}_2$. Its dual, $\hat{\mathbb{Z}}_n = \mathbb{Z}_n$, each character

$$\chi_j(a) = exp(2\pi\iota \tfrac{aj}{n});\ j{=}0; 1; ...n{-}1.$$

Group $\mathbb{Z}_2 = \{1; \epsilon\}$ and automorphism $\epsilon: \chi_j \to \chi_{n-j}$, and $\hat{\mathbb{Z}}_n$ splits into $\frac{n-1}{2}$ (or $\frac{n-2}{2}$) 2-point orbits $\{\omega_j = (\chi_j; \chi_{n-j})\}$, (depending on the odd\even parity of n), and a 1-point orbit: $\omega_0 = \{0\}$, (or a pair of 1-point orbits: $\{0\}$; $\{\frac{n}{2}\}$, for even n). So the dual object $\hat{\mathbb{D}}_n$ consists of $(\frac{n-1}{2})$ 2-D representations, corresponding to 2-point $\{\omega_j\}$, and a pair (or two pairs) of 1-dimensional representations of the factor-group $\mathbb{Z}_2 = \mathbb{D}_n / \mathbb{Z}_n$, corresponding to $\{0\}$ (and/or $\{\frac{n}{2}\}$). The reader could check directly all the general results (Theorem 4 of §3.1) for the dihedral group: (i) *the correspondence between irreducible representations and # of conjugacy classes;* (ii) *orthogonality relations, and the identity*

$\sum d(\pi)^2 = \text{order}(\mathbb{D}_n)$; (iii) *divisibility of the order of* $\mathbb{D}_n$ *by irreducible degrees* $\{d(\pi)\}$.

Example 3: *The tetrahedral (alternating) group* $\mathbb{A}_4 = (\mathbb{Z}_2 \oplus \mathbb{Z}_2) \rhd \mathbb{Z}_3$. We studied this example in the previous section, and now revisit it in the light of Theorem 1. Here normal subgroup H consists of elements

$$\{e; a = (12)(34);\ b = (13)(24);\ c = (14)(23)\},$$

while $K = \mathbb{Z}_3 = \{e; (123); (132)\}$ permutes $\{a, b, c\}$ cyclically. Thus $\hat{H}$ has two orbits: $\omega_0 = \{\chi_0\}$ with 3 characters (irreducible representations) of $\mathbb{Z}_3$ "sitting above it"; and a 3-point orbit ω_1 with a 3-dimensional representation π^1. One can verify straightforward that π^1 is equivalent to a representation of problem 4 (§1.3 of chapter 1).

3.3. General semidirect products. Now we turn to a general case of $G = H \rhd K$, with an arbitrary (noncommutative) normal subgroup H, and establish the analog of Theorem 1. This discussion will bring in a few important concepts in representation theory: *projective representations; cocycles* and *central extensions.*

The first step in the derivation is similar to the commutative case. Namely, the representation space breaks into the direct sum of fibers over a single G-orbit $\omega \subset \hat{H}$:

$$\mathcal{V} = \bigoplus_{\chi \in \omega} \mathcal{V}(\chi).$$

Here $\chi \in \hat{H}$ denote equivalence classes of irreducible representations of H. Notice, that any automorphism $\alpha \in Aut(H)$ can be lifted to space $\hat{H}$,

$$\alpha: T \to T^{\alpha}_h = T_{h^{\alpha}};\ h \in H.$$

Thus we get a natural action of G, respectively factor-group K on $\hat{H}$. As in the commutative case the restriction $S^{\chi} = T \mid \mathcal{V}(\chi)$ of stabilizer $G_{\chi} = H \rhd K_{\chi}$, is irreducible on the fiber-space $\mathcal{V}(\chi)$. However, the ensuing analysis of S^{χ} becomes more involved, since now we have to deal with equivalence classes of H-representations, rather than characters $\{\chi \in \hat{H}\}$.

Cocycles and central extensions. Given a representation $\chi \in \hat{H}$, and a group K of automorphisms of H, that leave "class χ" invariant, each element u of K defines an intertwining operator W_u in the representation space $\mathcal{V}_0$ of χ. Of course, such W_u is not unique, but is determined only modulo scalars (Schur's Lemma). However, once operators $\{W_u\}$ are chosen, they form a so called *projective representation* of K,

$$W_u W_v = \alpha(u, v) W_{uv};\ \text{for all } u, v \in K, \tag{3.2}$$

with scalar factor $\alpha(u, v)$.

The associative (group) law immediately yields a *cocycle condition*[5] on function

$\alpha(u;v)$,

$$\alpha(u;v)\alpha(uv;w) = \alpha(u;vw)\alpha(v;w), \text{ for all triples } u,v,w. \tag{3.3}$$

Cocycle α is called *trivial (coboundary)*, if

$$\alpha(u;v) = \frac{\beta(u)\beta(v)}{\beta(uv)}, \text{ for a function } (cochain)\ \beta(u).$$

A trivial cocycle implies that W could be transformed into an honest representation of K by multiplying each operator W_u with $\beta(u)$. Thus a projective representation is determined up to equivalence by the 2-nd *cohomology class* of α, i.e. element of the quotient group "2-cocycles"/"2-coboundaries".

All notions of representation theory: irreducibility, equivalence, etc., extend to projective representations. In fact, any projective representations of group K can be made an honest representation of a certain *central extension* of K, by group $\mathbb{T}$, $\mathbb{C}^*$, or more general Abelian group. For a given group K and a cocycle α satisfying (3.3) on K with values in $\mathbb{T} = \{z = e^{i\theta}: 0 \leq \theta < 2\pi\}$, we define an *$\alpha$-central extension*[6] of K, $G_\alpha = K \underset{\alpha}{\times} \mathbb{T}$, as the set of pairs $\{(z;u): z \in \mathbb{T}; u \in K\}$ with multiplication law,

$$(z;u)\cdot(z';v) = (zz'\alpha(u;v);uv).$$

It is easy to see that any α-projective representation of K gives rise to a representation of G_α. Conversely, any representation of G_α, restricted on a subset $\{(1;u)\} \simeq K$, yields an α-projective representation of K. Thus there is 1-1 correspondence between dual objects: α-$\hat{K}$ (α-projective irreducible representations of K) and $\widehat{G}_\alpha$. A natural example of central extension, is given by the celebrated *Heisenberg group*, that will be discussed at the end of the section and in chapter 6.

[5]The terminology came from algebraic topology. The cocycle condition can be written in a more conventional form, via *coboundary operator* ∂ on spaces $\mathfrak{C}^p(G)$ of *cochaines*, i.e. functions

$$\alpha(x_1;\dots x_p): \underbrace{G\times\dots\times G}_{p-times} \to \text{"numbers"}.$$

Operator ∂ takes a p-chain α into a $p+1$-chain,

$$\partial:\alpha \to (\partial\alpha)(x_0;x_1;\dots x_p) = \sum_{j=0}^{p}(-1)^j\alpha(x_0;\dots x_jx_{j+1};\dots x_p).$$

Here we write the group operation on numbers in the additive form. Linear operator ∂ sends vector space $\mathfrak{C}^p \to \mathfrak{C}^{p+1}$, and has the property $\partial^2 = \partial\cdot\partial = 0$. Hence, space of null-vectors $\mathfrak{Z}^p = \{\alpha \in \mathfrak{C}^p: \partial(\alpha) = 0\}$, called *cocycles*, contains the image of ∂, $\mathfrak{B}^p = \{\beta = \partial(\alpha): \alpha \in \mathfrak{C}^{p+1}\}$, called *coboundaries* (or trivial cocycles). The quotient $\mathfrak{Z}^p/\mathfrak{B}^p$ forms the p-th cohomology group, $\mathfrak{H}^p(G)$. Cocycles arise naturally in group extensions, and in off-diagonal matrix-entries of group representations (see ch.6).

[6]For finite K one can show that, modulo coboundary corrections, each cocycle α takes values in a finite subgroup $\mathbb{Z}_n$ of $\mathbb{T}$ (n-th roots of unity). So we can always use a finite group $\mathbb{Z}_n$ instead of $\mathbb{T}$ in the definition of the α-central extension. More general notion of the central extension arises, when circle $\mathbb{T}$ is replaced by an Abelian group A, and α represents an A-valued cocycle.

We proceed to analyze semidirect products. With each point $\chi \in \hat{H}$ (a class of irreducible representations of H, acting in space $\mathcal{V}_0$), we associate its stabilizer K_χ, a cocycle $\alpha = \alpha_\chi$ on K_χ and an α-projective representation W of K_χ, that results from its action on $\mathcal{V}_0$:

$$\chi(h^u) = W_u\, \chi(h) W_u^{-1};\ h \in H;\ u \in K_\chi.$$

On the other hand representation T of the product $H \triangleright K$ also yields a χ-primary subspace $\mathcal{V}(\chi)$ of $T \mid H$, on which stabilizer G_χ acts irreducibly by $S = S^\chi$. Space $\mathcal{V}_\chi$ factors into the tensor product $\mathcal{V}_0 \otimes \mathcal{V}_1$, and each $T_h \mid \mathcal{V}(\chi) = \chi(h) \otimes I$. Conjugating the latter with operators $\{S_u : u \in K_\chi\}$ we get

$$S_u^{-1}(\chi(h) \otimes I) S_u = \chi(h^u) \otimes I = (W_u^{-1} \otimes I)(\chi(h) \otimes I)(W_u \otimes I).$$

Therefore, operators $\{Q_u = S_u(W_u^{-1} \otimes I) : u \in K_\chi\}$ commute with all $\{\chi(h) \otimes I : h \in H\}$. By Schur's Lemma (see Theorem 3.5 of §1.3), operators $\{Q_u\}$ must be of the form $I \otimes \sigma_u$, which yields a representation σ of K_χ in $\mathcal{V}_1$. Since W was α-projective it follows that σ must be $\frac{1}{\alpha}$-projective in order for the product $S_u = W_u \otimes \sigma_u$, $u \in K_\chi$, to make an honest representation of K_χ. It is also obvious that irreducibility of $S \mid G_\chi$ is equivalent to irreducibility of σ.

The results of our analysis can now be summarized in the following

Theorem 4: *Let $G = H \triangleright K$ be a semidirect product. We take a representation $\chi \in \hat{H}$, and denote by ω a K-orbit through point $\{\chi\}$, by K_χ and G_χ - stabilizers of χ in groups K and G respectively. The action of group K_χ on "space of χ" determines a 2-cocycle (rather 2-nd cohomology class) $\alpha \in \mathcal{H}^2(K_\chi)$. We also pick a pair of representations: α-projective W in space $\mathcal{V}_0 = \mathcal{V}(\chi)$, and irreducible, $\frac{1}{\alpha}$ - projective σ in $\mathcal{V}_1$. Then*

(i) irreducible representations T of G are parametrized by all pairs $\{(\omega;\sigma)\}$, of K-orbit $\omega \subset \hat{H}$, and $\frac{1}{\alpha}$-projective representations σ of K_χ, at some $\chi \in \omega$.

(ii) the dual object of G can be viewed as a fibered space over the set of orbits $\Omega = \{\omega \subset \widehat{H}\}$, with fibers labeled by $\frac{1}{\alpha}$-projective irreducible representations of K_χ, the $\frac{1}{\alpha}$-dual object $(\frac{1}{\alpha}\text{-}K_\chi)^\frown$,

$$\widehat{G} = \bigcup_{\omega \subset \widehat{H}} \{(\omega; (\tfrac{1}{\alpha}\text{-}K_\chi)^\frown\}.$$

Furthermore, T is induced by a representation S of G_χ, whose restrictions on subgroups H and K_χ are:

$$S \mid H \simeq \chi \otimes I,\ S \mid K_\chi \simeq W \otimes \sigma.$$

Remark: Theorem 4 obviously includes all previous results for direct products

(§1.3) as well as semidirect products with Abelian normal subgroups. In both cases representation W becomes trivial, which simplifies the final result.

The following examples illustrate Theorems 1 and 4, as well as the notion of *central extension.*

Example 5: *The affine group*, $G = \{g: x \to ax+b\}$ is a semidirect product $\mathbb{F} \rhd \mathbb{F}^*$, with the action $a: b \to b^a = ab$. The dual group $\hat{\mathbb{F}}$ of a field $\mathbb{F}$ is identified with $\mathbb{F}$ by means of a nontrivial character $\chi_1: \mathbb{F} \to \mathbb{C}^*$, so that each $a \to \chi_a(b) \overset{def}{=} \chi_1(ab)$, and the action of $\mathbb{F}^*$ on $\hat{\mathbb{F}}$ becomes a multiplication, $a: \chi_b \to \chi_{ab}$. Clearly, $\hat{\mathbb{F}}$ splits into the union of two orbits, $\{0\} \bigcup \omega$, $\omega = \mathbb{F}^*$. The stabilizer of ω is trivial, so there is only one irreducible representation, "sitting above" ω, $\pi^\omega = ind(\chi_b \mid \mathbb{F}; G)$, where $b \in \mathbb{F}^*$. Thus the dual object $\hat{G}$ consists of an irreducible representation π^ω of degree $= \#\mathbb{F}^*$, and all 1-D characters of the Abelian factor-group

$$G/\mathbb{F} \simeq \mathbb{F}^*; \hat{G} = \{\pi^\omega\} \bigcup \mathbb{F}^*.$$

Example 6: *Symmetric group* $\mathbb{W}_4$ is a semidirect product of $\mathbb{A}_4 \rhd \mathbb{Z}_2$; $\mathbb{Z}_2$-factor could be represented by any odd permutation of order 2, e.g. $\sigma = (12)$. We already know all irreducible representations of $\mathbb{A}_4$ (example 6, §3.1), which is itself a semidirect product (problem 7). These are

(i) $\{\chi_0 = 1; \chi_1; \chi_2\}$ - 1-D characters of the commutative quotient $\mathbb{A}_4/[\text{commutator}] \simeq \mathbb{Z}_3$;

(ii) χ^3 - a 3-D irreducible representation.

Automorphism σ acting on the dual object $\hat{\mathbb{A}}_4$ leaves $\{1\}$ and $\{\chi^3\}$ invariant (one-point orbits), and permutes $\{\chi_1; \chi_2\}$ (a two-point orbit). Theorem 4 applies now, and we get a pair of representations of $\mathbb{W}_4$ over each one-point orbit (labeled by the dual object of $\mathbb{Z}_2$):

$\{1\} \to \{1; sgn\}$ - 1-D characters of the quotient $\mathbb{W}_4/\mathbb{A}_4 \simeq \mathbb{Z}_2$;

$\{\chi^3\} \to \{\pi^{3+}; \pi^{3-}\}$ - a pair of 3-D representations, $\chi^3 \mid \mathbb{A}_4$, extended by $\sigma \to I$, or $-I$.

A 2-point orbit $\omega = \{\chi_1; \chi_2\}$ yields a 2-D irreducible (induced) representation π^2 of $\mathbb{W}_4$, whose elements are represented by matrices,

$$\pi^2_h = \begin{bmatrix} \chi_1(h) & \\ & \chi_2(h) \end{bmatrix}; \text{ for } h \in \mathbb{A}_4; \text{ and } \pi^2_\sigma = \begin{bmatrix} & 1 \\ 1 & \end{bmatrix}.$$

Of course, the count of degrees: $1^2+1^2+2^2+3^2+3^2 = 24$, agrees with the general results of §3.1. Group $\mathbb{W}_4$ makes up the cubo-octahedral symmetries (§1.1), so it has a number of other interesting "geometric" representations, discussed in problem 8.

Example 7: *Symmetric group* $\mathbb{W}_5 = \mathbb{A}_5 \triangleright \mathbb{Z}_2$. Once again we know the dual object of $\mathbb{A}_5$: $\{1\}$; $\{\chi^{3+};\chi^{3-}\};\{\chi^4\};\{\chi^5\}$ (here we call irreducibles of $\mathbb{A}_5$ χ's, rather than π's). The dual space clearly breaks into 4 orbits: 3 of them one-pointed, and 1 a two-pointed orbit. Each one-point orbit yields a pair of $\mathbb{W}_5$-representations of the same degree (with $\mathbb{Z}_2$-generator σ being represented by $\pm I$), while a 2-orbit yields a degree-6 irreducible representation:

$$\begin{aligned} 1 &\to \quad \{1\};\ \{sgn\}; \\ \chi^4 &\to \quad \{\pi^{4+}\};\{\pi^{4-}\}; \\ \chi^5 &\to \quad \{\pi^{5+}\};\{\pi^{5-}\}; \\ \{\chi^{3+};\chi^{3-}\} &\to \quad \{\pi^6\}. \end{aligned}$$

Thus we recover all 7 irreducibles of $\mathbb{W}_5$, and their degrees (problem 8, §3.1).

Example 8: *The Heisenberg group* $G = \mathbb{H}_1$ consists of all 3×3 upper-triangular matrices

$$\left\{ g = \begin{bmatrix} 1 & a & c \\ & 1 & b \\ & & 1 \end{bmatrix} : a,b,c \in \mathbb{F} \right\}.$$

Group G is nilpotent of step 2, i.e. the commutator subgroup $[G;G]$ is equal to the center,

$$Z = \left\{ \begin{bmatrix} 1 & 0 & c \\ & 1 & 0 \\ & & 1 \end{bmatrix} : c \in \mathbb{F} \right\} \simeq \mathbb{F}.$$

Group G can be viewed as a semidirect product, $\mathbb{F}^2 \triangleright \mathbb{F}$, where $\mathbb{F}^2$ is identified with the normal subgroup

$$\left\{ h = \begin{bmatrix} 1 & 0 & c \\ & 1 & b \\ & & 1 \end{bmatrix} : b,c \in \mathbb{F} \right\};$$

and subgroup

$$\mathbb{F} = \left\{ u = \begin{bmatrix} 1 & a & 0 \\ & 1 & 0 \\ & & 1 \end{bmatrix} : a \in \mathbb{F} \right\},$$

acts on $\mathbb{F}^2$ by linear transformations

$$a: (b;c) \to (b; c+ab).$$

The dual group $\hat{\mathbb{F}}^2$ is isomorphic to $\mathbb{F}^2$, where character

$$\chi_{\lambda\beta}(c;b) = \chi_1(\lambda c + \beta b),$$

and the dual action of $\mathbb{F}$ on $\hat{\mathbb{F}}^2$ becomes,

$$a:(\lambda;\beta)\to(\lambda;\beta+a\lambda).$$

$\mathbb{F}$-orbits in $\hat{\mathbb{F}}^2$ are identified with lines parallel to the β-axis through $(\lambda;0)$, for each nonzeros λ,

$$\omega_\lambda=\{(\lambda;b):b\in\mathbb{F}\}$$

and points $\{(0;\beta)\colon \beta\in\mathbb{F}\}$ for $\lambda=0$. Stabilizers of the first family are trivial, while for $\omega=\{(0;\beta)\}$ they coincide with the center Z. The resulting irreducible representations, become

$$T^\lambda= ind(\chi_{\lambda,0}\,|\,\mathbb{F}^2;G),\text{ for }\lambda\neq 0, \tag{3.4}$$

and all characters of the factor-group $G/Z\simeq\mathbb{F}^2$. The reader could write down explicit formula for T^λ (problem 3 and chapter 6),

$$T^\lambda{}_g f(x)=\chi_0(\lambda c+\lambda b\cdot x)f(x+a), \tag{3.5}$$

where $\chi_0(a)$ is a nontrivial character of the additive group $\mathbb{F}$, a finite-field analog of e^{ia} on $\mathbb{R}$. So the dual object $\hat{G}$ consists of two families of representations: $\{T^\lambda\}_{\lambda\in\mathbb{F}^*}$ of degree $\#\mathbb{F}$, and 1-D characters $\{\chi\in\hat{\mathbb{F}}^2\}$. On the other hand group G can be viewed as a *central extension* of an Abelian group

$$\mathbb{F}^2=\left\{h=(a;b)\sim\begin{bmatrix}1&a&0\\&1&b\\&&1\end{bmatrix}\right\},$$

by a cocycle,

$$\alpha(h;h')=ab'. \tag{3.6}$$

Indeed, multiplying matrices

$$\begin{bmatrix}1&a&0\\&1&b\\&&1\end{bmatrix}\begin{bmatrix}1&a'&0\\&1&b'\\&&1\end{bmatrix}=\begin{bmatrix}1&a+a'&0\\&1&b+b'\\&&1\end{bmatrix}\begin{bmatrix}1&0&\alpha(h;h')\\&1&0\\&&1\end{bmatrix}.$$

So irreducible representations (3.4) of G can be viewed as α-projective representations of the commutative group $\mathbb{F}^2$, with cocycle α (3.6)!

More general (finite) Heisenberg group H_n represents a central extension of F^{2n} by $Z=\mathsf{F}$, i.e. elements $\{g\}$ are given by triples $\{(a;b;c)\colon a,b\in\mathsf{F}^n;\ c\in\mathsf{F}\}$, with multiplication

$$\alpha(g;g')=a\cdot b'. \tag{3.7}$$

H_n is also a semidirect product, $\mathsf{F}^{n+1};\mathsf{F}^n$, and has a family of irreducible representations $\{T^l\colon\lambda\in\mathsf{F}^*\}$ (3.5) in spaces of functions on F^n,

$$T^l{}_g f(x)=\chi_0(\lambda c+\lambda b\cdot x)f(x+a).$$

Those along with characters $\{\chi_{\alpha\beta}\}$ of the commutative factor-group $\mathsf{H}_n/Z\simeq\mathsf{F}^{2n}$

comprise all irreducible representations of $\mathbf{H}_n$ (see chapter 6 for more details). We shall describe now an interesting topological applications of the finite Heisenberg group.

Example 9: Vector fields on spheres. The problem is to *find the maximal number of tangent vector fields on spheres* $S^{n-1} \subset \mathbb{R}^n$, *linearly independent at each point* $x \in S^{n-1}$.

Let $D(n)$ denotes their number. If fields $\{\Xi_1;...\Xi_D\}$ form such a system, then the standard Gram-Schmidt orthogonalization yields an orthonormal basis of fields $\{\xi_1(x);...;\xi_D(x)\}$,

$$\xi_j(x)\cdot x = 0;\ \xi_j(x)\cdot\xi_k(x) = \delta_{jk};\ \text{for all } j,k;\ \text{at each } x. \tag{3.8}$$

Taking linearization of fields $\{\xi_j\}$ we get D matrices $\{A_1;...A_D\}$, so that

$$\xi_j(x) \simeq A_j x. \tag{3.9}$$

We shall apply the representation theory to solve a more simple linear problem (3.9), although the final answer, number $D(n)$, turned out to be the same for both (3.9) and (3.8). The orthogonality conditions (3.8) imply that matrices $\{A_j\}$ are antisymmetric and satisfy

$$A_j^2 = -I;\ A_jA_k = -A_kA_j,\ \text{for all } j \neq k; \tag{3.10}$$

(problem 3). We want to find the maximal number D of such matrices in $\mathbb{R}^n$. In the language of representation theory we deal with a group $G = G_D$ in $D+1$ generators $\{a_1;...a_D;b\}$, that obey the following relations,

$$a_j^2 = b;\ b^2=1;\ a_ja_k = a_ka_jb; \tag{3.11}$$

and ask for a minimal degree representation of G, that takes generator b into $-I$. We claim that G is a finite Heisenberg group $\mathbb{H}_n$ over the field[7] $\mathbb{F} = \mathbb{Z}_2$. Precisely,

(i) G contains center

$$Z = Z(G) = \begin{cases} \mathbb{Z}_2;\ \text{for even } D \\ \mathbb{Z}_2\times\mathbb{Z}_2;\ \text{for odd } D \end{cases};$$

and the quotient $G/Z \simeq \mathbb{Z}_2^D$ is commutative. Center Z is spanned by $\{b\}$ in the former (even) case, and by $\{b\}$ and $a = a_1...a_D$ in the latter case, so (problem 2),

$$G = \begin{cases} \mathbb{H}_{D/2};\ \text{even } D \\ \mathbb{H}_{\frac{D-1}{2}}\times\mathbb{Z}_2;\ \text{odd } D \end{cases}.$$

[7] Its official name is the *Clifford group*, as it generates the *Clifford algebra*, that plays an important role in representation theory, geometry, topology and Physics. It serves the basis for so the called spinor-groups (2-fold simply connected covers of orthogonal groups), gives rise to spinor structures on manifolds and the Dirac operator.

We shall state a few other simple properties of G, which the reader is asked to verify (problem 4):

(ii) any element of G is uniquely represented either as $a_{i_1}...a_{i_p}$, or as $ba_{i_1}...a_{i_p}$ $(i_1 < ... < i_p)$;

(iii) square of any element g in G belongs to the center, moreover,

$$g^2 = b^{\frac{p(p+1)}{2}}, \text{ for elements of length } p.$$

Form the previous example 8 (see problem 3) we already know all irreducible representations of the Heisenberg group. They are either 1-D characters of the commutative quotient G/Z (trivial on Z), or an irreducible $T = T^\lambda$, which takes central generator b to $-I$ (the only possible nontrivial character on $\mathbb{Z}_2$!). Representation T^λ can be realized on functions on $\mathbb{Z}_2^{D/2}$ (even D), or $\mathbb{Z}_2^{\frac{D-1}{2}}$ (odd D), (problem 3). Hence it has degree:

$$n = 2^{D/2}; \text{ or } 2^{\frac{D-1}{2}}.$$

It might seem we have essentially completed the argument. But there is a small catch, not to be overlooked! Construction (3.4) gives *complex irreducible representations* of $\mathbb{H}_n$, whereas our goal was a representation (3.10) of *real type*[8]. Of course, any complex T can be thought of a real representation of the double-degree, but our goal was to find the *exact minimal* D to represent (3.10).

So the argument would be completed after we determine the type of representation T^λ (real, complex or quaternionic[9]). The latter could be accomplished by a *type-criterion* due to Schur.

Schur's Criterion: *An irreducible representation T of a finite/compact group G belongs to a real, complex or quaternionic type, depending on the sign of the integral,*

$$Schur(T) = \int_G \chi_T(g^2)dg = \begin{cases} +; \ real-type \\ 0; \ complex \\ -; quaternionic \end{cases}. \tag{3.12}$$

The proof of Schur's criterion is outlined in problem 6. We apply it in our setup. Remembering that all squares $\{g^2 \in G\}$ are $\pm I$, depending on parity of $\frac{p(p+1)}{2}$ (for elements of length p, (iii)), we find the Schur's number (3.12) to be

[8] Real type T means that the representation space $\mathcal{V}$ has a real invariant subspace $\mathcal{V}_0$, so $T \mid \mathcal{V}$ becomes a complexification of $T \mid \mathcal{V}_0$.

[9] We remind the reader that an irreducible unitary representation T in complex space $\mathcal{V}$ belongs to the *quaternionic type*, if $T \sim \overline{T}$ (complex conjugate, or contragredient of T), i.e. T has real character $\chi_T = \overline{\chi}_T$, but there is no real invariant subspace in $\mathcal{V}$.

$$\chi(e)\left\{1-\binom{D}{1}-\binom{D}{2}+\binom{D}{3}+\binom{D}{4}-\ldots\right\}, \tag{3.13}$$

the "double-alternating" sum of binomial coefficients, the mth term corresponds to elements of length m in G. The latter can be written as $\Re(1+i)^D-\Im(1+i)^D$. Thence, one finds that the sign of $Schur(T)$ depends on the remainder of D modulo 8, and is given by the following table

D mod 8	0	1	2	3	4	5	6	7
sign	+	0	−	−	−	0	+	+

Combining the table and the known degrees of $\{T^\lambda\}$ we get the final result: *minimal dimension of* $\mathbb{R}^n$ *that realizes relations (3.10),*

$$n=\begin{cases} 2^{D/2}, \text{ if } D=8m, \text{ or } 8m+6; \\ 2^{\frac{D+1}{2}}; \text{ if } D=8m+1,\ 8m+3,\ 8m+4 \\ 2^{\frac{D+2}{2}}; \text{ if } D=8m+2, \text{ or } 8m+4 \\ 2^{\frac{D-1}{2}}; \text{ if } D=8m+7 \end{cases}$$

As an off-shot of our analysis we see that the number $D(n)$ depends only on the maximal factor 2^p in n !

Our exposition closely followed the Kirillov's book [Kirl].

Problems and Exercises.

1. (i) Find all irreducible characters of the affine group and show their orthogonality.

   (ii) Do the same for the finite Heisenberg groups $\mathbf{H}_1$ and $\mathbf{H}_n$.

2. Show that the Clifford group G of example 9 in D generators $\{a_i{:}1 \le i \le D; a_i^2 = -1; a_ia_j = -a_ja_i\}$ is isomorphic to a finite Heisenberg group over $\mathbf{Z}_2$: $\mathbf{H}_{D/2}$ (for even D) and $\mathbf{H}_{(D-1)/2} \times \mathbf{Z}_2$ (for odd D). Hint: for even $D = 2n$, elements $\{a_1a_2; a_3a_4; ...; a_{2n-1}a_{2n}\}$ generate a maximal commutative subgroup, $A \simeq \mathbf{Z}_2^{D/2}$; find another subgroup $B \simeq \mathbf{Z}_2^{D/2}$ to get a $\mathbf{Z}_2$-cocycle $\alpha(a;b)$. For odd $D = 2n+1$, we still have subgroups $A, B \simeq \mathbf{Z}_2^n$, but the center is now spanned by $b = -I$ and element $c = a_1 ... a_D$. Show that c commutes with all $\{a_j\}$.

3. i) Show that irreducible representations $\{T^\lambda\}$ of $\mathbf{H}_n$ can be realized in space of functions $\{f(x)\}$ on $\mathbf{F}$, by operators,
$$T^\lambda{}_g f(x) = \chi_0(\lambda c + \lambda b \cdot x) f(x+a),$$
where $g = (a,b;c)$; $\lambda \in \mathbf{F}^*$, and $\chi_0(a)$ – a nontrivial character from the additive group $\mathbf{F}$ to complex numbers, an analog of exponential function $\{e^{ia}\}$ on $\mathbf{R}$. It is well known (and could be easily verified) that any character $\chi = \chi_\lambda$ on $\mathbf{F}$ is obtained from χ_0 by a "λ-factor":
$$\chi_\lambda(a) = \chi_0(\lambda a);\ \lambda \in \mathbf{F}.$$
an analog of $\chi_\lambda(a) = e^{i\lambda a}$, on $\mathbf{R}$.

   ii) establish a similar formula for the n-D Heisenberg group $\mathbf{H}_n$ (the characters of $\mathbf{F}^n$ have the form: $\chi_\mu(a) = \chi_0(\mu \cdot a); \mu \in \mathbf{F}^n$).

4. Verify the relations (3.10) for matrices $\{A_j\}$ (Use a characterization of antisymmetric matrices, as $\{A\colon Ax \cdot x = 0, \text{for all } x \in \mathbf{R}^n\}$).

5. Verify properties (i)-(iii) of group G_D.

6. **Schur's "type-Criterion".**
   i) We denote by $S^2(T)$, $\wedge^2(T)$ *symmetric and antisymmetric tensor-extensions* of an operator/representation T in space $\mathcal{V}$. Show
$$\chi_T(g^2) = tr\, S^2(T_g) - tr \wedge^2(T_g), \text{ for any } g.$$

   ii) Observe that tensor-product space $\mathcal{V} \otimes \mathcal{V}$ is identified with the matrix-space $Mat(\mathcal{V})$, whereas subspaces of symmetric and antisymmetric tensors turn into symmetric and antisymmetric matrices $\{A = (a_{jk})\colon a_{kj} = \pm a_{jk}\}$, and group G acts on $Mat(\mathcal{V})$ by conjugation,
$$g{:}\, A \to {}^{\mathrm{T}}T_g A T_g. \tag{3.14}$$

   iii) Representation (3.14) on $Mat(\mathcal{V})$ is also equivalent to a G-action, $\overline{T}_g^{-1} A T_g$, on space $Mat(\mathcal{V};\mathcal{V}^*)$. Integrating the latter over G,
$$A \to \overline{A} = \int_G \overline{T}_g^{-1} A T_g,$$
on yields G-invariants $\overline{A}{:}\mathcal{V} \to \mathcal{V}^*$, intertwining representations T in $\mathcal{V}$ and $\overline{T}$ in $\mathcal{V}^*$. Note that $Mat(\mathcal{V}) \simeq S^2(\mathcal{V}) \oplus \wedge^2(\mathcal{V})$ - direct sun of symmetric and antisymmetric spaces, and show that Schur integral, $\int \chi_T(g^2)dg$, gives the difference of two intertwining numbers:
$$\boxed{m(T;\overline{T}) \text{ on symmetric space } S^2} - \boxed{m(T;\overline{T}) \text{ on antisymmetric} \wedge^2}.$$

   iv) Conclude that for complex-type T, $Schur(T) = 0$ (space $\mathrm{Int}(T;\overline{T}) = 0$!). For real-type T, the only intertwining operator $\{A\}$ lies in the symmetric subspace, $S^2(\mathcal{V})$, so $Schur(T) = 1$, while quaternionic T has $Int(T;T) \simeq Com(T)$ inside antisymmetric space $\wedge^2(\mathcal{V})$, so $Schur(T) = -1$!

   In the former case, $A = I$, and representation operators $\{T_g\}$ belong to the orthogonal group $O(n)$, in the latter case $A = \begin{bmatrix} & I \\ -I & \end{bmatrix}$, and $\{T_g\}$ belong to symplectic group $Sp(n)$.

7. Find all irreducible representations of alternating group $A_4 \simeq (Z_2 \times Z_2) \rhd Z_3$ by the method of Theorem 1.

8. Find irreducible characters of W_4, and make its character table.

9. Group W_4 acts naturally by symmetries of the cube (each $h \in A_4$ rotates 2 antipodal tetrahedral into themselves, while $\sigma \notin A_4$ flips both). There 3 different function-space on the cube:
a) $\mathcal{C}_F \simeq \mathbf{C}^6$ (functions on faces);
b) $\mathcal{C}_E \simeq \mathbf{C}^{12}$ (functions on edges);
c) $\mathcal{C}_V \simeq \mathbf{C}^8$ (functions on vertices).

So we get 3 associated regular representations of $G = W_4$: R^F; R^E; R^V.

i) Write each of them as the induced one, $ind(1 \mid H; G)$ (find subgroups H);

ii) Decompose each R into the sum of irreducible by two different methods, using characters (problem 8), or applying Frobenius reciprocity.

Chapter 4. Lie groups SU(2) and SO(3).

Groups of unitary and orthogonal matrices: *SU*(2) and *SO*(3) provide the simplest examples of the classical compact Lie groups, yet they exhibit many essential features of the general theory. Both appear in many different contexts in Mathematics and Physics, not surprising due to the fundamental rôle of 3-D Euclidian space and its translational and rotational symmetries. After a brief survey of groups *SU*(2), *SO*(3), their Lie algebras and Haar measures in §4.1, we construct and analyze (§4.2) their irreducible representations in different realizations: polynomials in 1 and 2 variables, infinitesimal construction. In the next section (§4.3) we compute matrix entries and characters of irreducible representations $\{\pi^m\}$ of *su*(2), and establish their connections to the classical orthogonal polynomials: Legendre and Jacobi. Then (§4.4) we turn to representations of *SO*(3) realized in spaces of spherical harmonics $\{\mathcal{H}_m\}$ on the 2-sphere. Here we shall see the connection between spectral theory of the spherical Laplacian and the regular representation of *SO*(3). The last section §4.5 will extend the *SO*(3)-theory to higher-dimensional spheres and orthogonal groups.

§4.1. *Lie groups SU*(2) *and SO*(3) and their Lie algebras.

1.1. Lie group *SU*(2) consists of all 2×2 complex unitary matrices of $det = 1$,

$$\left\{u = \begin{bmatrix} \alpha & \beta \\ -\bar{\beta} & \bar{\alpha} \end{bmatrix} : det\, u = |\alpha|^2 + |\beta|^2 = 1\right\};$$

while *SO*(3) is made of 3×3 real orthogonal matrices u of $det = 1$. Both are three dimensional Lie groups, which means they have the structure of smooth (real analytic) 3-manifolds with differentiable group operations of the product and inverse.

The two groups are closely related, *SU*(2) forms a two-fold cover of *SO*(3):

$$SO(3) \simeq SU(2)/\{\text{center} \pm \mathrm{I}\}. \tag{1.1}$$

One way to show (1.1) is to realize *SU*(2) as a multiplicative subgroup of quaternions,

$\mathrm{Q} = \{x + \boldsymbol{i}y + \boldsymbol{j}z + \boldsymbol{k}u; x, y, z, u\text{-real}; \boldsymbol{i}, \boldsymbol{j}, \boldsymbol{k}\text{ - imaginary units}\} = \{\xi = \alpha + \boldsymbol{j}\beta : \alpha, \beta\text{-complex}\}.$

In the (noncommutative) field of quaternions there exists the norm,

$$|\xi| = \sqrt{x^2+y^2+z^2+u^2},\ \text{with the property: } |\xi\eta| = |\xi||\eta| \text{ for all } \xi;\eta.$$

Hence the set of unit quaternions (unit sphere) $\boldsymbol{U} = \{\xi : |\xi| = 1\}$ forms a subgroup isomorphic to *SU*(2). Conjugation, $\xi \to u^{-1}\xi u$, with $u \in \boldsymbol{U}$ leaves invariant a 3-D real subspace of pure imaginary quaternions $\{\xi = y\boldsymbol{i} + z\boldsymbol{j} + u\boldsymbol{k}\}$. So it defines a map from $\boldsymbol{U} \simeq SU(2)$ onto *SO*(3) (problems 1;2).

1.2. Local coordinates (Euler angles). There are many possible choices of local coordinates on $SU(2)$ or $SO(3)$, for instance, a pair $(\alpha;\Re e\beta)$ or $(\alpha;\Im m\beta)$, on $SU(2)$, or any three independent entries of $u \in SO(3)$. Another (better) choice of coordinates is given by so called *Euler angles* (ϕ,θ,ψ). They arise in a factorization formula: $u = u_\phi v_\theta u_\psi$; $0 \le \theta < \pi$; $0 \le \phi,\psi < 2\pi$, where

$$u_\phi = \begin{bmatrix} e^{i\phi/2} & \\ & e^{-i\phi/2} \end{bmatrix}; v_\theta = \begin{bmatrix} \cos\frac{\theta}{2} & \sin\frac{\theta}{2} \\ -\sin\frac{\theta}{2} & \cos\frac{\theta}{2} \end{bmatrix}; u_\psi = \begin{bmatrix} e^{i\psi/2} & \\ & e^{-i\psi/2} \end{bmatrix}; \tag{1.2}$$

Indeed, writing matrix-entries of $u \in SU(2)$, as $\alpha = \cos\frac{\theta}{2} e^{i(\phi+\psi)/2}$; $\beta = \sin\frac{\theta}{2} e^{i(\phi-\psi)/2}$, we get (1.2). For $SO(3)$ the corresponding one-parameter subgroups can be identified with rotations by angles ϕ,ψ about the z-axis, and by angle θ about the x-axis. The Haar measure (invariant volume element) on both groups is expressed in the Euler-angles as

$$\boxed{du = \frac{1}{8\pi^2} \sin\theta \, d\theta \, d\phi \, d\psi} \tag{1.3}$$

Formula (1.3) is a special case of a general *invariant volume-element formula* on compact Lie groups to be derived in the next chapter. But translational invariance of du can be verified directly.

We observe that the quotient-space of $G = SO(3)$, modulo one-parameter subgroup $H = \{u_\phi : 0 \le \phi < 2\pi\}$, is isomorphic to the 2-sphere, $H\backslash G \simeq S^2$, with the natural action of G on S^2. Given a point $x = x(\theta;\psi) \in S^2$, we identify x with an element $u_x = v_\theta u_\psi$ of $SU(2)$, via the Euler decomposition. Then the product $dS(x) = \sin\theta \, d\theta \, d\psi$, represents the natural (rotation-invariant) area element on $S^2 \simeq H\backslash G$. We take $u \in SU(2)$ factor it as $u_\phi u_x$, then multiply on the right by $v \in G$. The product $(u_\phi u_x)v$ is factored once again according to (1.2) into $uv = u_{\phi'} u_{x'}$; where the new Euler coordinates are: $x' = x^v$ ($x \in S^2$, rotated by $v \in G$), and ϕ shifted by an angle depending on $(x;v)$, $\phi' = \phi + \phi(x;v)$. Since $dS(x)$ is invariant under all rotations $x \to x^v$; $v \in SO(3)$, and $d\phi$ is invariant under all translations: $\phi \to \phi + \phi_0$, it follows that $du = dS d\phi$ is the right-invariant Haar measure.

1.3. Lie algebras $su(2)$ and $so(3)$. Lie algebra $su(2)$ consists of all 2×2 hermitian antisymmetric traceless matrices:

$$A = \begin{bmatrix} ix & y+iz \\ y-iz & -ix \end{bmatrix}; \; x,y,z \in \mathbb{R}$$

while $so(3)$ is made of all 3×3 real antisymmetric matrices. Both have a basis of three elements:

$$su(2) = Span\left\{ H = \begin{bmatrix} i & 0 \\ 0 & -i \end{bmatrix}; V = \begin{bmatrix} 0 & 1 \\ -1 & 0 \end{bmatrix}; W = \begin{bmatrix} & i \\ i & \end{bmatrix} \right\} \tag{1.4}$$

$$so(3) = Span\left\{ H = \begin{bmatrix} & 1 & \\ -1 & & \\ & & 0 \end{bmatrix}; V = \begin{bmatrix} 0 & & \\ & & 1 \\ & -1 & \end{bmatrix}; W = \begin{bmatrix} & & 1 \\ & 0 & \\ -1 & & \end{bmatrix} \right\}$$

The basic elements satisfy the commutation relations:

$$[H;V] = 2W;\ [V;W] = 2H;\ [W;H] = 2V. \tag{1.5}$$

So both Lie algebras are isomorphic, but Lie groups are not! In physics $\{H; V; W\}$ are commonly called *Pauli (spin) matrices*, and denoted by $\sigma^0; \sigma^1; \sigma^2$. The corresponding one-parameter subgroups of $SU(2)$ generated by $\{H;V;W\}$ are

$$exp(tH) = \begin{bmatrix} e^{it} & \\ & e^{-it} \end{bmatrix}; exp(tV) = \begin{bmatrix} cost & sint \\ -sint & cost \end{bmatrix}; exp(tV) = \begin{bmatrix} cost & isint \\ isint & cost \end{bmatrix};$$

while for $SO(3)$ they consist of rotations about 3 coordinate axes.

Problems and Exercises:

1. Quaternions $\mathbf{Q}$ can be described in two different ways:
   i) 2×2 complex matrices $\left\{ \begin{bmatrix} \alpha & \beta \\ -\bar{\beta} & \bar{\alpha} \end{bmatrix} : \alpha; \beta \in \mathbf{C} \right\} \subset Mat_{2 \times 2}(\mathbf{C})$
   ii) all pairs $\{u = (\alpha; \boldsymbol{x}) : \alpha \in \mathbf{R}; \boldsymbol{x} \in \mathbf{R}^3\}$, made of real scalars and 3-vectors, with multiplication given by the dot and cross-products:
   $$u \cdot v = (\alpha\beta - \boldsymbol{x} \cdot \boldsymbol{y}; \alpha \boldsymbol{y} + \beta \boldsymbol{x} + \boldsymbol{x} \times \boldsymbol{y}).$$
   Check both statements.
2. Calculate explicitly the homomorphism $SU(2) \rightarrow SO(3)$. Hint: consider the adjoint action: $u \rightarrow Ad_u(X) = u^{-1}Xu$ of $SU(2)$ on its Lie algebra $su(2) \simeq \mathbf{R}^3$. Show that $\langle X; Y \rangle = -trXY$, is a positive definite inner product on $su(2)$, and $\{Ad_u : u \in SU(2)\}$ preserve this product. Thus $SU(2)$ act by orthogonal transformations on $su(2) \simeq \mathbf{R}^3$. Compute matrix of Ad_u in the basis $\{H; V; W\}$.

   Another way to establish the correspondence $\Im m\mathbf{Q} \simeq su(2)$ is via the cross-product in $\mathbf{R}^3$ (problem 1): $(u \cdot v)_{\Im m} \simeq \boldsymbol{x} \times \boldsymbol{y}$.

§4.2. *Irreducible representations of SU(2).*

2.1. Irreducible representations of group $SU(2)$ can be constructed starting from its natural action on the complex 2-space $\mathbb{C}^2 = \{(z,w)\}$. We shall call this representation $\pi = \pi_g^1$,

$$\pi_g\begin{pmatrix} z \\ w \end{pmatrix} = \begin{pmatrix} \alpha z + \beta w \\ -\bar{\beta}\bar{z} + \bar{\alpha} w \end{pmatrix}; \text{ for } g = \begin{pmatrix} \alpha & \beta \\ -\bar{\beta} & \bar{\alpha} \end{pmatrix}.$$

The action of $SU(2)$ *on* $\mathbb{C}^2$ extends to a representation π^m on the space of all homogeneous polynomials in (z,w) of degree m, $\mathcal{P}_m = \left\{f(z,w) = \sum_{0 \le j \le m} a_j z^j w^{m-j}\right\}$, as[1]

$$(\pi_g^m f)(z,w) = f(\bar{\alpha} z - \beta w; \bar{\beta} z + \alpha w) \tag{2.1}$$

It will be convenient to write polynomials $f \in \mathcal{P}_m$, in the form

$$f(z;w) = \sum_{-m/2 \le j \le m/2} a_j z^{m/2+j} w^{m/2-j}.$$

Let us observe that space $\mathcal{P}_m$ can be identified with the mth *symmetric tensor power of* $\mathbb{C}^2$,

$$\mathcal{P}_m \simeq \mathcal{S}^m(\mathbb{C}^2) \subset \underbrace{\mathbb{C}^2 \otimes \mathbb{C}^2 \otimes \ldots \otimes \mathbb{C}^2}_{m-times} = \mathcal{T}^m(\mathbb{C}^2)$$

by assigning monomials $z^j w^{m-j}$ to symmetrized tensors

$$\tilde{\xi}_j = Sym(\underbrace{e_1 \otimes \ldots \otimes e_1}_{j\ copies} \otimes \underbrace{e_2 \otimes \ldots \otimes e_2}_{(m-j)\ copies}) = \frac{1}{m!}\sum_{\sigma}(e_1 \otimes \ldots \otimes e_2)^{\sigma}; \tag{2.2}$$

sum over all permutations $\sigma \in \mathrm{W}_m$. Here $\{e_1;e_2\}$ form the natural basis of $\mathbb{C}^2$. All tensor-product spaces $(\mathbb{C}^2)^{\otimes m}$, consequently symmetric tensors $\mathcal{S}^m$ and polynomials $\mathcal{P}_m$, inherit the natural inner product of $\mathbb{C}^2$: $\langle \xi \mid \eta \rangle = \xi \cdot \bar{\eta}$;

$$\langle \xi_1 \otimes \ldots \otimes \xi_m \mid \eta_1 \otimes \ldots \otimes \eta_m \rangle = \prod_j \langle \xi_j \mid \eta_j \rangle.$$

So for symmetrized tensors (2.2) we find,

$$\langle \tilde{\xi}_j \mid \tilde{\xi}_k \rangle = \begin{cases} 0; & j \ne k \\ \frac{j!(m-j)!}{m!}; & j = k \end{cases}$$

which in the $\mathcal{P}_m$-space takes the form,

$$\langle z^j w^{m-j} \mid z^k w^{m-k} \rangle = \begin{cases} 0; & j \ne k \\ \frac{j!(m-j)!}{m!}; & j = k \end{cases} \tag{2.3}$$

Since the original π was unitary the resulting representation π^m (2.1) also

[1] We remind the reader that the group-action on manifold $g: x \to g(x)$ is transformed into the left action (representation) on functions $\{f(x)\}$, as $R_g f(x) = f(g^{-1}(x))$.

becomes unitary with respect to the product (2.3). Space $\mathcal{P}_m$ can be identified with polynomials in one-variable $x = \frac{z}{w}$ by writing

$$f(z,w) = \sum a_j z^j w^{m-j} \to F(x) = \sum a_j x^j; \text{ or } f(z,w) = w^m F(\tfrac{z}{w}).$$

Then representation π^m takes the form

$$(\pi_g^m F)(x) = (\bar{\beta} x + \alpha)^m F\Big(\frac{\bar{\alpha}\, x - \beta}{\bar{\beta} x + \alpha}\Big).$$

It will be convenient to write polynomials F as

$$F = \sum_{-m \le j \le m} a_j x^{m+j};$$

and relabel spaces $\mathcal{P}_m$, so that $\mathcal{P}_m$ will consists of $(2m+1)$-degree polynomials $(dim\mathcal{P}_m = 2m+1)$, with m taking on integer or half-integer values: $m = 0; \frac{1}{2}; 1; ...$

We define an inner product on space $\mathcal{P}_m = \{F = \sum a_j x^{m+j}\}$, according to (2.3)

$$\boxed{\langle F \mid \Phi \rangle = \sum_{-m \le j \le m} a_j\, \bar{b}_j \frac{(m+j)!(m-j)!}{(2m)!}} \tag{2.4}$$

and write representation π^m as

$$\boxed{(\pi_g^m F)(x) = (\bar{\beta} x + \alpha)^{2m} F\left(\frac{\bar{\alpha}\, x - \beta}{\bar{\beta} x + \alpha}\right)} \tag{2.5}$$

Theorem 1: *Representations* $\{\pi^m\}$ *are unitary and irreducible. Any unitary irreducible representation of* ***SU***(2) *is equivalent to* $\{\pi^m\colon m = 0; \frac{1}{2}; 1; ...\}$.

Proof: We shall use the infinitesimal method of section 1.4 and study the associated representation of Lie algebra $\mathbf{su}(2) = Span\{H; V; W\}$.

The one-parameter subgroups of the Lie algebra generators (1.4)

$$exp(tH) = \begin{bmatrix} e^{it} & \\ & e^{-it} \end{bmatrix};\ exp(tV) = \begin{bmatrix} cos\,t & sin\,t \\ -sin\,t & cos\,t \end{bmatrix};\ exp(tW) = \begin{bmatrix} cos\,t & i\,sint \\ i\,sin\,t & cost \end{bmatrix}.$$

after substitution in (2.5) yield

$$\begin{aligned} (\pi^m_{exptH} F)(x) &= e^{-i2mt}\, F(e^{2it} x) \\ (\pi^m_{exptV} F)(x) &= (x\, sin + cos)^{2m}\, F\Big(\frac{x\, cos - sin}{x\, sin + cos}\Big) \\ (\pi^m_{exptW} F)(x) &= (x\, isin + cos)^{2m} F\Big(\frac{x\, cos + i\, sin}{x\, isin + cos}\Big) \end{aligned} \tag{2.6}$$

Differentiating the RHS of (2.6) in t at $t = 0$, we find infinitesimal generators of $\{H; V; W\}$,

$$\pi_H^m = i(-2m + 2x\partial);\ \pi_V^m = [2mx - (1+x^2)\partial];\ \pi_W^m = i[2mx + (1-x^2)\partial]; \tag{2.7}$$

where ∂ denotes the derivative $\frac{d}{dx}$. Formula (2.7) could be directly verified to realize $\mathfrak{su}(2)$- commutation relations (1.5) by differential operators (problem 1).

Next we shall represent generators $\{H;V;W\}$ by matrices in the natural basis of monomials $\{x^{m+j}\}_{j=-m}^{m}$ in $\mathcal{P}_m$. The direct calculation with (2.7) yields

$$\pi_H^m = i\begin{bmatrix} 2m & & & \\ & 2(m-1) & & \\ & & \dots & \\ & & & -2m \end{bmatrix};\ \pi_V^m = \begin{bmatrix} 0 & 1 & & \\ -2m & 0 & 2 & \\ & -2 & 0 & 2m \\ & & -1 & 0 \end{bmatrix};\ \pi_W^m = i\begin{bmatrix} 0 & 1 & & \\ 2m & 0 & 2 & \\ & 2 & 0 & 2m \\ & & 1 & 0 \end{bmatrix}. \tag{2.8}$$

So operator π_H^m is represented by the diagonal matrix, diag$(2m;2m-2;\dots-2m)$, while π_V^m and π_W^m are given by tridiagonal matrices with 0's on the main diagonal, $\{1;2;\dots;2m\}$ above the diagonal and $\{2m;2m-1;\dots;1\}$ below the diagonal. Notice, that matrices π_V^m and π_W^m are not skew symmetric in $\mathbb{C}^{2m+1} \simeq \mathcal{P}_m$ with its natural hermitian product. But if we change it to the product (2.4), i.e.

$$\langle e_j | e_k \rangle = (2m-j)!j!\,\delta_{jk},$$

by conjugating $\{\pi_V;\pi_W\}$ with diagonal matrix $Q = diag\left\{\dots[(2m-j)!j!]^{-1/2}\dots\right\}$, this will bring both operators to a skew symmetric form (problem 6).

Irreducibility. By Schur's Lemma we need to show that $Com(\pi^m)$ is scalar. Matrix $A = \pi_H^m$ was shown to be diagonal with distinct eigenvalues. Commutator of any such A consists of diagonal matrices. But any diagonal matrix $B = diag(b_1;\dots;b_{2m+1})$ that commutes with $\{\pi_V^m;\pi_W^m\}$, also commutes with their linear combination. We take $X = \pi_V^m \pm i\pi_W^m$ - an upper/lower triangular matrix with a single sub/super-diagonal row of non-vanishing pairwise disjoint entries. It is easily to show that any diagonal B that commutes with such X must be scalar (problem 2). This proves the first statement.

To show that $\{\pi^m : m = 0;\frac{1}{2};1;\dots\}$ comprise the entire "irreducible list" of $SU(2)$, we shall apply once again the infinitesimal method. Let us note that any representation of Lie algebra $\mathfrak{K} = \mathfrak{su}(2)$ can be lifted (exponentiated) to a global representation of $SU(2)$, since $SU(2) \simeq S_3$ is simply connected.

Lie algebra $\mathfrak{K} = \mathfrak{su}(2)$ is contained in the complex Lie algebra $\mathfrak{G} = \mathfrak{sl}(2)$, and forms a so called *real (compact) form*. This means that $\mathfrak{G} = \mathfrak{K} \oplus i\mathfrak{K}$, as any matrix $A \in \mathfrak{G}$ can be decomposed into its symmetric ($i\mathfrak{K}$) and antisymmetric ($\mathfrak{K}$) parts. So any representation of $\mathfrak{K}$ extends to a complex representation of $\mathfrak{G}$,

$$\pi_{A+iB} = \pi_A + i\pi_B, \; A, B \in \mathfrak{R},$$

and this also applies to Lie group-representations, so (2.5) extends to SL_2 as

$$\boxed{(\pi_g^m F)(x) = (-cx+a)^{2m} F\left(\frac{dx-b}{-cx+a}\right)} \text{ for } g = \begin{bmatrix} a & b \\ c & d \end{bmatrix} \tag{2.9}$$

In fact there exists a natural 1-1 correspondence between representations of $su(2)$, respectively $SU(2)$, and *complex representations*[2] of sl_2, which respects irreducibility and decomposition. This correspondence, known as *Weyl "unitary trick"*, will be examined closely in chapter 5 for more general simple and semisimple groups. Representations of sl_2 are easier to deal with. We introduce the so called *Cartan basis* in sl_2:

$$h = -iH = \begin{bmatrix} 1 & \\ & -1 \end{bmatrix}; \; X = \tfrac{1}{2}(V - iW) = \begin{bmatrix} 0 & 1 \\ & 0 \end{bmatrix}; \; Y = -\tfrac{1}{2}(V + W) = \begin{bmatrix} 0 & \\ 1 & 0 \end{bmatrix}$$

and verify the commutation relations

$$[h;X] = 2X; \; [h;Y] = -2Y; \; [X;Y] = h. \tag{2.10}$$

Let π be an irreducible representation of sl_2 in a complex vector space $\mathcal{V}$. We denote by $\widehat{h} = \pi_h$; $\widehat{X} = \pi_X$; $\widehat{Y} = \pi_Y$ its generators, respectively $\widehat{H} = \pi_H$; $\widehat{V} = \pi_V$; $\widehat{W} = \pi_W$ - the generators of $su(2)$. Operator $\widehat{h}$ is diagonalizable with real eigenvalues, since $\widehat{H} = i\widehat{h}$ is equivalent to a skew-symmetric operator (problem 3). Let $\lambda_0 > \lambda_1 > \dots$ denote the eigenvalues of $\widehat{h}$, and $\{E(\lambda_j)\}$ the corresponding eigenspaces. We shall use the following

Lemma 2: *i) All eigenspaces of π_h are one dimensional, $E(\lambda_j) = span\{v_j\}$; v_j- the the j-th eigenvector.*

ii) the highest eigenvalue $\lambda_0 = 2m$ is an integer, and the j-th eigenvalue $\lambda_j = 2(m-j)$; *for $0 \le j \le 2m$. So dim $\mathcal{V} = 2m+1$.*

iii) operator $\widehat{X}$ is raising eigenvalues: $\widehat{X}(v_j) = a_j v_{j-1}$; while $\widehat{Y}(v_j) = b_j v_{j+1}$ lowering them.

iv) sequences of coefficients $\{a_j\}$, $\{b_j\}$, obtained by applying $\widehat{X}$ and $\widehat{Y}$ to the j-th eigenvectors satisfy the relations

$$\boxed{\lambda_j = 2(m-j) = a_{j+1}b_j - b_{j-1}a_j}. \tag{2.11}$$

To prove the Lemma we pick an eigenvector v_j of an eigenvalue λ_j and apply the lowering operator π_Y to v_j. It will move v_j to another eigenvector of π_h, due to the

[2] $sl_2(\mathbf{C})$ can be considered as either real or complex Lie algebra, so one can talk about its real or complex (holomorphic) representations. On the Lie-group level this corresponds to an additional holomorphic (algebraic) structure on $G = SL_2$, which turns it into a complex (algebraic) manifold. In this context one can talk about the usual class of continuous representations T, or impose stronger assumptions, like holomorphic (algebraic) T, i.e. representations with holomorphic (algebraic) matrix entries: $t(x) = \langle T_x \xi \mid \eta \rangle$.

commutation relation (2.10),

$$\hat{h}\hat{Y}(v_j) = (\hat{Y}\hat{h} + [\hat{h};\hat{Y}])(v_j) = (\lambda_j - 2)\hat{Y}(v_j).$$

So $\hat{Y}$ maps $E(\lambda_j)$ into $E(\lambda_j-2)$, and in a similar fashion, $\hat{X}: E(\lambda_j)\to E(\lambda_j+2)$. This shows that the eigenvalue spectrum of $\hat{h}$ is invariant under the shift $\lambda\to\lambda\pm 2$. To get relation (2.11) we apply the commutator $[\hat{X};\hat{Y}]$ *to* v_j. Next we pick the highest eigenvector v_0 and form a sequence $\{v_j = \hat{Y}^j(v_0)\in E(\lambda_0-2j)$. The span $\mathcal{V}_0 = Sp\{v_0;...v_j...\}$ is invariant under all three generators $\{\hat{h};\ \hat{X};\ \hat{Y}\}$, as

$$\hat{X}(v_j) = j[\lambda_0 - 2(j-1)]v_j. \tag{2.12}$$

Hence (irreducibility) $\mathcal{V}_0 = \mathcal{V}$. So spectrum of $\hat{h}$ is $\{\lambda_0;\ \lambda_0-2;...\}$, as claimed, and all eigenspaces are 1-dimensional. In order for a sequence of coefficients $a_j = j[\lambda_0 - 2(j-1)]$ in (2.12) to vanish after finitely many steps λ_0 must be an integer. The symmetric form of sequence $\{\lambda_0;(\lambda_0-2);...;-\lambda_0\}$ results from an involution on $\mathfrak{G}$, that interchanges X and Y.

2.2. Matrix realization. We shall apply Lemma 2 to show that any irreducible representation π of $\mathfrak{su}(2)$ (respectively $\mathfrak{sl}_2$) is equivalent to π^m of (2.1). The goal is to construct a basis in the representation space $\mathcal{V}(\pi)$, where the generators $\{H;V;W\}$, or $\{h;X;Y\}$ could be brought to the form (2.8).

We start with the highest eigenvector v_0 of operator $\hat{h}$, according to Lemma 2, and iterate it by lowering operator $\hat{Y}$, to get a sequence of eigenvectors $v_j = \hat{Y}^j(v_0)$. Here all coefficients $b_j = 1$, and $\{a_j\}$ can be calculated by (2.11),

$$a_j = \lambda_0 + ... + \lambda_{j-1} = j(2m-j+1).$$

Thus, in the basis $\{v_j = \hat{Y}^j(v_0)\}$, operators $\hat{h};\ \hat{X};\ \hat{Y}$ are represented by matrices

$$\hat{h} = \begin{bmatrix} 2m & & & \\ & 2(m-1) & & \\ & & \ddots & \\ & & & -2m \end{bmatrix};\ \hat{X} = \begin{bmatrix} 0 & 2m & & \\ & 0 & 2(2m-1) & \\ & & 0 & 2m \\ & & & 0 \end{bmatrix};\ \hat{Y} = \begin{bmatrix} 0 & & & \\ 1 & 0 & & \\ & 1 & 0 & \\ & & 1 & 0 \end{bmatrix}.$$

Renormalizing the basis: $v_j \to \sqrt{j(2m-j+1)}v_j$, we can bring π into the symmetric form: $\hat{Y} = \hat{X}^*$, i.e.

$$\hat{X} = \begin{bmatrix} 0 & \sqrt{2m} & & \\ & 0 & \sqrt{2(2m-1)} & \\ & & 0 & \sqrt{2m} \\ & & & 0 \end{bmatrix};\ \hat{Y} = \begin{bmatrix} 0 & & & \\ \sqrt{2m} & 0 & & \\ & \sqrt{2(2m-1)} & 0 & \\ & & \sqrt{2m} & 0 \end{bmatrix}. \tag{2.13}$$

In this form one easily recognizes the symmetrized form of generators $\hat{V} = \hat{X} - \hat{Y}$ and $\hat{W} = i(\hat{X} + \hat{Y})$ of $\mathfrak{su}(2)$ derived in (2.8) (problem 6). So representation $\pi \sim \pi^m$ of

(2.5), and we complete the proof of the Theorem.

Remarks: 1) The construction of irreducible **su**(2)-representations $\{\pi^m\}$ in (2.5) gave at once all finite-dimensional representations of $sl_2(\mathbf{R})$. Both series were linked via complex/analytic continuation through their joint complex hull $sl_2(\mathbf{C})$. Next we would like to determine all irreducible finite-D representations of $sl_2(\mathbf{C})$, considered as *real Lie algebra*, respectively representations of the *real Lie group* $G = SL_2(\mathbf{C})$. Both (group and algebra) have an involution $g \to \bar{g}$ (complex conjugation), and two series of representations, arising from **su**(2): *holomorphic* $\{\pi_g^m\}$ and *antiholomorphic* $\{\bar{\pi}_g^m = \pi_{\bar{g}}^m\}$. It turns out that all other representations are obtained by tensor/Kronecker products of two series. Namely,

$$\pi_g^{n,m} = \pi_g^n \otimes \bar{\pi}_g^m;\ g \in G;$$

and

$$\pi_z^{n,m} = \pi_z^n \otimes \mathrm{I} + \mathrm{I} \otimes \bar{\pi}_z^m$$

for Lie algebra elements z (problem 7), a consequence of general properties of complexifications and representations of direct products (Theorem 4 of §1.3). Representations $\pi^{n,m}$ of $SL_2(\mathbf{C})$ could be realized in polynomials of holomorphic and antiholomorphic variables: $F(z,w) = \sum a_{kj} z^k \bar{w}^j$, by an extension of (2.5),

$$(\pi_g^{m,n} F)(z;w) = (cz+d)^{2m}(c\bar{w}+d)^{2n} F\Big(\frac{dz-b}{-cz+a};\frac{d\bar{w}-b}{-c\bar{w}+a}\Big); \tag{2.14}$$

where matrix $g = \begin{bmatrix} a & c \\ b & d \end{bmatrix} \in SL_2(\mathbf{C})$. All representations $\{\pi^m;\ \bar{\pi}^n;\ \pi^{n,m}\}$ of SL_2 are non-unitary (problem 5).

2) The infinitesimal construction of irreducible representations π^m in terms of the eigendata of $\widehat{H}$, for a specially chosen (Cartan) basis $\{h; X; Y\}$ of sl_2, gives the simplest example of the *highest weight construction* of irreducible representations of simple and semisimple Lie algebras, to be studied in the next chapter.

Problems and Exercises:

1. Verify the commutation relation of $su(2)$ for differential operators $\{\pi_H; \pi_V; \pi_W\}$ in (2.7).

2. Check: if a diagonal matrix $A = \mathrm{diag}(a_1;...;a_n)$ commutes with an upper/lower diagonal matrix B with nonzeros entries $b_1;...;b_{n-1}$ above/below the diagonal, then A is scalar.

3. Show that for any representation T of sl_2 the operator T_h is diagonalizable (Hint: a representation T_g of a Lie group G is unitary iff all infinitesimal generators $\{T_X : X \in \mathfrak{G}\}$ are skew symmetric. Any representation of a compact group is unitarizable!).

4. If a pair of matrices $(A;X)$ satisfies the commutation relation $[A;X] = \alpha X$, for $\alpha \neq 0$, then X is nilpotent, i.e. $X^m = 0$ for some m (Study the eigenspaces and root-subspaces of X).

5. Apply problem 4 to show that SL_2 has no nontrivial unitary representations in finite dimension.

6. Conjugate matrices (2.8) with $Q = \mathrm{diag}\left\{..\,[(2m-j)!j!]^{-1/2}...\right\}$, to bring them to a skew-symmetric form obtained from (2.13).

7. i) Given a complex Lie algebra $\mathfrak{G}$, show that its complexification $\tilde{\mathfrak{G}} = \mathfrak{G} \otimes \mathbb{C} = \{X+iY;\ X,Y \in \mathfrak{G}\}$ is isomorphic to $\mathfrak{G} \oplus \mathfrak{G}$ (direct sum of Lie algebras).

   ii) Complex conjugation $g \to \bar{g}$ in $\mathfrak{G}$ induces a *conjugation* of representations: $T \to \overline{T}_g = T_{\bar{g}}$; which takes holomorphic representations into antiholomorphic

   ii) Conclude (via Theorem 4 of §1.3), that any irreducible representation T of $\mathfrak{G}$, as a real Lie algebra, is equivalent to a tensor/Kronecker product of holomorphic and antiholomorphic irreducible representations: $T \simeq \pi \otimes \bar{\sigma}$.

§4.3. Matrix entries and characters of irreducible representations: Legendre and Jacobi polynomials.

In this section we shall compute matrix entries and characters of irreducible representations $\{\pi^m\}$ of $su(2)$, constructed in the previous section and establish their connection to classical orthogonal polynomials: Legendre and Jacobi.

We shall find matrix entries: $\pi^m_{kn} = \langle \pi^m_g e_k \mid e_n \rangle$ in the natural basis of $\mathcal{P}_m$

$$\{e_k = \frac{1}{\sqrt{(m-k)!(m+k)!}} x^{m-k}: -m \le k \le m\},$$

Notice that the inner product in $\mathcal{P}_m$ can be expressed by means of the differential operations: $(m+k)!\partial^{m-k}$, associated to each monomial $\{x^{m-k}\}$, where $\partial = \frac{d}{dx}$. In other words each polynomial $f = \sum a_k x^{m+k}$ is assigned a differential operator

$$D_f = f(\partial) = \sum (m-k)! a_k \partial^{m+k}.$$

The inner product (4.7) in $\mathcal{P}_m$ is then given by

$$\langle f \mid g \rangle = D_f[\bar{g}]\Big|_{x=0}; \text{ where } f = \sum_{-m}^{m} a_m x^{m+k},\ g = \sum_{-m}^{m} b_k x^{m+k}. \tag{3.1}$$

Using this new form of the inner product and formula (3.1) we can write the kn-th matrix entry of π^m as

$$\pi^m_{kn}(u) = \sqrt{\frac{(m+n)!}{(m-n)!(m-k)!(m+k)!}}\, \partial^{m-k}\Big\{(dx-b)^{m-k}(-cx+a)^{m+k}\Big\}\Big|_{x=0}; \tag{3.2}$$

for any $u = \begin{bmatrix} a & b \\ c & d \end{bmatrix} \in SL_2(\mathbb{C})$, or $SU(2)$. Changing variable, $x \to z = c(dx-b)$, and derivative $\partial_x \to \partial_z = \frac{1}{cd}\partial_x$, we can rewrite (3.2) as

$$\pi^m_{kn}(u) = \sqrt{\frac{(m+n)!}{(m-n)!(m-k)!(m+k)!}}\, \frac{(cd)^{m-n}}{(c^{m-k}d^{m+k})}\, \partial_z^{m-n}\Big\{z^{m-k}(1-z)^{m+k}\Big\}\Big|_{z=-cb} \tag{3.3}$$

Next we shall recast (3.3) in terms of *Euler angles* on $SU(2)$, introduced in §4.1,

$$u = \begin{bmatrix} \alpha & \beta \\ -\bar{\beta} & \bar{\alpha} \end{bmatrix} = \begin{bmatrix} e^{i\phi/2} & \\ & e^{-i\phi/2} \end{bmatrix} \begin{bmatrix} cos\theta/2 & sin\theta/2 \\ -sin\theta/2 & cos\theta/2 \end{bmatrix} \begin{bmatrix} e^{i\psi/2} & \\ & e^{-i\psi/2} \end{bmatrix}; \tag{3.4}$$

where the entries $\alpha = cos\frac{\theta}{2} e^{i(\phi+\psi)/2}$; $\beta = sin\frac{\theta}{2} e^{i(\phi-\psi)/2}$. Remembering that all $\Big\{e_k = \frac{x^{m-k}}{\sqrt{\dots}}\Big\}$ are eigenvectors of diagonal elements $\{exp\, tH\}$, that appear in the right and left factors of (3.4),

$$\pi^m_{exp\,tH}(e_k) = e^{ikt} e_k;$$

we get

$$\pi^m_{kn}(\phi,\theta,\psi) = e^{i(k\psi+n\phi)} \pi^m_{kn}(0,\theta,0) = \sqrt{\frac{\dots}{\dots}}\, e^{i(k\psi+n\phi)} \Big(\frac{sin^{k-n}\theta/2}{cos^{k+n}\theta/2}\Big) \partial_z^{m-n}\Big\{z^{m-k}(1-z)^{m+k}\Big\};$$

the RHS to be evaluated at $z = sin^2\theta/2$. Making another substitution: $z \to x = cos\theta$ (the new x bears no relation to the original variable of $\mathcal{P}_m$), and the substitution

$$1-z = \frac{1+x}{2} = cos^2\theta/2;\ z = \frac{1-x}{2} = sin^2\theta/2\,;\ \text{and}\ \partial_z = -2\partial_x;$$

we bring π^m_{kn} to the form

$$\pi^m_{kn}(\phi,\theta,\psi) = e^{i(k\psi+n\phi)}P^m_{kn}(cos\,\theta),$$

with a single variable function

$$P^m_{kn}(x) = \frac{(-1)^{m-n}}{2^{m+k+n}}\sqrt{\frac{(m+n)!}{(m-n)!(m-k)!(m+k)!}}(1-x)^{\frac{k-n}{2}}(1+x)^{-\frac{k+n}{2}}\partial_x^{m-n}\{(1-x)^{m-k}(1+x)^{m+k}\}. \quad (3.5)$$

We shall express functions P^m_{kn} in terms of the classical *Jacobi polynomials* [Erd], [Leb]. There are many different ways to introduce Jacobi polynomials, one of them is through the *Rodriguez formula:*

$$P^{(\alpha,\beta)}_m = \frac{(-1)^m}{2^m m!}(1-x)^{-\alpha}(1+x)^{-\beta}\Big(\frac{d^m}{dx^m}\Big)\{(1-x)^{\alpha+m}(1+x)^{\beta+m}\}. \quad (3.6)$$

The Jacobi polynomials are orthogonal on $[-1;1]$, relative to the *weighted* L^2–*product:*

$$\langle f \mid g\rangle = \int_{-1}^{1} f(x)\overline{g(x)}w(x)dx, \quad (3.7)$$

with weight $w = (1-x)^\alpha(1+x)^\beta$. In fact, orthogonalization of the natural basis $\{1;x;x^2;...x^m;...\}$ vie product (3.7) provides an alternative definition of $\{P^{(\alpha,\beta)}\}$. Furthermore, Jacobi polynomials are known to satisfy a second order differential equation: $L[y]+\lambda_m y = 0,\ y = P^{(\alpha,\beta)}_m(x)$, where

$$L = \partial(1-x^2)\partial + [\alpha-\beta+(\alpha+\beta)x]\partial,\ \text{and}\ \lambda_m = m(m+\alpha+\beta+1). \quad (3.8)$$

In other words, $\{P^{(\alpha,\beta)}_m\}_m$ represent eigenfunctions of the *Jacobi differential operator* L (3.8) with eigenvalues $\{\lambda_m\}$. This gives yet another characterization of Jacobi polynomials. Returning to functions $\{P^m_{kn}\}$ that appear in formulae (3.5)-(3.6) for matrix entries π^m_{kn} we find them to be

$$P^m_{kn}(x) = C(1-x)^{\frac{n-k}{2}}(1-x)^{\frac{n+k}{2}}P^{(n-k;\,n+k)}_{m-n}(x),$$

with coefficients

$$C = C(n;m;k) = 2^{-(n+k)}\sqrt{\frac{(m+n)!(m-n)!}{(m+k)!(m-k)!}}.$$

Thus we get a representation of matrix entries of $\{\pi^m\}$ in Euler angles as products of exponentials, trigonometric terms and Jacobi polynomials,

$$\boxed{\pi^m_{kn}\,(\phi,\theta,\psi) = C(...)e^{\,i(k\phi+n\psi)}(sin\,\tfrac{\theta}{2})^{n-k}(cos\,\tfrac{\theta}{2})^{n+k}P^{(n-k;\,n+k)}_{m-n}(cos\theta)}$$

A special class of Jacobi polynomials called *Legendre* are given by

$$P_m(x) = \frac{(-1)^m}{2^m m!}\left\{(1-x^2)^m\right\}^{(m)}.$$

They correspond to special values $\alpha = \beta = 0$ in formula (3.6). So Legendre polynomials are orthogonal on $[-1;1]$ relative to constant weight $w(x) = 1$,

$$\langle f \mid g \rangle = \int_{-1}^{1} f(x)\bar{g}(x)dx;$$

and satisfy the differential equation:

$$L[P_m] + m(m+1)P_m = 0, \text{ where } L = \partial(1-x^2)\partial,$$

in other words $\{P_m\}$ are eigenfunctions of the *Legendre differential operator* L. Let us remark that many other classes of orthogonal polynomials on $[-1;1]$ (*Tchebyshev; Gegenbauer; etc.)* are also special cases of Jacobi, corresponding to certain values of $\alpha;\beta$. Legendre polynomials give a special set of matrix entries,

$$\pi^m_{00}(\phi,\theta,\psi) = \text{Const}\, P_m(cos\theta),$$

called *spherical functions* on $SU(2)$. More general entries of the form π^m_{0n} are given by the so called *associated Legendre functions*:

$$P^n_m(x) = \frac{(-1)^m}{2^m m!}(1-x^2)^{-n/2}\left\{(1-x^2)^m\right\}^{(m-n)}.$$

Namely,

$$\pi^m_{0n}(\phi,\theta,\psi) = \frac{1}{\sqrt{(m+n)!(n-m)!}}P^n_m(cos\theta)e^{in\phi}.$$

Family $\{P^\nu_m\}_m$ also forms an orthogonal system of eigenfunctions in $L^2[-1;1]$ for the so called *ν-th associated Legendre operator:*

$$L_\nu = \partial(1-x^2)\partial - \frac{\nu^2}{1-x^2}, \text{ with eigenvalues } \{\lambda_m = m(m+1)\colon m \geq \nu\}. \tag{3.9}$$

We shall examine them closely in the next section.

Problems and Exercises:

1. Verify the orthogonality relations for Jacobi polynomials and find the norming constants $\| P^{(\alpha,\beta)}_m \|^2$.
2. Show that the Jacobi differential operator L is symmetric w.r. to the product (3.7): $\langle Lf \mid g\rangle = \langle f \mid Lg\rangle$.
3. Use orthogonality of Jacobi polynomials and the fact that the Haar measure on $SU(2)$, $du = sin\theta d\theta d\phi d\psi$, in the Euler-angle coordinates, to show that matrix entries $\{\pi^m_{kn}\}$ are orthogonal. Find their norms $\| \pi^m_{kn} \|_{L^2(G)}$.

§4.4. Representations of $SO(3)$: Angular momentum and Spherical harmonics.

Irreducible representations of the orthogonal group $SO(3)$ will be realized in spaces of spherical harmonics $\mathcal{H}_m \subset L^2(S^2)$. Subspaces $\{\mathcal{H}_m\}$ can be characterized in different ways: harmonic polynomials of degree m on $\mathbb{R}^3$, eigenspaces of spherical Laplacian Δ_S on S^2, irreducible subspaces of $SO(3)$. We shall give several proofs of the main decomposition result. One of them exploits important concepts of *Weyl algebra* and *dual pairs* (due to R. Howe). It finds an interesting link between irreducible representations of $SO(3)$ and those of SL_2. Another (computational) argument provides a group-theoretic base for some well-known recurrence relations of associated Legendre functions. In conclusion we apply the analysis of the regular representation on S^2 to the spherical Radon transform.

In the preceding sections we constructed irreducible representations $\{\pi^m\}$ of $SU(2)$, labeled by the *spin* parameter $m = 0;\frac{1}{2};1;\frac{3}{2};...$ (integers and half-integers), which at the same time comprised all complex (holomorphic) irreducible representations of SL_2. Only half of $\{\pi^m\}$, those of integer-valued m, factors through the representation of the orthogonal group $SO(3) = SU(2)/\{1;-1\}$. Indeed, operators representing the diagonal (Cartan) subgroup $\{exp\, tH\}$ of $SU(2)$,

$$\pi^m_{exp\, tH} = \begin{bmatrix} e^{i2mt} & & \\ & e^{i(2m-2)t} & \\ & & \ddots \end{bmatrix};$$

become trivial at $t=\pi$ (i.e. at $u=-I$), iff m is integer.

Here we shall give another construction of representations $\{\pi^m\}$ of integral spin realized in *spherical harmonics* on S^2. We denote by $\mathcal{P}_m$ the space of all homogeneous polynomials in 3 variables of degree m,

$$dim\,\mathcal{P}_m = C^{m+2}_{m} K = \frac{(m+2)(m+1)}{2},$$

and consider its subspace $\mathcal{H}_m$ of *harmonic polynomials*, i.e. solutions of the Laplace's equation,

$$\Delta f = (\partial_1^2 + \partial_2^2 + \partial_3^2) f = 0.$$

Group $G = SO(3)$ acts on $\mathbb{R}^3$ by rotations and this action extends to polynomials,

$$T_u f(x) = f(x^u),\ u \in SO(3),\ x \in \mathbb{R}^3,$$

and $x \to x^u$ -the action of $SO(3)$ on $\mathbb{R}^3$. There is a natural inner product on polynomials $\mathcal{P} = \bigoplus_0^\infty \mathcal{P}_m$, which makes T unitary. It can be introduced via differentiation on $\mathbb{R}^3$, $D = (\partial_1;\partial_2;\partial_3)$. Namely, to each polynomial $p = \sum a_\alpha x^\alpha$, $\alpha = (\alpha_1;\alpha_2;\alpha_3)$, we assign a constant-coefficient differential operator $p(D) = \sum a_\alpha D^\alpha$; for instance radial polynomial $r^2 = |x|^2$ yields the Laplacian Δ. Then we set

$$\langle p \mid q \rangle = [p(D)\bar{q}](0), \text{ i.e. } \langle x^\alpha \mid x^\beta \rangle = \begin{cases} 0; & \text{if } \alpha \ \beta \\ \alpha!\,; & \text{if } \alpha = \beta \end{cases} \tag{4.1}$$

for any pair of multi-indices α,β. This product coincides with the product induced on symmetric tensor powers of $\mathbb{R}^3$, $\mathcal{S}^m(\mathbb{R}^3) \simeq \mathcal{P}_m$ by the natural $\mathbb{R}^3$-product. Indeed, given a pair of m-tensors $\xi = \xi_1 \otimes \ldots \otimes \xi_m$, $\eta = \eta_1 \otimes \ldots \otimes \eta_m$; with the usual product $\langle \xi \mid \eta \rangle = \prod \langle \xi_i \mid \eta_i \rangle$, and the operation of symmetrization,

$\mathrm{Sym}_m : \xi \to \mathcal{S}^m(\xi) = \frac{1}{m!} \sum \xi_{\sigma(1)} \otimes \ldots \otimes \xi_{\sigma(m)}$ - sum over all permutations σ,

the product of symmetrized tensors,

$$\langle \mathcal{S}^m(\xi) \mid \mathcal{S}^m(\eta) \rangle = \frac{1}{m!} \langle \xi \mid \mathcal{S}^m(\eta) \rangle.$$

Applying the latter formula to decomposable tensors of the form

$$\xi_\alpha = \underbrace{e_1 \otimes \ldots \otimes e_1}_{\alpha_1 - times} \otimes \underbrace{e_2 \otimes \ldots \otimes e_2}_{\alpha_2 - times} \otimes \ldots$$

whose symmetrizers $\mathcal{S}^m(\xi_\alpha)$ are identified with monomials $x^\alpha \in \mathcal{P}_m$, we get

$$\langle \mathcal{S}^m(\xi_\alpha) \mid \mathcal{S}^m(\xi_\beta) \rangle = \frac{1}{m!} \langle \xi_\alpha \mid \mathcal{S}^m(\eta_\beta) \rangle = \begin{cases} 0; & \text{if } \alpha \neq \beta \\ \frac{\alpha!}{m!}; & \text{if } \alpha = \beta \end{cases}.$$

Theorem 4.1: (i) *Space $\mathcal{P}_m$ is decomposed into the direct orthogonal sum*

$$\mathcal{P}_m = \mathcal{H}_m \oplus r^2 \mathcal{H}_{m-2} \oplus \ldots \tag{4.2}$$

where $\mathcal{H}_k$ denotes the space of harmonic polynomials of degree k. Hence the entire space $\mathcal{P}$ splits into the tensor product $\mathcal{R} \otimes \mathcal{H}$ of the algebra of radial polynomials $\mathcal{R} = \{f = \sum a_k r^{2k}\}$, and the subspace of harmonic polynomials $\mathcal{H} = \bigoplus_m \mathcal{H}_m$.

(ii) Each subspace $\mathcal{H}_m \subset \mathcal{P}_m$ is $SO(3)$-invariant and irreducible, furthermore the restriction of the regular representation $R \mid \mathcal{H}_m \simeq \pi^m$.

From the orthogonal expansion (4.2) we easily compute the dimension of $\mathcal{H}_m$

$$\dim \mathcal{H}_m = \dim \mathcal{P}_m - \dim \mathcal{P}_{m-2} = 2m+1.$$

Theorem 4.1 gives the spectral (primary) decomposition of the regular representation R on $L^2(S^2)$, and at once the spectral resolution of the Laplacian Δ_{S^2}.

Its first statement follows from product formula (4.1). Indeed, given any pair of functions $g \in \mathcal{P}_m$; $r^2 f$ with $f \in \mathcal{P}_{m-2}$, the orthogonality relation

$\langle r^2 f \mid g \rangle = 0$, for all $f \in \mathcal{P}_{m-2}$; is equivalent to

$$[f(D)\Delta g](0) = \langle f \mid \Delta g\rangle = 0, \quad \text{all } f \in \mathcal{P}_{m-2} \Leftrightarrow \Delta g = 0\,, \text{ i.e. } g\text{ - harmonic.}$$

Therefore, space $\mathcal{H}_m$, an orthogonal complement of G-invariant subspaces $\{r^{2k}\mathcal{P}_{m-2k}\}_{k=0;1;\dots}$, is also G-invariant. It remains to prove irreducibility of $R \mid \mathcal{H}_m$ and equivalence to π^m.

We shall give 3 arguments, based on very different but equally important ideas. One of them will exploit *Weyl algebra* $\mathcal{W} = \mathcal{W}_3$, and so called *Dual (commuting) pairs* in $\mathcal{W}$. As a byproduct we shall find an interesting connection to (∞-D) representations of SL_2. The second argument is fairly straightforward (though somewhat tedious), it will link representations $\{\pi^m\}$ to *eigenfunctions of the 2-sphere Laplacian, spherical harmonics* and *Legendre functions.* The third and shortest utilizes irreducible characters of $SO(3)$, and uniqueness of the map: $T \leftrightarrow \chi_T$ on compact groups, explained in chapter 3.

Proof 1: The *Weyl algebra* $\mathcal{W}$ on $\mathbb{R}^n$ consists of all differential operators with polynomial coefficients $\mathcal{W} = \{A = \sum a_\alpha(x)D^\alpha\}$. Algebra $\mathcal{W}$ acts naturally on space $\mathcal{P}$ of polynomial functions, as well as other function-spaces, e.g. $L^2(\mathbb{R}^n)$, and this action is irreducible (problem 1), since $\mathcal{W}$ contains generators of all translations and multiplications on $\mathbb{R}^n$, and anything commuting with translations and multiplications must be scalar.

The Weyl algebra in $\mathbb{R}^3$ contains a Lie subalgebra $\mathfrak{so}(3)$ of 1-st order differential operators (vector fields), generators of rotations about x, y, z-axis,

$$\partial_H = y\partial_z - z\partial_y;\quad \partial_V = z\partial_x - x\partial_z;\quad \partial_W = x\partial_y - y\partial_x, \tag{4.3}$$

which correspond to the Cartan basis $\{H;V;W\}$ (1.4). Using vector (cross-product) notation we can write triple (3), as

$$(\partial_H;\partial_V;\partial_W) = x \times \nabla.$$

Vector derivative $J = x \times \nabla$ is called in Quantum mechanics the *angular momentum operator,* by analogy with the *classical angular momentum* $x \times p$ (the meaning and the role of *angular momentum* will be explored in depth in chapter 7 (see [BL]).

Another interesting subalgebra of $\mathcal{W}$ is generated by 2-nd order operators:

$$X = \Delta;\ h = x\cdot\nabla;\ Y = |\,x\,|^2. \tag{4.4}$$

One can show $\{h;X;Y\}$ to obey the commutation relations (2.10) of $\mathfrak{sl}_2$ (problem 2)[3]. We denote by $\mathcal{A}_1 \subset \mathcal{W}$ the associative algebra (hull), generated by $\mathfrak{sl}_2$ (4.4), and $\mathcal{A}_0$ will denote the associative hull of $\mathfrak{so}(3)$ in $\mathcal{W}$. Obviously, algebras $\mathcal{A}_0$ and $\mathcal{A}_1$ commute in $\mathcal{W}$, since all three generators (4.4) are rotation invariant. Furthermore, using Weyl's "orthogonal polynomial invariants", one can show (problem 3) that $\mathcal{A}_0$, $\mathcal{A}_1$ form a *maximal commuting* pair in $\mathcal{W}$, called *dual pair*, according to R. Howe [How]. We shall use the following general result.

Proposition: *If an irreducible algebra of operators $\mathcal{W}$ in space $\mathcal{V}$ has a maximal (dual) pair $(\mathcal{A}_0;\mathcal{A}_1)$, then space $\mathcal{V}$ is decomposed into the direct sum of tensor-products*

$$\mathcal{V} \simeq \bigoplus_m \mathcal{V}_m \otimes \mathcal{K}_m, \tag{4.5}$$

so that $\mathcal{A}_0$ acts irreducibly on each $\mathcal{V}_m$, while $\mathcal{A}_1$ on each $\mathcal{K}_m$. Moreover, all representations $\{\mathcal{A}_0 \mid \mathcal{V}_m\}$, respectively $\{\mathcal{A}_1 \mid \mathcal{K}_m\}$, are pair-wise different.

The argument was essentially outlined in Theorem 4 of §1.3 (irreducible representations of product-groups $G = H \times K$). As a consequence, we get a 1-1 correspondence between irreducible representations of $\mathcal{A}_0$ and $\mathcal{A}_1$ that appear in the decomposition of $\mathcal{W}$. In our case ($\mathcal{V} = \mathcal{P}$ or L^2), the decomposition takes the form:

$$\mathcal{P} = \bigoplus_{m=0}^{\infty} \mathcal{R} \otimes \mathcal{H}_m; \text{ similarly } L^2 = \bigoplus_{m=0}^{\infty} \mathcal{R} \otimes \mathcal{H}_m;$$

where $\mathcal{R}$ denotes the radial functions $\{f(r)\}$ (polynomial or L^2), and $\mathcal{H}_m$ - harmonic polynomials of degree m. Both algebras $\mathfrak{K} = so(3)$ and $\mathfrak{H} = sl_2$ respect this decomposition, $\mathfrak{K}:\mathcal{H}_m \to \mathcal{H}_m$, and $\mathfrak{H}:\mathcal{R} \to \mathcal{R}$. Hence, due to the uniqueness of components (4.5), the resulting pairs of representations: $\pi = R \mid \mathcal{H}_m$ of $SO(3)$; and T^m of $SL_2(\mathbb{R})$ on $\mathcal{R}$ must be irreducible, QED.

Remark: The "dual pair" approach yields more than was claimed in Theorem 4.1, not only we are able to realize irreducibles of $SO(3)$ in spaces of harmonic polynomials $\mathcal{H}_m$, but as a byproduct we obtained a series of irreducible representations of $SL(2;\mathbb{R})$, realized in various radial components of the decomposition $\mathcal{P} = \bigoplus_m \mathcal{R} \otimes \mathcal{H}_m$. These are so called *discrete series representations,* studied in chapter 7.

Proof 2. Harmonic polynomials $\{f(x)\} \subset \mathcal{H}_m$ can be restricted on the unit $S^2 \subset \mathbb{R}^3$, due to homogeneity. The resulting functions $\{Y(x) = f \mid S^2 : f \in \mathcal{H}_m\}$, are called *spherical harmonics of degree m.* It turns out that spaces $\mathcal{H}_m(S^2)$ form a complete orthogonal system in $L^2(S^2)$ with the natural (invariant) L^2-product,

$$L^2 = \bigoplus_0^{\infty} \mathcal{H}_m.$$

[3]The same clearly holds in any space $\mathbb{R}^n$, $n \geq 2$. In fact, lurking behind the scene there sits a much larger symplectic algebra $sp(n)$, and its natural "oscillator" representation in $L^2(\mathbb{R}^n)$ (see chapter 6).

In fact, each $\mathcal{H}_m$ will be shown to coincide with an eigenspace of the spherical Laplacian Δ_S. We shall find a basis in spaces of spherical harmonics $\mathcal{H}_m$, and then compare two actions of Lie algebra generators $\{H,V,W\}$ in $R\,|\,\mathcal{H}_m$ and in π^m.

The $\mathbb{R}^3$-Laplacian Δ in polar coordinates $(r;\phi;\theta)$ takes the form

$$\Delta = \partial_r^2 + \tfrac{2}{r}\partial_r + \tfrac{1}{r^2}\Delta_S;$$

where Δ_S denotes its spherical part,

$$\Delta_S = \partial_\theta^2 + \cot\theta\,\partial_\theta + \frac{1}{\sin^2\theta}\partial_\phi^2. \tag{4.6}$$

Given a homogeneous function of degree m, restricted on the sphere,

$$f(x) = |\,x\,|^m Y(x'),\ x' = x/|\,x\,|,$$

the Laplacian of f,

$$\Delta(f) = r^{m-2}[m(m+1) + \Delta_S]Y,$$

is reduced to the spherical Laplacian of Y. This shows that $\mathcal{H}_m$-spherical harmonics on S^2 are eigenfunctions of the Δ_S:

$$\Delta_S Y + m(m+1)Y = 0, \tag{4.7}$$

of eigenvalue $\lambda = m(m+1)$. The standard separation of variables θ,ϕ in (4.6)-(4.7), $Y = F(\phi)G(\theta)$, yields a pair of ODE's for functions $G(\theta)$ and $F(\phi)$

$$F'' + k^2F = 0 \text{ (periodic in } \phi) \Rightarrow F = e^{\pm ik\phi}$$

$$G'' + \frac{1}{\sin\theta}G' + \left\{m(m+1) - \frac{k^2}{\sin^2\theta}\right\}G = 0. \tag{4.8}$$

Changing variable: $z = \cos\theta$, equation (4.8) is transformed into the k–*th associated Legendre equation* (3.9)

$$L_k[G] = -\partial(1-z^2)\partial G + \frac{k^2}{(1-z^2)}G = m(m+1)G. \tag{4.9}$$

Solutions of (9) is the *k-th associated Legendre function (of degree m)* $G = P_m^k$; the m-th eigenfunction of the associated Legendre operator L_k. Functions $\{Y_m^k(\theta,\phi) = e^{ik\phi}P_m^k(\cos\theta)\}_{k=-m}^m$ solve the eigenvalue problem (4.7), and are easily seen to form a orthogonal basis in the the m-th eigensubspace $\mathcal{E}_m$ of Δ_S. One can directly verify that $\mathcal{E}_m = \mathcal{H}_m\,|\,S^2$ (problem 4). The one-parameter group of azimuthal rotations $\{\exp\phi H\}$ (stabilizer of the north pole) is diagonalized in the basis $\{Y_m^k\}$, the eigencharacters being $\{\dots\, e^{ik\phi} \dots\}$, as in (2.9). It remains to show that two other basic elements of $so(3)$ are represented by suitable tridiagonal matrices (2.9).

A straightforward, though somewhat tedious way, is to compute 3 generators $\{H;V;W\}$ of $so(3)$ (angular momentum) in polar coordinates on the 2-sphere. In $\mathbb{R}^3$ they are given by

$$J_x = y\partial_z - z\partial_y;\ J_y = z\partial_x - x\partial_z;\ J_z = x\partial_y - y\partial_x,$$

Passing to polar coordinates we find (problem 5),

$$J = x \times \nabla_x = r u \times (...) = w\partial_\theta - \frac{v}{sin\theta}\partial_\phi = \begin{bmatrix} -sin\phi \\ cos\phi \\ 0 \end{bmatrix}\partial_\theta + \begin{bmatrix} cot\theta\, cos\phi \\ cot\theta\, sin\phi \\ 1 \end{bmatrix}\partial_\phi;$$

where $(u; v; w)$ are unit orthogonal vectors:

$$u = \begin{bmatrix} sin\theta\, cos\phi \\ sin\theta\, sin\phi \\ cos\theta \end{bmatrix};\ v = \begin{bmatrix} cos\theta\, cos\phi \\ cos\theta\, sin\phi \\ -sin\theta \end{bmatrix};\ w = \begin{bmatrix} -sin\phi \\ cos\phi \\ 0 \end{bmatrix}; \tag{4.10}$$

columns of the Jacobian matrix $A = \frac{\partial(x; y; z)}{\partial(r; \theta; \phi)}$ (fig.1). Whence we get three components of J on the sphere,

$$J_x = -sin\phi\, \partial_\theta - cot\theta\, cos\phi\, \partial_\phi;\ J_y = cos\phi\, \partial_\theta - cot\theta\, sin\phi\, \partial_\phi;\ J_z = \partial_\phi. \tag{4.11}$$

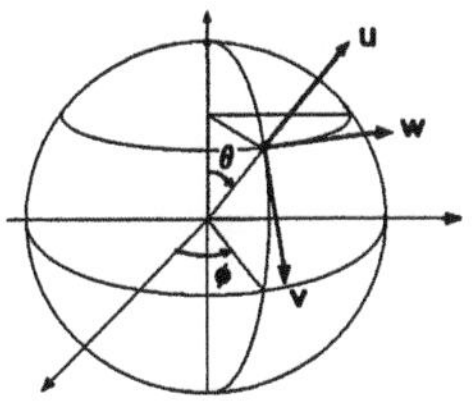

Fig. 1. *The orthogonal unit frame* $\{u; v; w\}$ *on the sphere.*

Next we apply momentum operators (11) to the basic spherical harmonics

$$\{Y_m^k = e^{ik\phi} P_m^k(cos\theta)\}_{k=-m}^m, \text{ of } \mathcal{H}_m.$$

The generator of azimuthal rotations about the z-axis, J_z is diagonalized in the basis $\{Y_m^k\}$, $J_z[Y_m^k] = ik\, Y_m^k$. To compute $J_x; J_y$ we change θ to variable $z = cos\theta$, $\partial_z = \frac{1}{sin\theta}\partial_\theta$, as in (4.8), and introduce operators Q_+, Q_- on S^2,

$$Q_\pm = \sqrt{1-z^2}\,\partial_z \mp \frac{iz}{\sqrt{1-z^2}}\partial_\phi. \tag{4.12}$$

The x and y-components of J are expressed in terms of $Q_\pm$

$$J_x = \frac{1}{2i}(e^{i\phi}Q_+ - e^{-i\phi}Q_-);\ J_y = \frac{1}{2}(e^{i\phi}Q_+ - + e^{-i\phi}Q_-).$$

Operators $Q_\pm$ are simply related to the raising and lowering generators $\{\hat{X}; \hat{Y}\}$ of sl_2 (2.10). Indeed,

$$\hat{X} = J_x + iJ_y = e^{i\phi}Q_+;\ \hat{Y} = J_x - iJ_y = e^{-i\phi}Q_-. \tag{4.13}$$

Now we can easily evaluate $J_x; J_y$ on spherical harmonics,

$$J_x[Y_m^k] = \frac{i}{2}e^{i(k+1)\phi}\underbrace{\left\{\sqrt{1-z^2}\,\partial + \frac{kz}{\sqrt{1-z^2}}\right\}}_{Q_k^+}P_m^k + \frac{i}{2}e^{i(k-1)\phi}\underbrace{\left\{\sqrt{1-z^2}\,\partial - \frac{kz}{\sqrt{1-z^2}}\right\}}_{Q_k^-}P_m^k \tag{4.14}$$

and similarly for J_y

$$J_y[Y_m^k] = \tfrac{1}{2}e^{i(k+1)\phi}Q_k^+[P_m^k] - \tfrac{1}{2}e^{i(k-1)\phi}Q_k^-[P_m^k].$$

The new (reduced) raising/lowering operators $Q_k^{\pm}$ are obtained by restricting $Q_{\pm}$ of (4.12) to functions of the form $\{e^{ik\phi}f(z)\}$. Now it remains to apply the well-known recurrence relations for associated Legendre functions (problem 6),

$$Q_k^+(P_m^k) = a_k P_m^{k+1};\ \text{and}\ Q_k^-(P_m^k) = b_k P_m^{k-1}; \tag{4.15}$$

with coefficients $\{a_k; b_k\}$, depending on a particular normalization of $\{P_m^k\}$ to bring J_x and J_y (or $\widehat{X};\widehat{Y}$ (4.13)), to the familiar form of tridiagonal (raising/lowering) operators (2.9)-(2.13) in basis $\{Y_m^k\}$.

Summary: The argument of part 2 provided more than mere realization of irreducibles $\{\pi^m\}$ in spherical harmonics. Among other results we established

• solution of the eigenvalue problem for the spherical Laplacian Δ_S: *eigenvalues* $\lambda_m = m(m+1)$; $m = 0; 1; ...$; *eigenspaces are spherical harmonics* $\mathcal{H}_m$.

• we showed that the basis $\{Y_m^k = e^{ik\phi}P_m^k(cos\theta)\}$ gives a solution of the joint eigenvalue problem for operators Δ_S and J_z:

$$\begin{cases} \Delta_S[Y_m^k] = m(m+1)Y_m^k \\ J_z[Y_m^k] = ik\,Y_m^k \end{cases}$$

• the azimuthal angular momentum J_z is shown to decompose $L^2(S^2)$ into the direct sum of eigenspaces: $\bigoplus_{-\infty}^{\infty}\mathcal{E}_k$; each $\mathcal{E}_k = \{F = e^{ik\phi}f(cos\theta)\} \simeq L^2[-1;1]$. The Laplacian restricted on $\mathcal{E}_k$ becomes the k-th associated Legendre operator, $\Delta_S \mid \mathcal{E}_k \simeq L_k$. So the $km-$th spherical harmonics factors into the product of the k-th harmonics in ϕ and the km-th associated Legendre function P_m^k (the eigenfunction of L_k).

• the raising/lowering operators $\widehat{X};\widehat{Y}$ given by (4.13) take Y_m^k into $Y_m^{k\pm 1}$. Thus one gets the recurrence relation (4.16) for associated Legendre functions:

$$P_m^{k+1} = \left(\sqrt{\cdots}\,\partial + \frac{kz}{\sqrt{\cdots}}\right)P_m^k;\ P_m^{k-1} = \left(\sqrt{\cdots}\,\partial - \frac{kz}{\sqrt{\cdots}}\right)P_m^k; \tag{4.17}$$

as well as the formula relating the k-th associated Legendre function to the Legendre polynomial P_m (problem 6),

$$P_m^k(z) = (1-z^2)^{k/2}\{(1-z^2)^m\}^{(m+k)}. \tag{4.18}$$

Proof 3: The shortest proof of Theorem 4.1, however, comes from the general representation theory of chapter 3. To prove equivalence of π^m to $R \mid \mathcal{H}_m$, it is enough to compare the characters of two representations: T and S are equivalent, iff their characters are equal, $\chi_T = \chi_S$. Since characters $\{\chi(u)\}$ are constant on conjugacy classes of $u \in G$, it suffices to compute them on class-representatives. In case of $SU(2)$ each u is conjugate (equivalent) to a diagonal matrix,

$$u = v^{-1}\begin{bmatrix} e^{i\phi/2} & \\ & e^{-i\phi/2} \end{bmatrix} v = v^{-1} exp\phi H\, v;$$

where $\{e^{\pm i\phi/2}\}$- eigenvalues of u. Similarly each $u \in SO(3)$ is conjugate to

$$e^{\phi H} = \begin{bmatrix} 1 & & \\ & cos\phi & sin\phi \\ & -sin\phi & cos\phi \end{bmatrix}.$$

In both cases operators $\pi^m_{exp\phi H}$ are digonalized in the natural basis, $\pi^m_H \simeq \mathrm{diag}\{e^{im\phi}; e^{i(m-1)\phi}; ...; e^{-im\phi}\}$, and we readily find

$$\boxed{\chi_m(u) = \sum_{-m}^{m} e^{ik\phi} = \frac{e^{i(m+\frac{1}{2})\phi} - e^{-i(m+\frac{1}{2})\phi}}{e^{i\phi/2} - e^{-i\phi/2}} = \frac{sin(m+\frac{1}{2})\phi}{sin\,\phi/2}} \tag{4.19}$$

Since characters of $\{\pi^m\}$ on diagonal elements coincide with those of $\{R \mid \mathcal{H}_m\}$, it follows that the representations are equivalent, QED.

Remark: Irreducible characters $\{\chi_m = tr\pi_m\}$ are real and orthogonal with respect to the *reduced Haar measure* (problem 7) on conjugacy classes,

$$\langle \chi_m \mid \chi_k \rangle = \int \chi_m(\phi)\, \chi_k(\phi)\, sin^2(\tfrac{\phi}{2})\, d\phi = \delta_{mk}.$$

Using this relation one can decompose the product of any two characters

$$\chi_m \chi_k = \sum_{j=|m-k|}^{|m+k|} \chi_j.$$

The product of characters gives the character of the tensor (Kronecker) product of two representations, $\chi_{T\otimes S} = \chi_T \chi_S$, and similar relation holds for direct sums: $\chi_{T\oplus S} = \chi_T + \chi_S$. As a consequence we get the *Clebsch-Gordon* (decomposition) formula for tensor products of irreducible $SU(2)$ - representations,

$$\boxed{\pi^m \otimes \pi^k = \bigoplus_{j=|m-k|}^{|m+k|} \pi^j}$$

Formula (4.19) is a special case of the Weyl character-formula, that will be studied in the next chapter.

Application to the Radon transform on the sphere. The Radon transform is defined by integrating function $f(x)$ on S^2 along all great circles (geodesics) γ (see §2.5),

$$\mathcal{R}: f \to \hat{f}(\gamma) = \frac{1}{2\pi}\oint_\gamma f ds.$$

Each circle γ can be identified with its north pole $\xi = \xi(\gamma)$, so one can think of $\mathcal{R}$, as transforming functions $\{f(x)\}$ into functions $\{\hat{f}(\xi)\}$ on S^2. Clearly, operator $\mathcal{R}$ commutes with the regular representation R of $SO(3)$ on S^2, so by Theorem 4.1 it must be scalar on each subspace $\mathcal{H}_k$ of spherical harmonics, $\mathcal{R}|\mathcal{H}_m = c_m I$. We could also say that $\mathcal{R}$ must be a "function of the Laplacian Δ on S^2". Instead of Δ it is often convenient to use operator $L = \sqrt{-\Delta + \frac{1}{4}} - \frac{1}{2}$, whose eigenvalues are integers $\{m = 0; 1; ...\}$, $L \mid \mathcal{H}_m = m$. Then we can write $\mathcal{R} = F(\Delta)$, where $F(m) = c_m$. To find function F (or coefficients $\{c_m\}$) it suffices to evaluate $\mathcal{R}$ of a special set of spherical harmonics. We shall chose the zonal ones, $Y^0 = Y^0{}_m(\theta) = P_m(cos\theta)$. The $\mathcal{R}$-image of such Y^0 is another zonal harmonics, $c_m Y^0$, whence we find the coefficient c_m,

$$c_m = \frac{\hat{Y}^0(0)}{Y^0(0)}. \tag{20}$$

It remains to compute the numerator and the denominator of (20). The former is obtained by integrating Y^0 along any meridian-circle of S^2. In the standard normalization of the Legendre polynomials,

$$P_k = \frac{1}{2^k k!}[(1-x^2)^k]^{(k)}, \tag{21}$$

we get for even $k = 2m$ ($\mathcal{R}$ vanishes on all odd functions!),

$$\text{numerator} = \frac{1}{2\pi}\oint_\gamma Y^0 ds = \frac{1}{\pi}\int_0^\pi P_{2m}(\cos\theta)d\theta = \frac{1}{2^{4m}}\binom{2m}{m}^2,$$

(see [Erd], ch.10), while

$$\text{denominator} = P_{2m}(0) = \frac{(-1)^m\binom{2m}{m}}{2^{2m}}.$$

Hence follows the ratio,

$$c_{2m} = \frac{(-1)^m}{2^{2m}}\binom{2m}{m}.$$

The latter can be expressed in terms of the ratio of Γ-functions, namely,

$$F(z) = \frac{\Gamma(z+1)}{2^z\Gamma(z/2+1)^2},$$

evaluated at all integers. So

$$\boxed{c_{2m} = F(2m)}\ m = 0; 1; ... \text{ or } \boxed{\mathcal{R} = F(\sqrt{-\Delta + \tfrac{1}{4}} - \tfrac{1}{2})}$$

The standard Stirling formula for $\Gamma(z)$, yields the following asymptotics of $F(z)$,

$$F(z) \sim \frac{1}{2}\sqrt{\frac{z}{\pi}}, \text{ as } z \to \infty.$$

So one could say that the Radon transform is equivalent to a fractional power

$$\mathcal{R} \sim (-\Delta)^{\frac{1}{4}}, \text{ restricted on "even functions".}$$

Problems and Exercises:

1. If operator Q commutes with all multiplications $\phi: f(x) \to \phi(x)f(x)$, and translation on $\mathbf{R}^n$, then Q must be scalar. Use it to show that Weyl algebra $\mathcal{W}_n$ acts irreducibly on $L^2(\mathbf{R}^n)$ (or polynomials $\mathcal{P}$).

2. Verify the commutation relations (2.10) for operators $\{h; X; Y\}$ of (4.4).

3. **Orthogonal invariants:** group $G = SO(3)$ acts on $\mathbf{R}^3$, and on its "cotangent space" $\mathbf{R}^6 \simeq \mathbf{R}^3 \oplus \mathbf{R}^3$,

$$u: (x; \xi) \to (u(x); u(\xi)); \; x; \xi \in \mathbf{R}^3. \tag{4.22}$$

This action results from a general "cotangent map" defined by a linear transformation $g \in GL_3$

$$g: (x; \xi) \to (g(x); {}^{\mathrm{T}}g^{-1}(\xi)). \tag{4.23}$$

Both actions on $\mathbf{R}^3$ and $\mathbf{R}^6$ extend to polynomial functions: $\mathcal{P}_3 = \{f(x)\}$ and $\mathcal{P}_6 = \{f(x; \xi)\}$.

i) Show that algebra $\mathcal{A}_3$ of $SO(3)$ – invariants in $\mathcal{P}_3$ consists of radial functions $\{f = \sum a_j r^{2j}\}$, i.e. $\mathcal{A}_3$ is generated by a single invariant: $r^2 = |x|^2$.

ii) Algebra $\mathcal{A}_6$ of G-invariants in $\mathcal{P}_6$ is generated by 3 elements: $\{|x|^2;\; x \cdot \xi;\; |\xi|^2\}$ (a special case of Weyl's Theorem on invariants [We]).

iii) Elements $\{A = \sum a_{\alpha\beta} x^\alpha \partial^\beta\}$ of the Weyl algebra $\mathcal{W}_3$ can be identified with polynomials $\{f(x; \xi) = \sum a_{\alpha\beta} x^\alpha \xi^\beta\}$, called *symbols* of differential operators A. Check that the action (4.22)-(4.23) of GL_3 and $SO(3)$ on $\mathcal{P}_6$ and $\mathcal{W}_3$ coincide. Conclude that the only $SO(3)$ – invariants of $\mathcal{W}_3$ are generated by operators (4.4).

4. Check that $\{f = r^m Y^k_m(\phi; \theta)\}$ are harmonic polynomials in $\mathbf{R}^3$, so $\mathcal{H}_m \mid S^2 = \mathcal{E}_m$ - the m–th eigenspace of Δ_S.

5. Find the gradient ∇_x and the angular momentum operator $x \times \nabla$ in polar coordinates $x = r\, sin\theta sin\phi$; $y = r\, sin\theta\, cos\phi$; $z = r\, cos\theta$, of $\mathbf{R}^3$.

Compute the Jacobian matrix: $A = \dfrac{\partial(x, y, z,)}{\partial(r, \theta, \phi)} = (u; rv; r sin\theta\, w)$,

where $\{u; v; w\}$ are three unit orthogonal vectors (4.10) obtained from columns of A; invert A using orthogonality, and show that in polar coordinates,

$$\nabla_x = u\partial_r + \frac{v}{r}\partial_\theta + \frac{w}{r sin\theta}\partial_\phi.$$

6. **Associated Legendre functions:**

i) Derive the recurrence relations (4.17) for associated Legendre functions (Use raising/lowering operators $Q_\pm$ (4.14) related to sl_2 generators $\widehat{X}; \widehat{Y}$, via (4.13)).

ii) Apply recurrence relations (4.17) to derive formula (4.18) for associated Legendre functions. Show that operator $R_k = (1 - z^2)^{k/2} \partial^k = \rho^k \partial^k$, that takes P_m into P^k_m, factors into the product of 1st order operators:

$$R_k = (\rho\partial + (k-1)\rho')(\rho\partial + (k-2)\rho')...(\rho\partial + \rho')\,\rho\partial; \; \rho = \sqrt{1-z^2};$$

and the j-th factor coincides with the *reduced raising operator* Q^+_j of (4.14).

Establish the intertwining relation: $L_k R_k = R_k L_0$; i.e. R_k intertwines the 0-th and the k – th (associated) Legendre operators, and complete the argument.

7. Derive the reduced Haar measure on conjugacy classes of $SU(2)$: $d\mu(\phi) = sin^2(\frac{\phi}{2})d\phi$.

§4.5. Laplacian on the n-sphere.

Spherical Laplacian Δ_S on S^{n-1} is invariant under orthogonal rotations $\{u \in SO(n)\}$, its eigenspaces consist of spherical harmonics $\mathcal{H}_k = \{Y^k$: of degree $k\}$, the eigenvalues being: $\lambda_k = k(k+n-2) = (k+\frac{n-2}{2})^2 - (\frac{n-2}{2})^2$. In this section we shall exploit the spectral theory of Δ_S, to find the Green's functions (fundamental solutions) of the heat and wave problems on the sphere, and establish interesting relations with special functions (Gegenbauer-Legendre), and the Poisson kernel in the unit ball $B \subset \mathbb{R}^n$.

5.1. Harmonic polynomials and decomposition of $L^2(S^{n-1})$. The Laplacian Δ in polar coordinates $\{r;\theta\}$ of $\mathbb{R}^n$ has the form,

$$\Delta = \partial_r^2 + \tfrac{n-1}{r}\partial_r + r^{-2}\Delta_S, \tag{5.1}$$

where Δ_S denotes the spherical Laplacian on S^{n-1}. We shall introduce subspaces of spherical harmonics in $L^2(S^{n-1})\}$, by taking homogeneous harmonic polynomials of degree k on $\mathbb{R}^n$,

$$\mathcal{H}_k = \Big\{f = \sum_{|\alpha|=k} a_\alpha x^\alpha : \Delta f = 0\Big\},$$

and restricting them on S^{n-1}. Spectral decomposition of Δ_S is analogous to the 2-D case of the previous section (Theorem 4.1). Namely,

Theorem 5.1: (i) Space $L^2(S^{n-1})$ is decomposed into the direct orthogonal sum of spherical harmonics $\{\mathcal{H}_k\}$;

(ii) $\{\mathcal{H}_k\}$ are eigenspaces of the Laplacian Δ_S, of eigenvalues $\lambda_k = k(k+n-2)$, *and also irreducible subspaces of the regular representation* $R_u f = f(x^u)$, of group $G = SO(n)$ *on* $L^2(S^{n-1})$.

The eigenvalues are easily obtained from the polar form (5.1). Indeed, for any harmonic function f, homogeneous of degree k (which restricts to a spherical harmonics $Y = f \mid S$), we have

$$0 = \Delta[f]|_S = \{k(k-1) + (n-1)k + \Delta_S\}[Y],$$

hence,

$$\Delta_S[Y] = -k(k+n-2)Y; \text{ on } Y \in \mathcal{H}_k.$$

The irreducibility part will follow from the general results of chapter 5 (JJ5.3; 5.7). More elementary argument exploits a characterization of $\{\mathcal{H}_k\}$ as eigenspaces of Δ_S, and the fact, that each $\mathcal{H}_k$ contains a unique $SO(n-1)$-invariant (axi-symmetric function) Y_0. Indeed, such Y_0 becomes a function of a single angle θ, between $x \in S^{n-1}$, and the axis of rotation. Decomposing Δ_S in spherical angles $\{\theta;...\}$ on S^{n-1},

$$\Delta_S = \partial_\theta{}^2 + (n-2)cot\theta\partial_\theta + \frac{1}{sin^2\theta}\Delta',$$

where Δ' is an S^{n-2} ("equatorial") Laplacian, we see that Y_0 solves an ODE,

$$Y_{\theta\theta} + (n-2)cot\theta Y_\theta + k(k+n-2)Y = 0. \tag{5.2}$$

Changing variable, $\theta \to z = cos\theta$, we bring (5.2) to the standard (Gegenbauer) form, and get a unique solution,

$Y_0(\theta) = \mathrm{C}_k^m(\cos\theta)$, - a Gegenbauer polynomial of degree k, and order $m = \frac{n-2}{2}$.

Once all $\{\mathcal{H}_k\}$ are shown to possess a unique K-invariant ($K = SO(n-1)$), they must be irreducible under the regular action R of $G = SO(n)$ on $L^2(S)$, QED.

5.2. The heat and wave kernels on S^{n-1}. Next shall study the wave and heat-kernels on S^{n-1}. Our main tool will be the Dirichlet problem: $\Delta u = 0; u \mid_S = f$; in the unit ball $B \subset \mathbb{R}^n$. In chapter 2 we derived the Poisson kernel of the ball,

$$P(x;y) = C_n \frac{1-r^2}{(1-2rcos\theta + r^2)^{n/2}}; \tag{5.3}$$

where $r = \mid x \mid$; θ - angle between x and y, and constant $C_n = \dfrac{\Gamma(n/2)}{2\pi^{n/2}}$. Kernel P can be viewed as a formal solution of an operator-valued ODE,

$$u'' + \frac{n-1}{r}u' + \frac{1}{r^2}\Delta[u] = 0;\ u\mid_{r=1} = f. \tag{5.4}$$

Here Δ denotes the Laplacian Δ_S. Indeed, solving formally (5.4) we get

$$u = r^{s(\Delta)}[f], \tag{5.5}$$

where exponential $s(\Delta)$ is found from the characteristic equation of (5.5),

$$s(s-1) + (n-1)s + \Delta = 0, \Rightarrow \boxed{s = \frac{2-n}{2} \pm \sqrt{(\frac{2-n}{2})^2 - \Delta}} \tag{5.6}$$

We shall drop the negative root, since u is required to be regular at $\{0\}$, and denote square-root term by L,

$$L = \sqrt{(\frac{2-n}{2})^2 - \Delta}, \tag{5.7}$$

a positive self-adjoint operator with spectral resolution,

$$L = \bigoplus_k \mu_k E_k. \tag{5.8}$$

To define (5.8) we take eigenvalues $\{\lambda_k = k(k+n-2)\}$ of Δ_S, call the corresponding spectral projections $E_k: L^2(S) \to \mathcal{H}_k$ (spherical harmonics of degree k), and set eigenvalue

$$\mu_k = \sqrt{\lambda_k + (\frac{n-2}{2})^2} = k + \frac{n-2}{2},$$

in accord with (5.6) and the general notion of operator "$\phi(\Delta)$" (Appendix A, and §2.3).

and write

$$u(r,\omega) = r^{L-(\frac{n-2}{2})}[f];\ \omega \in S^{n-1}. \tag{5.9}$$

In other words Poisson kernel $P(r;\omega;\omega')$, represents a "function of operator L" (5.9), $\phi(L)$, where

$$\phi(\lambda) = r^{(\lambda - \frac{n-2}{2})}.$$

From the Poisson kernel $\{P_r\}$ (5.3) we can easily pass to a semigroup of L, $\{e^{-tL}\}$. Indeed,

$$P = C_n r^{1-n/2} \frac{(1/r - r)}{[\frac{1+r^2}{r} - 2cos\theta]^{n/2}}.$$

Comparing it with operator $r^{(L-\ldots)}$ of (5.9) and dropping $r^{1-n/2}$ in both, we get the integral kernel of r^L,

$$r^L(\omega;\omega') = C_n \frac{(1/r - r)}{[\frac{1+r^2}{r} - 2cos\theta]^{n/2}};\quad cos\theta = \omega\cdot\omega';\quad \omega,\omega' \in S^{n-1}.$$

Now a substitution $r = e^{-t}$, brings it into the semigroup-form,

$$\boxed{e^{-tL} = C_n \frac{sinh\, t}{[2(cosh\, t - cos\,\theta)]^{n/2}}} \tag{5.10}$$

From (5.10) we also get a Neumann semigroup-kernel,

$$\boxed{L^{-1}e^{-tL} = \int_t^{\infty} e^{-tL}dt = \frac{2C_n}{(n-2)[2(cosh\, t - cos\,\theta)]^{\frac{n-2}{2}}}} \tag{5.11}$$

Let us remark that formulae (5.10)-(5.11) can be viewed as the Dirichlet/Neumann Poisson kernels on the semi-infinite cylinder $\{(\omega;t)\} = S^{n-1} \times [0;\infty)$.

Next we shall analytically continue them in the complex range of parameter t, $Re\,t \geq 0$, to get wave-propagators $\{\cos(tL); \frac{sin(tL)}{L}\}$. As in §2.3 we formally substitute $(it - \epsilon)$ in both kernels and let $\epsilon \to 0$. So

$$\begin{aligned} L^{-1}e^{itL} &= \lim_{\epsilon\to 0} \frac{C_n'}{[2(cosh(it-\epsilon) - cos\theta)]^{(n-2)/2}} \\ e^{itL} &= \lim_{\epsilon\to 0} \frac{C_n sint}{[2(cosh(it-\epsilon) - cos\theta)]^{n/2}}; \end{aligned} \tag{5.12}$$

and $cos-$ $sin-$propagators are the real and imaginary parts of corresponding unitary groups. In the special case of S^2, it gives,

$$L^{-1}e^{itL} = \frac{1}{2\pi\sqrt{2(\cos t - \cos\theta)}};$$

hence

$$L^{-1}\sin(tL) = \begin{cases} \frac{1}{2\pi\sqrt{2(\cos t - \cos\theta)}}; & |\theta| \leq t \\ 0; & |\theta| \geq t \end{cases}; \tag{5.13}$$

which then extends periodically in time t with period 2π.

Formula (5.13) gives an S^2-version of the 2-D Euclidian propagator (chapter 2),

$$\frac{1}{2\pi\sqrt{t^2 - |x|^2}}.$$

It obeys, as the former, the finite-propagation-speed principle, but due to time-periodicity[4], the out-spreading wave-front reassembles after $t = 2\pi$ at s single source, and the whole pattern repeats again.

In higher even dimensions ([Ta2], chapter 4) one finds the propagators to be

$$\boxed{L^{-1}\sin(tL) = \frac{1}{2(2\pi)^{(n-2)/2}}\left(\frac{1}{\sin\theta}\frac{\partial}{\partial\theta}\right)^{(n-4)/2}\left[\sin^{n-3}\theta(M^\theta f)(x)\right]\Big|_{\theta = t}}$$

$$\boxed{\cos(tL) = \frac{1}{2(2\pi)^{(n-2)/2}}\sin\theta\left(\frac{1}{\sin\theta}\frac{\partial}{\partial\theta}\right)^{(n-2)/2}\left[\sin^{n-3}\theta(M^\theta f)(x)\right]\Big|_{\theta = t}} \tag{5.14}$$

where $M^\theta f(x)$ denotes the mean value of f on the geodesic sphere , centered at x of radius θ, $S_\theta(x) = \{y: x \cdot y = \cos\theta\}$

$$M^\theta f(x) = \int_{S_\theta(x)} f(y)dS(y).$$

The even-D formulae clearly demonstrate the strict Huygens principle: at each time t, distributional kernels of $L^{-1}\sin(tL)$, and $\cos(tL)$ are supported on the geodesic sphere $S_t = \{x \cdot y = \cos t\}$, rather than the ball $\{x \cdot y \geq \cos t\}$.

The heat-propagator can be constructed by Fourier-transforming the wave-group (5.12), or cos-propagator (5.14),

$$e^{-tL^2} = \int_{-\infty}^{\infty} \frac{1}{\sqrt{4\pi t}} e^{-s^2/4t}\cos(sL)\,ds.$$

Using 2π-periodicity of $\cos(tL)$ and the Poisson summation formula of §2.1 for the 1-D Gaussian,

[4]Notice that spectrum of L, $\{\lambda_k(L) = k + \frac{n-2}{2}\}$, consists of integers for even n's, and half-integers for odd n's. So a unitary group of L, $\{e^{itL}\}$ is either 2π or 4π-periodic.

$$G(z;t) = \sum_m exp(-\tfrac{1}{2}tm^2+imz) = (2\pi t)^{-1}\sum_\nu exp(-\frac{(z-\nu)^2}{2t}) = \Theta(z;t) \text{ - theta-function,}$$

we get the spherical Gaussian for odd-dimensional S^{n-1},

$$\boxed{e^{-t\Delta} = (\frac{1}{2\pi\, sin\theta}\frac{\partial}{\partial\theta})^{(n-2)/2}\Theta(\theta;t)}$$

Remark: One can derive similar formulae for hyperbolic Laplacians on spaces $\mathbb{H}^n$ by analytically extending metric tensor (see [Ta2], chapter 4).

5.3. Spherical harmonics. We described eigenspaces $\{\mathcal{H}_k\}$ of L as degree-k harmonic polynomials on S^{n-1}. To describe them explicitly we can choose a single element $Y_0 \in \mathcal{H}_k$ and apply all rotations $\{u \in SO(n)\}$ to it. The simplest choice is $Y_{\boldsymbol{a}} = (\sum a_j x_j)^k = (\boldsymbol{a}\cdot\boldsymbol{x})^k$ - a linear form on $\mathbb{R}^n$, with complex coefficients $\{a_j\}$, raised to the k-th power. One easily verifies,

$$\Delta[Y_{\boldsymbol{a}}] = k(k-1)(\boldsymbol{a}\cdot\boldsymbol{a})(\boldsymbol{a}\cdot\boldsymbol{x})^{k-2},$$

hence $Y_{\boldsymbol{a}}$ is harmonic, iff $\boldsymbol{a}\cdot\boldsymbol{a} = \sum a_j^{\,2} = 0$. Since $SO(n)$ acts irreducibly on $\mathcal{H}_k$ (as will be explained in §5.7), the latter coincides with the linear span of all such Y's[5],

$$\mathcal{H}_k = \text{Span}\{Y_{\boldsymbol{a}}\colon \boldsymbol{a}\cdot\boldsymbol{a} = 0\}.$$

Another possible choice is an axi-symmetric (zonal) function $Y_0 = P_k(\cos\theta)$, i.e. an $SO(n-1)$-invariant Y_0. Since regular representation R on S^{n-1} is induced by the stabilizer subgroup $K = SO(n-1)$, we have shown via the Frobenius reciprocity, that each $\pi^k \subset R$ has a unique K-invariant. Function $Y(\theta) = P_k(\cos\theta)$ satisfies a "radially reduced" Laplace's condition,

$$Y'' + (n-2)cot\theta\, Y' + k(k+n-2)Y = 0,$$

or in terms of variable $z = cos\theta$, $Y = P(z)$ solves,

$$(1-z^2)^{-m}\frac{d}{dz}(1-z^2)^m\frac{d}{dz}P + k(k+n-2)P = 0,\; m = \frac{n-2}{2}.$$

The solutions are well known *Gegenbauer (ultraspherical) polynomials* $\{P_k = C_k^m(z)\colon k = 0;1;...\}$ of order $m = \frac{n-2}{2}$ (see [Erd, vol.2]).

An important feature of $Y_0 = P_k(\cos\theta)$ is described by the following Lemma.

Lemma 2: *Integral kernel $E_k(x;y) = P_k(\cos\theta)$, $\cos\theta = x\cdot y$, properly normalized, defines an orthogonal projection $E_k\colon L^2 \to \mathcal{H}_k$.*

Indeed, any G-invariant (integral) operator $K(x;y)$ on S^{n-1}, a rank-one symmetric space (see §5.7), must be of the form $K(x\cdot y)$ (problem 1), i.e. K must depend only on the

---

[5] Later (chapter 5) we shall see that $\{Y_{\boldsymbol{a}}\}$ defines the so called highest weight-vector of an irreducible representation π^k of $SO(n)$ on h_k.

geodesic distance between x and y: $d(x;y)=cos^{-1}(x\cdot y)$. So E_k commutes with the G-action on L^2. It clearly takes L^2 into $\mathcal{H}_k$,

$$u(x)=\int_S P_k(x\cdot y)f(y)dS(y)\Rightarrow \Delta[u]=\int \Delta_x[P_k(...)]f\,dS=\lambda_k u.$$

Hence, by irreducibility of π^k, $E_k|\mathcal{H}_k=$ Const, and the constant is found from the normalizing condition, Const $=E_k[P_k](1)=\|P_k\|^2=1$.

Now we can compare the Neumann semigroup, $L^{-1}e^{-tL}$, in two different representations: in terms of the integral kernel (5.11), and through the spectral decomposition via projections $\{E_k\}$,

$$\frac{C_n'}{[cosh\,t-cos\,\theta]^{(n-2)/2}}=\sum_0^\infty \frac{1}{\mu_k}e^{-t\mu_k}P_k(cos\theta),$$

where $\{\mu_k=k+\frac{n-2}{2}\}$ are eigenvalues of L. Returning to the original variable $r=e^{-t}$, we get the expansion of the conjugate (Neumann) Poisson kernel of the sphere,

$$\boxed{\frac{C'}{(1-2rcos\theta+r^2)^{(n-2)/2}}=\sum_0^\infty r^kP_k(cos\theta)} \tag{5.15}$$

Here we renormalized polynomials $\{P_k\}$, by dividing them with μ_k. Relation (5.15) gives a *generating function*[6] for the Gegenbauer polynomials $\{C_k^m\}$ $(m=\frac{n-2}{2})$. The latter in turn yields the Cauchy-integral formulae for $\{P_k=C_k^m\}$,

$$P_k(t)=\frac{1}{2\pi i}\oint_\gamma \frac{1}{(1-2t\zeta+\zeta^2)^m}\frac{d\zeta}{\zeta^{k+1}}\text{ - contour integral in } \mathbb{C}.$$

In special case $n=3$ we get the familiar Legendre polynomials $P_k(t)=\frac{1}{2^kk!}[(1-t^2)^k]^{(k)}$, whose generating function,

$$\boxed{\sum_0^\infty r^kP_k(t)=\frac{1}{\sqrt{1-2rt+r^2}}.}$$

Further results, details and references could be found in the book [Ta2] by M.Taylor, whose approach we closely followed, and [Hel2].

[6] A generating function of an orthogonal family $\{\phi_k(z)\}$ is defined as a function $F(r;z)$, that admits a Taylor/Laurent expansion in variable r, with coefficients $\{\phi_k\}$, $F(r;z)=\sum r^k\phi_k(z)$. All classical orthogonal polynomials are known to have generating functions ([BE]), in particular Gegenbauer $\{C^\alpha\}$ are generated by

$$(1-2rt+r^2)^{-a}=\sum_0^\infty r^kC_k^a(t).$$

Problems and Exercises:

1. Show that the commutator of the regular representation R of $SO(n)$ on S^{n-1} consists of integral operators $K(x;y) = F(x \cdot y)$, i.e. kernel K depends only on the geodesic distance $\theta = \cos^{-1}(x \cdot y)$, between x and y.

   Hint: the n–sphere is a 2-point symmetric space, i.e. any pair $(x;y)$ is taken into any other equidistant pair $(x';y')$ by an orthogonal $u \in SO(n)$ (compare to problem 4 of §1.3).

Chapter 5. Classical compact Lie groups and algebras.

In this chapter we shall develop the structure and the representation theory of ***classical compact Lie groups*** and associated ***complex and real Lie algebras***. There are four types of classical ***simple*** compact Lie groups and algebras: unitary, orthogonal (odd and even) and symplectic. Together with three ***exceptional*** algebras they provide a complete list of all simple Lie algebras, due to the celebrated ***Cartan classification Theorem***.

We shall start this chapter with a brief introduction to the basic structural theory of simple and semisimple Lie algebras: ***Cartan subalgebra, root system, Weyl group***. These results will lay ground for a subsequent study of representations of semisimple algebras. We shall give several constructions of irreducible representations, including ***highest weight*** realization and the ***Weyl's theory of tensor powers*** and ***Young symmetrizers***. Finally, we proceed to the heart of the classical compact Lie group theory: the celebrated ***Weyl character formulae***. The last section applies results of preceding parts (§§5.1-6) to study Laplacians on compact symmetric space $K\backslash G$.

§5.1 Simple and Semisimple Lie Algebras, Weyl unitary trick.

We introduce basic types of Lie algebras, solvable, nilpotent, simple and semisimple, and the Cartan list of simple algebras. Then we outline the *Weyl unitary trick*, that relates representations of compact Lie groups to those of complex semisimple algebras $\{\mathfrak{G}\}$. This makes available the machinery of chapter 3 to study representations of semisimple algebras. It implies, in particular, that any finite-D representation T of $\mathfrak{G}$ is completely reducible (direct sum of irreducible components), also any semisimple $\mathfrak{G}$ is itself decomposed into the direct sum of simple ideals.

1.1. Definitions. We remind the reader the basic notions of *simple, semisimple, solvable and nilpotent* Lie algebras and groups, introduced in §1.4 (chapter 1). Solvable and nilpotent algebras are defined through an *upper/lower derived series* of $\mathfrak{B}$,

$$\text{(I)}\ \mathfrak{B}\supset\mathfrak{B}^1=[\mathfrak{B};\mathfrak{B}]\supset\mathfrak{B}^2=[\mathfrak{B}^1;\mathfrak{B}^1]\supset\ldots\supset[\mathfrak{B}^n;\mathfrak{B}^n]=\mathfrak{B}^{n+1}\supset\ldots$$

$$\text{(II)}\ \mathfrak{B}\supset\mathfrak{B}_1=[\mathfrak{B};\mathfrak{B}]\supset\mathfrak{B}_2=[\mathfrak{B};\mathfrak{B}_1]\supset\ldots\supset[\mathfrak{B};\mathfrak{B}_n]=\mathfrak{B}_{n+1}\supset\ldots$$

Here $[\mathfrak{X};\mathfrak{Y}]$ denotes $Span\{[X;Y];X\in\mathfrak{X},Y\in\mathfrak{Y}\}$. All terms of the upper/lower series $\{\mathfrak{B}_k\subset\mathfrak{B}^k\}$ are *ideals* of $\mathfrak{B}$, $[\mathfrak{B};\mathfrak{B}_k]\subset\mathfrak{B}_k$; moreover the *commutant* $\mathfrak{B}_1$ is the smallest ideal so that $\mathfrak{B}/\mathfrak{B}_1$ is commutative.

Solvable algebras have the upper derived series (I) terminated by $\{0\}$, $\mathfrak{B}^{n+1}=0$, while *nilpotent* (a subclass of solvable) have the lower series (II) terminates by $\{0\}$.

Typical examples are upper/lower triangular matrices:

$$\left\{\begin{bmatrix} a & * & * \\ & \ddots & * \\ & & c\end{bmatrix}\right\}\text{solvable};\quad \left\{\begin{bmatrix} 0 & * & * \\ & \ddots & * \\ & & 0\end{bmatrix}\right\}\text{nilpotent (0's on the main diagonal !).}$$

Obviously, the last term $\mathfrak{B}^n$ of (I) is a commutative ideal, and all quotients $\mathfrak{B}^k/\mathfrak{B}^{k+1}$ are commutative ideals of $\mathfrak{B}/\mathfrak{B}^{k+1}$ (solvable case), while $\mathfrak{B}_n \subset \text{center}\mathfrak{B}$, and any $\mathfrak{B}_k/\mathfrak{B}_{k+1}$ belongs in the center of the factor – algebra $\mathfrak{B}/\mathfrak{B}_{k+1}$ in the nilpotent case. All ideals $\{\mathfrak{B}_k\}$ and $\{\mathfrak{B}^k\}$ are *characteristic,* i.e. invariant under all automorphisms (respectively derivations) of $\mathfrak{B}$. If $\mathfrak{B}$ is solvable, then $\mathfrak{B}^1$ is known to be nilpotent (cf. problem 1, §5.3). Also algebra $\mathfrak{B}$ is nilpotent iff all operators $\{ad_X\}$ are nilpotent, $ad_X^n = 0$.

Semisimple Lie algebras $\mathfrak{G}$ can be described in many different ways:

(i) $\mathfrak{G}$ *has no solvable, hence, abelian ideals;*

(ii) the commutator subalgebra $[\mathfrak{G};\mathfrak{G}] = \mathfrak{G}$.

The third characterization is given in terms of the *Killing form,* an inner product on $\mathfrak{G}$, defined by

$$\langle X \mid Y \rangle \overset{def}{=} tr(ad_X ad_Y). \tag{1.1}$$

(iii) algebra $\mathfrak{G}$ *is semisimple iff the Killing form is non-degenerate.*

One can show that ideal and factors $\mathfrak{G}/\mathfrak{H}$ of semisimple algebras are also semisimple.

We will not provide a complete detail of the equivalence proof, but just mention that the argument is based on *invariance properties* of the Killing form relative to the adjoint action:

$$\begin{aligned} \langle ad_X Y \mid Z \rangle + \langle Y \mid ad_X Z \rangle &= 0, \text{ all } X, Y, Z \in \mathfrak{G} \\ \langle Ad_g\, Y \mid Ad_g\, Z \rangle &= \langle Y \mid Z \rangle, \text{ all } Y, Z \in \mathfrak{G} \text{ and } g \in G. \end{aligned} \tag{1.2}$$

The invariance relations (1.2) imply that the null – space of the Killing form, $\mathfrak{N} = \{Y : \langle X \mid Y \rangle = 0, \text{ for all } X \in \mathfrak{G}\}$ is an ideal, and the Killing form on $\mathfrak{N}$ is identically $= 0$. Any such Lie algebra $\mathfrak{N}$ could be shown to be solvable, which contradicts semi-simplicity of $\mathfrak{G}$, and proves (i).

The reader is invited to check (i)-(iii) for the classical *simple* Lie algebras, listed below (problem 6).

Algebra $\mathfrak{A}$ is called *simple,* if it has no ideals. All simple (complex) Lie algebras were classified by Eli Cartan. There are four classical series, and 5 exceptional algebras. The four classical series in the standard designation consist of

(A_n) *Special linear algebras:* $\mathfrak{sl}(n+1;\mathbb{C}) = \{\text{all } (n+1)\times(n+1) \text{ matrices } A \text{ of } trA = 0\}$.

(B_n) *Odd-dimensional orthogonal algebras:* $\mathfrak{so}(2n+1;\mathbb{C}) = \{$complex matrices $A\vec{x}\cdot\vec{y}+\vec{x}\cdot A\vec{y}=0\}$, with the standard symmetric dot product $\vec{x}\cdot\vec{y}=\sum x_k y_k$ on $\mathbb{C}^{2n+1}$,

(C_n) *Symplectic algebras:* $\mathfrak{sp}(n;\mathbb{C}) = \{2n\times 2n$ matrices A, satisfying $\langle JA\vec{x}\mid\vec{y}\rangle+\langle J\vec{x}\mid A\vec{y}\rangle=0\}$, where $J=\begin{bmatrix}0 & I_n\\ -I_n & 0\end{bmatrix}$ denotes the symplectic form on $\mathbb{C}^{2n}$.

(D_n) *Even-D orthogonal algebras:* $\mathfrak{so}(2n;\mathbb{C})$.

The 5 exceptional Lie algebras are designated as G_2; F_4; E_6; E_7; E_8 subscript indicating their *rank*, and will be described in the next section.

The structure of any Lie algebra could be reduced to its "elemental blocks". Namely,

Theorem 1: *i) Any semisimple Lie algebra splits into the direct sum of simple ideals:* $\mathfrak{G}=\oplus\mathfrak{G}_k$ *- irreducible components of the adjoint action* $ad_X\mid\mathfrak{G}$.

ii) An arbitrary algebra admits a Levi-Malćev decomposition (§1.4) *into a semidirect product,* $\mathfrak{B}=\mathfrak{R}\rhd\mathfrak{G}$, *of the maximal solvable ideal* $\mathfrak{R}$ *(radical of* $\mathfrak{B}$*), and a semisimple subalgebra* $\mathfrak{G}$.

The first statement will follow from the *complete reducibility* of any (finite – D) representation T of a semisimple Lie algebra, via the *Weyl unitary trick.*

1.2. The Weyl unitary trick. The Weyl "unitary trick" allows to reduce representations of complex semisimple Lie algebras to those of compact groups. The key element of the Weyl reduction is based on an important notion of *Cartan involution.* All simple and semisimple complex Lie algebras are known to have an *involutive automorphism* σ, with properties

(i) $\sigma[X;Y]=[\sigma(X);\sigma(Y)]$, (ii) $\sigma(cX)=\bar{c}X$, for any complex c, (iii) $\sigma^2=1$.

The set of fixed points of σ, $\mathfrak{K}=\{X\colon \sigma(X)=X\}$ forms a *real subalgebra* $\mathfrak{K}$ of $\mathfrak{G}$, called *real compact form* of $\mathfrak{G}$, i.e.

$$\mathfrak{G}=\mathfrak{K}\oplus\mathfrak{P} \tag{1.3}$$

direct sum of the subalgebra $\mathfrak{K}$ and the subspace $\mathfrak{P}=i\mathfrak{K}$, the subgroup $K\subset G$ of algebra $\mathfrak{K}\subset\mathfrak{G}$, is compact. Decomposition (1.3) is crucial in the Weyl's method. It means that algebra $\mathfrak{G}$ is a *complexification* of the compact real Lie algebra $\mathfrak{K}$. Therefore, any complex representation of $\mathfrak{G}$ (respectively *analytic* representation of its Lie group G) is uniquely determined by its restriction on the compact form $\mathfrak{K}$ (respectively *compact Lie subgroup* K). Thus we get at our disposal the whole machinery of representations of *compact groups,* developed in chapter 3. Before we shall sketch the general argument for

(1.3) let us give a few examples.

Examples: 1) Algebra $\mathfrak{G} = sl(n)$, with involution $\sigma(X) = -X^*$, has real compact form $\mathfrak{K} = su(n)$;

2) Orthogonal algebra: $so(n;\mathbb{C})$ has an involution: $X \to -{}^T X$, and compact form $\mathfrak{K} = so(n;\mathbb{R})$;

3) Symplectic: $sp(n;\mathbb{C}) = \left\{ \begin{bmatrix} a & c \\ b & -a^* \end{bmatrix}; \ a \in gl_n(\mathbb{C});\ b, c \text{ - hermitian symmetric} \right\}$ has involution: $X \to -X^*$, and the compact form

$$\mathfrak{K} = \left\{ \begin{bmatrix} a & c \\ -c^* & a \end{bmatrix}; a \in su(n);\ c \text{ - hermitian symmetric} \right\} = sp(n),$$

called *compact symplectic algebra.*

The existence proof of the *Cartan involution* and *real compact form* (1.3) semisimple algebras is based on certain properties of the Killing form (1.1). This form will also yield a simple characterization of compact (real) Lie algebras.

Theorem 2: *A real semisimple algebra* $\mathfrak{K}$ *is compact iff the Killing form is negative definite* $\langle X \mid X \rangle < 0$, *for all* $X \neq 0$.

To establish (ii) we observe that all Lie algebra adjoint operators $\{ad_X\}$ are antisymmetric with respect to the Killing form, hence Lie group $\{Ad_g\}$ are orthogonal. The definiteness (positive or negative) of the Killing form implies that the adjoint map $g \to Ad_g$ takes G into the compact orthogonal group $O(n)$, $n = dimG$. Since G is semisimple, it implies that the factor-group G/Z (modulo discrete center Z) is compact. Hence Z is finite and G is itself compact.

Conversely, given a compact (simple) group G, there exists a definite (positive/negative) product $X \cdot Y$ on $\mathfrak{G}$ s.t. adjoint maps $\{\mathrm{Ad}_g\}$ is orthogonal (the latter is easily achieved by "averaging" any product with respect to the Haar measure on G). Any other bilinear form, including Killing's, is given then by a suitable symmetric operator

$$\langle X \mid Y \rangle = QX \cdot Y.$$

The invariance condition (1.2) implies that Q commutes with the adjoint representation Ad_g, and since the latter is irreducible (simplicity of $\mathfrak{G}$!), $Q = \lambda I$, a multiple of the identity, by Schur's Lemma. Obviously, constant $\lambda < 0$, since for any antisymmetric matrix A, $trA^2 \leq 0$, QED.

After Theorem 2 one can easily find compact real forms (1.3) in $\mathfrak{G}$, any maximal real subalgebra $\mathfrak{K}$ in $\mathfrak{G}$ with negative definite Killing form $\{\langle Y \mid Y \rangle < 0; Y \in \mathfrak{K}\}$. The orthogonal complement of such $\mathfrak{K}$, $\mathfrak{K}^{\perp} = \{Y : \langle X \mid Y \rangle = 0;\ X \in \mathfrak{K}\}$, must be a null-space of the Killing form $\{\langle Y \mid Y \rangle = 0;\ Y \in \mathfrak{K}^{\perp}\}$, otherwise, subalgebra $\mathfrak{K}$ could be extended by adding $Y_0 \in \mathfrak{K}^{\perp}$ to it. But degeneracy of the Killing form on $\mathfrak{K}^{\perp}$ contradicts once again semi-simplicity of $\mathfrak{G}$. So $\mathfrak{K}^{\perp} = 0$, and the complex hull of $\mathfrak{K}$, $\mathfrak{K} \oplus i\mathfrak{K} = \mathfrak{G}$, QED.

Finally, the proof of Theorem 1 (complete reducibility for representations, T_X and ad_X, of semisimple Lie algebras) amounts by the Weyl unitary trick to complete reducibility of representations of compact Lie groups, established in chapter 3 (§3.1).

Problems and Exercises:

1. Find the derived series for the algebra of all upper triangular matrices.

2. Show that the ideals $\{\mathfrak{B}_k; \mathfrak{B}^k\}$ of the upper/lower derived series are characteristic, i.e. invariant under derivations,
$$Der(\mathfrak{B}) = \{A:\mathfrak{B}\to\mathfrak{B}; A[X;Y] = [A(X);Y]+[X;A(Y)]; X;Y\}.$$

3. Find the algebra of derivations $Der(\mathfrak{B})$ for the Heisenberg Lie algebra
$$\mathfrak{B} = \left\{\begin{bmatrix} 0 & a & c \\ & 0 & b \\ & & 0 \end{bmatrix} : \ all\ a,b,c\right\}$$

4. Verify the invariance relations (1.2) for the Killing form.

5. Show that $K = \{$fixed elements of $\sigma\}$ is the maximal compact subgroup of G.

6. Check that classical algebras $\mathfrak{G} = SL_n;\ SO(n);\ Sp(n)$ are simple, i.e. the adjoint action $\{ad_X: X \in \mathfrak{G}\}$ is irreducible; the Killing form on all of them is non-degenerate.

7. Show that the real orthogonal group $\boxed{SO(4;\mathbb{R}) \simeq SU(2)\times SU(2)/\mathbb{Z}_2}$.

   i) Use a representation of 3-sphere by unit quaternions, §4.1 (chapter 4),
   $$S^3 \simeq \mathbb{Q}_1^* = \{\xi = \alpha+\beta\mathbf{k}: |\xi|^2 = |\alpha|^2+|\beta|^2 = 1\} \simeq SU(2);$$

   ii) define the action of group $G = SU(2)\times SU(2)/\{\pm I\}$ on S^3 via quaternionic multiplication,
   $$(\xi;\eta): z \to \xi^{-1}z\eta;\ \ \xi,\eta,z \in \mathbb{Q}_1^*$$

   iii) verify that the stabilizer of point $\{1\} \in \mathbb{Q}_1^*$, $H \simeq SU(2)/\{\pm I\} = SO(3)$;

   iv) use a characterization of $SO(4)$, as group acting transitively on S^3 with a fixed point stabilizer $G_x \simeq SO(3)$.

§5.2. Cartan subalgebra. Root system. Weyl group.

In this section we shall introduce and study three most important structural elements of semisimple Lie algebras: *Cartan subalgebra, root system* (roots and root vectors), and the *Weyl group*. Most of the results will be illustrated with examples, *sl(n)* and *so(n)*. The structure theory presented here will lay ground for the subsequent study of representations of semisimple algebras.

2.1. Any semisimple Lie algebra will be shown to possess a maximal abelian subalgebra $\mathfrak{H}$, called *Cartan subalgebra* of $\mathfrak{G}$, such that the family of adjoint operators $\{ad_H : H \in \mathfrak{H}\}$ are simultaneously diagonalized

$$ad_H = \begin{bmatrix} 0 & & \\ & \alpha_1(H) & \\ & & \alpha_2(H) \end{bmatrix} \tag{2.1}$$

The 0-diagonal block here corresponds to $\mathfrak{H} =$ "null-space of $\{ad_H : H \in \mathfrak{H}\}$". All other (nonzeros) diagonal blocks are one-dimensional, the corresponding entries $\alpha_1(H); \alpha_2(H); ...$, (linear functionals on $\mathfrak{H}$) are called *roots* of $\mathfrak{H}$, and the eigenvectors $\{X_\alpha\}$ - *root vectors,*

$$ad_H(X_\alpha) = \alpha(H) X_\alpha;\ H \in \mathfrak{H}.$$

To find a Cartan subalgebra in $\mathfrak{G}$ one picks a regular element $H \in \mathfrak{G}$, whose adjoint map ad_H has the maximal possible *rank* $= dim\, ad_H(\mathfrak{G})$. The commutator of such H is the Cartan subalgebra.

The *root system* $\Sigma = \{\alpha\}$ can always be *ordered,* so for any pair α,β we have either $\alpha < \beta$, or $\alpha > \beta$, furthermore Σ is divided into two opposite halves: *positive roots* Σ_+ and *negative roots* Σ_- (the splitting depends, of course, on a particular choice of ordering in Σ). Having chosen some ordering we call a positive root α *simple*, if α can not be decomposed into the sum of other positive roots $\beta + \gamma$. Simple roots are linearly independent and form a basis of $\mathfrak{H}$, $dim \mathfrak{H} = \#\{\text{simple roots}\}$ is called the *rank* of Lie algebra $\mathfrak{G}$. We shall not provide the general argument (see [Ser]; [Jac]; [Hel]), but rather illustrate all concepts with the example of Lie algebra $\mathfrak{sl}(n)$:

1) *Cartan subalgebra* $\mathfrak{H}$ consists of diagonal matrices: $h = diag(...h_j...)$; $trh = \sum h_j = 0$.

2) *Root system* Σ is labeled by pairs of indices: jk $(j \neq k)$, where $\alpha_{jk}(h) = h_j - h_k$.

3) *Positive roots:* $\{\alpha_{jk}\}$ correspond to $j<k$, *negative* to $j>k$.

This comes from the natural lexicographical ordering of Cartan elements (real diagonal matrices): $H=(h_1;h_2;...)>H'=(h'_1;...)$, *if the first* j, *where* $h_j \neq h'_j$, *has* $h_j > h'_j$. It follows that two roots $\alpha_{jk}>\alpha_{j'k'}$ iff $j\le j'$, and (in case $j=j'$) $k>k'$.

4) *Simple roots* are of the form $\{\alpha_{jj+1}=\alpha_j\}$.

Any other positive roots are sums of simple roots: $\alpha_{ik}=\sum_i^{k-1}\alpha_j$; root α_{1n} is the *highest*, while α_{n1} - the *lowest (negative!)*.

5) *Root vectors* $\{X_\alpha = X_{jk}\}$ are Kronecker $\delta-$matrices with 1 at the jk^{th} place and 0 at the rest. We shall denote by $\{X_{jk}\}$ $(j<k)$ *positive root vectors*, and by $\{Y_{jk}=X_{kj}\}$ $(j<k)$ *negative* root vectors, so jk will always mean pair $j<k$.

7) *The commutation relations:*

$$[h;X_\alpha]=\alpha(h)X_\alpha;\ \ [h;Y_\alpha]=-\alpha(h)Y_\alpha;\ [X_\alpha;Y_\alpha]\in\mathfrak{H}; \tag{2.2}$$

hold for any $h\in\mathfrak{H}$; and root $\alpha=jk$.

Returning to the general situation, Cartan algebra $\mathfrak{H}\subset\mathfrak{G}$ has a natural inner product, the *Killing form* of $\mathfrak{G}$:

$$\langle h\mid h'\rangle = tr(ad_h ad_{h'}) = \sum_{\alpha\in\Sigma}\alpha(h)\alpha(h') \ - \text{ sum over all roots,} \tag{2.3}$$

in terms of its *root-system*. Product (2.3) allows us to identify each root α (a linear functional on $\mathfrak{H}$) with an element $H_\alpha\in\mathfrak{H}$, turning Σ into a system of vectors $\{H_\alpha\}$ in $\mathfrak{H}$.

Let us compute roots $\{H_{ij}\}$ for Lie algebra sl_n. Take a set of basic diagonal matrices $\{E_{ij}=diag(...\ 1;\ ...\ -1;...)\}$, with 1 on the i-th, (-1) on the j-th place, and the rest zeros, and compute the product (problem 2)

$$\langle E_{ij}\mid h\rangle = 2n(h_i-h_j). \tag{2.4}$$

So roots H_{ij} are diagonal matrices E_{ij} divided by 2 (we ignore the unessential factor n in (2.4). In special cases $\mathfrak{sl}(2)$ and $\mathfrak{sl}(3)$, the positive root system is made of matrices,

$\mathfrak{sl}(2)$: $H=diag(\frac12;-\frac12)$;

$\mathfrak{sl}(3)$: $H_1=diag(\frac12;-\frac12;0)$; $H_2=(0;\frac12;-\frac12)$; $H_{12}=(\frac12;0;-\frac12)=H_1+H_2$;
here $\{H_1;H_2\}$ form a basis of simple roots of $\mathfrak{sl}(3)$.

Root system $\{H_\alpha\}$ along with all (positive/negative) root vectors $\{X_\alpha;Y_\alpha\}$ forms the so called *Cartan basis* of Lie algebra $\mathfrak{G}$. One can easily verify the following commutation relations for the Cartan basis of $\mathfrak{sl}(3)$:

$$[H_1; X_{12}] = X_{12}\ ;\ [H_1; Y_{12}] = -Y_{12};\ [X_{12}\ ; Y_{12}] = 2\,H_1; \tag{2.5}$$

which generalize the commutation formulae (2.10) *for* $\mathfrak{sl}(2)$ in chapter 4.

Similarly relations are verified for $\mathfrak{sl}(n)$,

$$[H_{jk}; X_{jk}] = X_{jk};\ [H_{jk}; Y_{jk}] = -Y_{jk};\ [X_{jk}; Y_{jk}] = 2H_{jk};\ \text{ for all pairs } jk.$$

Let us now compute the (Killing) inner product on the Cartan subalgebra $\mathfrak{H} \subset \mathfrak{sl}(3)$,

$$\langle h \mid H_1\rangle = \tfrac{3}{2}(h_1 - h_2);\ \langle h \mid H_2\rangle = \tfrac{3}{2}(h_2 - h_3);\ \langle h \mid H_{12}\rangle = \tfrac{3}{2}(h_1 - h_3), \tag{2.6}$$

for $h = (h_1; h_2; h_3)$.

Introducing new coordinates on $\mathfrak{H}$: $t_1 = \frac{1}{2}(h_1 - h_2)$, $t_2 = \frac{1}{2}(h_2 - h_3)$, i.e. writing $h = t_1H_1 + t_2H_2$, we find

$$\| H_j \| = \tfrac{3}{2};\ \langle H_1 \mid H_2\rangle = -\tfrac{3}{4};\ \langle H_1 \mid H_{12}\rangle = \tfrac{3}{4};\ \langle h \mid H_1\rangle = 3t_1;\ \langle h \mid H_{12}\rangle = 3(t_1 + t_2);\ \text{etc.}$$

The inner-product structure on the real vector-space $\mathfrak{H}$ allows one to associate to any root $h_\alpha \in \Sigma$ a reflection $s_\alpha: \mathfrak{H} \to \mathfrak{H}$, w.r.t. to the plane orthogonal to h_α,

$$s_\alpha : h \to h - \frac{\langle h \mid h_\alpha\rangle}{\langle h_\alpha \mid h_\alpha\rangle} h_\alpha; \tag{2.7}$$

It turns out that all reflections $\{s_\alpha\}$ map the root system Σ into itself. Thus the family $\{s_\alpha\}$ generates a finite group $\mathbb{W}$ of isometries of $\mathfrak{H}$ called the *Weyl group* of $\mathfrak{G}$.

For algebras $\mathfrak{sl}(n)$ reflections $\{s_\alpha : \alpha = jk\}$, can be explicitly calculated by (2.6) – (2.7) (problems 1,2). One can show that s_{ij} transposes the ith and j^{th} entry of h. Therefore the Weyl group of $\mathfrak{sl}(n)$ coincides with the permutation group of n elements $\mathbb{W} = \mathbb{W}_n$.

2.2. Root systems for simple and semisimple Lie algebras. An alternative way to introduce simple and semisimple Lie algebras is to start with a *root system* $\Sigma = \{\alpha\}$ in space $\mathfrak{H} = \mathbb{R}^m$, equipped with an inner product $\langle\,|\,\rangle$, a finite system of vectors, invariant under all reflections:

$$s_\alpha : \beta \to \beta - 2\frac{\langle \beta \mid \alpha\rangle}{\langle \alpha \mid \alpha\rangle}\alpha. \tag{2.8}$$

Symmetries (2.8) generate the Weyl group of root-system: $\mathbb{W} = \mathbb{W}(\Sigma)$. Under some additional (minor) constraints such systems could be completely classified and give rise to root systems of simple and semisimple Lie algebras. The former are called *indecomposible root-systems* (i.e. Σ can not be broken into orthogonal pieces: $\Sigma' \cup \Sigma''$ with $\langle \alpha \mid \beta\rangle = 0$, for all $\alpha \in \Sigma'$; $\beta \in \Sigma'$), while the latter (semisimple) are made of orthogonal *simple root systems*, $\Sigma_1 \bigcup \Sigma_2 \bigcup ... \bigcup \Sigma_p$, according to a decomposition of $\mathfrak{G}$ into the direct sum of simple components $\bigoplus_1^p \mathfrak{G}_k$. Geometrically, all root systems of low ranks: $r = 1;2;3$ are sketched below (fig.2;3). For higher ranks we adopt a description of [Ser2] (chapter 5), in terms of an orthonormal basis $\{e_1;...e_n\} \subset \mathbb{R}^n$, and the lattice $\Lambda = \Lambda_n$ spanned by $\{e_j\}$.

Series A_n ($n \geq 1$): We take $\mathfrak{H}$ to be a hyperplane in $\mathbb{R}^{n+1}$ orthogonal to $\sum_1^{n+1} e_j$. Then Σ consists of all vectors in $\mathfrak{H} \bigcap \Lambda_{n+1}$ of norm 2, i.e. all $\{e_j - e_k\colon j \neq k\}$, the basis could be chosen as $\{e_j - e_{j+1}\colon 1 \leq j \leq n\}$, and the Weyl group $\mathbb{W} = \mathbb{W}_{n+1}$ is made of all permutations of $\{1;2;...;n+1\}$.

Series B_n ($n \geq 1$): In space $\mathfrak{H} = \mathbb{R}^n$ we consider all lattice points of norm 1 and $\sqrt{2}$; $\Sigma = \{\alpha \in \Lambda_n\colon \langle \alpha \mid \alpha \rangle = 1, \text{ or } 2\}$. Clearly, Σ is made of $\{\pm e_i \pm e_j\colon i \neq j\}$ and $\{\pm e_j\}$ the basis of Σ: $\{e_1 - e_2;\ e_2 - e_3;\ ...;\ e_{n-1} - e_n;\ e_n\}$; and the Weyl group $\mathbb{W}$ is generated by all permutations of $\{1;2;...n\}$, and all sign changes (multiplications): $e_j \to \pm e_j$. So $\mathbb{W} = (\underbrace{\mathbb{Z}_2 \times ... \times \mathbb{Z}_2}_{n-times}) \triangleright \mathbb{W}_n$ - semidirect product.

For $n = 1$ algebras A_1 and B_1 are isomorphic, i.e. $\mathfrak{sl}(2) \simeq \mathfrak{so}(3)$ (chapter 4).

Series C_n ($n \geq 1$): The root system of C_n-type is *dual* to the B_n-root system, i.e. $\Sigma(C) = \{\alpha^* = 2\frac{\alpha}{\langle \alpha \mid \alpha \rangle};\ \alpha \in \Sigma(B)\}$. So $\Sigma(C)$ consists of $\{\pm e_i \pm e_j\colon i \neq j\}$ and $\{\pm 2e_j\}$, has a basis $\{e_1 - e_2;\ e_2 - e_3;\ ...;\ e_{n-1} - e_n;\ 2e_n\}$; and the same Weyl group as B_n.

Series D_n ($n \geq 2$): The D_n-root system consists of all $\alpha \in \Lambda_n$ of $\langle \alpha \mid \alpha \rangle = 2$, i.e. all $\{\pm e_i \pm e_j\colon i \neq j\}$ with the basis $\{e_1 - e_2;\ e_2 - e_3;\ ...;\ e_{n-1} - e_n;\ e_{n-1} + e_n\}$. The Weyl group is made of all permutations and all sign changes: $e_j \to \pm e_j$; with even number of $\{-\}$, the latter group being isomorphic to $Z_n \simeq (\mathbb{Z}_2)^{n-1} \subset (\mathbb{Z}_2)^n$. So Weyl group $\mathbb{W} = \mathrm{Z}_n \triangleright \mathbb{W}_n$.

Algebra G_2: This root-system was sketched in fig.2 (iv). It could be described as the set of algebraic integers $\{z = a + b\omega + c\omega^2\colon a, b, c \in \mathbb{Z}\}$ of the cyclotomic field $\mathbb{Q}(\omega) \subset \mathbb{C}$, generated by the cubic root of identity $\omega = e^{i2\pi/3}$, of norm $|z| = 1$, or 3.

For other exceptional algebras we refer to [Ser]. In low ranks there are many overlaps between 4 series:

$\boxed{n = 1}$ $C_1 \simeq A_1 \simeq B_1$: $\mathfrak{sp}(1) \simeq \mathfrak{sl}(2) \simeq \mathfrak{so}(3)$ (chapters 1,4);

$\boxed{n = 2}$ $C_2 \simeq B_2$: $\mathfrak{sp}(2) \simeq \mathfrak{so}(5)$; and $D_2 \simeq A_1 \oplus A_1$: $\mathfrak{so}(4) \simeq \mathfrak{sl}(2) \oplus \mathfrak{sl}(2)$ (problem 5)

$\boxed{n = 3}$ $D_3 \simeq A_3$: $\mathfrak{so}(6) \simeq \mathfrak{sl}(4)$ (problem 6).

Given a root system $\Sigma = \{\alpha\}$ in $\mathbb{R}^n = \mathfrak{H}$ with a chosen basis $\{\alpha_1;...\alpha_n\}$ one can construct the corresponding semisimple Lie algebra. We denote by $\Sigma_\pm$ the set of positive/negative roots, and set

$$\mathfrak{G} = \mathfrak{H} \oplus_{\alpha > 0} (\mathfrak{G}_\alpha \oplus \mathfrak{G}_{-\alpha})$$

with 1-dimensional subspaces: $\mathfrak{G}_\alpha = \mathrm{span}\{X_\alpha\}$; $\mathfrak{G}_{-\alpha} = \mathrm{span}\{Y_\alpha = X_{-\alpha}\}$. One has the following commutation relations,

(i) $[h;X_\alpha] = \alpha(h)X_\alpha$; where $\alpha(h) = \langle \alpha^* \mid h\rangle = 2\frac{\langle \alpha \mid h\rangle}{\langle \alpha \mid \alpha\rangle}$;
and $\alpha^* = 2\frac{\alpha}{\langle \alpha \mid \alpha\rangle} = H_\alpha$ denotes the dual vector to α.

The root-vectors could be normalized so that

(ii) $[X_\alpha;Y_\alpha] = H_\alpha$; $[X_\alpha;X_\beta] = \begin{cases} N_{\alpha\beta}X_{\alpha+\beta}; & \text{if } \alpha+\beta \in \Sigma \\ 0; & \text{otherwise} \end{cases}$

The coefficients $\{N_{\alpha\beta}\}$ depend on the choice of basis in Σ.

The construction can also be implemented in terms of generators of $\mathfrak{G}$, i.e. a chosen basis $\{\alpha_1;...\alpha_n\}$ in $\mathfrak{H}$, respectively the dual basis $\{H_1;...H_n\}$, and the corresponding root vectors $\{X_1;...X_n\}$ (positive), and $\{Y_1;...Y_n\}$ (negative). The generators satisfy *Weyl commutation relations:*

$$\begin{aligned} &[H_i;H_j] = 0; \\ &[X_i;Y_i] = H_i;\ [X_i;Y_j] = 0, \quad \text{for } i \neq j; \\ &[H_i;X_j] = n(i;j)X_j;\ [H_i;Y_j] = -n(i;j)Y_j; \end{aligned} \tag{2.9}$$

furthermore,

$$ad_{X_i}^{-n(i;j)+1}(X_j) = 0;\ ad_{Y_i}^{-n(i;j)+1}(Y_j) = 0; \tag{2.10}$$

Here numbers $\{n_{ij} = n(\alpha_i;\alpha_j)\}$ denote pairings of basic roots $\{\alpha_1;...\alpha_n\}$,

$$n(\alpha;\beta) = 2\frac{\langle \alpha \mid \beta\rangle}{\langle \alpha \mid \alpha\rangle}.$$

The set of numbers $\{n(i;j)\}$ forms the *Cartan matrix* of Σ. Numbers $n(\alpha;\beta)$ are known to take on integer values $\{0;\pm 1;\pm 2;\pm 3\}$. They have a simple geometric interpretation,

$$n(\beta;\alpha) = 2\frac{|\beta|}{|\alpha|}\cos\theta;$$

hence

$$n(\alpha;\beta)n(\beta;\alpha) = 4\cos^2\theta;$$

where θ denotes the angle between roots α and β. In particular, angle θ could take on only 4 possible sets of values: $\theta = \{\frac{\pi}{2}\};\{\frac{\pi}{3};\frac{2\pi}{3}\};\{\frac{\pi}{4};\frac{3\pi}{4}\};\{\frac{\pi}{6};\frac{5\pi}{6}\}$, accordingly we get 7 possible configurations of (non-colinear) pairs $\{\alpha;\beta\}$ (see fig.1 and the table).

Table 1.

1) $n(\alpha;\beta)=0$	$n(\beta;\alpha)=0$	$\theta=\frac{\pi}{2}$	orthogonal roots
2) $n(\alpha;\beta)=1$	$n(\beta;\alpha)=1$	$\theta=\frac{\pi}{3}$	$\lvert\beta\rvert=\lvert\alpha\rvert$
3) $n(\alpha;\beta)=-1$	$n(\beta;\alpha)=-1$	$\theta=\frac{2\pi}{3}$	$\lvert\beta\rvert=\lvert\alpha\rvert$
4) $n(\alpha;\beta)=1$	$n(\beta;\alpha)=2$	$\theta=\frac{\pi}{4}$	$\lvert\beta\rvert=\sqrt{2}\lvert\alpha\rvert$
5) $n(\alpha;\beta)=-1$	$n(\beta;\alpha)=-2$	$\theta=\frac{3\pi}{4}$	$\lvert\beta\rvert=\sqrt{2}\lvert\alpha\rvert$
6) $n(\alpha;\beta)=1$	$n(\beta;\alpha)=3$	$\theta=\frac{\pi}{6}$	$\lvert\beta\rvert=\sqrt{3}\lvert\alpha\rvert$
7) $n(\alpha;\beta)=-1$	$n(\beta;\alpha)=-3$	$\theta=\frac{5\pi}{6}$	$\lvert\beta\rvert=\sqrt{3}\lvert\alpha\rvert$

Let us remark that basis $\{\alpha_1;...\alpha_n\}$ in any root system Σ is chosen in such a way, that the Cartan numbers $\{n(i;j)\}$ are negative, so all angles θ_{ij} between pairs $\{\alpha_i;\alpha_j\}$ are $>\frac{\pi}{2}$. This also explains the choice of negative exponentials $-n(i;j)$ in (2.10). Finally, any set of Weyl generators (2.9)-(2.10) yields a semisimple Lie algebra with root system Σ.

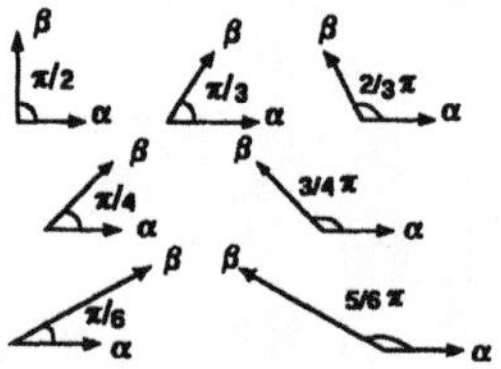

Fig.1: *illustrates all possible angles between pairs of root vectors $\{\alpha;\beta\}$ and their relative length.*

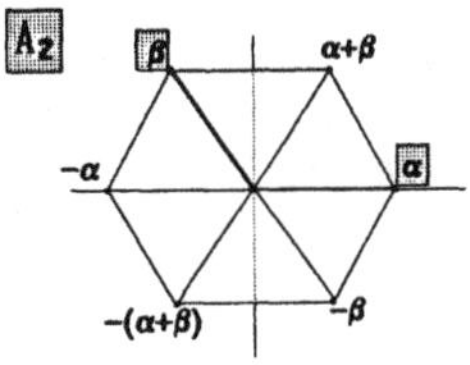

Fig. 2: *shows all root systems in rank 2. There are 4 different cases:*

(i) A_2*: algebra* $\mathfrak{sl}(3)$ *has 6 roots in hexagonal arrangement; roots* α, β, $\alpha+\beta$ *correspond to diagonal triples:* $(1;-1;0)$; $(0;1;-1)$ *and* $(1;0;-1)$;

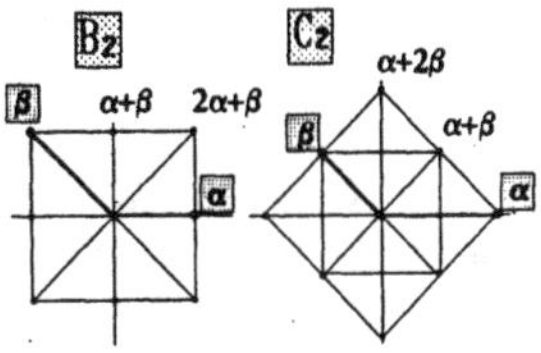

(ii) B_2 *and* C_2*: algebras* $\mathfrak{so}(5)$ and $\mathfrak{sp}(2)$ *are isomorphic, as evidenced from their root diagrams; those become identical when roots* α *and* β *are interchanged.*

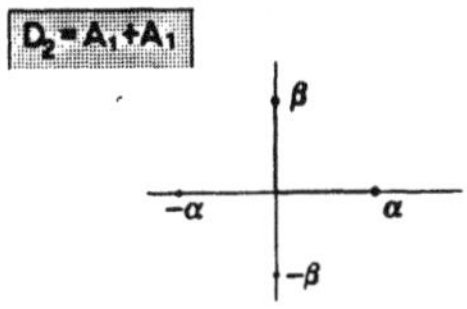

(iii) D_2*: algebra* $\mathfrak{so}(4)$ *is not simple, but breaks into the direct sum of 2 orthogonal* A_1*- diagrams. We have already mentioned that*

$$\mathfrak{so}(4) \simeq \mathfrak{su}(2) \oplus \mathfrak{su}(2),$$

hence complex $\mathfrak{so}(4) \simeq \mathfrak{sl}(2) \oplus \mathfrak{sl}(2)$.

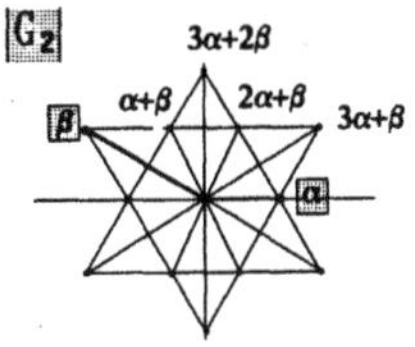

(iv) Exceptional Lie algebra G_2 *has 12 roots arranged in a king-David star.*

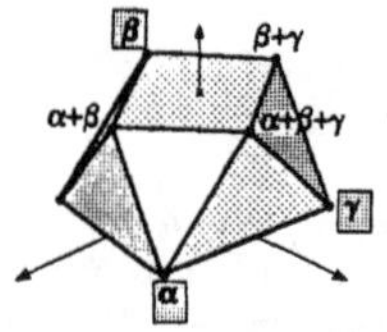

Fig.3(a): *Root system $A_3 = D_3$ ($sl_4 \simeq so(6)$) consists of 12 vertices of cubo-octahedron. The figure shows 6 positive roots, spanned by a triple of simple roots: $\alpha; \beta; \gamma$. All roots have equal length.*

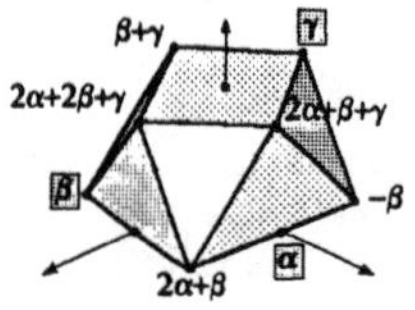

Fig.3 (b): *Root system B_3 (algebra $sp(3)$) consists of 12 vertices and 6 centers of square faces. We have shown 9 positive roots, spanned by simple roots: $\alpha;\beta;\gamma$.*

The C_3 - root system is made of the same 12 points, but the "long roots" (vertices) and short roots (centers) interchange. So α becomes long, while β,γ short.

Dynkin diagrams. Root systems Σ are conveniently labeled by certain graphs, called Dynkin diagrams. The vertices of diagrams are simple (positive) roots of a fixed Cartan basis. Two roots α, β are connected by a *single, double* or *triple* bar, if they make angles $\frac{2\pi}{3}$; $\frac{3\pi}{4}$ and $\frac{5\pi}{6}$ respectively (the orthogonal roots are disconnected). Furthermore, the lines between uneven roots are equipped with arrows from the short to the long root.

Figure 4 below sketches Dynkin diagrams of all simple Lie algebras, classical and exceptional.

Fig.4: *Dynkin diagrams in ranks 2 and 3.*

Problems and Exercises:

1. Use formula (2.6) to compute reflections $\{s_{jk}\}$ of $\mathfrak{sl}(3)$. Show that the Weyl group of $\mathfrak{sl}(3)$ is $\mathbf{W}_3$.
2. Check the product formula (2.4); calculate $\|E_{jk}\|$ and $\langle E_{jk} \mid E_{j'k'}\rangle$ for the roots $E_{jk} = \mathrm{diag}(\ldots 1;\ldots -1\ldots)$ of $\mathfrak{sl}(n)$. Find angles between roots. Show that the Weyl group of $\mathfrak{sl}(n)$ is $\mathbf{W}_n$.
3. Find the Cartan subalgebra, root system and the Weyl group of the Lie algebra $\mathfrak{so}(4)$.
4. Check that a Weyl reflection
$$\sigma_\alpha\colon h \to h - 2\frac{\langle h \mid \alpha\rangle}{\langle \alpha \mid \alpha\rangle}\alpha,$$
defined on the Cartan subalgebra $\mathfrak{H} \subset \mathfrak{G}$, extends to an automorphisms of the whole Lie algebra $\mathfrak{G}$, by $\sigma_\alpha\colon X_\beta \to X_{\sigma_\alpha(\beta)}$.
5. Show the real orthogonal group $SO(4;\mathbb{R}) \simeq SU(2) \times SU(2)/\mathbb{Z}_2$.
i) Use the representation of 3-sphere by unit quaternions, §4.1 (chapter 4),
$$S^3 \simeq \mathbb{Q}_1^* = \{\xi = \alpha + \beta\mathbf{k}\colon |\xi|^2 = |\alpha|^2 + |\beta|^2 = 1\} \simeq SU(2);$$
ii) define the action of group $G = SU(2) \times SU(2)$ on S^3 via quaternionic multiplication,
$$(\xi;\eta)\colon z \to \xi^{-1} z \eta;\ \xi, \eta, z \in \mathbb{Q}_1^*$$
iii) verify that the stabilizer of point $\{1\} \in \mathbb{Q}_1^*$, $H \simeq SU(2)/\{\pm I\} = SO(3)$;
iv) use a characterization of $SO(4)$, as group acting transitively on S^3 with a fixed point stabilizer $G_x \simeq SO(3)$.
6. Show that Lie algebras $\mathfrak{sl}(4)$ and $\mathfrak{so}(6)$ are isomorphic, by comparing their compact forms: $\mathfrak{su}(4)$ and $\mathfrak{so}(6;\mathbb{R})$, and choosing an appropriate Cartan basis for both.
7. Compute the Cartan matrix $\{n(i;j)\}$ for 4 classical series: $A_n;B_n;C_n;D_n$ and for G_2.
8. Use a description of Cartan numbers $\{n(i;j)\}$ and angles in table 1 to get all root systems in rank 2 and 3.

§5.3. Highest weight representations.

The main result in the representation theory of semisimple Lie algebras describes their structure in terms of *weights,* linear functionals on the Cartan subalgebra $\mathfrak{H}$ of $\mathfrak{G}$. Weights are linear combinations of roots, $\beta = \sum n_\alpha \alpha$, with integral coefficients, so they belong to a lattice Γ spanned by roots. Each irreducible $\pi \in \widehat{G}$ is labeled by its *highest weight* λ. We shall give a detailed description of the weight-diagram of an irreducible representation, and its relation to the Weyl group of $\mathfrak{G}$, emphasizing the example of $SL(n)$. Then we construct some explicit realization of the *highest weight representations* π^λ of $G = SL(n)$ in function spaces over G, and its quotients. This construction underlines the fundamental role of 2 algebraic operations: symmetrization and antisymmetrization, in the representation theory of $SL(n)$. It also exhibits $\{\pi^\lambda\}$, as "induced representations" in spaces of polynomial (holomorphic) functions on G.

3.1. General construction. We shall use the previous notations: $\mathfrak{G}$ - a semisimple Lie algebra, $\mathfrak{H}$ its Cartan subalgebra, $\{H_\alpha\}$ - the root system in $\mathfrak{H}$; $\{X_\alpha;Y_\alpha\}$ - Cartan basis of root vectors, and $\mathbb{W}$ - the Weyl group. In case of $\mathfrak{G} = \mathfrak{sl}(n)$, $\mathfrak{H}$ consists of diagonal matrices, and roots and root vectors are

$$H_\alpha = H_{jk} = \mathrm{diag}(\cdots\tfrac{1}{2};\cdots-\tfrac{1}{2}\cdots);\ X_{jk} = \delta_{jk};\ Y_{jk} = \delta_{kj}\ \text{- Kronecker d, for } j<k.$$

The Weyl group $\mathbb{W} = \mathbb{W}_n$ consists of all permutations $\{\sigma\}$, acting on diagonal $h \in \mathfrak{H}$, and also on the whole Lie algebra $\mathfrak{G}$ itself by conjugations, $X \to \sigma^{-1} X \sigma$.

The following Theorem reveals the structure of irreducible representations of terms of weights $\{\beta\}$ (linear functionals on $\mathfrak{G}$), and weight-vectors.

Theorem 1: *i) Any irreducible representation T of algebra $\mathfrak{G}$ in a finite-dimensional space $\mathcal{V}$, restricted on the Cartan subalgebra breaks down into the direct sum of eigenspaces $\mathcal{V} = \bigoplus_\beta \mathcal{V}_\beta$, where*

$$T_h | \mathcal{V}_\beta = \beta(h) I = \langle H_\beta | h \rangle I.$$

The system of eigen-functional $\{\beta\}$, called weights of T, has the follo[...] properties:

ii) all $\{\beta\}$ belong to the lattice spanned by simple roots, i.e. $\beta = \sum n_j \alpha_j$, with integral coefficients $\{n_j\}$.

iii) the weight diagram $\{\beta\}$ of T is invariant under the Weyl group, acting on the Cartan algebra, furthermore multiplicities of eigenspaces, $d(\beta) = \dim \mathcal{V}_\beta = d(\beta$[...] are equal for all $\{\beta\}$ in the same Weyl-orbit.

iv) There exists a unique highest weight vector $\xi_0 = \xi(\beta_0)$: $T_h\xi_0 = \beta_0(h)\xi_0$, for all h, whose weight $\beta_0 \geq \beta$, for all β in the weight diagram. Vector ξ_0 is annihilated by all raising operators: $T_X\xi_0 = 0$ for all $X = X_\alpha$. The span of ξ_0 by all lowering operators $\{T_{Y_1}T_{Y_2}...(\xi_0)\}$ generates the entire space $\mathcal{V}$.

Its proof will consists of several steps.

Step 1: Existence of the highest-weight vector ξ_0 follows from a general theorem in the representation theory of solvable Lie algebras, so called *Lie's Theorem.* We observe that the algebra $\mathfrak{B}_+$ spanned by $\mathfrak{H}$ and all raising operators $\mathrm{Span}\{\mathfrak{H};X_\alpha\}_{\alpha>0}$ (called *Borel subalgebra* of $\mathfrak{G}$) is solvable. Indeed, $ad_h(X_\alpha) = \alpha(h)X_\alpha$ and $[X_\alpha;X_\beta] \subset X_{\alpha+\beta}$; for all pairs of roots $\alpha;\beta$. In case of $\mathfrak{sl}(n)$ the Borel subalgebra $\mathfrak{B}_+$ is made of all upper-triangular matrices !

Lie's Theorem: *Any solvable Lie algebra of operators $\mathfrak{B} \subset End(\mathcal{V})$ has an eigenvector, $X\xi_0 = \lambda(X)\xi_0$, for all X in $\mathfrak{B}$. Consequently, algebra $\mathfrak{B}$ can be brought into the upper-triangular form, i.e.all operators $\{X\}$ can be represented by the upper-triangular matrices in $\mathcal{V}$.*

The proof is inductive, either in dimension of $\mathfrak{B}$, or in its the *height,* the length of the derived series. Let v be an eigenvector of an ideal $\mathfrak{A} = [\mathfrak{B};\mathfrak{B}] = \mathfrak{B}_1$ (inductive assumption !) and μ – the corresponding eigenweight: $Yv = \mu(Y)v$, for all Y in $\mathfrak{A}$. For any X in $\mathfrak{B}$ we denote by $\xi_X = e^X(v)$, and show that ξ_X is also an eigenvector of $\mathfrak{A}$ of an eigenweight $\mu^X(Y) = \mu(exp\,ad_X(Y))$. Indeed,

$$Ye^X(v) = e^X Ad_{expX}(Y)\ v = \mu(Ad_{expX}(Y))\, e^X(v).$$

It remains to observe that the number of weights $\{\mu\}$ of a finite-D representation is always finite. Hence a weight can not be continuously deformed by the family of operators $\{X\colon \mu \to \mu^X\}$ and we get $\mu^X = \mu$, for all X. The latter means that the eigenspace $\mathcal{V}_\mu = \{v\colon Yv = \mu(Y)v\}$ is invariant under the entire Lie algebra $\mathfrak{B}$.

Now we can restrict operators $\{X \mid \mathcal{V}_\mu\}$, and reduce the problem to a two – step solvable factor – algebra $\mathfrak{B}/ker\mu = \mathfrak{B}_1$. Algebra $\mathfrak{B}_1$ has at most a one-D commutator,

$$[\mathfrak{B}_1;\mathfrak{B}_1] = Span\{Y_0\};\ [X;Y_0] = \alpha(X)Y_0,$$

and the operator Y_0 is scalar: $Y_0 = \mu_0 I = \mu(Y_0)I$. But

$$\mathrm{tr}Y_0 = \mu_0 \dim\mathcal{V} = \frac{1}{\alpha(X)}\,tr[X;\ Y_0] = 0.$$

Therefore, $Y_0 = 0$ and we get the commutative algebra of operators $\mathfrak{B}_1/[...;...] \simeq \mathfrak{B}/[..;..]$ acting on the eigenspace $\mathcal{V}_\mu$. So weight μ must be $= 0$, and the problem is reduced to a commutative family of operators, acting on $\mathcal{V}$, which is well known to have a joint eigenvector, QED.

Existence of the highest-weight vector readily follows now (for other applications of Lie's Theorem see problem 1).

Step 2: Weight diagram. Other statements of Theorem 1 will be demonstrated in the case of algebra $sl(n)$. By the Weyl "unitary trick" the latter is equivalent to the study of representations of $SU(n)$, a *compact form of* $SL(n)$.

We denote by $D \simeq \mathbf{T}^{n-1}$ the diagonal (Cartan) subgroup of $G = SU(n)$. All operators T_h ($h \in D$) can be diagonalized,

$$T_h = \begin{bmatrix} \ddots & & \\ & e^{im\cdot\theta} I_m & \\ & & \ddots \end{bmatrix}; \text{ where } h = diag(e^{i\theta_1}; e^{i\theta_2};...);\ \theta = (\theta_1;\theta_2;...);\quad \theta_j = 0;$$

and weight $m = (m_1; m_2;...)$ represents a tuple of integers. Obviously, any weight $m = (m_1;m_2;...)$ is uniquely determined, modulo a diagonal shift:

$$(m_1;m_2;...) \to (m_1+m_0;\ m_2+m_0;...).$$

Moreover, any weight $m^\sigma = (m_{\sigma(1)};\ m_{\sigma(2)};\ ...)$, with σ in the Weyl group of $sl(n)$, belongs to the weight diagram (spectrum) of representation $h \to T_h$ of D and multiplicity $d(m) = d(m^\sigma)$, for all m and σ. Indeed, $\mathbf{W}$ acts by automorphisms of $SU(n)$ and $SL(n)$, σ: $X \to \sigma^{-1} X\ \sigma$, that leave Cartan subgroup D invariant. So the $\mathbf{W}$-orbit of any m in the weight diagram contains $m^\sigma = (m_1;\ m_2;\ ...)$ with $m_1 \geq m_2 \geq ...$ We shall use this ordering of weights, and choose the highest weights m. The corresponding vector ξ_0 is annihilated by all raising operators: $T_{X_{jk}} \xi_0 = 0$. Indeed, for any weight β and root α, operator $T_{X_\alpha}: \mathcal{V}_\beta \to \mathcal{V}_{\beta+\alpha}$, i.e. X_α raises weights, while $T_{Y_\alpha}: \mathcal{V}_\beta \to \mathcal{V}_{\beta-\alpha}$, lowers them. I claim that $\mathcal{V}_0 = \text{span}\{v = product(T_{Y_\alpha}^{n_\alpha})(\xi_0)$: $\alpha = jk\}$, is an invariant subspace of T. Indeed, all v's are the eigenvectors of D, so the Cartan subalgebra takes $\mathcal{V}_0$ into itself. It remains to show that all operators $\{T_X\}$ take $\mathcal{V}_0$ into itself. Given any product of lowering operators $Y_1 Y_2 ... Y_N$ ($Y_j = Y_\beta$), and a raising $X = X_\alpha$, we use the commutation relation

$$XY_1Y_2 ...Y_N = Y_1 ...Y_N X + \mathrm{Sum}_{1 \leq j \leq N} Y_1 ...[X;Y_j] ...Y_N.$$

Applying both sides to ξ_0, the first term in the RHS vanishes ($X\xi_0 = 0!$). To analyze the remaining terms we observe that $[X_\alpha;Y_\beta]$ is either a multiple of $Y_{\beta-\alpha}$ (if $\alpha<\beta$), or belongs to the Cartan subalgebra h (if $\alpha = \beta$), or a multiple of $X_{\alpha-\beta}$ (if $\alpha>\beta$). In the first two cases we obviously remain in $\mathcal{V}_0$. In the later case we continue the process, i.e. write each of the remaining products

$$X_{\alpha-\beta}Y_{j+1}\dots Y_N \text{ as } Y_{j+1}\dots Y_N X_{\alpha-\beta}+\sum \dots,$$

and argue as above. The process will terminate after all terms will transform into products of Y's. This shows that $\mathcal{V}_0$ is invariant under the whole group (algebra) G, hence by irreducibility $\mathcal{V}_0 = \mathcal{V}$.

Step 3: Uniqueness of the highest weight vector is easier to prove by extending representation T to the Lie group G of the algebra $\mathfrak{G}$. We shall prove it for $G = SL(n)$, by embedding representation T into a function-space on G, i.e. realizing it as part of the regular representation.

i) Any irreducible representation T of G can be realized on a subspace of functions on G by picking a vector ξ_0 in $\mathcal{V}$, a functional η_0 in $\mathcal{V}^*$ (the dual space) and mapping $v = T_g\xi_0$ into a function (matrix entry) $f(g) = \langle T_g\xi_0 \mid \eta_0\rangle$. Obviously, W: $v = \sum a_j T_{g_j}(\xi_0) \to f = \langle T_g v \mid \eta_0\rangle$ maps $\mathcal{V}$ into a finite-D invariant subspace of the regular representation R on G. Furthermore, since $X \to T_X$ is a complex representation of the complex Lie algebra, all entries are regular (holomorphic) functions on G (G has a natural complex structure !), in fact, for an algebraic groups, like $SL(n)$, they all are polynomial functions .

ii) We choose now ξ_0 be the highest weight vector of T, η_0 be the lowest weight vector of the contragredient representation $\tilde{T}_g = T'_{g^{-1}}$ (inverse of the dual operator T'_g to T_g). So $T_{expX}(\xi_0) = \xi_0$, $T'_{expY}(\eta_0) = \eta_0$ for all raising X and lowering Y, and $T_h\,\xi_0 = \lambda(h)\xi_0$, $T'_h\eta_0 = \mu(h)\eta_0$. The corresponding matrix entry $f(g) = \langle T_g\xi_0 \mid \eta_0\rangle$ has the properties that

$$f(ygx) = f(g);\ \text{for all } x \in N^+;\ y \in N^-$$

the upper and lower triangular subgroups of G)

$$f(h) = \lambda(h) = \mu(h^{-1}),\ \text{all diagonal } h. \tag{3.1}$$

Indeed,

$$\langle T_h\xi_0 \mid \eta_0\rangle = \lambda(h)\langle\xi_0 \mid \eta_0\rangle = \langle\xi_0 \mid T'_{h^{-1}}\eta_0\rangle = \mu(h^{-1})\langle\xi_0 \mid \eta_0\rangle,$$

and $\langle T_{ygx}\xi_0 \mid \eta_0\rangle = \langle T_g\xi_0 \mid \eta_0\rangle$. It remains to see that $\langle\xi_0 \mid \eta_0\rangle \neq 0$. Here we shall use the Gauss decomposition of any (generic) matrix into the product: $g = y\cdot h\cdot x$ (lower-triangular, diagonal, and upper-triangular). Then

$$f(g) = \langle T_h\xi_0 \mid \eta_0\rangle = \lambda(h)\langle\xi_0 \mid \eta_0\rangle.$$

But $f(g)$ can not be identically 0, due to irreducibility.

iii) Obviously, uniqueness of the function $f(g)$ satisfying properties (3.1) implies uniqueness of the highest (lowest) weight vectors.

To complete the proof of Theorem 1 it remains to observe that weights $\{m\}$ of the representation T of $\mathfrak{sl}(n)$ described as tuples of integers $(m_1; m_2; ...)$ are integral multiples of roots. In other words we have to express $m\cdot\theta = \sum_1^{n-1} m_j\theta_j$ (m_n can always be taken 0!) as $\sum_1^{n-1}\nu_j(\theta_j-\theta_{j+1})$.

Corollary: *An irreducible (analytic) representation π of a semisimple Lie group G has a unique vector v_0 that satisfies: $\pi_x(\xi_0)=\xi_0$ for all $x\in N^+$, and $\pi_h(\xi_0)=\lambda(h)\xi_0$ for all $h\in D$. Such ξ_0 must be the highest weight vector and λ the corresponding highest weight.*

Indeed, the proof shows that Span$\{\pi_y(\xi_0): y\in N^-\}$ is an invariant subspace of π.

Example 2: We shall illustrate the foregoing with an example of group $SL(3)$ and two its representations: the natural (smallest) representation π^1 in $\mathbb{C}^3$ and the adjoint representation π^2 on its Lie algebra $\mathfrak{sl}(3)\simeq\mathbb{C}^8$. For π^1 the weight diagram consists of three weights:

$m=(1,0,0)\to matrix\,(m_{jk})=\begin{bmatrix}1 & 1 & \\ & 0 & \end{bmatrix}\to\beta=\frac{1}{3}(2,1);$

$m=(0,1,0)\to matrix\,(m_{jk})=\begin{bmatrix}-1 & 0 & \\ & 1 & \end{bmatrix}\to\beta=\frac{1}{3}(-1,1);$

$m=(0,0,1)\to\begin{bmatrix}0 & -1 & \\ & -1 & \end{bmatrix}\to\beta=\frac{1}{3}(-1;-2)$. All three of them have multiplicity 1.

For π^2 the weight- diagram coincides with the root-diagram, so it has 6 roots: $\alpha=H_{jk}$ $(j\neq k)$ (3 *positive*: $j<k$, and 3 negative: $j>k$), each of multiplicity 1 and the weight $\alpha=0$ of multiplicity 2.

Remark: a general formula for multiplicities $\{d(\beta;\lambda): -\lambda\le\beta\le\lambda\}$ in a weight-diagram of an irreducible representation π^λ of semisimple compact Lie group G is due to Kostant (see [CT]),

$$\boxed{d(\beta;\lambda)=\sum_{\sigma\in W}(\det\sigma)P((\lambda+\rho)^\sigma-(\beta+\rho))}$$

sum over all Weyl transformations $\{\sigma\}$. Here $P(\gamma)$ denotes a partition function, which measures the number of positive weights "less than γ" in the root lattice

$$P(\gamma)=\{\gamma'=\sum m_\alpha\alpha\le\gamma: m_\alpha\ge 0\}.$$

3.2. Explicit form of highest weight representations of $\mathfrak{sl}(n)$. We denote by D the diagonal (Cartan) subgroup of $G=SL(n)$, by $N_\pm$- the subgroups of upper/lower triangular matrices with $\{1\}$ on the main diagonal, by $B_\pm=DN_\pm$- the corresponding *Borel subgroups* (all upper/lower triangular matrices).

Any (almost any) matrix $g\in SL(n)$ can be uniquely factored into the product of lower-, diagonal, and upper- triangular matrices,

$$g = yhx,\ y \in N_-;\ x \in N_+;\ h \in D,$$

the well known *Gauss decomposition* of the Linear algebra.

Spaces $\mathfrak{F}_\lambda$. We pick the highest/lowest weight vectors ξ_0 and η_0 of representations $T=\pi^\lambda$ and $\tilde{T}$ (contragredient), and assign to each $v \in \mathcal{V} = \mathcal{V}(T)$ its matrix entry:

$$v \to \langle T_g v \mid \eta_0 \rangle = f(g).$$

Functions $\{f_v = f(g) = \langle T_g v \mid \eta_0 \rangle\}$ form a subspace $\mathfrak{F}_\lambda$ in space $\mathfrak{C}(G)$, satisfying :

i) $f(yg) = f(g)$; for all $y \in N_-$ (the lower-triangular subgroup)

ii) $f(hg) = \lambda(h) f(g)$; for all diagonal h.

Furthermore, the *highest weight function* $f_0(g) = \langle T_g \xi_0 \mid \eta \rangle$ is invariant under the right multiplication with $x \in N_+$: $f_0(gx) = f_0(g)$. By the Gauss decomposition, $g = yhx$, any $f \in \mathfrak{F}_\lambda$ is uniquely determined by its values on N_+, and the highest weight function $f_0(x) = 1$ on N_+. Thus $f_0(g) = \lambda(h)$, in terms of the Gauss decomposition $g = yhx$. We also have

iii) $\mathfrak{F}_\lambda = \mathrm{Span}\{R_g(f_0) = f_0(xg) = \lambda(h(x,g)) f_0(x^g) \colon g \in G\}$.

Properties (i-ii) show that T is embedded into the induced representation: $R = ind(\lambda \mid B_-;G)$, where $\lambda \mid B$ means an obvious extension of character λ from D to the Borel subgroup $B_- = N_- D$, $\lambda(yh) = \lambda(h)$. Property (iii) completely characterizes $\mathfrak{F}_\lambda$ in the representation space of R. So we can think of $\mathfrak{F}_\lambda$, either as a subspace of $\mathfrak{C}(G)$, satisfying (i-iii), or a subspace of $\mathfrak{C}(P_-)$. In the former case $\mathfrak{F}_\lambda$ is spanned by all right translates of $f_0 = \lambda(h)$, in the latter by all operators $R_g(f_0)$ applied to $f_0=1$ in $\mathfrak{F}(P_-)$ - realization.

Proposition 3: *Representation $T = R \mid \mathfrak{F}_\lambda$ is irreducible.*

To show irreducibility of $\mathfrak{F}_\lambda$ we apply once again the unitary trick and the 1-1 correspondence between (analytic) representations of $SL(n)$ and those of $SU(n)$. Obviously, $T = R \mid \mathfrak{F}_\lambda$ is analytic, so if $T \mid_{SU(n)}$ were reducible, each component of it would contain a highest-weight vector, by Theorem 1, which contradicts uniqueness of the function f_0.

We shall illustrate this construction with the example of $SL(2)$. The Gauss factorization:

$$\begin{bmatrix} a & b \\ c & d \end{bmatrix} = \begin{bmatrix} 1 & 0 \\ c/a & 1 \end{bmatrix} \begin{bmatrix} a & 0 \\ 0 & 1/a \end{bmatrix} \begin{bmatrix} 1 & b/a \\ 0 & 1 \end{bmatrix}$$

identifies $B_-\backslash G$ with the group

$$N_+ = \left\{ \begin{bmatrix} 1 & x \\ 0 & 1 \end{bmatrix} : x \in \mathbb{C} \right\}.$$

Space $\mathcal{F}_\lambda$ consists of all functions $f(x)$. Weight $\lambda = m$ (integer), i.e. $\lambda(h) = h^m$, for $h = diag(h; h^{-1})$, and the induced representation $R_g f(x) = \lambda(h(x;g))f(x^g)$, where the cocycle $h(x,g)$ and the action of G on $\{x\}$ can be explicitly calculated:

$$\begin{bmatrix} 1 & x \\ 0 & 1 \end{bmatrix}\begin{bmatrix} a & b \\ c & d \end{bmatrix} = \begin{bmatrix} 1 & 0 \\ \frac{c}{a+cx} & 1 \end{bmatrix}\begin{bmatrix} a+cx & \\ & \frac{1}{a+cx} \end{bmatrix}\begin{bmatrix} 1 & \frac{b+dx}{a+cx} \\ & 1 \end{bmatrix} = y(x,g)h(x,g)x^g.$$

Thus $h(x,g) = diag(a+cx; \frac{1}{a+cx})$, element g acts on x by fractional linear transformations $g: x \to x^g = \frac{dx+b}{cx+a}$; the resulting representation

$$R_g f(x) = (a+cx)^m \ f(\frac{b+dx}{a+cx}).$$

Space $\mathcal{F}_\lambda$ spanned by $f_0 = 1$ consists of all polynomial functions of degree m. Thus we get a familiar from chapter 4 realization of representations $\{\pi^m\}$ of $SL(2)$, or $SU(2)$ in polynomial functions.

Basic representations $\{\pi^k\}$. Returning to $SL(n)$ we shall first construct a special family of representations $\{\pi^{(k)}\}_{k=1}^{n-1}$ acting in function-spaces $\mathcal{F}_k$, that will provide a *basis* of all irreducible $\{\pi^\lambda\}$, in the sense that each space $\mathcal{F}_\lambda$ will be built from $\mathcal{F}_k$.

To this end it will be convenient to write weights λ of a representation T of $SL(n)$ as tuples of integers $m = (m_1; m_2; \dots m_n)$, i.e. $\lambda(h) = h_1^{m_1} h_2^{m_2} \dots$ for any $h = (h_1; h_2; \dots h_n)$ in the diagonal subgroup D. The n-tuples of integers $(m_1; m_2; \dots)$ are *lexicographical ordered,* with the highest weight (character) λ satisfying $m_1 \geq m_2 \geq \dots \geq m_n = 0$. Such n-tuples are often called *signatures* of T. We shall alternate two notations: π^λ (λ-highest weight), or π^m (m-signature) for irreducible representations of $SL(n)$. Often it will be convenient to change the natural coordinates (diagonal entries) $(h_1; h_2; \dots h_n)$ on D to another set: $\delta_1 = h_1$; $\delta_2 = h_1 h_2$; ...; $\delta_k = h_1 h_2 \dots h_k$; ... . Then $\lambda(h) = \delta_1^{r_1} \delta_2^{r_2} \dots$ with a tuple of integers: $\alpha = (r_1; r_2; \dots)$, $r_1 = m_1 - m_2$; ...; $r_k = m_k - m_{k+1}$; ... . So a third label for irreducible representations of $SL(n)$ will be $\{\pi^{(\alpha)}\}$.

All irreducible representations of $SL(n)$, consequently $SU(n)$, will be realized in certain spaces of polynomial functions on G. Notice that matrix entries $\{x_{ij}\}$ of any $g = (x_{ij})$, as well as the minors:

$$\left\{ M_{j_1 j_2 \dots j_k}^{i_1 i_2 \dots i_k}(g) = det(x_{i_p j_q})_{pq=1}^k \right\}$$

"cut-off" by rows: $i_1, i_2, \dots i_k$; and columns: $j_1, j_2, \dots j_k$), are polynomial functions on $SL(n)$

of degrees 1,2,... Let us introduce the following function spaces on G:

$\mathfrak{F}_1 = \mathrm{Span}\{x_{1j}\}_{j=1}^{n}$ - all entries of 1-st row;

$\mathfrak{F}_2 = \mathrm{Span}\{ M^{12}_{j_1 j_2}\}$ - all minors made of the 1-st and 2-nd row entries

$\mathfrak{F}_k = \mathrm{Span}\{M^{12...k}_{j_1 j_2 ... j_k} = M_{j_1 j_2 ... j_k}\}$ - all minor made of the first k rows, $\dim \mathfrak{F}_k = \binom{n}{k}$.

Spaces $\mathfrak{F}_k$ are invariant under the right multiplication with elements $g \in SL(n)$ as a consequence of the general product formula for minors of any pair of $n \times n$ matrices A and B

$$M^{i_1 i_2 ... i_k}_{j_1 j_2 ... j_k}(AB) = \sum_{p_1<p_2<...p_k} M^{i_1 i_2 ... i_k}_{p_1 p_2 ... p_k}(A)\, M^{p_1 p_2 ... p_k}_{j_1 j_2 ... j_k}(B). \tag{3.2}$$

Abbreviating notations for tuples of multi-indices $\boldsymbol{j} = (j_1<j_2<...<j_k)$, we can write (3.2), as

$$M_{\boldsymbol{j}}^{\boldsymbol{i}}(AB) = \sum_{\boldsymbol{p}} M^{\boldsymbol{i}}_{\boldsymbol{p}}(A)\, M^{\boldsymbol{p}}_{\boldsymbol{j}}(B), \tag{3.3}$$

the form similar to the standard matrix-multiplication $A \cdot B$, but with k-tuples $\{\boldsymbol{i}; \boldsymbol{j}\}$ in place of indices $\{i; j\}$, in matrix entries.

We fix a k-tuple of row indices $\boldsymbol{i} = (i_1 i_2 ... i_k)$, and consider a subspace spanned by all minors with the given $\boldsymbol{i}$, $\mathfrak{F}_{\boldsymbol{i}} = \mathrm{span}\{M^{\boldsymbol{j}}_{\boldsymbol{i}}$: all $\boldsymbol{j}\}$. It follows from (3.3), that $\mathfrak{F}_{\boldsymbol{i}}$ is invariant under right multiplication with $B \in SL(n)$. Thus we get a family of representations $\{\pi^{(k)}\}_{k=1}^{n-1}$ in spaces $\mathfrak{F}_k = \mathfrak{F}_{(1;2...k)}$, all subrepresentations of the regular/induced representation R.

Representations $\{\pi^{(k)}\}$ can be constructed from the simplest natural action $\pi^1 \simeq R \mid \mathfrak{F}_1$ of $SL(n)$ on $\mathbb{C}^n \simeq \mathfrak{F}_1$, by extending π^1 to all antisymmetric (exterior) tensor powers. Indeed, spaces $\mathfrak{F}_k$ are identified with k^{th} exterior powers of $\mathbb{C}^n$, $\wedge^k(\mathbb{C}^n)$, by mapping the natural basis:

$$e_{j_1} \wedge e_{j_2} \wedge ... \wedge e_{j_k} \to M^{12...k}_{j_1 j_2 ... j_k}(g).$$

Functions $f \in \mathfrak{F}_k$ satisfy $f(yhg) = \alpha_k(h) f(g)$, for all y, h, g, where

$$\alpha_k(h) = \delta_k = h_1 h_2 ... h_k,$$

in particular, $\mathfrak{F}_k$ can be realized as a subspace of $\mathfrak{C}(P_-)$. We need to show that representations $\{\pi^{(k)}\}$ are irreducible and find their signatures $\{\alpha_k\}$, i.e.$\mathfrak{F}_k = \mathfrak{F}_{\alpha_k}$, as defined earlier. We notice that each minor $M_{j_1 j_2 ... j_k}$ is an eigenvector of the diagonal group D acting by the right translation of weight $\lambda(h) = h_{j_1} h_{j_2} ... h_{j_k}$, the corresponding tuple $m = (0;...;1;0; ...1;...)$, with 1's in the j_1, j_2, ... j_k spots and zeros at the rest. Obviously, the highest weight λ has signature $m = (1; 1; ...1; 0; 0; ...)$, the first k entries being 1, while the rest 0. Its signature in terms of r-parameters becomes (0; ...;1;0;...) (1 at the kth spot). We shall called it α_k.

It remains to show that $M_{12\ldots k}(g)$ is a unique function in $\mathfrak{F}_k$ satisfying $f(gx) = f(g)$, for all $x \in N_+$. Notice that all highest weight λ's belong to a lattice in the vector space $\mathfrak{H}$ (Cartan subalgebra) of $\mathfrak{G} = \mathfrak{sl}(n)$, spanned by the positive roots of $\mathfrak{G}$. So one can consider formal sums of highest weights: $(\lambda,\mu) \to \lambda+\mu$. It turns out that this operation extends to representation spaces $\mathfrak{F}_\lambda$, namely the product $\mathfrak{F}_\lambda \mathfrak{F}_\mu = \mathfrak{F}_{\lambda+\mu}$. Indeed, $\mathfrak{F}_{\lambda+\mu} \supset \mathfrak{F}_\lambda \mathfrak{F}_\mu$, since all products $f = f_1(g) f_2(g)$ satisfy:

$$f(yhg) = \lambda(h)\mu(h)f(g), \text{ for all } y, h, g,$$

and $\mathfrak{F}_{\lambda+\mu}$ is spanned by $f_0 = 1 = 1 \cdot 1$ (in $\mathfrak{F}(P)$ -realization). But $f_0 = 1$ is the only highest weight function in $\mathfrak{F}_\lambda \mathfrak{F}_\mu$, as all highest weight functions are characterized by the property $f(gx) = f(g)$, or $f(x) = \text{Const}!$

Starting with the basis $\{\mathfrak{F}_k\}$ one can construct all other spaces $\mathfrak{F}_\lambda$, $\lambda = (m_1, m_2, \ldots) \leftrightarrow \alpha = (r_1, r_2, \ldots r_k)$, by taking $\mathfrak{F}_\lambda = \mathfrak{F}_1^{r_1} \mathfrak{F}_2^{r_2} \ldots \mathfrak{F}_k^{r_k}$. Thus we established the following

Theorem 4: (i) *For any pair of highest weights λ and μ the product of irreducible subspaces $\mathfrak{F}_\lambda \mathfrak{F}_\mu$ is an irreducible subspace $\mathfrak{F}_{\lambda+\mu}$ of the regular/induced representation.*

(ii) All irreducible representations $\{\pi^\lambda\}$ are generated by the representations $\{\pi^k\}_{k=1}^{n-1}$ in spaces $\mathfrak{F}_k = \text{Span}\{X_{j_1 j_2 \ldots j_k}\} \simeq \Lambda^k(\mathfrak{F}_1)$. Moreover, if $\lambda = \sum r_k \alpha_k$, then $\mathfrak{F}_\lambda$ coincides with the product

$$\mathfrak{F}_\lambda = \mathfrak{F}_1^{r_1} \ldots \mathfrak{F}_k^{r_k} \ldots$$

Remark: The reader should not confuse the multiplication of spaces $\{\mathfrak{F}_\lambda\}$ with the standard operation of tensor (Kronecker) product of representations $\{\pi^\lambda\}$. The latter are typically highly reducible. But the above construction of irreducible representations of $SL(n)$ based on "products" of antisymmetric tensor powers $\pi^k = \Lambda^k(\pi^1)$, is closely related to tensor product construction of irreducible representations, examined in the next section.

Example 5. We conclude with an example of $SL(3)$. Two basic representations of $SL(3)$ are $\pi^1: \mathbb{C}^3 \to \mathbb{C}^3$ (the natural) and $\pi^2: \Lambda^2(\mathbb{C}^3) \simeq \mathbb{C}^3 \to \Lambda^2(\mathbb{C}^3)$. Although both act in a 3-dim space they are different, in fact π^2 is contragredient of π^1. Indeed, space $\Lambda^2(\mathbb{C}^3) \simeq \mathfrak{F}_2$ has a basis made of minors $M_{12}(g)$; $M_{13}(g)$; $M_{23}(g)$. Formula (3.2) shows that $M_{jk}(ga)$ are linear combinations of $M_{pq}(g)$ with coefficients that are 2×2 minors of

$a \in SL(3)$, so the matrix of π_a^2 in the basis $\{M^1 = M_{23};\ M^2 = M_{13};\ M^3 = M_{12}\}$ is $(A_{jk}) = {}^{T}(a_{jk})^{-1}$.

Any irreducible representation of $SL(3)$ with signature $\alpha = (p, q)$ is realized in the space $\mathcal{F}_{pq}$ of polynomial functions

$$f(x; y) = \sum_{|\beta| = p;\ |\gamma| = q} c_{\beta\gamma} x^{\beta} y^{\gamma},$$

where $x = (x_1, x_2, x_3)$; $y = (y_1, y_2, y_3)$ denote the first two rows of matrix g, and $Y_1 = x_2 y_3 - x_3 y_2$; etc. are the corresponding minors. Group G acts on space $\mathcal{F}_{p,q}$ by the change of variable: $f \to f(xg; g^{-1}y)$, i.e. the left multiplication of a row vector x by matrix g and the right of a column y by g^{-1}.

We get as special cases representations $\pi^{(p,0)}$, which represent all symmetric tensor powers of π^1, $\mathcal{S}^p(\pi^1)$, and $\pi^{(0,q)} = \mathcal{S}^q(\pi^2)$, i.e. q-th symmetric tensor power of the basic antisymmetric representation π^2. Indeed, the corresponding function-spaces are

$$\mathcal{F}_{p0} = \{\text{polynomials of degree } p \text{ in } x_1, x_2, x_3\} \simeq \mathcal{S}^p(\mathbb{C}^3),$$

and

$$\mathcal{F}_{0q} = \{\text{polynomials in } y_1, y_2, y_3 \text{ of degree } q\} \simeq \mathcal{S}^q(\mathbb{C}^3),$$

but the G-action on the x- and y-spaces are different (contragredient one to the other!).

Problems and Exercises:

1. Apply Lie's Theorem to show (i) commutator $[\mathfrak{B};\mathfrak{B}]$ of a solvable Lie algebra $\mathfrak{B}$ is nilpotent; (ii) algebra $\mathfrak{B}$ is nilpotent iff all adjoint operators $\{\mathrm{ad}_X : X \in \mathfrak{B}\}$ are nilpotent, i.e. $\mathrm{ad}_X^n = 0$, for some $n = n(X)$.

2. Introduce variables $\{t_j = \theta_j - \theta_{j+1}\}$, and show that $\{\theta_1; \dots \theta_{n-1}\}$ are expressed in terms of $\{t_1; \dots t_{n-1}\}$ as

$\theta_1 = \frac{1}{n}\{(n-1)t_1 + (n-2)t_2 + \dots + t_{n-1}\}$

$\theta_2 = \frac{1}{n}\{\ -t_1 + (n-2)t_2 + \dots + t_{n-1}\}$

$\theta_{n-1} = \frac{1}{n}\{-t_1 - 2t_2 - \dots - (n-2)t_{n-2} + t_{n-1}\}$

Get as a corollary,

$$\sum_1^{n-1} m_j\theta_j = \frac{1}{n}\sum_1^{n-1} \nu_j(\theta_j - \theta_{j+1})$$

with integral coefficients $\{\nu_j\}$, given by the following relations in terms of the differences $\{m_{jk} = m_j - m_k\}_{j<k}$

$\nu_1 = \sum_{j>1} m_{1j}$

$\nu_2 = \sum_{j>2} m_{1j} + \sum_{j>2} m_{2j}$

$\nu_{n-1} = m_{1;n-1} + m_{2;n-1} + \dots + m_{n-2;n-1}.$

$$\begin{bmatrix} m_{12} & m_{13} & \dots & m_{1;n-1} \\ & m_{23} & \dots & m_{2;n-1} \\ & & & m_{n-2;n-1} \end{bmatrix}$$

3. A symmetric tensor power of the representation π^1 of $SL(3)$ acts in a 6-D space $S_2(\mathbf{C}^3) = \mathbf{C}^3 \otimes_S \mathbf{C}^3$. Show that $\pi^1 \otimes_S \pi^1$ is irreducible and calculate its weight diagram.

4. Prove (3.2).

5. Show that $\pi^{(k)} \simeq \wedge^k(\pi^{(1)})$-the kth exterior power of π^1 in the space $\wedge^k(\mathbf{C}^n)$, i.e.$\mathfrak{F}_k \simeq \wedge^k(\mathbf{C}^n)$ and $\pi^{(k)}(\xi_1 \wedge \dots \wedge \xi_k) = (\pi^1\xi_1) \wedge \dots \wedge (\pi^1\xi_k)$, for all products $\xi_1 \wedge \dots \wedge \xi_k$.

6. Show that $M_{12\dots k}$ is the only N_+-invariant minor (i.e.$M(gx) = M(g)$ for x in N_+) among all $\left\{ M_{j_1 j_2 \dots j_k} \right\}$.

7. Show that representations π^k and π^{n-k} of $SL(n)$ are contragredient. Hint: spaces $\wedge^k$ and $\wedge^{n-k}$ are dual each to the other by the pairing:

$(\tilde{\xi} = \xi_1 \wedge \xi_2 \dots \wedge \xi_k ; \tilde{\eta} = \eta_1 \wedge \dots \wedge \eta_{n-k}) \rightarrow \xi_1 \wedge \dots \wedge \eta_{n-k} = \det(\xi_1;\xi_2;\dots\xi_n) \bigwedge_1^n e_k = \langle \tilde{\xi} \mid \tilde{\eta} \rangle \bigwedge_1^n e_k$

So $\langle \pi_g^k \tilde{\xi} \mid \pi_g^{n-k} \tilde{\eta} \rangle = \langle \tilde{\xi} \mid \tilde{\eta} \rangle \det g$.

8. Let A^k denote the kth exterior power a matrix $A \in GL(n)$, i.e. entries A_{ij}^k are $k \times k$ minors of A. Find the inverse of A^k.

§5.4. Tensors and Young tableaux.

§5.4 outlines the Weyl theory of tensor powers, invariants and representations of classical compact groups. It turns out that all representations of compact groups are obtained from the simplest (natural) representation π (e.g. $su(n)$ acting in $\mathbb{C}^n$), by taking all tensor powers $\{\mathcal{T}^m(\pi): m=0;1;2...\}$ of π, and decomposing them according to the action of permutation group $\mathbf{W}_m$. The latter gives rise to various types of $\mathbf{W}_m$-tensorial symmetries (statistics), of which symmetric and antisymmetric tensors form two special (extreme) cases. All other (intermediate) symmetries are described by the so called *Young tableaux*. The theory of Young tableaux and symmetrizers yields irreducible representations of both $SL(n)$ and $\mathbf{W}_m$. In fact, two sets are naturally paired, as $SL(n)$ and $\mathbf{W}_m$ form a maximal commuting pair (Howe's "dual pair") in the corresponding tensor-power space.

4.1. Invariants. Let us start with a simple example of rank-2 tensors. Space $\mathcal{T}^2(\mathbb{C}^n) = S_2 \oplus A_2$, breaks into the direct sum of its symmetric and antisymmetric parts, both subspaces being invariant under $\pi^{(m)}$ and irreducible (problem 1). The corresponding projections:

$$\chi^s: \xi \otimes \eta \to \tfrac{1}{2}(\xi \otimes \eta + \eta \otimes \xi) \ (\textit{symmetrization}),$$

and

$$\chi^a: \xi \otimes \eta \to \tfrac{1}{2}(\xi \otimes \eta - \eta \otimes \xi) \ (\textit{anti-symmetrization}),$$

provide a canonical decomposition of $\mathcal{T}^2$. A possible approach to a general (higher-rank) problem, due to H.Weyl [We3], lies through the study of all invariants of the upper triangular group N_+ in the tensor algebra $\mathcal{T} = \bigoplus_0^\infty \mathcal{T}^m$ (with tensor multiplication $\mathcal{T}^k \otimes \mathcal{T}^m \subset \mathcal{T}^{k+m}$). Thinking of tensors as multilinear forms over $\mathbb{C}^n$, the product of two forms (k-form Ψ and j-form ϕ) becomes a $(k+j)$-form $\Psi(\xi_1;..\xi_k)\phi(\eta_1;..\eta_j)$. Once all N_+ invariants are known it remains to identify eigenvectors of the diagonal group D among them, i.e. find the highest weight tensors.

One natural set of N_+- invariants, we have encountered so far, were exterior products: $\omega_k = e_1 \wedge \ldots \wedge e_k$, where $\{e_1; \ \ldots \ e_n\}$ is the natural basis of $\mathbb{C}^n$. Written as multilinear forms on $\mathbb{C}^n$ they become

$$\omega_k(\xi;\eta; \ \ldots \ \zeta) = det \begin{bmatrix} \xi_1 & \xi_2 \cdots & \xi_k \\ \eta_1 & \eta_2 \cdots & \eta_k \\ \zeta_1 & \zeta_2 \cdots & \zeta_k \end{bmatrix}$$

$k \times k$ determinants formed by the first k entries of vectors $\xi = (\xi_1;..\xi_n)$; $\eta = (\eta_1;..\eta_n)$; ... Next we can take their tensor powers $\{\omega_k^r = \omega_k \otimes ... \otimes \omega_k\}$, i.e. kr-multilinear forms:

$$\omega_k(\xi_1;\xi_2;..\xi_k)\,\omega_k(\eta_1;..\eta_k)\ldots\omega_k(\zeta_1;..\zeta_k),$$

in vector variables $\{\xi_1..\xi_k; \eta_1..\eta_k; ..; \zeta_1..\zeta_k: \ \xi_i; \eta_j; \zeta_m \in \mathbb{C}^n\}$, which are also invariant under N_+, as well as their tensor products:

$$\omega_\alpha = \omega_k^{r_k}\omega_{k-1}^{r_{k-1}}\dots\omega_1^{r_1}, \text{ of rank } m = \sum_{1\le j\le k} j r_j. \tag{4.1}$$

Permutation group $\mathbf{W}_m$ acts naturally on the tensor-product space $\mathcal{T}^m$ by interchanging tensorial components,

$$\sigma\colon \omega = \xi_1 \otimes \dots \otimes \xi_m \to \sigma\omega = \xi_{\sigma(1)} \otimes \dots \otimes \xi_{\sigma(m)}$$

and this action clearly commutes with the representation $\pi^{(m)}$ of $SL(n)$. Therefore, all permutations of forms (4.1) $\{\omega_{\alpha s} = s\omega_\alpha\colon s \in \mathbf{W}_m\}$ are also N_+ invariants. One of the main result of the Weyl SL_n- invariant theory asserts: *the space of all N_+ invariants of $\mathcal{T}^m$ is generated by tensors $\{\omega_{\alpha s}\}$, i.e. each invariant is a linear combination of $\{\omega_{\alpha s}\}$, where α denotes a partitions of $m = \sum jr_j$, $\alpha = (r_1;\dots r_k)$, and $s \in \mathbf{W}_m$.*

Obviously, each m-tensor $\omega_k^{r_k}\omega_{k-1}^{r_{k-1}}\dots\omega_1^{r_1}$ is an eigenvector of the diagonal group D, of a highest (eigen) weight $\lambda(h) = h_1^{m_1}\dots h_k^{m_k}$; with $m_j = \sum_{i\ge j} r_i$. Thus by counting all highest weight tensors (with their multiplicities) we obtain a decompositions of $\pi^{(m)}$ into the sum of irreducible representations $\{\pi^\alpha\}$, of signature $\alpha = (m_1; m_2; \dots m_k)$, as well as multiplicities of $\{\pi^\alpha\}$ in $\pi^{(m)}$.

In special cases (tensors of low rank) this procedure yields a complete and explicit answer. However, to decompose higher rank tensor spaces $\mathcal{T}^m$ one needs to classify different types of *tensorial symmetries,* in other words to study the action of the permutation group $\mathbf{W}_m$ on $\mathcal{T}^m$.

Two examples of $\mathbf{W}_m$-symmetries were already mentioned, the *symmetric* and *antisymmetric* parts of $\mathcal{T}^m$: $\mathcal{S}^m(\mathbb{C}^n)$ and $\Lambda^m(\mathbb{C}^n)$, given by two projections, the *symmetrizer*

$$\chi^s\colon \xi_1 \otimes \dots \otimes \xi_m \to \frac{1}{m!}\sum_{\sigma \in \mathbf{W}_m} \xi_{\sigma(1)} \otimes \dots \otimes \xi_{\sigma(m)}; \tag{4.2}$$

and *anti-symmetrizer*

$$\chi^a\colon \xi_1 \otimes \dots \otimes \xi_m \to \frac{1}{m!}\sum_{\sigma \in \mathbf{W}_m} (-1)^\sigma \xi_{\sigma(1)} \otimes \dots \otimes \xi_{\sigma(m)} = \frac{1}{m!}\xi_1 \wedge \dots \wedge \xi_m. \tag{4.3}$$

Both subspaces $\mathcal{S}^m$ and Λ^m are invariant under $\pi^{(m)}$, and the resulting subrepresentations: $\pi^{(m)} \mid \mathcal{S}^m$ and $\pi^{(m)} \mid \Lambda^m$, are irreducible with signatures:

$$\alpha_{sym} = (m;0\dots), \text{ and } \alpha_{ant} = (\underbrace{1,1,\dots 1}_{m-times},0,\dots) \text{ (see problem 1).}$$

But unlike the case of rank 2, the higher rank tensor spaces $\mathcal{T}^m$ $(m \ge 3)$ are not made up of their symmetric and antisymmetric parts: $\mathcal{S}^m \oplus \Lambda^m$,

$$\dim\mathcal{S}^m + \dim\Lambda^m = n^2 < n^m = \dim T^m.$$

There exists a variety of intermediate *mixed symmetries.* We shall see that all such symmetries could be built of two basic operations of *symmetrization and anti-symmetrization* (4.2)-(4.3). The idea is to construct a family of minimal projections $\{\chi\}$ in the group algebra of the symmetric group. The construction will be based on the important notion of *Young tableaux* and the corresponding *Young symmetrizer.*

4.2. Young tableaux and symmetrizers. Given an integer m, we form a partition of m into the sum of non-increasing integers $(m_1 \geq m_2 \geq \ldots \geq m_k \geq 0)$:

$$\alpha = (m_1;...;m_k);\ m = m_1 + m_2 + \ldots + m_k\ (k \leq n-1).$$

A Young tableaux α is defined as a diagram of k rows, with m_1 entries in first row, m_2 - in the second, $\ldots m_k$ in the kth row (fig. 5).

(*a*)

1	6	9	11	13
2	7	10	12	
3	8			
4				
5				

(*b*)

2	5	8	11	1
3	6	9	12	
4	7	10		

Fig.5: *Young tableaux of signatures:* $\alpha = (5;4;2;1;1)$, $m = 13$, *in the standard allocation* (a); *and* $\alpha = (5;4;3)$, $m = 12$ *with a cyclic permutation* $s = (1;...;12)$ (b).

An alternative set of parameters to describe the αth *tableau,* are k-tuples of integers $\{r_1;r_2...;r_k\colon r_j = m_j - m_{j+1}\}$, r_k being the number of columns of largest size "k" $r_{k-1} = \#$ columns of size "$k-1$", etc.

So Young tableaux label naturally conjugacy classes in $\mathbf{W}_m$. At the same time they will serve to build irreducible characters $\{\chi_\alpha\}$ of $\mathbf{W}_m$, via *Young symmetrizers.* From the general theory of chapter 3, we already know that characters of finite groups correspond in a 1-1 way to conjugacy classes.

Each tableau α could be filled with numbers $\{1,\ldots m\}$ in $m!$ different ways, according to all permutations $\sigma \in \mathrm{W}_m$. Our convention will be to fill in the tableaux column-wise, starting from the upper left corner. This would be called *the standard allocation,* all others can be obtained from the standard one by permutations $s \in \mathrm{W}_m$ (fig. 5).

A simple example of Young tableau is given by a rectangular $k \times r$ - box $(m = kr)$ with an arbitrary distribution of integers $\{1;2;\ldots;m\}$,

$$\begin{bmatrix} 1 & k{+}1 \dots \\ 2 & k{+}2 \dots \\ k & 2k \quad m \end{bmatrix} \text{(standard)}, \begin{bmatrix} i_1 & i_{k+1} \dots \\ i_2 & \dots \dots \\ i_k & i_{2k} \ i_m \end{bmatrix} \text{(permuted)}.$$

Given a diagram α we can introduce two subgroups of $\mathbf{W}_m$: $P = \{p\}$ made of all horizontal (row) permutations, and $Q = \{q\}$ made of all vertical (column) permutations, in the standard allocation. If (α, s) denotes any other Young tableau, based on α, then the corresponding subgroups P', Q' are conjugates of P and Q: $P' = sPs^{-1}$, $Q' = sQs^{-1}$.

In examples (a)-(b) of fig.5 we get

(a) the columnar subgroup: $Q \simeq \mathbf{W}_5 \times \mathbf{W}_3 \times \mathbf{W}_2 \times \mathbf{W}_2 \times \mathbf{W}_1$, is generated by permutations of $\{1;2;3;4;5\} \simeq \mathbf{W}_5$; $\{6;7;8\} \simeq \mathbf{W}_3$; $\{9;10\}$ and $\{11;12\} \simeq \mathbf{W}_2$; and $\{13\} \simeq \mathbf{W}_1$- trivial;
the row subgroup: $P \simeq \mathbf{W}_5 \times \mathbf{W}_4 \times \mathbf{W}_2 \times \mathbf{W}_1 \times \mathbf{W}_1$, permutes subsets

$$\{1;6;9;11;13\} \bigcup \{2;7;9;12\} \bigcup \{3;8\} \bigcup \{4\} \bigcup \{5\}.$$

(b) subgroup $Q \simeq \mathbf{W}_3 \times \mathbf{W}_3 \times \mathbf{W}_3 \times \mathbf{W}_2 \times \mathbf{W}_1$ permutes $\{2;3;4\} \bigcup \{5;6;7\} \bigcup \{8;9;10\}$..
while $P \simeq \mathbf{W}_5 \times \mathbf{W}_4 \times \mathbf{W}_3$ permutes $\{2;5;8;11;1\} \bigcup$...

One can associate with any allocated Young tableau (α,s) the corresponding *Young symmetrizer* $\chi = \chi_{\alpha s}$

$$\chi = s^{-1}\Big(\sum (-1)^q qp \Big) s, \tag{4.4}$$

sum over all $q \in Q$, $p \in P$, where $(-1)^q$ denotes the *signature* of q. We call the reader's attention to the order of two operations: *columnar anti-symmetrizers* $\{(-1)^q q\}$ *of diagram* α, *followed by row - symmetrizers* $\{p\}$.

Symmetrizer χ (4.4) could be understood as an element of the group algebra $\mathbb{C}(\mathbf{W}_m)$, supported on the product of two subgroups[1]: $QP \subset \mathbf{W}_m$, or alternatively as an operator in tensor space $\mathcal{T}^m$, representing a suitable combination of tensorial permutations $p, q \in \mathbf{W}_m$.

Examples: 1) Tableaux$\begin{bmatrix} 1 & 3 \\ 2 & 4 \end{bmatrix}$ in $\mathcal{T}^4$ has two subgroups: $Q, P \simeq \mathbb{Z}_2 \times \mathbb{Z}_2$; generated by pairs: $\{q_1 = (12); q_2 = (34)\}$, and $\{p_1 = (13);\ p_2 = (24)\}$. So its symmetrizer

$$\chi_\alpha = (1 - q_1)(1 - q_2)(1 + p_1)(1 + p_2),$$

as an element of group algebra $\mathbb{C}(\mathbf{W}_4)$. The corresponding operator on tensors $\mathcal{T}^4$, abbreviated by diagrams $\begin{bmatrix} x_1 & x_2 \\ y_1 & y_2 \end{bmatrix}$, with vectors $\{x_i; y_j\}$ filling in appropriate slots of tableaux α, becomes

$$\chi_\alpha\!: \begin{bmatrix} x_1 & x_2 \\ y_1 & y_2 \end{bmatrix} \to x_1 \wedge y_1 \otimes x_2 \wedge y_2 + x_2 \wedge y_1 \otimes x_1 \wedge y_2 + x_1 \wedge y_2 \otimes x_2 \wedge y_1 + x_2 \wedge y_2 \otimes x_1 \wedge y_1.$$

[1]Notice that the product-set PQ is not the direct group-product $P \times Q$, as two subgroups do not commute in $\mathbf{W}_m$.

In other words χ is made up of two operations: anti-symmetrization of columns followed by symmetrization of rows. Later we shall see that symmetrizer χ maps $\mathfrak{T}^4$ onto an irreducible subspace of $SL(n)$ of signature $\alpha = (2,2,0\ldots)$.

2) Tableau $\begin{bmatrix} 1 & 3 & 5 \\ 2 & 4 & \end{bmatrix}$ has $P \simeq \mathbf{W}_3 \times \mathbf{Z}_2$; with elements $\{p_1 \in \mathbf{W}_3(1;3;5);\ p_2 = (24)\}$, and $Q \simeq \mathbf{Z}_2 \times \mathbf{Z}_2$ generated by $\{q_1 = (12);\ q_2 = (34)\}$. So

$$\chi_\alpha = (1-q_1)(1-q_2)(1+p_2)(\sum_{\mathbf{W}_3} p_1),$$

a combination of $8 \cdot 3! = 48$ group elements in $\mathbf{W}_5$.

In general, Young symmetrizers will provide a family of projections onto irreducible components of $\pi^{(m)}$. To state precisely the result we shall adopt the following terminology:

i) *Lexicographical ordering of Young diagrams*: $\alpha = (m_1, m_2, \ldots)$ is said to be greater than $\alpha' = (m_1'; m_2'; \ldots)$, $\alpha > \alpha'$, if the first coordinate $m_j \neq m_j'$ satisfies $m_j > m_j'$ so one has $m_1 = m_1'; \ldots m_{j-1} = m_{j-1}'$; but $m_j > m_j'$.

ii) we call elements $\{s = qp : q \in Q; p \in P\}$ *simple*, and say that $s, \sigma \in \mathbf{W}_m$ are *simply related*, $s \sim \sigma$, if $s = \sigma(qp)$, for a simple qp. We shall see (Lemma 3, corollary 5), that such relation defines an equivalence in $\mathbf{W}_m$.

Theorem 2: *i) Any symmetrizer $\chi = \chi_{\alpha\sigma}$, considered as an element of group algebra $\mathbb{C}(\mathbf{W}_m)$ has $\chi * \chi = \mu\chi$, with constant $\mu = \mu(\alpha)$, depending on α only, so $\frac{1}{\mu}\chi$ becomes an idempotent (projection) in $\mathbb{C}(\mathbf{W}_m)$.*

ii) *Two symmetrizers $\chi = \chi_{\alpha s}$ and $\chi' = \chi_{\alpha'\sigma}$ are mutually disjoint $\chi * \chi' = 0$, if $\alpha \neq \alpha'$ ($\alpha > \alpha'$), or for $\alpha = \alpha'$, if σ and s are not simply related, $\sigma \neq s(qp)$. So mutually disjoint projections $\chi;\chi'$ correspond to different diagrams α, α', or to unrelated s and σ.*

iii) If $\alpha = \alpha'$ and $\sigma;s$ are simply related, $\sigma = s(qp)$, then

$$\chi * \chi' = \mu s\Big(\sum (-1)^q qp\Big)\sigma^{-1}.$$

The proof is based on a simple combinatorial Lemma.

Lemma 3: *If (α,s) and (α',σ) are two Young diagrams with either $\alpha > \alpha'$, or $\alpha = \alpha'$, but s and σ non-simply related, then there exist two indices j,k that belong to the same row in (α,s) and the same column in (α',σ).*

Proof of the Lemma: We take indices in the first (longest !) row of (α,s) and call them $i_1, i_2, \ldots$ If the first row of α', $m_1' < m_1$, then at least two of i's must necessarily lie in the same column of (α',σ), so the Lemma's conclusion would hold. Assuming $m_1' = m_1$ and no i's belong to the same column in α', we can bring all of them by a vertical permutation q_1 of α' to the first row, and within the first row we rearrange them by horizontal p_1 in the same order as in (α,s). Thus 1st rows of (α,s) and (α',σ) become identical,i.e.$s = \sigma q_1 p_1$ (modulo remaining rows). We continue the process with the remaining rows, each time seeking the Lemma's conclusion to fail. This yields at the end $\alpha = \alpha'$ and $s = \sigma\ q_1 p_1\ q_2 p_2 \dots q_k p_k$. Since each p_j is a permutation of the jth row, it commutes with all remaining terms of the product, and the latter becomes σ $(q_1 q_2 \dots)(p_1 p_2 \dots)$. So the only way for the Lemma's conclusion to fail is when $\alpha = \alpha'$ and s being simply related to σ, QED.

Corollary 4: Given any pair $\{(\alpha,s);(\alpha',\sigma)\}$ with either $\alpha > \alpha'$, or $\alpha = \alpha'$, but s unrelated to σ, there exists a row transposition $p_0 \in P$, and a column transposition $q_0 \in Q$, so that $s p_0 s^{-1} = \sigma q_0 \sigma^{-1}$. In particular, for each non-simple $s \neq qp$, there exist a pair p_0, q_0, s.t. $s p_0 = q_0 s$.

Another result could be deduced along the lines of Lemma 3.

Corollary 5: *For any pair of elements $p \in P$, $q \in Q$ in a fixed Young tableau α, there exists another pair $p' \in P$; $q' \in Q$, so that simple element $qp = p'q'$.*

Hence, *simple relation*, $s \sim \sigma$, is indeed an equivalence, although product QP is not a subgroup of $\mathbf{W}_m$!

Proof of Theorem 2: We observe that symmetrizer χ of Young tableau (α,s) has certain invariance properties w.r. to the left multiplication with elements $q \in sQs^{-1}$ and right multiplication with $p \in sPs^{-1}$,

$$\chi(q\sigma p) = q\chi p = (-1)^q \chi(\sigma), \text{ all } q, p.$$

The same holds for $\chi^2 = \chi * \chi$, or $\chi * \chi'$,

$$\chi * \chi'(qtp) = (-1)^q \chi * \chi'(t), \text{ for all } q \in sQs^{-1};\ p \in \sigma P \sigma^{-1}. \tag{4.5}$$

Now we take a pair q_0, p_0 of the corollary and get: $\chi^2(q_0 t p_0) = -\chi^2(t)$, by (4.5), and $\chi^2(q_0 t p_0) = \chi^2(t)$, whenever t is non-simple by the corollary. Thus

$$\chi^2(t) = \begin{cases} 0; \text{ for non-simple } t; \\ \mu = \chi^2(e); \text{ for simple } t = qp \end{cases}.$$

Similarly one verifies other statements of the Theorem.

As a corollary of Theorem 2 we get a family of idempotents (projections) $\{\chi = \chi_{\alpha s}\}$ in the group algebra $\mathbb{C}(\mathbf{W}_m)$. Each of them is *minimal*, so χ projects group algebra $\mathbb{C}(\mathbf{W}_m)$ onto an irreducible subspace $\mathcal{V} = \mathcal{V}_{\alpha s} = \chi * \mathbb{C}$. Indeed, the argument of Theorem 2 shows that $\chi * f * \chi = \mu(f)\chi$, for any f in the group algebra $\mathbb{C}(\mathbf{W}_m)$. In particular, any subprojection ψ of χ satisfies $\chi*\psi*\chi = \psi = \mu\chi$, hence $\psi = \mu\chi$ (a multiple of χ). Furthermore, all irreducible representations $\{\tau = \tau^{\alpha,s} = R \mid \mathfrak{T}\}$ are equivalent, so $\tau = \tau^{\alpha}$ (depends only on signature α). The multiplicity of τ^{α} in R, $d(\alpha)$, and the norming coefficients $\mu = \mu(\alpha)$ of symmetrizer $\chi = \chi_{\alpha s}$ are related by

$$\mu(\alpha)d(\alpha) = m! = |\mathbf{W}_m|.$$

Indeed, trace of the convolution operator $\chi:\phi \to \chi*\phi$, in the group algebra $\mathbb{C}(\mathbf{W}_m)$,

$$\operatorname{tr}\chi = \chi(e)m! = m!,$$

is equal to $\mu \cdot d(\tau)$, but the degree of each irreducible τ coincides by the Frobenius reciprocity with its multiplicity $m(\tau)$ in the regular representation R,

$$d(\tau) = m(\tau) = \#\{\text{disjoint projections } \chi_{(\alpha s)} \text{ of signature } \alpha\}.$$

Minimal projections $\chi_{(\alpha s)}$ of algebra $\mathbb{C}$ give rise to central projections χ^{α} on primary components $\mathfrak{T}_{\alpha}$ of the regular representation,

$$\chi^{\alpha} = \frac{1}{\mu^2}\sum_{s \in \mathbf{W}_m} \chi_{(\alpha,s)} = \frac{1}{\mu^2}\sum_{s} s\Big(\sum (-1)^q qp\Big)s^{-1}.$$

Projections $\{\chi^{\alpha}\}$ are orthogonal, and the family $\{\chi^{\alpha}\}_{\alpha}$ is complete in the center $\mathfrak{Z}(\mathbf{W}_m)$ of the group algebra. Indeed, their number is equal to the number of partitions of m, which is the same as the number of conjugacy classes in $\mathbf{W}_m$, or equivalence classes of irreducible representations of $\mathbf{W}_m$! Thus we get at once a complete characterization and explicit construction of irreducible representations $\{\tau^{\alpha}\}$ of $\mathbf{W}_m$.

Next we shall apply Young symmetrizers to our main problem: decomposition of tensor products $\pi^{(m)}$ of $SL(n)$.

Theorem 6: *(i) Each operator $\chi = \chi_{\alpha s}$ projects $\mathfrak{T}^m(\mathbb{C}^n)$ onto an irreducible subspace $\mathfrak{T}_{\alpha s}$ of $\pi^{(m)}$, and the restriction $\pi^{(m)} \mid$ "range χ" $\simeq \pi^{\alpha}$, an irreducible representation of $SL(n)$ of signature α.*

(ii) The multiplicity of π^{α} in $\pi^{(m)}$ is equal to $d(\alpha) =$ degree of τ^{α}. Moreover, the central projection χ^{α} of W_m projects T^m onto the primary subspace $\mathfrak{T}_{\alpha}$ of π^{α}, invariant under both groups.

(iii) The joint action of $\mathrm{W}_m \times SL(n)$ on $\mathfrak{T}_m$ factors into the product of irreducible representations $\tau^{\alpha} \otimes \pi^{\alpha}$, and $\mathfrak{T}_{\alpha} \simeq \mathrm{Space}(\tau^{\alpha}) \otimes \mathrm{Space}(\pi^{\alpha})$.

In other words, if $\tau^{(m)}$ denotes the action of W_m on $\mathfrak{T}^m$ ($\tau^{(m)} = R \otimes I_n$ - the n-th multiple of the regular representation), and $\pi^{(m)}$ the action of $SL(n)$ on T^n, then the tensor-product action of $\mathrm{W}_m \times SL(n)$ is decomposed as

$$\tau^{(m)}\cdot\pi^{(m)} = R\otimes\pi^{(m)} \simeq \underset{\alpha}{\oplus}\tau^{\alpha}\otimes\pi^{\alpha},$$

sum over all partitions α: $m=\sum m_j$.

The last statement means that groups $SL(n)$ and W_m form a *dual (maximal commuting) pair* in $\mathcal{T}^m$, by analogy with pairs: $\{SO(3);SL_2\}$, or $\{SO(n);SL_2\}$ in $L^2(\mathbb{R}^n)$, studied in chapter 4. So the algebras spanned by SL_n and W_m in $\mathcal{T}^{(n)}$ are mutual commutants:

$$Com(\pi^{(m)}\mid SL(n)) = Alg\,(W_m) \text{ and } Com(W_m) = Alg\,(\pi^{(m)}).$$

Proof: i) Irreducibility of subspaces $\mathcal{T}_{\alpha s}$ is established by finding all Weyl (N_-) invariants in $\mathcal{T}_{\alpha s}$ and identifying the unique highest weight vector. Let $\chi=\chi_{\alpha s}$ be a Young symmetrizer of partition α. I claim that χ maps all but one element of the Weyl basis of invariants $\{\omega_{\alpha s} = s\ \omega_1^{r_1}\omega_2^{r_2}\dots\omega_k^{r_k}\}$ into 0,

$$\chi_{\alpha s}(\omega_{\alpha'\sigma}) = \begin{cases} 0;\ if\ \alpha\neq\alpha' \text{ or } \alpha=\alpha' \text{ but } \sigma\neq s \\ \omega_{\alpha s};\ \text{if } \alpha=\alpha' \text{ and } \sigma=s\end{cases}$$

Indeed, the basis of Weyl invariants written in the tensorial notation consists of

$$\omega_\alpha = (e_1\wedge e_2\dots\wedge e_k)^{\otimes r_1}\otimes(e_1\wedge\dots\wedge e_{k-1})^{\otimes r_2}\dots\otimes\ e_1^{\otimes r_1},$$

or their permutations $s\omega_\alpha$, $s\in W_m$. The corresponding Young tableau is filled by elements $\{e_1;\dots;e_k\}$, according to the following allocation scheme,

$\boxed{e_1e_1\dots\dots e_1}$ m_1 times

$\boxed{e_2e_2\dots\ \ e_2}$ m_2 times

$\boxed{e_ke_k..e_k}$ m_k times

Obviously, all row permutations $p\in P$ of diagram α leave ω_α fixed while each q multiplies it by $(-1)^q$. Thus $\chi_\alpha(\omega_\alpha)=\omega_\alpha$, as claimed. For any diagram α' different from α, or unrelated s and σ there will be a pair of basic elements $\{e_j;e_j\}$, that lie in the same column of diagram $(\alpha', s^{-1}\sigma)$. So the corresponding determinant (wedge product) will be 0. We have thus shown each subspace $\mathcal{T}_{\alpha s}=\chi_{\alpha s}[\mathcal{T}^m]$ to have a unique highest weight vector, hence follows irreducibility.

ii) Next we take the primary subspace $\mathcal{T}_\alpha=\chi_\alpha[\mathcal{T}^m]$ and want to establish irreducibility of the joint action of $W_m\times SL(n)$ on $\mathcal{T}_\alpha$. This would imply the proper factorization of space $\mathcal{T}_\alpha$ and representation $(\tau^{(m)}\cdot\pi^{(m)})\mid\mathcal{T}_\alpha$ into the product $\tau^\alpha\otimes\pi^\alpha$. But each minimal projection $\chi_\alpha\in\mathbb{C}(W_m)$, maps $\mathcal{T}^m$ onto an irreducible subspace of $G=SL(n)$. From the general result for commuting pairs $(W_m;G)$ we get a tensor product decomposition of part (iii), QED.

Remark 7: Irreducible representations of $G=W_m$ can be realized in a more direct way, using partitions (Young tableaux) α: $m=\sum m_j$ ($m_1\geq m_2\geq\dots$). With any α

we associate a subgroup $H = H_\alpha \simeq \mathbf{W}_{m_1} \times ... \times \mathbf{W}_{m_k}$ of G, and a consider two natural 1-D representations of H_α: trivial **1**, and signature $\lambda_s = \operatorname{sgn} s$. Let T_α and T'_α be the corresponding induced representations of G: $T = \operatorname{ind}(\mathbf{1} \mid H;G)$; $T' = \operatorname{ind}(\operatorname{sgn} \mid H;G)$. All partitions $\{\alpha\}$ split into dual pairs: $\alpha \leftrightarrow \alpha^*$, by interchanging rows and columns (so $m_j^* = \#\{m_k \geq j\}$) (fig.6).

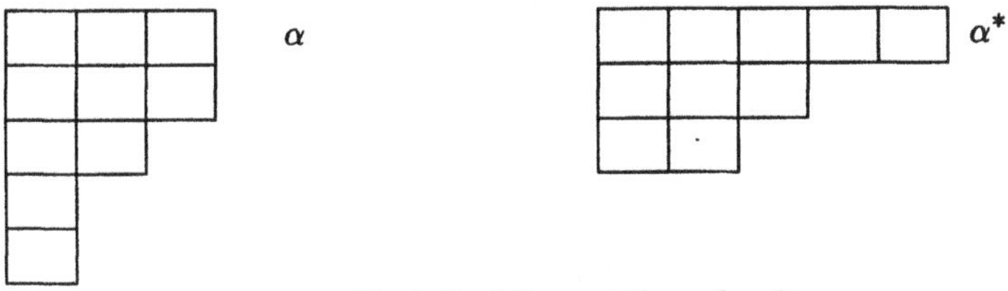

Fig.6: *Dual Young tableaux* $\{\alpha;\alpha^*\}$.

The key result (Weyl; von Neumann) gives the intertwining numbers of representations $\{T_\alpha\}$ and $\{T'_\alpha\}$:

$$\mu(T_\alpha;T'_\beta) = \begin{cases} 1; \text{ for } \beta = \alpha^* \\ 0; \text{ all other} \end{cases}.$$

Hence, T_α and T'_{α^*} have a joint irreducible component τ^α of multiplicity 1. The degree of τ^α can be derived by careful analysis of intertwining operators, and is equal to

$$\boxed{\dim \tau^\alpha = \frac{m! \prod_{i<j} (\ell_i - \ell_j)}{\ell_1! ... \ell_k!}}$$

where $\ell_i = m_i + i - 1$.

Remark 8: In the previous section we realized irreducible representations of $SL(n)$ in function spaces $\mathcal{F}_\lambda$ (of small degree). Tensors and Young tableau give another realization in "large spaces" $\mathcal{T}^m$. There exists an intermediate construction, where the role of $\mathbf{W}_m$ is played by the group $SL(k)$ (see [Ze]). Namely, we take a vector space $\mathfrak{X}$ of all $k \times n$ matrices with the natural actions of two groups $H = SL(k)$, and $G = SL(n)$ by the left/right multiplication:

$$(h,g): X = (x_{ij}) \rightarrow h^{-1} X g = \pi^1_{h,g}(X).$$

This action extends to all polynomial functions

$$\mathcal{P}_m = \{f(X) = \sum c_\beta X^\beta : \beta = (\beta_{ij}), \ \sum \beta_{ij} = m\},$$

of degree m, and gives rise to a representation

$$\pi^m_{(hg)} f(X) = f(h^{-1} X g).$$

The polynomial space $\mathcal{P}_m \simeq \mathcal{S}^m(\mathfrak{X})$ is the mth symmetric tensor power of $\mathfrak{X}$, so π^m is the mth symmetric power of π^1. The reader is asked to show that representation π^m of $G \times H$

is irreducible, groups G and H form a dual (maximal) pair in any space $\mathcal{P}_m$ (problem 5), hence there is 1-1 correspondence between signatures of H and G.

Problems and Exercises:

1. Show that both are irreducible subrepresentations of $\pi^{(2)}$(Hint: identify $\mathcal{S}^2$ with the space of symmetric matrices and Λ^2 with antisymmetric, with the action of G on both spaces by $g: X \to {}^{\mathrm{T}}gXg$. Find all eigenvectors of the diagonal subgroup).
2. Show that $\mathcal{T}^3$ has a decomposition with signatures $\alpha = (3,0,\ldots)$ (symmetric); $\alpha = (1,1,1,0\ldots)$ (antisymmetric); and 2 copies of $\alpha = (2,1,0\ldots)$ (Hint: count Weyl invariants, observe that $\omega_2(\xi,\zeta)\omega_1(\eta) - \omega_2(\eta,\zeta)\omega_1(\xi) = \omega_2(\xi,\eta)\omega_1(\zeta)$ for any triple of vectors ξ,η,ζ in $\mathbf{C}^n$!).
3. Compute symmertizers of Young tableaux: $\begin{pmatrix} 1 & 3 \\ 2 & \end{pmatrix}$ and $\begin{pmatrix} 3 & 2 \\ 1 & \end{pmatrix}$.
4. Show that π^m splits into the direct sum of irreducible products $\tau^\alpha \otimes \pi^\alpha$, where τ^α and π^α are irreducible representations of H and G respectively of the same signature $\alpha = (m_1 \geq m_2 \geq \ldots \geq m_j \geq \ldots)$. (Hint: use a version of Gauss decomposition for $k \times m$ matrices and show that the highest weight vectors for both groups are the familiar products of minors:
$$M_{12\ldots p}(X)^{m_p} M_{12\ldots p-1}(X)^{m_{p-1}} \ldots M_1(X)^{m_1},$$
constructed from the first p columns of X, then first $p-1$ columns, etc.).
5. Show that the natural action of groups $SL(n) \times SL(m)$ in space $\mathfrak{X} = Mat_{n \times m}$ is irreducible; groups $SL(n)$ and $SL(m)$ form a maximal dual pair in any polynomial (symmetric tensor) space $\mathcal{P}_k(\mathfrak{X})$.
6. Use Young tableau and formula (4.5) to classify irreducible representations of symmetric groups $\mathbf{W}_3$; $\mathbf{W}_4$; $\mathbf{W}_5$. Find their characters and degrees. Compare the results for $\mathbf{W}_4$ with §3.3 (example 3.7), identify Young tableau of $\{1; sgn; \pi^2; \pi^{3+}; \pi^{3-}\}$.

§5.5. Haar measure on compact semisimple Lie groups.

We compute the Haar measure on compact Lie group, reduced to conjugacy classes of G, taking as a model group $SU(n)$. This result will lay the ground for the celebrated Weyl character formulae in §5.6, that lie at the heart of the representation theory of classical compact groups (§5.6).

We shall assume G to be a subgroup of $GL_n(\mathbb{C})$, so its Lie algebra $\mathfrak{G} = T_e$ - tangent space at $\{e\}$, could be identified with a subalgebra of $Mat_n = gl(n)$. The tangent space at any point $g \in G$ is then $T_g = g\mathfrak{G}$. Each element $X \in \mathfrak{G}$ (tangent vector at $\{e\}$), extends to a left-invariant vector field $\xi(g) = gX$ (matrix product).

To construct an *invariant volume element* on G, we can choose a basis of left invariant vector fields, or the corresponding left-invariant differential 1-forms: $\omega_j = \sum a_{jk}(x)dx_k$, where $\{x_1;...;x_N: N = dimG\}$ are local coordinates on G (vector fields could be identified with 1-forms via a left-invariant metric). Then we define an invariant *volume element*, as a differential $N-$form

$$\omega = \bigwedge_1^N \omega_k = \det(a_{jk})dx_1 \wedge ... \wedge dx_N. \tag{5.1}$$

A canonical way to construct $\{\omega_k\}$ is to consider a Lie-algebra-valued differential 1-form, called *gauge potential*

$$\Omega = g^{-1}dg = \sum g^{-1}\partial_k(g)dx_k.$$

At each point g in the group, Ω takes on values in the cotangent space at $\{g\}$, and is clearly left-invariant, $(g_0g)^{-1}d(g_0g) = g^{-1}dg$. Choosing a basis $\{e_1;... e_N\}$ in $\mathfrak{G}$ and expanding Ω in this basis, we get a basis of scalar G-invariant 1-forms,

$$\Omega = \sum \omega_k e_k;$$

needed for the volume element (5.1). It is an easy exercise to directly verify (via the group identity $g^*g = I$) that for unitary/orthogonal groups $SU(n)$, $SO(n)$ matrix-differential $g^{-1}dg$ is skew-symmetric, $dg_{jk} = -d\bar{g}_{kj}$, so dg takes on values in the Lie algebra $su(n)$ or $so(n)$.

The above procedure provides a general construction of the Haar measure dg on matrix Lie groups. However, for the purposes of analysis we often need an explicit form of dg (or ω) in a suitably chosen coordinate system on G.

Here we shall do it for the unitary groups $U(n)$, $SU(n)$, and more general compact Lie groups G, based on the structure theory of §5.2. These results will be needed in the section on *Weyl character formulae* for irreducible representations of compact groups G.

Since characters depend only on conjugacy classes, it suffices to compute the *reduced Haar measure* on conjugacy classes of G. We denote by H the space of conjugacy classes: $K_h = \{g^{-1}hg: g \in G\}$, and choose coordinates $h = (h_1;...;h_m)$ on H, as well as transversal variables $y = (y_1;...)$ on each class $K_h \simeq G/G_h$ (G modulo the stabilizer of h, $G_h = \{g: g^{-1}hg = h\}$). We would like to decompose the Haar measure on G, as

$$dg = \rho(h)dh\, d_h y, \tag{5.2}$$

where $d_h(y)$ denotes a G-invariant measure on K_h; dh - some natural measure on H, and the density factor $\rho(h)$ measures the relative size of conjugacy classes $\{K_h: h \in H\}$. Our goal is to compute factor $\rho(h)$ in (5.2). We shall do it first for group $\boldsymbol{SU}(n)$. Each unitary matrix u is conjugate to a diagonal (Cartan) matrix

$$h = \begin{bmatrix} e^{i\theta_1} & & \\ & e^{i\theta_2} & \\ & & e^{i\theta_n} \end{bmatrix}, \quad u = vhv^{-1}. \tag{5.3}$$

Matrix h is unique, modulo all permutations of diagonal entries, i.e the *Weyl group* action on $\boldsymbol{SU}(n)$. Thus a generic conjugacy class is identified with a point in a "positive cone" in the torus, $H = \{\theta_1 > \theta_2 > ... > \theta_n\} = \mathbb{T}^n/\text{modulo } \mathbb{W}_n$, the so called *Weyl chamber*[2].

Factor v in (5.3) is also non unique, $v \in \boldsymbol{SU}(n)/\mathbb{T}^n$, since stabilizer of a generic diagonal element h is $\mathbb{T}^n$. So one can think of G (or rather a dense open subset in G), as decomposed into the product: $G \simeq H \times G/\mathbb{T}^n$. To find invariant integral in coordinates (h,v) we differentiate the relation: $u = v^{-1}hv$,

$$du = (-v^{-1}dv\,v^{-1})hv + v^{-1}dh\,v + v^{-1}h\,dv,$$

and multiply it on the right with $u^{-1} = v^{-1}h^{-1}v$

$$du\,u^{-1} = v^{-1}(dh\,h^{-1} + h\,dv\,v^{-1}h^{-1} - dv\,v^{-1})v.$$

Introducing right-invariant cotangent vectors $\delta u = duu^{-1}$, the latter can be written, as

$$\delta u = v^{-1}\{\delta h + (Ad_h - I)\delta v\}v. \tag{5.4}$$

Equation (5.4) represents the Jacobian map Φ' of the coordinate transformation: $\Phi:(h;v) \to u$. We need to compute the determinant of Φ'. Modulo unessential v-

[2] In general a Weyl chamber $A \subset \mathfrak{H}$ in a semisimple algebra $\mathfrak{G}$ with a fixed system of positive roots $\Sigma_+ = \{\alpha\}$, consists of all $\{H \in \mathfrak{H}: \langle \alpha \mid H \rangle > 0, \text{ all } \alpha\}$ (fig.7).

conjugation (which has no effect on $\det\Phi'$), the Jacobian map consists of an identity block in h-variables ($\dim = n-1$), and a h-dependent linear block $Ad_h - I$, in the complementary variables $\delta v = \{\delta v_{ij}\}$.

Variables $\{v_{ij}\}$ can be chosen so that the diagonal part $\delta v_{jj} = 0$, which follows from the decomposition of the Lie algebra $\mathfrak{su}(n)$ into the sum of its "diagonal" (Cartan) part, and "off-diagonal" (root) parts $\mathfrak{H} \oplus \mathfrak{B}$. Regarding matrix entries $\{\delta u_{jk}: j \leq k\}$, as generators of independent coordinates on G, and writing $h_k = e^{i\theta_k}$, we get the formula for the Haar measure on G

$$\boxed{du = \prod_{j \leq k} \delta u_{jk} = |\, det(Ad_h - \mathrm{I})\,|^2 d\theta_1 ... d\theta_n \prod_{j<k} \delta v_{jk} = |\, det(...)\,|^2 d\theta dv} \tag{5.5}$$

Here dv represents the invariant volume on the quotient-space $SU(n)/\mathbb{T}^n$, the determinant of $(Ad_h - \mathrm{I})$ being squared, since each off-diagonal entry v_{jk} represents a complex variable. Finally, remembering that all off-diagonal elements $\{\delta u_{ij} = \delta v_{ij}\}$ are eigenvectors of the adjoint map Ad_h (root vectors in $\mathfrak{G}$),

$$(Ad_h - I)\delta v_{jk} = (h_j/h_k - 1)\delta v_{jk} = \left(e^{i(\theta_j - \theta_k)} - 1\right)\delta v_{jk};$$

we can write the determinant factor in (5.5) explicitly in terms of parameters θ,

$$\prod_{j<k} \left|\left(e^{i(\theta_j - \theta_k)} - 1\right)\right|^2 = \prod_{j<k} \left(e^{i(\theta_j - \theta_k)} - 1\right)\left(e^{-i(\theta_j - \theta_k)} - 1\right) = \prod_{j<k} 4\sin^2\left(\frac{\theta_j - \theta_k}{2}\right). \tag{5.6}$$

In this form the result easily extends to all simple/semisimple compact Lie groups G, the role of diagonal matrices being played by the Cartan subalgebra $\mathfrak{H} \subset \mathfrak{G}$, and pairs (ij) being replaced by positive roots $\{\alpha\}$ of $\mathfrak{G}$.

Precisely, we decompose G into the product $H_+ \times G/H$, where $H = exp\mathfrak{H}$ is a Cartan subgroup (maximal torus) in G, $\mathfrak{H}_+$ denotes a positive *Weyl chamber* in $\mathfrak{H}$, a cone made of all elements $\{X \in \mathfrak{H}: \langle \alpha \mid X \rangle > 0\}$, for all positive roots $\alpha \in \Sigma_+$ (see fig.7), and $H_+ = \exp \mathfrak{H}_+$- the image of the Weyl chamber in H. The key to a decomposition lies in the fact, that generic $u \in G$ can be conjugated to a Cartan element h, $u = vhv^{-1}$ determined uniquely, modulo Weyl-group[3]. The relation (5.4) is derived the same way as for $SU(n)$, and determinant is evaluated in terms of positive roots. The final result can be stated in terms of variables $h = exp\, i\theta$, $\theta \in \mathfrak{H}$, and $v \in G/H$,

$$\boxed{du = \left|\det(\mathrm{Ad}_h - \mathrm{I})\right|^2 dh\, dv = \prod_{\alpha > 0} \left|e^{i\langle \alpha \mid \theta \rangle} - 1\right|^2 d\theta dv.}$$

[3] A Lie group analog of diagonalization of unitary matrices.

In this form the result will be used in subsequent sections.

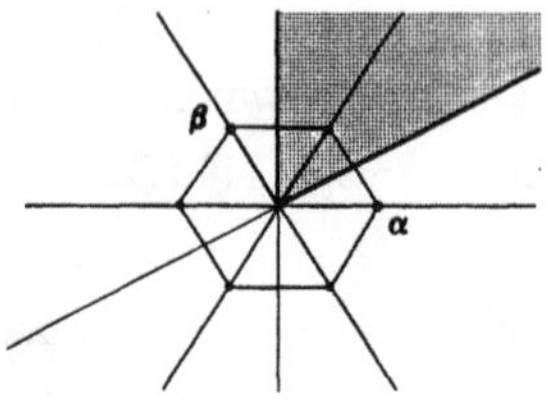

Fig.7: *The Weyl chamber of* $sl(3)$ *(shaded) consists of all weights* λ, *that make positive products with the positive basic roots:* $\langle\lambda\,|\,\alpha\rangle > 0$; $\langle\lambda\,|\,\beta\rangle > 0$.

Problems and Exercises:

1. a) Compute the Haar measure for other classical compact groups: $SO(n)$ and $Sp(n)$.
   b) Compare (5.6) with formulae of §4.1 for groups $SU(2)$ and $SO(3)$.

§5.6. The Weyl character formulae.

Our main goal in this section is to derive the celebrated Weyl character formulae for irreducible representations of compact semisimple Lie groups. Most of the discussion below will concentrate on groups $SU(n)$; $U(n)$ (respectively SL_n and GL_n, but whenever possible we shall indicate connections and extensions to other compact groups.

6.1. First Weyl formula. Irreducible representations $\{\pi^{\alpha}\}$ of group $G = SU(n)$, or $U(n)$ were described in §5.3 in terms of their *signatures* (highest weights) $\alpha = (m_1; m_2; \dots m_n)$. The latter means an ordered n-tuple of integers[4]: $m_1 \geq m_2 \geq \dots \geq m_n$. We denote the character of π^{α} by $\chi_{\alpha} = \mathrm{tr}\pi^{\alpha}$. Since function $\chi = \chi_{\alpha}$ is conjugate-invariant on G, and any element $u \in SU(n)$ is conjugate to a diagonal matrix $h = \mathrm{diag}(h_1; \dots; h_n)$, it suffices to evaluate χ^{α} on the diagonal (Cartan) subgroup $D \simeq \mathbb{T}^{n-1}$, or $\mathbb{T}^n$ (for $SU(n)$ and $U(n)$, respectively).

We shall start with two special cases: the symmetric $\mathcal{S}^k(\pi)$ and antisymmetric $\Lambda^k(\pi)$ tensor powers of the natural representation π in $\mathbb{C}^n$. Their characters will be denoted p_k and q_k respectively. In both cases we know explicitly all weights $\{\beta\}$ and weight vectors:

• For symmetric space $\mathcal{S}^k(\mathbb{C}^n)$ realized by polynomials of degree k in n variables, $\mathcal{P}_k(x_1; \dots; x_n)$, the weight vectors are monomials $\{x^{\beta}: \beta = (k_1; \dots; k_n)\}$, and the corresponding weights are tuples of multi-indices β of norm $|\beta| = k$. Each has multiplicity 1, hence

$$p_k(h) = \sum_{|\beta| = k} h^{\beta} = \sum_{k_1 + \dots + k_n = k} h_1^{k_1} \dots h_n^{k_n}. \tag{6.1}$$

• Similarly for antisymmetric $\Lambda^k(\mathbb{C}^n)$ weights are labeled by ordered tuples $\boldsymbol{i} = (i_1 < i_2 < \dots < i_k)$, $h^{\boldsymbol{i}} = h_{i_1} \dots h_{i_k}$, with the corresponding weight vectors: $e_{i_1} \wedge \dots \wedge e_{i_k}$. Hence,

$$q_k(h) = \sum_{\boldsymbol{i}} h^{\boldsymbol{i}} = \sum_{i_1 < \dots < i_k} h_{i_1} \dots h_{i_k}.$$

In general, each character $\chi(h) = \chi_{\alpha} = \mathrm{tr}\pi^{\alpha}$ of $SU(n)$, restricted on the diagonal (Cartan) subgroup $D = \{h: h = (e^{i\theta_1}; \dots; e^{i\theta_n})\}$ becomes a trigonometric polynomial $\sum a_m e^{im \cdot \theta}$, with integer coefficients a_m = multiplicity of the mth weight in the weight diagram of π^{α}. Unfortunately the multiplicities $\{a_m\}$ are not easy to compute, so we

[4]We remind the reader that different n-tuples α yield different representations of $U(n)$, but for $SU(n)$, only the differences $\{(\dots; m_i - m_{i+1}; \dots)\}$ matter. Indeed, shifting all m_j by any p, $m_j \to m_j + p$, will multiply π^{α} by a central character: $\{e^{ipz} I\}$ for $z = zI$, which is trivial on $SU(n)$. So signatures α can always be chosen, so that $\sum m_j = 0$, or $m_n = 0$. We shall use both choices, depending on the context.

shall proceed differently.

Let us observe that each character is invariant under all Weyl transformations $\sigma \in \mathbb{W}$, which in case of $SU(n)$; $U(n)$ coincides with the symmetric group $\mathbb{W}_n$. Indeed, all Weyl elements σ belong to G itself, so $\sigma h \sigma^{-1} = h^{\sigma}$, for any diagonal (Cartan) element h, which yields

$$\chi(h^{\sigma}) = \sum a_m e^{im\cdot\theta^{\sigma}} = \sum a_{m^{\sigma}} e^{im\cdot\theta} = \chi(h).$$

Therefore all Fourier coefficients $\{a_m\}$ with m in the same $\mathbb{W}$-orbit $\omega = \{m^{\sigma}: \sigma \in \mathbb{W}\}$ are equal,

$$a_m = a_{m^{\sigma}} = a_{\omega}, \text{ for all } m, \sigma.$$

Thus we can rewrite χ as the sum of *orbital terms:* $\chi_{\omega} = \sum_{m \in \omega} e^{im\cdot\theta}$, i.e. $\chi = \sum a_{\omega}\chi_{\omega}$, summation over all $\mathbb{W}$-orbits ω in the weight diagram, with integer coefficients a_{ω}.

Next we shall apply the orthogonality relation for irreducible characters:

$$\langle \chi \mid \chi \rangle = \int_G \chi\, \bar{\chi} dg = 1, \; dg \text{ - normalized Haar measure on } G = U(n). \tag{6.2}$$

Due to conjugate-invariance of $f = |\chi|^2$ integral (6.2) can be reduced from G to the set of conjugacy classes. The latter as we already know are parametrized by points of the quotient space $D/\mathbb{W}$ (diagonal subgroup $D = \mathbb{T}^{n-1}$, modulo Weyl group $\mathbb{W}$), or by a "positive cone" $D_+ = \mathbb{T}^{n-1}_+ = \{\theta: \theta_1 \geq \theta_2 \geq \ldots \geq \theta_n\}$ in D.

The reduced Haar measure was calculated in §5.6,

$$du = \left| \prod_{j<k} (e^{i\theta_j} - e^{i\theta_k}) \right|^2 d\theta_1 \ldots d\theta_{n-1} dv = \Delta(h) d\theta dv; \tag{6.3}$$

where $\theta = (\theta_1;\ldots\theta_n)$ parametrize the Cartan (maximal) torus $D \subset G$ (any $u \in G$ can be conjugated into a diagonal matrix $h = h(u)$, with eigenvalues $\{e^{i\theta_k}\}$, ordered lexicographically: $\theta_1 > \theta_2 > \ldots$, and $v \in G/D$ implements the conjugation $u = v^{-1}hv$). We shall recast it in a more convenient form using the well known expansion of the vandermond determinant,

$$\Delta(h) = \prod_{j<k} (e^{i\theta_j} - e^{i\theta_k}) = \det \begin{bmatrix} 1 \; e^{i\theta_1} \ldots e^{i(n-1)\theta_1} \\ \ldots\ldots\ldots\ldots \\ 1 \; e^{i\theta_n} \ldots e^{i(n-1)\theta_n} \end{bmatrix}.$$

The determinant can be expanded in the usual way

$$\sum_{\sigma \in W} (-1)^{\sigma} e^{i[(n-1)\theta_{\sigma(1)} + \ldots + 0\cdot\theta_{\sigma(n)}]}, \tag{6.4}$$

sum over all permutations σ. In case of $SU(n)$ product $(\prod_1^n e^{i\theta_k})^{\frac{n-1}{2}}$ can be pulled out from each term of (6.4) and get it into a familiar sum of exponentials (cf. chapter 4),

$$\exp\tfrac{i}{2}[(n-1)\theta_{\sigma(1)} + (n-3)\theta_{\sigma(2)} + \ldots + (1-n)\theta_{\sigma(n)}]. \tag{6.5}$$

Exponentials of each term (6.5) represent a particular positive weight ρ on D, (transformed by $\sigma \in \mathbb{W}$), namely the *one-half sum of positive roots:*

$$\rho = \tfrac{1}{2}\sum_{j<k} H_{jk} = \tfrac{1}{2}(n-1; n-3; \ldots; 1-n).$$

On $su(n)$ weight ρ can also be represented by a tuple $(n-1; n-2; \ldots 1; 0)$, due to the zero-trace cancellation. Thus the reduced Haar measure can be written as

$$du = |\Delta(h)|^2 d\theta = \left|\sum_{\sigma \in \mathbf{W}} (-1)^\sigma e^{i\langle \rho^\sigma \mid \theta\rangle}\right|^2 d\theta. \tag{6.6}$$

Next we substitute character $\chi = \sum a_\omega \chi_\omega$, written as the sum of orbital terms

$$\chi_\omega = \sum_{m \in \omega} e^{im\cdot\theta},$$

and the Haar measure (6.6) into the orthogonality relation (6.2). Let us observe that the density of the reduced Haar measure (6.6) factors into the product $\Delta \cdot \bar{\Delta}$, with $\Delta(h)$ being equal to an alternating orbital sum,

$$\tilde{\chi}q_{\omega(\rho)} = \sum (-1)^\sigma e^{i\langle \rho^\sigma \mid h\rangle}.$$

So the integrand of (6.2), $\chi \cdot \bar{\chi}\Delta \cdot \bar{\Delta}$, combines the products of symmetric and antisymmetric orbital sums: $\chi_\omega \tilde{\chi}_{\omega'}$, corresponding to various orbits ω in the weight diagram of α, and a special orbit $\omega' = \omega(\rho)$ of weight ρ.

Let us compute the product of two such orbital terms: $\chi_\omega \tilde{\chi}_{\omega'}$, where orbit $\omega = \omega(\lambda)$ comes from a positive weight λ, and $\omega' = \omega(\rho)$. We find

$$\chi_\omega \tilde{\chi}_{\omega'} = \sum_{s,\sigma} (-1)^s e^{i(\lambda^\sigma + \rho^s)} = \sum_\sigma \pm \tilde{\chi}_{\omega(\lambda^\sigma + \rho)}.$$

Orbits $\{\omega(\lambda^\sigma + \rho)\}$ with different weights λ may coincide, but picking the highest weight α in the original ω, the orbit $\omega(\alpha^\sigma + \rho)$ will be "highest" among all $\{\omega(\lambda^\sigma + \rho)$: $\lambda, \sigma\}$ and unique! In particular, taking the highest weight α of a representation, the product

$$\chi_{\omega(\alpha)} \tilde{\chi}_{\omega'} = \tilde{\chi}_{\omega(\alpha+\rho)} + \sum b_\omega \chi_\omega;$$

breaks into the sum of the "highest orbital term" $\tilde{\chi}_{\omega(\alpha+\rho)}$, and some lower terms $\{\chi_\omega\}$ with integer coefficients b_ω. It remains to observe that different orbital terms $\{\tilde{\chi}_\omega\}$ are orthogonal on torus $D = \mathbb{T}^{n-1}$ with respect to the usual "flat" Haar measure $d\theta$,

$$\langle\tilde{\chi}_\omega \,|\, \tilde{\chi}_{\omega'}\rangle = \int_{\mathbf{T}^{n-1}} \tilde{\chi}_\omega \bar{\tilde{\chi}}_{\omega'} d\theta = \begin{cases} 0; \text{ if } \omega \neq \omega' \\ |\,\omega\,|; \text{ if } \omega = \omega' \end{cases};$$

where $|\,\omega\,| = n!$, for the highest-weight orbit ω. Therefore,

$$\langle\chi \,|\, \chi\rangle_{L^2(G)} = \frac{1}{n!}\left\|\tilde{\chi}_{\omega(\alpha+\rho)} + \sum b_\omega \tilde{\chi}_\omega\right\|^2_{L^2(\mathbf{T}^{n-1})} = 1 + \frac{1}{n!}\sum |\,\omega\,|\, b_\omega^2$$

must be equal 1. But the latter is impossible unless all lower orbital terms $\{\chi_\omega\}$ cancel each other, i.e. all coefficients $b_\omega = \sum \pm a_{\omega'}$ are zero. This means that the product $\chi \cdot \Delta$ consists of a single term $\tilde{\chi}_{\omega(\alpha+\rho)}$, where α is the highest weight of the representation π, and $\rho = \frac{1}{2}$"sum of positive roots". As a consequence we get

1-st Weyl formula:

$$\chi_\alpha(\theta) = \frac{\sum_{\sigma \in \mathbf{W}} \det(\sigma) e^{i\langle(\alpha+\rho)^\sigma \,|\, \theta\rangle}}{\sum_{\sigma \in \mathbf{W}} \det(\sigma) e^{i\langle\rho^\sigma \,|\, \theta\rangle}} \tag{6.7}$$

The Weyl formula was written here in a general form, valid for any semisimple compact Lie group, with product $\langle.\,|\,.\rangle$ given by the Killing form on the Cartan subalgebra of $\mathfrak{G}$. From (6.7) we shall deduce degrees of irreducible representations $\{\pi^\alpha\}$. Let us observe that (6.7) represents the ratio of two vandermond determinants. Indeed, passing to variables $\{h_k = e^{i\theta_k}\}_{k=1}^n$

$$\chi_\alpha(h) = \frac{det\begin{bmatrix} h_1^{m_1+n-1} & \dots & h_1^{m_n} \\ h_n^{m_1+n-1} & \dots & h_n^{m_n} \end{bmatrix}}{det\begin{bmatrix} h_1^{n-1} & \dots & 1 \\ h_n^{n-1} & \dots & 1 \end{bmatrix}}; \tag{6.8}$$

where $\alpha = (m_1; m_2; \dots m_n)$. Here and henceforth it will be convenient to replace signature $\alpha = (m_1; \dots m_n)$ with another n-tuple ℓ, whose components $\{\ell_j = m_j + n - j\}$, so ℓ and α differ by ρ. We shall also introduce a column vector

$$\vec{h}^\ell = \begin{bmatrix} h_1^\ell \\ \vdots \\ h_n^\ell \end{bmatrix} \text{-the } \ell^{th} \text{ power of } \vec{h}. \tag{6.9}$$

Then (6.8) can be written as

$$\chi_\alpha(h) = \frac{det\left|\vec{h}^{\ell_1} \dots \vec{h}^{\ell_n}\right|}{det\left|\vec{h}^{n-1} \dots \vec{1}\right|}. \tag{6.10}$$

Once character χ is known the degree of the corresponding representation π is obtained by evaluating χ at $\{e\}$. However, formal substitution $\theta = 0$ (respectively $h_1 = h_2 = \dots = h_n = 1$) in (6.8), results in indeterminacy $\frac{0}{0}$. To resolve it we shall make a substitution: $h_1 = z^{n-1}$; $h_2 = z^{n-2}$; $\dots h_n = 1$, in terms of parameter z. The numerator of (6.8) now turns into the vandermondian in new variables $w_1 = h_1^{m_n+n-1}$; $w_2 = h_2^{m_n+n-2}$; $\dots$ $w_n = h_n^{m_n}$, while the denominator is equal to the vandermondian of $\{z^{n-1}; \dots;1\}$. Thus

$$\text{numerator} = \prod_{j<k} (z^{m_j+n-j} - z^{n_k+n-k}); \text{ denominator} = \prod_{j<k} (z^{n-j} - z^{n-k}) .$$

Pulling out a suitable power of z from both products we rewrite the ratio as

$$z^{N-M}\frac{\prod(z^{m_j-m_k+k-j} - 1)}{\prod(z^{k-j}-1)};$$

Passing to the limit as $z\to 1$ we obtain the degree of π^α

$$\boxed{d(\pi^\alpha) = \frac{\prod_{j<k} (m_j - m_k + k-j)}{\prod_{j<k} (k-j)} = \frac{\prod_{\beta>0} \langle\alpha+\rho \mid \beta\rangle}{\prod_{\beta>0} \langle\rho \mid \beta\rangle}} \tag{6.11}$$

Notice that the RHS of (6.11) consists of the Killing products of positive weights $\alpha+\rho$ and ρ with all positive roots β of $\mathfrak{sl}(n)$. In such form (6.11) extends to all semisimple Lie groups and algebras.

6.2. The second Weyl character formula. The derivation will be based on a remarkable matrix identity due to Cauchy.

Lemma 1: *The determinant of the $n\times n$ matrix depending on $2n$ parameters $(h_1,\dots,h_n)$ and $(z_1,\dots,z_n)$*

$$det\left[\frac{1}{1-h_jz_k}\right] = \frac{\Delta(h)\,\Delta(z)}{\prod_{jk}(1-h_jz_k)}. \tag{6.12}$$

where $\Delta(h)$, $\Delta(z)$ denote vandermondian difference-products $\prod_{j<k}(h_j-h_k)$ and $\prod_{j<k}(z_j-z_k)$.

Proof of the Lemma proceeds by induction in n.

Step 1°. Subtract the 1st row from all the others, the resulting jk^{th} entry becomes

$$\frac{z_k(h_j-h_k)}{(1-h_jz_k)(1-h_1z_k)}.$$

Pulling common factors (h_j-h_1) from j^{th} row and $\frac{1}{1-h_1z_k}$ from k^{th} column we express the $n\times n$ determinant as

$$\frac{\prod_{j>1}(h_j-h_1)}{\prod_{k=1}^{n}(1-h_1z_k)}\begin{vmatrix} 1 & & 1 \\ \frac{z_1}{1-h_2z_1} & \cdots & \frac{z_n}{1-h_2z_n} \\ \frac{z_1}{1-h_nz_1} & & \frac{z_n}{1-h_nz_n}\end{vmatrix}.$$

Step 2°. Now we subtract the 1st column from all the rest. The resulting jk^{th} entry is then

$$\frac{z_k - z_1}{(1-h_j z_k)(1-h_j z_1)}.$$

Pulling out common factors: $(z_k - z_1)$ from the kth column and $\frac{1}{1-h_j z_1}$ from the jth row, we reduce the $n-th$ determinant to the $(n-1)-st$,

$$det_n = \frac{\prod_{j>1}(h_j - h_1)\prod_{k>1}(z_k - z_1)}{\prod_{j=1}^{n}(1-h_j z_1)\prod_{k=1}^{n}(1-h_1 z_k)}\, det_{n-1}$$

whence the result easily follows.

The Cauchy determinant $K(h;z) = det\left[\frac{1}{1-h_j z_k}\right]$ can be expanded in two different ways. On the one hand we shall see that

$$K(h;z) = \sum \left|\vec{h}^{\ell_1} \ldots \vec{h}^{\ell_n}\right| \left|\vec{z}^{\ell_1} \ldots \vec{z}^{\ell_n}\right|, \tag{6.13}$$

summation over all tuples of indices $\ell_1 > \ell_2 > \ldots > \ell_n$. On the other hand the RHS of (6.12) will be expanded into a similar series in "determinantal powers of $\{\vec{z}^{\ell}\}$", but with different coefficients: symmetric characters $\{p_k(h)\}$ (6.1). Precisely, the RHS of (6.12), divided by $\Delta(h)$, will be shown to consist of

$$\sum \left|\vec{z}^{\ell_1}\, \vec{z}^{\ell_2} \ldots \vec{z}^{\ell_n}\right| \begin{vmatrix} p_{\ell_1 - n+1} & \ldots & p_{\ell_1 - 1} & p_{\ell_1} \\ \vdots & & \vdots & \vdots \\ p_{\ell_n - n+1} & \ldots & p_{\ell_n - 1} & p_{\ell_n} \end{vmatrix}; \tag{6.14}$$

as above the summation extends over all ordered n-tuples $\ell = (\ell_1 > \ell_2 > \ldots > \ell_n)$. Using symbolic notations $\vec{p}_{\ell}; \vec{p}_{\ell-1}; \ldots$ for column-vectors $\begin{bmatrix} p_{\ell_1 - k} \\ \vdots \\ p_{\ell_n - k} \end{bmatrix}$ of (6.14) $(k = 0, 1, \ldots)$ we can rewrite it in the following compact form

$$\frac{\Delta(z)}{\prod_{jk}(1-h_j z_k)} = \sum_{\ell} \left|\vec{p}_{\ell-n+1} \cdots \vec{p}_{\ell-1}\, \vec{p}_{\ell}\right| \left|\vec{z}^{\ell_1}\, \vec{z}^{\ell_2} \ldots \vec{z}^{\ell_n}\right| \tag{6.15}$$

Now it remains to compare the coefficients of z-terms in expansions (6.13) and (6.15), and to remember the 1-st character formula (6.10), to get

2-nd Weyl formula:
$$\boxed{\chi_\alpha(h) = \begin{vmatrix} p_{\ell_1 - n+1} & \ldots & p_{\ell_1 - 1} & p_{\ell_1} \\ \vdots & & \vdots & \vdots \\ p_{\ell_n - n+1} & \ldots & p_{\ell_n - 1} & p_{\ell_n} \end{vmatrix}} \tag{6.16}$$

an $n \times n$ determinant whose entries are symmetric characters:

$$p_k(h) = \sum_{|\beta| = k} h^\beta, \text{ and signature } \alpha = (m_1; \ldots m_n)$$

is related to an n-tuple $\ell = (\ell_1; \ldots \ell_n)$ through the weight $\rho = (n-1; n-2; \ldots 0)$ -

$\frac{1}{2}\times$ *sum of positive roots:* $\ell_1{=}m_1+n{-}1; m_2=\ell_2{+}n{-}2;\dots;m_n{=}\ell_n$.

Warning! Some indices $m=\ell_k-n{+}j$ in (6.16) could turn negative. The convention here is that the negative-order symmetric characters p_m $(m<0)$ are assumed to be 0, so they won't enter expansion (6.16).

Formula (6.16) blends once again two fundamental tensorial operations of *symmetrization* and *anti-symmetrization* to produce in an intricate and somewhat mysterious pattern all irreducible characters of $SU(n)$. Curiously the role of two operations is now interchanged compared to the previous sections (§5.4-5). There we built all irreducible $\{\pi^\alpha\}$, based on *symmetrization* of all *antisymmetric tensor-powers* $\{\pi^{(k)}$ in $\Lambda^k(\mathbb{C}^n)$: $1\le k\le n\}$, as generators (spaces $\mathfrak{F}_\lambda$). Here irreducible $\{\pi^\alpha\}$ arise from "*anti-symmetrizers*" (6.16) of various *symmetric tensor-representations.*

To show (6.13) we note that each column of matrix $\left[\frac{1}{1-h_jz_k}\right]$ has a geometric series expansion in powers of z_k, whose coefficients are powers (6.9) of column $\vec{h}$

$$\left|(\vec{1}+z_1\vec{h}+z_1^2\vec{h}^2+\dots);\ (\vec{1}+z_2\vec{h}+z_2^2\vec{h}^2+\dots);\dots;\ (\vec{1}+z_n\vec{h}+\dots)\right| \tag{6.17}$$

Then we expand (6.17) in multi-powers of $(z_1;\dots z_m\dots)$ and collect terms with the identical h-determinants. This results in a double-determinant expansion (6.13). Since h-terms of (6.13) are precisely the numerators in the 1-st character formula, the *Cauchy determinant* $\left[\frac{1}{1-h_jz_k}\right]$ could be called a *generating function* of irreducible characters $\{\chi_\alpha\}$. To derive (6.15) we shall expand the RHS of the Cauchy identity (6.12). First observe that for each $z=z_k$ the product

$$\prod_{j=1}^{n}\frac{1}{1-zh_j}=\sum_{m=0}^{\infty}z^m\sum_{|\alpha|=m}h^\alpha=\sum z^m p_m(h), \tag{6.18}$$

with symmetric characters $\{p_m(h)\}$ (6.1). Multiplying expansions (6.18) for different $z=z_1;z_2;\dots;z_n$ we get

$$\prod_{jk=1}^{n}\frac{1}{1-h_jz_k}=\sum p_{m_1}p_{m_2}\cdots p_{m_n}\sum_{\tau\in W}z^{\tau(m)}; \tag{6.19}$$

where $m=(m_1;\dots;m_n)$ denotes an ordered tuple of integers (signature), and summation extends over all such tuples. We multiply (6.19) by the vandermondian difference-product

$$\Delta(z)=\sum_{s\in W}(-1)^s\,z^{s(\rho)},$$

with the "$\frac{1}{2}$-sum positive weight" $\rho=(n{-}1;\dots;0)$ to bring the RHS into the form

$$\sum_m p_{m_1}\cdots p_{m_n}\sum_{s,\tau}(-1)^s z^{\tau(m)+s(\rho)}. \tag{6.20}$$

Inner sum (6.20) consists of alternating orbital terms corresponding to various (*non ordered)* weights $\ell=(\ell_1;\dots;\ell_n)$ in the form $\ell=\rho+m^\sigma$. Notice that only non-degenerate

ℓ's (i.e. $\ell_j \neq \ell_k$, all j,k) will give a nonzeros contribution to (6.20). Using the symmetry of each product $p_{m_1}\dots p_{m_n}$ with respect to permutations τ: $m \to m^\tau = \tau(m)$, we can bring (6.20) in the form

$$\sum_{\ell}\sum_{s}(-1)^s z^{s(\ell)} \sum_{\tau} p_{\tau(m)}. \tag{6.21}$$

Here we abbreviated the product notation $p_{m_1}\dots p_{m_n}$ as $p_{(m)}$. The outer sum in (6.21) extends over all (non-ordered) ℓ-tuples, that are related to m-tuples via ρ and τ: $\ell = m^\tau + \rho$. We denote by $\ell' = (\ell_1;\dots\ell_n)$ the unique highest tuple in the orbit of ℓ, (i.e. $\ell_1 > \dots > \ell_n$), and by σ the permutation that takes ℓ to ℓ', and rewrite (6.21) as the sum over all ordered ℓ-tuples

$$\sum_{\ell'}\sum_{s\in\mathbf{W}}(-1)^s z^{s(\ell')}\sum_{\sigma\in\mathbf{W}}(-1)^\sigma p_{(\ell'-\sigma(\rho))}. \tag{6.22}$$

Here we utilized once again the total **W**- symmetry of the inner sum. Each of two inner sums (6.22) collapses into a $n\times n$ determinant in variables $\{z_k\}$ and $\{h_k\}$ (or $p_k(h)$), whence follows (6.15).

Remark 6.2: Characters have simple transformation properties under the operations of the direct sum and tensor product of representations: $\chi_{T\oplus S} = \chi_T + \chi_S$; $\chi_{T\otimes S} = \chi_T\chi_S$. The second Weyl formula gives a formal expression of arbitrary χ_α in terms of products of symmetric characters p_k. This expansion can be "lifted" to representations. In other words we get an expansion of any irreducible representation π^α of $SL(n)$ in terms of symmetric representations $\{\pi^k = \mathcal{S}^k(\pi)\}_{k=0}^\infty$. Namely,

$$\pi^\alpha = \begin{vmatrix} \pi^{\ell_1} & \dots & \pi^{\ell_1-n+1} \\ \dots & & \dots \\ \pi^{\ell_n} & & \pi^{\ell_n-n+1}\end{vmatrix}^{\otimes} = \sum_{\sigma\in\mathbf{W}}(-1)^\sigma \bigotimes_{j=1}^{n} \pi^{\ell_j+1-\sigma(j)}; \tag{6.23}$$

each entry in the formal determinant (6.23) stands for a symmetric representation π^k (in tensors of rank k) of signature $(k;0;\dots0)$, and $\pm$ signs mean the direct sum/difference of representations.

Example 8.3: A representation of signature $\alpha = (4,1,0)$ of $SU(3)$, respectively $\ell = (6,2,0)$ can be represented by the tensorial determinant

$$\begin{vmatrix} \pi^4 & \pi^5 & \pi^6 \\ \pi^0 & \pi^1 & \pi^2 \\ 0 & 0 & \pi^0 \end{vmatrix} = \pi^0\otimes\pi^4\otimes\pi^1 \ominus \pi^5\otimes\pi^0\otimes\pi^0 \simeq \pi^4\otimes\pi^1\ominus\pi^5. \tag{6.24}$$

Here π^0 denotes the trivial representation, two slots (3,1) and (3,2) in the matrix (6.24) are filled with zeros, since they have negative orders: $-2,-1$. Notice that $\pi^5 \subset \pi^4\otimes\pi^1$, since $\mathcal{S}^5(\mathbb{C}^n)$ is naturally embedded in $\mathcal{S}^4\otimes\mathcal{S}^1$. The degree of π^α can be computed in 2 different ways, by the 1-st character formula (6.11),

$$d(\pi^{\alpha}) = \frac{(6-2)(6-0)(2-0)}{(2-1)(2-0)(1-0)} = 24;$$

on the other hand from 2-nd formula (6.24) we get

$$d(\pi^{\alpha}) = d(\pi^4)d(\pi^1) - d(\pi^5) = 15 \cdot 3 - 21 = 24.$$

Additional comments.

Our exposition in §5.6 followed [We3] and [Ze]. We shall conclude this section with yet another interesting formula for irreducible characters and degrees $\{d(\alpha)\}$ (see [Že], [Ta2], [Sim]). It can be stated in terms of the highest weight matrix entry,

$$\psi_{\alpha}(g) = \langle \pi_g^{\alpha} v_{\alpha} \mid v_{\alpha} \rangle,$$

where v_{α} is the highest-weight vector. Functions $\{\psi_{\alpha}\}$ have the property,

$$\psi_{\alpha+\beta} = \psi_{\alpha}\psi_{\beta},$$

a consequence of the tensor multiplication rule:

$$\pi^{\alpha} \otimes \pi^{\beta} = \pi^{\alpha+\beta} \oplus \ldots \text{ (lower weight } \pi\text{'s)}.$$

If we choose a system of basic roots $\{e_1;\ldots e_n\}$, $n = rk(\mathfrak{G})$, call $\psi_j = \psi_{\alpha_j}$, and write any α as a linear combination of $\{e_j\}$ with integral coefficients,

$$\alpha = \sum m_j e_j;$$

then clearly,

$$\psi_{\alpha} = \prod \psi_j^{m_j}.$$

The function ψ_{α} plays a role in the study of quantum partition function for hamiltonians, associated to Yang-Mills potentials. But it also has an interesting connection to the character χ_{α}. Namely,

$$\chi_{\alpha}(x) = d(\alpha) \int_G \psi_{\alpha}(g^{-1}xg)\, dg;$$

also the degree of π^{α} is given in terms of ψ,

$$d(\alpha) = \left(\int_G \left| \int_G \psi_{\alpha}(g^{-1}xg) dg \right|^2 dx \right)^{-2}. \qquad (6.25)$$

There is no direct link of (6.25) to the classical 1-st and 2-nd Weyl formulae, so its meaning remains obscure.

Problems and Exercises:

1. Verify Weyl formula for characters p_k and q_k of the symmetric and antisymmetric tensor powers, $\mathcal{S}^k(\pi)$ and $\wedge^k(\pi)$.

2. Write the 1-st and 2-nd Weyl character formulae for the orthogonal and symplectic groups $SO(n)$, $Sp(n)$.

3. Compare the 1-st and 2-nd Weyl formulae with the results of chapter 4 for characters of $SU(2)$ and $SO(3)$. Verify formula (6.23) for $SU(2)$, using the Clebsch-Gordan decomposition of §4.4. (Caution: in chapter 4 we used a slightly different labeling of $SU(2)$ representations, so the present "integral" signature $\alpha = (m_1; m_2)$ would correspond to weight $m = \frac{1}{2}(m_1 - m_2)$ of chapter 4).

4. Find characters of all irreducible representations of $sl(n; \mathbb{C})$, considered as real Lie algebra (any such π is the product of a holomorphic and antiholomorphic representations $\pi^\alpha \otimes \overline{\pi}^\beta$).

§5.7. Laplacians on symmetric spaces.

Symmetric spaces are defined as Riemannian manifolds, whose curvature tensor is invariant under all parallel transports (see Appendix C). E. Cartan posed the problem to classify all such spaces. By two ingenious arguments he gave the problem a group-theoretic formulation. His first approach was based on the *holonomy group* of $\mathcal{M}$. The holonomy group K at point $x \in \mathcal{M}$ consists of the parallel transport operators on tangent space T_x along all closed path $\{\gamma\}$ through x. Clearly, different points $x \in \mathcal{M}$ give isomorphic groups K. The Riemannian metric is always invariant under K, and for locally symmetric $\mathcal{M}$, the curvature tensor is also K–invariant. Hence, it follows from the Cartan structural equations (Appendix C), that each holonomy $u \in K$ induces a local isometry in a neighborhood of x, that leaves x fixed. This yields some algebraic relations between Lie algebra $\mathfrak{K}$ of group K, and the metric and curvature tensors, g and R on $\mathcal{M}$,

$$g(AX;Y) + g(X;AY) = 0, \text{ for all } A \in \mathfrak{K};\ X, Y \in T_x.$$

$$[A; R_0(X,Y)] = R(AX,Y) + R(X,AY); \text{ and } R(X,Y) \in \mathfrak{K}.$$

Cartan showed that any R satisfying such relations and the standard symmetry condition of the Riemann tensor, defines a locally symmetric space. Thus he was able to reduce the problem to (i) classification of all possible holonomy groups of symmetric spaces; and (ii) derivation of the curvature tensor on symmetric spaces in terms of their holonomy groups.

The second Cartan's method is based on another important observation: *the curvature tensor is invariant under all parallel transports iff each geodesic symmetry* $\gamma(s) \to \gamma(-s)$, *at* $x \in \mathcal{M}$ *defines a local isometry on* $\mathcal{M}$. Here $\gamma = \gamma(s;\xi)$ denotes a geodesic through x in the direction $\xi \in T_x$. The latter gave meaning to the terminology "locally symmetric". This result becomes particularly significant for "*globally symmetric spaces*", where each local symmetry extends through the global symmetry of $\mathcal{M}$. Hence, any globally symmetric space possesses a transitive group G of isometries, and could be represented as a quotient $K \backslash G$. Subgroup $K \subset G$ stabilizes point x in $\mathcal{M}$, and itself consists of fixed points of an involutive automorphism θ (that results from the geodesic reflection). Group G becomes semisimple, when one drops a "trivial (Euclidian) factor". The problem is then reduced to the study of involutive automorphisms of semisimple Lie algebras.

7.1. Compact symmetric spaces. In this section we shall be mostly interested in *compact symmetric spaces* $\mathcal{M} = K \backslash G$ i.e. quotients of compact semisimple Lie groups G, modulo a subgroup K, which stabilizes an involutive (Cartan) automorphism θ of G,

$$K = \{u \in G: \theta(u) = u\},\ \theta^2 = 1.$$

Automorphism θ splits Lie algebra $\mathfrak{G}$ of group G into the direct (orthogonal) sum of the subalgebra $\mathfrak{K}$ of K, and a subspace $\mathfrak{P}$, identified with the tangent space at $x_0 = \{K\}$, other tangent spaces[5] $\{T_x : x = x_0^g\}$, being given by the adjoint action of G, $T_x \simeq \mathrm{Ad}_g(\mathfrak{P})$, applied to $\mathfrak{P}$. Furthermore, automorphism θ takes on value ± 1 ($\theta\ \mathfrak{K} \simeq I$ and $\theta\ \mathfrak{P} \simeq -I$), so the resulting Lie brackets between the $\mathfrak{K}$ and $\mathfrak{P}$ take the form:

$$[\mathfrak{K};\mathfrak{K}] \subset \mathfrak{K};\ [\mathfrak{K};\mathfrak{P}] \subset \mathfrak{P};\ [\mathfrak{P};\mathfrak{P}] \subset \mathfrak{K}. \tag{7.1}$$

Space $\mathfrak{P}$ contains a maximal abelian (*Cartan*) subalgebras $\mathfrak{T} \simeq \mathbb{R}^r$, whose dimension $n = \dim \mathfrak{T}$ is called the *rank* of $\mathcal{M}$. The image of $\mathfrak{T}$ under the exponential map $\mathfrak{G} \to G$, forms a maximal *geodesically flat* torus $\gamma = \exp \mathfrak{T} \simeq \mathbb{T}^n$ in $\mathcal{M}$. Space $\mathcal{O}_n$ of all flat n-tori $\{\gamma\}$ is itself a smooth manifold, whose dimension depends on $\dim G$ and its rank. Group G acts on $\mathcal{O}_n$, turning it into a homogeneous space $K_0 \backslash G$, K_0 - stabilizer of $\mathfrak{T}$ in G.

7.2. Restricted root system. The abelian part $\mathfrak{T}$ of $\mathfrak{P}$ could be embedded into a larger Cartan subalgebra $\mathfrak{H}$ of $\mathfrak{G}$, in fact, $\mathfrak{H}$ can be chosen as the sum $\mathfrak{T} \oplus \mathfrak{A}$ (Cartan parts of $\mathfrak{P}$ and $\mathfrak{K}$). Algebra $\mathfrak{T}$ breaks space $\mathfrak{G}$ into the sum of eigensubspaces $\oplus \mathfrak{G}_\alpha$, which lie in $\mathfrak{P} \oplus \mathfrak{K}$ (but not in either of two summonds). We want to define a *restricted root system* $\Sigma = \{\alpha\}$ in the $\mathfrak{P}$-component of $\mathfrak{G}$. To this end we square adjoint operators $\{\mathrm{ad}_H^2 : H \in \mathfrak{T}\}$, and note that squares map $\mathfrak{P}$ into itself by (7.2). Hence, $\mathfrak{P}$ is broken into the direct sum: $\bigoplus_\alpha (\mathfrak{P}_\alpha \oplus \mathfrak{P}_{-\alpha})$, where $\mathrm{ad}_H^2 \mid \mathfrak{P}_{\pm\alpha} = \alpha(H)^2$. As in the case of entire $\mathfrak{G}$, the restricted root system $\Sigma(\mathfrak{P})$ can be ordered, i.e. split into its positive and negative parts $\Sigma_+ \cup \Sigma_-$. It has a basis of positive roots $\{\alpha_1; \ldots \alpha_n\}$ $(n = \mathrm{rk}\mathfrak{P} = \dim \mathfrak{T})$, the (restricted) Weyl group $\mathbb{W}(\mathfrak{P})$, etc. In one respect, however, root systems $\Sigma(\mathfrak{G})$ and $\Sigma(\mathfrak{P})$ differ significantly, the former was shown in §5.2 to be multiplicity free, i.e. $\dim \mathfrak{G}_\alpha = 1$, for all $\alpha \neq 0$, whereas the latter has typically higher multiplicities $\{m_\alpha \geq 1\}$. In fact, each root α of $\mathfrak{P}$ can be extended through a root of $\mathfrak{G}$, and m_α measures the number of such expansions. Examples below will illustrate the general concepts.

7.3. Classification. Symmetric spaces (compact and noncompact) were completely classified by Cartan (see [Hel]), who discovered their close connection with semisimple Lie algebras and groups. Here we shall list the basic examples of so called *irreducible globally symmetric compact spaces.*

[5] One can associate with any *compact pair* $\{\mathfrak{G}; \mathfrak{K}\}$ *a noncompact (dual) pair* $\{\mathfrak{G}^*; \mathfrak{K}\}$ (in the terminology of [Hel]), where $\mathfrak{G} = \mathfrak{K} \oplus \mathfrak{P}$, $\mathfrak{G}^* = \mathfrak{K} \oplus i\mathfrak{P} \subset \mathfrak{G}_{\mathbb{C}}$ - a complexification of $\mathfrak{G}$. The standard example of a dual pair is Lie algebras $\mathfrak{G} = \partial\,(n) \supset \mathfrak{K} = \partial \sqcap (n)$, and $\mathfrak{G}^* = \partial](n; \mathbb{R}) \supset \partial \sqcap (n)$. The $\mathfrak{G}$-Kil form on $\mathfrak{P}$ (better to say, on its dual $i\mathfrak{P} \subset \mathfrak{G}_{\mathbb{C}}$), defines a G–invariant Riemannian metric on $\mathcal{M}$, whose curvature tensor is obtained from the Lie bracket,

$$R(X,Y) \cdot Z = -[[X;Y];Z], \text{ for } X,Y,Z \in \mathfrak{P} = \mathfrak{P}_x. \tag{7.2}$$

Notice that the curvature tensor is always anti-symmetric relative to the metric $\{g(x)\}$. So operator $R(X,Y)$ belongs to the Lie algebra of the holonomy group K of $\{x\}$, and the triple bracket (7.2) does obey the basic properties of the curvature tensor (Appedix C). Indeed, $[[\mathfrak{P};\mathfrak{P}];\mathfrak{P}] \subset \mathfrak{P}$, and operators $\{\mathrm{ad}_{[X,Y]} : X, Y \in \mathfrak{P}\}$ belong to the adjoint Lie algebra $\mathfrak{K}$, acting orthogonally on $\mathfrak{P}$.

Table of symmetric spaces.

Space	*Involution*	*Rank*
(i) $SU(n)/SO(n)$	$\theta: X \to \bar{X}$ (complex conjugation)	$r = n-1$
(ii) $SU(2n)/Sp(n)$	$\theta(X) = J_n \bar{X} J_n^{-1}$, where J_n is symplectic matrix $\begin{bmatrix} & I \\ -I & \end{bmatrix}$ in $\mathbb{C}^{2n}$	$r = n-1$
(iii) $SU(p+q)/S(U(p)\times U(q))$	$\theta(X) = I_{pq}\bar{X} I_{pq}$, where I_{pq} is the matrix of the indefinite $(p;q)$-form: $\sum_1^p x_j y_j - \sum_{p+1}^{p+q} x_j y_j$, in $\mathbb{C}^{p+q}$;	$r = min(p;q)$
(iv) $SO(p+q)/SO(p)\times SO(q)$	$\theta(X) = I_{pq} X I_{pq}$, same I_{pq} as above in $\mathbb{R}^{p+q}$;	$r = min(p;q)$
(v) $SO(2n)/U(n)$	$\theta(X) = J_n X J_n^{-1}$	$r = \sim\frac{n}{2}]$
(vi) $Sp(n)/U(n)$	$\theta(X) = \bar{X} = J_n X J_n^{-1}$	$r = n.$

The are a few other examples, related to exceptional simple Lie algebras. More general symmetric spaces can be decomposed into products of the irreducibles ones: $\mathcal{M}_1 \times \mathcal{M}_2 \times \ldots$ The simplest example is the product of 2-spheres $(S^2 \times S^2)/\mathbb{Z}_2$, a quotient $SO(4)/SO(2)\times SO(2)$ (problem 5).

7.4. Laplacians on manifold and symmetric spaces. We recall the definition of the Laplace-Beltrami operator Δ on a Riemannian manifold $\mathfrak{M}$. Let (g_{ij}) denote the Riemann metric tensor on tangent spaces, $\{T_x\}$, and (g^{ij}) - the dual metric on cotangent spaces $\{T_x^*\}$, matrix $\check{g} = (g^{ij}) = (g_{ij})^{-1}$. So the metric (arc-length) element,

$$ds^2 = \sum_{ij} g_{ij} dx^i dx^j,$$

in local coordinates $x = (x_i)$ on $\mathfrak{M}$, while the volume element of g has the form,

$$dV = \frac{dx}{\sqrt{det(g_{ij})}} = \sqrt{\det(g^{ij})}dx = J dx, \text{ where } J = \sqrt{det(g^{ij})}.$$

The g-gradient and g-divergence operations on functions $\{f\}$ and vector-fields

$\{F\}$ are given by

$$\nabla f = (\sum_j g^{ij}\partial_j f) = \check{g}\cdot\partial f;\ \nabla\cdot F = \tfrac{1}{J}\sum_k \partial_k(Jf_k) = \tfrac{1}{J}\partial\cdot(JF).$$

One can easily check that operations ∇ and $\nabla\cdot$ are dual (adjoint) one to the other relative to the L^2-inner product with the volume element,

$$\langle \nabla f \mid F\rangle = \int \nabla f\cdot F dV = -\int f\nabla\cdot F dV, \tag{7.3}$$

the dot-product in (7.3) clearly means the (g_{ij})-product on tangent-spaces $\{T_x\}$.

Let us remark that any coordinate change, $x=\phi(y)$, takes tensor (g_{ij}) into $g^{\phi} = {}^{\mathrm{T}}AgA$, and $\check{g}\to A^{-1}\check{g}({}^{\mathrm{T}}A^{-1})$, where $A=\phi'$ denotes the Jacobian of map ϕ. Hence one could check that all the above definition do not depend on a particular choice of local coordinates (x).

The g-Laplacian is a product of operations g-div and g-grad,

$$\Delta = \nabla\cdot\nabla = \tfrac{1}{J}\sum_{ij}\partial_i J g^{ij}\partial_j = \tfrac{1}{J}\partial\cdot J\check{g}\partial.$$

Once again the definition is coordinate-free.

For the purpose of analysis on symmetric spaces we shall need the so called *multi-polar* (or *generalized radial*) coordinates on $\mathfrak{X}$. A natural framework for introducing such coordinates is a manifold $\mathfrak{X}$ with a compact Lie group K of isometries, like symmetric space $\mathfrak{X} = K\backslash G$. We choose a transversal manifold $\mathfrak{T}\subset\mathfrak{X}$, that cuts across each K-orbit S at a single point t, and assume (without loss of generality) that $\mathfrak{T}$ is orthogonal to all orbits $\{S=S_t\}$. Manifold $\mathfrak{T}$ will play the role of multi-radial directions, while orbits $\{S_t\}$ will form a family of spheres of multi-radii t.

The following examples will illustrate the concept:

1) $\mathbb{R}^n$ with the natural $SO(n)$-action, where $\mathfrak{T}=\mathbb{R}_+$ (half-line), while $S=S_r$ are standard spheres of radii $r\geq 0$;

2) sphere $S^n\subset\mathbb{R}^{n+1}$, with $SO(n)$ acting by axi-symmetric rotations. Here $\mathfrak{T}$ is a great circle (in fact, semicircle) through the "North pole" of S^n, while family $\{S_t\}$ is made of horizontal (transverse) spheres of "spherical radii" $r(t)=t$ - angle, respectively Euclidian radii $r(t)=\sin t$ (see fig.8).

3) Hyperbolic space $\mathbb{H}^n$, is usually realized as a one-sheet hyperboloid $\{(x;x_j): x_0^2-\sum x_j^2 = 1\}$ in $\mathbb{R}^{n+1}$. Once again $SO(n)$ acts on $\mathbb{H}^n$ by axi-symmetric rotations, and foliates it into a one-parameter family of orbits $\{S_t\}$ of radii $r=\sinh t$.

Fig.8: *Orbit-spaces of the n-sphere and n-hyperboloid relative to $SO(n)$-action.*

We remark that all 3 represent model examples of rank-one symmetric spaces $\mathfrak{X} = K\backslash G$ with symmetry-groups: $\mathbb{M}_n = \mathbb{R}^n \triangleright SO(n)$; $SO(n{+}1)$; $SO(1;n)$, and the maximal compact subgroup $K = SO(n)$. Here radial manifold $\mathfrak{T}$ is one-dimensional. But higher rank symmetric spaces provide examples of multi-D radial part.

4) Group $SO(p)\times SO(q)$, $p \le q$, acting on the flat space[6] $\mathfrak{X} = \mathrm{Mat}_{p\times q}$,

$$(u,v)\colon X \to u^{-1}Xv,$$

has the radial (transverse) manifold, made of diagonal matrices

$$\left\{X_t = \begin{bmatrix} t_1 & & & 0 \\ & \ddots & & 0 \\ & & t_p & 0 \end{bmatrix};\ t = (t_1;\dots t_p);\ t_1 > t_2 > \dots > t_p\right\} \tag{7.4}$$

Given a manifold $\mathfrak{X}$ with compact Lie group K, acting by isometries, and a transverse radial manifold $\mathfrak{T}$, we pass to new variable $\{(t;\theta)\}$ on $\mathfrak{T}\times S$, (orbit $S \simeq K_x\backslash K$), and find,

$$ds^2 = dt^2 + \sum g'_{ij}(t;\theta)\, d\theta^i d\theta^j;$$

where g' denotes the restricted metric on K-orbits $\{S = S_t\}$. Hence, the volume density $J = \sqrt{det(g'_{ij})}$, will depend on t only (due to K-invariance of the metric g' on S_t). In fact, factor $J(t)$ will represent the Riemannian m-volume ($m = dimS$) of the orbit. The Laplacian in "multi-polar" coordinates,

$$\Delta = \tfrac{1}{J}\partial_t J \partial_t + \partial_\theta \cdot g'(t;\theta)\partial_\theta; \tag{7.5}$$

and its radial part,

$$\boxed{\Delta_R = \tfrac{1}{J}\Delta_{\mathfrak{T}} J - \tfrac{1}{J}\Delta_{\mathfrak{T}}(J)} \tag{7.6}$$

[6]Let us notice that space $\mathfrak{X}VMat_{p-q}$ represents a linearization (tangent) of the symmetric space $SO(p{+}q)/SO(p)\times SO(q)$ of type IV.

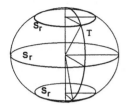

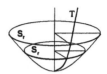

Here $\Delta_{\mathfrak{T}}$ denotes the natural Laplacian on $\mathfrak{T}$. We shall demonstrate both formulae with a few examples.

7.5. Rank-one spaces: $\mathbb{R}^n$; S^n; $\mathbb{H}^n$. Here the polar variable r measures the geodesic distance from the given point (pole) to $x = (r;\theta)$. The metric tensor in polar coordinates is

$$ds^2 = \begin{cases} dr^2 + r^2 d\theta^2; & \mathbf{R}^n \\ dr^2 + sin^2 r d\theta^2; & S^n \\ dr^2 + sinh^2 r d\theta^2; & \mathbf{H}^n \end{cases}$$

Hence the density factors and the radial Laplacians become,

$$J(r) = \begin{cases} r^{n-1}; \text{ on } \mathbf{R}^n \\ \sin^{n-1} r; \text{ on } S^n \\ \sinh^{n-1} r; \text{ on } \mathbf{H}^n \end{cases} ; \Delta_R = \begin{cases} \partial_r^2 + \frac{n-1}{r}\partial_r; \text{ on } \mathbf{R}^n \\ \partial_r^2 + (n-1) cot\, r \partial_r; \text{ on } S^n \\ \partial_r^2 + (n-1) coth\, r \partial_r; \text{ on } \mathbf{H}^n \end{cases}$$

7.6. Higher rank spaces. Turning to a general symmetric space $\mathfrak{X} = K\backslash G$, we choose a maximal abelian subalgebra $\mathfrak{T}$ in $\mathfrak{P}$ ($\dim\mathfrak{T} = n$), pick a system of positive roots in $\mathfrak{T}$, $\{\alpha \in \Sigma_+(\mathfrak{P})\}$, or $i\Sigma_+(\mathfrak{P})$, in the compact case, and consider a *Weyl chamber* in $\mathfrak{T}$ (fig.7),

$$A = \{H : \langle \alpha \mid H \rangle > 0, \text{ for all } \alpha \in \Sigma_+\}$$

Exponential map, $\exp: \mathfrak{P} \to \mathfrak{X}$, takes A into a flat (totally geodesic) manifold $\mathcal{T} \subset \mathfrak{X}$, that cuts transversely through all K-orbits. So here parameters $\{t = (t_1;...;t_n)\}$ are coordinates of element $H \in A$ in a fixed basis of simple roots $\{\alpha_1;...\alpha_n\}$, $t_j = \langle H \mid \alpha_j \rangle$.

To understand the role of the Weyl chamber, as "radial parameters", we first turn to the linearized case, i.e. space $\mathfrak{P} \subset \mathfrak{G}$, with the adjoint action of K, $\{\mathrm{ad}_K \mid \mathfrak{P}\}$. Here a generic K-orbit S punches the Weyl chamber at a single point $\alpha = \alpha(S)$. This follows from a characterization of K-invariant functions (polynomials) on $\mathfrak{P}$. Namely,

Theorem (Chevallet): (i) *Any K–orbit on $\mathfrak{P}$ coincides with the joint level set of the K–invariant functions/polynomials on $\mathfrak{P}$;*

(ii) *Any K–invariant function f is uniquely determined by its its restriction on the Cartan component $\mathfrak{T}$ of $\mathfrak{P}$, and $f \mid \mathfrak{T}$ is $\mathbb{W}$-invariant (here $\mathbb{W}$ means the Weyl group of $\mathfrak{P}$, generated by all symmetries/reflections of the restricted root system).*

Thus we get a 1-1 correspondence between K-orbits in $\mathfrak{P}$ and $\mathbb{W}$-orbits in $\mathfrak{T}$. So a generic K-orbit will hit Cartan part $\mathfrak{T}$ at exactly $\{\mathbb{W}(\alpha)\}$-locations, with only one point

$\sigma(\alpha)$ in the Weyl chamber, and the latter give the requisite transverse (polar) directions.

We shall illustrate the statement for the rank-one space S^n. Here $\mathfrak{P} \simeq \mathbf{R}^n$, $K = SO(n)$ acts on $\mathfrak{P}$ by rotations; Cartan $\mathfrak{T}$ is identified with the 1-st coordinate axis; K-invariants are all even polynomials of $r = |x|$, $f = \sum a_m |x|^{2m}$; $\mathbf{W}$ consists of 2 elements, $\pm : x_1 \to \pm x_1$, hence $\mathbf{W}$-invariants are all even polynomials in x_1, and the correspondence $f \to f \mid \mathfrak{T}$ is clearly 1-to-1. Moreover, any even ($\mathbf{W}$-invariant) $f = x_1{}^{2m}$ extends to a K-invariant by integration over K,

$$x_1{}^{2m} \to \int_K [u(x)]_1{}^{2m} du = \text{Const}\, r^{2m}.$$

Similar arguments apply in the general case.

Passing to a curved homogeneous space $\mathfrak{X} = K \backslash G$ the exponential map, $\exp: \mathfrak{P} \to \mathfrak{X}$, will take the K-orbits in $\mathfrak{P}$ into the K-orbits in $\mathfrak{X}$, and bend Cartan part $\mathfrak{P}$, and its Weyl chamber, into a flat torus (or a totally geodesic Euclidian leaf, in the noncompact case). Of course, the density-factors and the resulting radial Laplacians will look different in both cases. Precisely,

Symmetric spaces of complex semisimple groups $K \backslash G$. Here radial sector $\mathfrak{T}$ coincides with the real part of the complex Cartan subalgebra $\mathfrak{H} \subset \mathfrak{G}$, function

$$J(H) = \prod_{\alpha \in \Sigma_+} (e^{\alpha(H)} - e^{-\alpha(H)}) = \sum_{\sigma \in \mathbf{W}} \det(\sigma) e^{\rho^\sigma(H)}; \tag{7.7}$$

where $\rho = \frac{1}{2} \sum_{\alpha > 0} \alpha$ - the half-sum of positive roots. The radial Laplacian here is

$$\boxed{\Delta_R = \tfrac{1}{J}\left[\Delta_A - \langle \rho \mid \rho \rangle\right] J} \tag{7.8}$$

where Δ_A is the standard "flat Laplacian" on the real Cartan (Abelian) sector $\mathfrak{T}$.

Compact groups and symmetric spaces. The radial part of group G is made of its Cartan subalgebra $\mathfrak{H}$ (the Weyl chamber), and the density takes the form,

$$J(H) = \prod_{\alpha \in i\Sigma_+} 2 \sin \frac{\alpha(H)}{2} = \sum_{\sigma \in \mathbf{W}} \det(\sigma) e^{i\rho^\sigma(H)}; \tag{7.9}$$

The 2-nd formula was essentially derived in §7.5, and the first could be established in a similar way. For symmetric space $\mathfrak{X} = K \backslash G$ each root α will come with its multiplicity, so

$$J(H) = \prod_{\alpha \in i\Sigma_+} 2 \sin^{m_\alpha}\left(\tfrac{1}{2}\alpha(H)\right).$$

Hence the corresponding radial Laplacians takes the form

$$\boxed{\Delta_R = \Delta_A + \sum_{\alpha \in \Sigma_+} m_\alpha \cot\alpha(...)\partial_\alpha} \tag{7.10}$$

Here ∂_α means a partial derivative $\alpha \cdot \partial$, in the direction α.

In case of noncompact symmetric space all $\{\sin\alpha\}$ in J are replaced by hyperbolic $\{\sinh\alpha\}$, as in (7.7). By the same pattern all $\{\cot\alpha\}$ in (7.10) are replaced by hyperbolic $\{\coth\alpha\}$.

Once again, for compact groups expression (7.10) simplify as in (7.8) to

$$\boxed{\Delta_R = \tfrac{1}{J}\left[\Delta_A + \langle \rho^* \mid \rho^* \rangle\right] J} \tag{7.11}$$

Here $*$ indicates the half-sum of positive imaginary roots of $\mathfrak{G}$ (compact Lie algebras have purely imaginary roots!).

Flat spaces. We think of the adjoint action of K on the $\mathfrak{P}$-component in the Cartan decomposition, $\mathfrak{G} = \mathfrak{K} \oplus \mathfrak{P}$, $\mathfrak{G}$ is assumed to be a noncompact Lie algebra, $\mathfrak{K}$ its maximal compact subalgebra. Let us remark that any Cartan pair $\{\mathfrak{K} \subset \mathfrak{G}\}$ can be associated with a triple of algebras and the corresponding symmetric spaces: hyperbolic, spherical and flat-type. Namely,

(i) noncompact algebra $\mathfrak{G} = \mathfrak{K} \oplus \mathfrak{P}$ (here the Killing form $\langle X \mid Y \rangle$ is positive on $\mathfrak{P}$, due to "maximality" of $\mathfrak{K} \subset \mathfrak{G}$), yields hyperbolic-type symmetric space$\mathfrak{X} = K\backslash G$, with noncompact symmetry group G.

ii) the "dual" compact Lie algebra $\widetilde{\mathfrak{G}} = \mathfrak{K} \oplus i\mathfrak{P} \subset \mathfrak{G}_{\mathbf{C}}$ (the Killing form negative everywhere) yields a compact-type space $\widetilde{\mathfrak{X}} = K\backslash\widetilde{G}$, with compact symmetry group $\widetilde{G}$ of algebra $\widetilde{\mathfrak{G}}$.

iii) the motion algebra $\mathfrak{M} = \mathfrak{P} \oplus \mathfrak{K}$, where $\mathfrak{P}$ is considered as an abelian algebra, yields the flat symmetric space $\mathfrak{P} = K\backslash \mathsf{M}$, with the motion symmetry group, $\mathsf{M} = \mathfrak{P} \triangleright K$ (K acting by the adjoint representation on $\mathfrak{P}$).

Three model examples: $\mathfrak{G} = \mathbf{so}(1;n)$; $\widetilde{\mathfrak{G}} = \mathbf{so}(n+1)$; and $\mathfrak{M} = \mathbb{R}^n \triangleright \mathbf{so}(n)$, with $\mathfrak{K} = \mathbf{so}(n)$ and $\mathfrak{P} \simeq \mathbb{R}^n$, exemplify the general situation. Another illustrative example is furnished by the pair: $\mathbf{sl}(n;\mathbb{R}) \supset \mathbf{so}(n;\mathbb{R})$. Here $\mathfrak{P}$ consists of real symmetric matrices $SM_n(\mathbb{R})$, the dual compact algebra is $\widetilde{\mathfrak{G}} = \mathbf{su}(n) = \mathbf{so}(n) \oplus i\mathfrak{P}$, while the flat (motion) group is the semidirect product: $SM_n \triangleright SO(n)$, with $SO(n)$, acting by conjugation $X \to u^{-1}Xu$, on the abelian group SM_n.

In the general flat case, the Weyl chamber $A \subset \mathfrak{H}$ makes up the "cone of radial directions", the density-factor and the radial Laplacian become,

$$\boxed{J(H) = \prod_{\alpha \in \Sigma_+} \alpha(H)^{m_\alpha}; \ \Delta_R = \Delta_A + \sum_{\alpha \in \Sigma_+} \frac{m_\alpha}{\alpha(H)} \partial_\alpha}$$

7.7. Spectra of Laplacians on symmetric spaces. We shall solve the eigenvalue problem for Laplacians on symmetric spaces $K\backslash G$, in terms of the weight structure of $\{\mathfrak{K};\mathfrak{P}\}$ and the representation theory of group G, developed in the preceding sections.

Theorem 2: (i) *Eigenvalues* $\{\lambda_\alpha\}$ *of the Laplacian* Δ *on symmetric space* $\mathfrak{X} = K\backslash G$ *are labeled by the restricted weight lattice* $\{\alpha = \sum k_j \alpha_j; k_j \geq 0\}$ *in the Weyl chamber* $A \subset \mathfrak{T}$. *The* α*-th eigenvalue is equal to*

$$\boxed{\lambda_\alpha = (\alpha + \rho)^2 - \rho^2} \tag{7.12}$$

where $\rho = \frac{1}{2}\sum m_\alpha \alpha$ *- the half-sum of all positive roots taken with their multiplicities* $\{m_\alpha\}$.

(ii) *The multiplicity of* λ_α *is equal to the degree of an irreducible representation* π^α *of group* G.

The last statement requires some clarifications: in part (i) we talked about restricted roots in the Cartan sector $\mathfrak{T} \subset \mathfrak{P}$, while (ii) evidently refers to weights of $\mathfrak{G}$ itself, i.e. vectors of $\mathfrak{H}$. But the latter breaks into the direct sum $\mathfrak{A} \oplus \mathfrak{T}$ of the Cartan sectors of $\mathfrak{K}$ and $\mathfrak{P}$, hence weights $\{\alpha \in \mathfrak{T}\}$ can be identified with elements of $\mathfrak{H}$, that "vanish on the $\mathfrak{K}$-part". Indeed, we shall give a precise meaning to this statement in terms of the regular representation of G on $\mathfrak{X}$, induced by the trivial character **1** of subgroup K. The derivation of (7.12) will consists of 3 steps.

1°. We show that invariant Laplacian Δ on $\mathfrak{X}$ is the image of the so called *Casimir element* L in the center of the *enveloping algebra*[7] $\mathfrak{A}(\mathfrak{G})$. The latter gives the Laplacian on G itself and on all quotients $K\backslash G$. Since Casimir L, and its image Δ commute with the regular representation R of G on $L^2(\mathfrak{X})$, the eigenspaces of Δ are precisely the isotropic (primary) components of R,

$$R \simeq \bigoplus_{\alpha \in \widehat{G}} \pi^\alpha \otimes \mathrm{d}_\alpha.$$

2°. We show that representation R on symmetric $\mathfrak{X}$ has simple (multiplicity-free) spectrum, i.e. each π^α enters R with multiplicity 1, hence the intertwining algebra of

[7] Enveloping algebra $\mathfrak{A}(\mathfrak{G})$ means the associative hull of the Lie algebra $\mathfrak{G}$, i.e. all noncommutative polynomials in variables $\{X_1;...X_n\}$ (a basis of vector space $\mathfrak{G}$), where all products obey the Lie bracket relation, $X_iX_j - X_jX_i = \sum c_{ij}^k X_k$, where $\{c_{ij}^k\}$ are structure constants.

the representation R, is generated by a single element Δ. In other words, any intertwining operator $QR_u = R_uQ$, $u \in G$, is a function of Δ, $Q = f(\Delta)$. This implies that

$$L^2(\mathfrak{S}) = \bigoplus_\alpha \mathcal{H}_\alpha;$$

where eigensubspaces $\{\mathcal{H}_\alpha\}$ are at once the irreducible subspaces of G (the reader may compare it with the case of S^2-Laplacian and the orthogonal action of $SO(3)$ on S^2 in chapter 4).

3°. Finally we use the radially reduced Laplacians above, and the Weyl character formulae of the previous section to compute the eigenvalues (infinitesimal characters) of the Casimirs L on irreducible spaces of π^α.

1° **Lemma:** *Regular representation R on any compact symmetric space $\mathcal{M} = K\backslash G$ has simple spectrum, i.e. multiplicity of any irreducible $\pi \in \widehat{G}$ in R,*

$$m(\pi;R) \leq 1.$$

The general argument (cf. [GGP]) exploits the Cartan automorphism θ, and so called *K-spherical functions* on G. The latter consists of all functions on G, bi-invariant under all right and left translations with K,

$$\mathfrak{L}^K = \{f: f(uxv) = f(x);\ u,v \in K; x \in G\} \subset \mathfrak{L} = L(G).$$

Let π be an irreducible representation of group G (or algebra $\mathfrak{L}$) in space $\mathcal{V}$, and $\mathcal{V}_0$ be a subspace of K-invariants in $\mathcal{V}$. Then a simple argument,[8] due to Godement (problem 2) shows that $\mathcal{V}_0$ is invariant under the spherical subalgebra $\mathfrak{L}^K$,

$$\pi_f \mid \mathcal{V}_0 \subset \mathcal{V}_0, \text{ for all } f \in \mathfrak{L}^K,$$

and the restriction $\{\pi_f \mid \mathcal{V}_0 : f \in \mathfrak{L}^K\}$ is irreducible. But the spherical algebra on any semisimple Lie group G (compact or noncompact alike) equipped with a Cartan automorphism is always commutative (problem 2). Hence, $\dim \mathcal{V}_0 = 1$, and Frobenius reciprocity (§3.2) immediately implies simple spectrum of R.

2°. **Casimir element.** Let G be a semisimple Lie group with Lie algebra $\mathfrak{G}$ and $B(X;Y) = \mathrm{tr}(\mathrm{ad}_X \mathrm{ad}_Y)$ - the Killing form on $\mathfrak{G}$. We pick a basis $\{X_1;...X_n\}$ in $\mathfrak{G}$, form a symmetric tensor (metric) $g_{ij} = B(X_i;X_j)$, its inverse $(g^{ij}) = (g_{ij})^{-1}$, and define an element Δ of degree 2 in the enveloping algebra $\mathfrak{A}(\mathfrak{G})$,

$$\Delta = -g^{ij} X_i X_j.$$

[8]The argument applies to all pairs $\{G;K\}$, not necessarily compact G, and yet more general pairs of algebras $\mathfrak{L} \supset \mathcal{A} = \chi\mathfrak{L}\chi$, where χ is an idempotent (projection) in $\mathfrak{L}$, $\chi^2 = \chi$.

One can verify (problem 6), that

- Δ belongs to the center of $\mathfrak{A}(\mathfrak{G})$, i.e. $\mathrm{ad}_X(\Delta)=0$, for all $X\in\mathfrak{G}$.

- on simple algebras $\mathfrak{G}$ element Δ is unique and independent of a particular basis (since the Killing form is unique)!

A particular choice of the Cartan basis $\{H_j{:}X_\alpha;Y_\alpha\}$ (§5.2) yields the Casimir element of the form,

$$\Delta=\sum_j H_jH'_j-\sum_\alpha X_\alpha Y_\alpha;$$

where $\{H'_j\}$ are dual vectors to $\{H_j\}$: $\langle H_i\,|\,H'_j\rangle=\delta_{ij}$. The corresponding differential operator $\Delta=\sum g^{ij}\partial_{X_i}\partial_{X_j}$ on G commutes with all left and right translations (here $\{\partial_X\}$ denotes left-invariant vector fields on G). Let us also remark that coefficients $\{g_{ij}\}$ give rise to a left-invariant pseudo-Riemannian metric on G (Riemannian in the compact case), and Δ becomes the Laplace-Beltrami operator of this metric. Furthermore, any G-action on manifold $\mathcal{M}$, yields a representation of its Lie algebra $\mathfrak{G}$ by vector fields (1-st order differential operators) $\{\partial_X{:}X\in\mathfrak{G}\}$. Hence the enveloping algebra $\mathfrak{A}(\mathfrak{G})$ acts on $\mathcal{M}$ by higher order differential operators:

$$\mathfrak{A}\ni p(X)=\sum a_m X_1^{m_1}\ldots X_n^{m_n}\to p(\partial)=\sum a_m\partial_{X_1}^{m_1}\ldots\partial_{X_n}^{m_n}.$$

Clearly, Casimir element Δ, considered on a symmetric space $\mathcal{M}=K\backslash G$, turns into a G-invariant Laplacian on $\mathcal{M}$. Let us remark that a G-invariant metric tensor (or differential operator) on $\mathcal{M}$ is uniquely determined by it values at a single point $x_0\in\mathcal{M}$, and its value (symbol) at x_0 must be a ad_K-invariant polynomial on $\mathfrak{P}$. Here K denotes the stabilizer of point x_0 in G

The eigenvalues $\{\lambda_\alpha\}$ of Δ on $L^2(\mathcal{M})$ are labeled by highest weights $\{\alpha\}$ (irreducible components) of the corresponding regular representation $R=R^{\mathcal{M}}$. By Schur's Lemma, Δ restricted on any irreducible component $\mathcal{V}_\alpha$ of R, is scalar,

$$\Delta\,|\,\mathcal{V}_\alpha=\lambda_\alpha I.$$

3°. To compute eigenvalues $\{\lambda_\alpha\}$ it suffices to evaluate element $\Delta\in\mathfrak{A}(\mathfrak{G})$, on the irreducible character χ_α, a $\mathfrak{C}^\infty$-function on $\mathcal{M}$,

$$\Delta[\chi_\alpha]=\lambda_\alpha\chi_\alpha.$$

Here we shall utilize the Weyl character formula of the preceding section and the explicit formula for radially reduced Laplacians (7.8); (7.11). We recall the Weyl character formula,

$$\chi_\alpha(exp\,H) = \frac{\sum\limits_{\sigma\in W} \det(\sigma) e^{i\langle(\alpha+\rho)^\sigma \mid H\rangle}}{\sum\limits_{\sigma\in W} \det(\sigma) e^{i\langle\rho^\sigma \mid H\rangle}};\ H \in \mathfrak{H} \tag{7.13}$$

The denominator of (7.13), $J(H)$, represents the Haar density restricted on the maximal torus of G. Formulae (7.8); (7.11) show that radially-reduced Laplacian/Casimir on Lie group itself is conjugate to a flat Laplacian on the maximal abelian subgroup (torus) in G. Precisely,

$$\Delta_R = \frac{1}{J}\left[\Delta_{\mathbb{T}^r} + \langle\rho\mid\rho\rangle\right]J,$$

with weight $\rho = \frac{1}{2}\sum\limits_{\alpha>0}\alpha$. By (7.13) the "reduced Casimir" applied to a character χ_α, turns into the standard (flat) Laplacian on $\mathbb{T}^n$, evaluated on the trigonometric polynomial in the numerator of (7.13),

$$\Delta_R[\chi_\alpha] = \frac{1}{D}(\Delta_{\mathbb{T}^n} - \langle\rho\mid\rho\rangle)\left[\sum \det(\sigma) e^{i\langle\alpha+\rho\mid h\rangle}\right] = (\langle\alpha+\rho\mid\alpha+\rho\rangle - \langle\rho\mid\rho\rangle)\chi_\alpha,$$

Since all abelian characters $\{e^{i\langle\beta^\sigma\mid h\rangle}:\sigma\in \mathbb{W}\}$ on $\mathbb{T}^n$ have equal norms, we get the result,

$$\boxed{\lambda_\alpha = \|\alpha+\rho\|^2 - \|\rho\|^2} \tag{7.14}$$

The difference between regular representations of group G on itself and on the symmetric space $K\backslash G$ is that the latter selects a particular set of roots, "orthogonal to $\mathfrak{K}$". Also multiplicities of eigenvalues $\{\lambda_\alpha\}$ on G and on $\mathcal{M}$, are different. The former are always equal to $\deg^2(\pi^\alpha)$ by the Frobenius reciprocity, so spectrum of the group-Laplacian,

$$\operatorname{spec}(\Delta^G) = \{\lambda_\alpha = \|\alpha+\rho\|^2 - \|\rho\|^2 : \#(\lambda_\alpha) = d(\pi^\alpha)^2 : \alpha\in\widehat{G}\},$$

the latter have dimension equal to degree of π^α, since (Lemma 1°) spectrum of $R^{\mathfrak{X}}$ is multiplicity-free. This completes the proof of the Theorem.

Remark: In all the above compact cases (group G, or symmetric quotient $\mathcal{M}$) eigenvalues $\{\lambda_\alpha\}$ of the Laplacian (Casimir) are in 1-1 correspondence with irreducible representations (highest weights) $\{\alpha\}$. In fact, the lexicographic ordering of weights $\{\alpha\}$, via a choice of the positive (Cartan) roots in $\mathfrak{H}$ (ot $\mathfrak{T}$), yields the same ordering of eigenvalues:

$$\boxed{\alpha<\beta} \Rightarrow \boxed{\lambda_\alpha < \lambda_\beta};\ \text{for all } \alpha,\beta \text{ in the Weyl chamber.}$$

Indeed, by (7.14) the difference,

$$\lambda_\beta - \lambda_\alpha = \|\beta+\rho\|^2 - \|\alpha+\rho\|^2 = \langle \beta+\alpha+2\rho \mid \beta-\alpha\rangle > 0,$$

since $\langle \rho \mid \gamma\rangle > 0$, for any positive weight γ. We shall conclude the discussion with 2 examples.

7.8. Examples.

• The n-sphere $S^n \simeq SO(n+1)/SO(n)$. The n-sphere Laplacian was analyzed in §4.5. We shall reexamine it in the light of previous discussion. Here subalgebra $\mathfrak{K} \simeq so(n)$, subspace $\mathfrak{P}$ consists of matrices:

$$X = X_b = \begin{bmatrix} 0 & b \\ -{}^{T}b & 0 \end{bmatrix}$$

with columnar vector $b \in \mathbb{R}^n$; isotropy subgroup $K = SO(n)$ acts on $\mathfrak{P}$ by rotations:

$$u^{-1} X_b u = X_{u(b)}.$$

Maximal abelian subalgebras $\mathfrak{T} \subset \mathfrak{P}$ are 1-dimensional (rank 1!), and the *exp*-image of $\mathfrak{T}$ becomes a closed geodesics (great circle γ) in S^n. Lattice Λ consists of integers k, and the basic positive root $\{1\}$ has multiplicity $n-1$, hence $\rho\ \frac{n-1}{2}$. The k-th eigenvalue of the Laplacian Δ_{S^n},

$$\boxed{\lambda_k = k(k+n-1) = (k+\rho)^2 - \rho^2;\ \rho = \tfrac{n-1}{2}} \tag{7.15}$$

the eigensubspace $\mathcal{H}_k$ being made of all *spherical harmonics* of degree k, restrictions on $S^n \subset \mathbb{R}^{n+1}$, of all harmonic polynomials (§4.4-5 of ch.4),

$$\mathcal{H}_k = \Big\{ f = \sum_{|\alpha|=k} a_\alpha x^\alpha : \Delta f = 0 \Big\} \text{ in } \mathbb{R}^{n+1}.$$

Furthermore, the $\mathcal{H}_k$-component of the regular representation R of group $G = SO(n+1)$ on space $L^2(S^n)$, is an irreducible representation π^k of weight $k = (k;0;0;...)$ (cf. Theorem 5.1 of §4.5). So

$$R \simeq \bigoplus_0^\infty \pi^k;$$

each π^k entering R with multiplicity 1.

• **Symmetric space** $SU(n)/SO(n)$. Here subalgebra $\mathfrak{K}$ consists of real antisymmetric matrices, while subspace $\mathfrak{P}$ is realized by real symmetric matrices. Indeed, any $su(n)$-element $Z = X+iY$, whose real part $X \in so(n) = \mathfrak{K}$, and imaginary part $Y \in SM_n$. Group $K = SO(n)$ acts on $\mathfrak{P}$ by conjugations:

$$u: X \to u^{-1} X u.$$

Cartan subgroup $\mathfrak{T} \subset \mathfrak{P}$ consists of diagonal matrices, $\mathfrak{T} \simeq \mathbb{R}^{n-1}$ (rank $= n-1$!), and $exp\,\mathfrak{T}$ spans a geodesically flat torus $\gamma \simeq \mathbb{T}^{n-1}$ in $\mathcal{M}$. Positive roots in $\mathfrak{T}$

$\mathfrak{H}\cap\mathfrak{P}\simeq\mathbb{R}^{n-1}$ are of the form

$$H_{ij}=\operatorname{diag}(0;\ldots 1;\ldots -1;\ldots);\ \text{on the } i\text{-th and } j\text{-th place,}$$

the basis consists of

$$H_j=\operatorname{diag}(\ldots 1;-1;\ldots);\ j=1;\ldots n-1;$$

while weights are made of all integral combinations of the basic roots:

$$\alpha=\sum_j k_j H_j=\operatorname{diag}(k_1;\,k_2-k_1;\ldots;-k_{n-1}).$$

The half sum of positive roots:

$$\tfrac{1}{2}\rho=\tfrac{1}{2}(n-1;n-3;\ldots;-n+1),$$

while the inner product

$$\langle H\mid H'\rangle=\operatorname{tr}(ad_H\,ad_{H'})=\sum_{i<j}(h_i-h_j)(h'_i-h'_j).$$

So the eigenvalues of Δ are equal to

$$\boxed{\lambda_\alpha=\sum_{i<j}\left\{(k_{ij}+\rho_{ij})^2-\rho_{ij}^2\right\}}\tag{7.16}$$

where the components of weight $\alpha=(k_{ij})$ are given by

$$k_{ij}=(k_i-k_{i-1})-(k_j-k_{j-1});\ \rho_{ij}=\rho_i-\rho_j=2(j-i).$$

Instead of $(n-1)$-tuple $\alpha=(k_1;\ldots k_{n-1})$ we could use the entries of the diagonal matrix, representing it, $\alpha=\operatorname{diag}(m_1;\ldots m_n)$: where $m_j=k_j-k_{j-1}$; satisfies $\sum m_j=0$; and $m_1\geq m_2\geq\ldots\geq m_n$ (condition of "highest weight", element of the Weyl chamber). Then components $k_{ij}=m_i-m_j$, and we get

$$\boxed{\lambda_\alpha=\sum_{i<j}\left\{[(m_i-m_j)+(j-i)]^2-(j-i)^2\right\}}$$

Let us remark that the rank of symmetric space $\mathcal{M}=SU(n)/SO(n)$ is equal to the rank of group G itself. So both spaces have the identical set of weights. As a consequence, we find that the regular representation $R^{\mathcal{M}}$, consists of all irreducibles $\pi\in\widehat{G}$, but unlike R^G, each π enters $R^{\mathcal{M}}$ with multiplicity 1. So multiplicities, rather than "spectrum of irreducible $\{\pi\}$", mark the difference between regular representations on $\mathcal{M}$ and G).

Problems and Exercises:

1. Show that any $n\times m$ matrix X has a unique representation of the form: $X=uX_tv$, with $u\in SO(n)$; $v\in SO(m)$, and diagonal matrix X_t of (7.4).

2. Use Cartan automorphism θ on group G to show that spherical algebra $\mathcal{L}^K$ is commutative.

   Hint: Cartan θ define an automorphism of functions on $G: f\to f^\theta=f(x^\theta)$, $(f*h)^\theta=f^\theta h^\theta$, while the inversion, $x\to x^{-1}$, yields an anti-automorphism:
   $$f\to \check{f},\ (f*h)\check{}=\check{h}*\check{f}.$$
   But on spherical functions $\{f\}$, both θ and $\check{}$ are equal: $f^\theta(x)=f(x^{-1})$!.

3. Let $\mathcal{L}$ be an algebra with an idempotent (projection) element χ, and let $\mathcal{A}=\chi\mathcal{L}\chi$ denote the corresponding "spherical" subalgebra of $\mathcal{L}$. If representation T of $\mathcal{L}$ in space $\mathcal{V}$ is irreducible, and subspace $\mathcal{V}_0=T_\chi(\mathcal{V})$ is the image of T_χ, *then* $\mathcal{V}_0$ is invariant under $\mathcal{A}$ and the restriction $T_f\,|\,\mathcal{V}_0$ $(f\in\mathcal{A})$ is an irreducible representation of $\mathcal{A}$.

   Hint: assume that $\mathcal{V}_0$ contains a proper $\mathcal{A}$-invariant subspace $\mathcal{V}'_0$, take the $\mathcal{L}$-span of $\mathcal{V}'_0$, $\mathcal{V}'=\{T_f(\mathcal{V}'_0): f\in\mathcal{L}\}$, and show that $T_\chi(\mathcal{V}')\subset\mathcal{V}'_0$, hence $\mathcal{V}'\subsetneq\mathcal{V}$, contradiction.

   Apply this result to a pair $G\supset K$-compact, where algebra $\mathcal{L}=L^1(G)$, χ any (irreducible) character of K (e.g. $\chi=1_K$), and $\mathcal{A}=\mathcal{L}^K=\chi*\mathcal{L}*\chi$ - the spherical subalgebra.

4. (i) Show that the radially reduced Laplacian on rank-one spaces $\mathbb{R}^n$; S^n; $\mathbb{H}^n$ has the form $\Delta=\frac{1}{J}\partial J\partial$, where J denotes the radially reduced Haar density.

   (ii) Conjugate Δ with $\sqrt{J}$, $L\to L^*=\sqrt{J}L\frac{1}{\sqrt{J}}$, and derive the resulting flat Schrödinger operator,
   $$L^*=\partial^2-\left[\frac{1}{2}\left(\frac{J'}{J}\right)'+\frac{1}{4}\left(\frac{J'}{J}\right)^2\right]$$
   (iii) Use general form of L^* to compute radial part of the Laplacian on rank-one symmetric spaces $\mathbb{R}^n$; S^n; $\mathbb{H}^n$.

   (iv) Show that potential
   $$V=\begin{cases}\dfrac{(n-1)(n-3)}{4r^2};\ \text{on } \mathbb{R}^n\\[2ex] \dfrac{(n-1)^2}{4}-\dfrac{(n-1)(n+1)}{4\sin^2 r};\ \text{on } S^n\\[2ex] \dfrac{(n-1)^2}{4}+\dfrac{(n-1)(n-3)}{4\sinh^2 r};\ \text{on } \mathbb{H}^n\end{cases} \tag{7.17}$$

   Remark: Dimension $n{=}3$ is exceptional for the flat (Euclidian) and hyperbolic cases. Here V turns into a constant, so the radially-reduced Laplacian is conjugate to a flat Laplacian (2-nd derivative) on $\mathbb{R}_+$, hence solutions (Green's functions) of such problems are expressed in terms of elementary functions (cf. §2.3).

5. Apply problem 7 of §5.1 and results of chapters 3,4, to decompose the regular representation R of $SO(4)$ on the 3-sphere, and derive the spectrum of he Laplacian on S^3. Show that
   $$R\simeq\bigoplus_0^\infty \pi^k\otimes\pi^k,$$
   where $\{\pi^k: k{=}0;1;...\}$ are standard irreducible representations of $SU(2)$ of §4.2, and $SO(4)=SU(2)\times SU(2)/\{\pm I\}$. Hint: identify R with the double left-right action $(u,v): f\to f(u^{-1}xv)$, on $L^2(G)$, $G=SU(2)$, and use a decomposition of $L^2(G)$ into the sum of matrix algebras $\bigoplus_\pi Mat_{d(\pi)}$ (chapter 3).

6. **Casimirs and invariant polynomials on $\mathfrak{G}^*$:** (i) The enveloping algebra $\mathfrak{U}(\mathfrak{G})$ can be identified with all polynomials on $\mathfrak{G}^*$- the dual to Lie algebra $\mathfrak{G}$, by taking symmetrized products,

$$\Sigma: X = X_1 \dots X_m \to \bar{X} = \frac{1}{m!}\sum_{\sigma} X_{\sigma(1)} \dots X_{\sigma(m)};\ X_j \in \mathfrak{G} \text{ (sum over all permutations).}$$

Indeed, any symmetrized element $\bar{X} \in \mathfrak{U}(\mathfrak{G})$ gives rise to a polynomial function,

$$p_X(\xi) = \sum c_{1\dots m} \langle X_1 \mid \xi \rangle \dots \langle X_m \mid \xi \rangle;\ \xi \in \mathfrak{G}^*.$$

One should think of $\{p_X(\xi)\}$ as "noncommutative" generalizations of symbols of differential operators.

(ii) Furthermore, the *co-adjoint* action of G on both spaces, $\mathfrak{U}(\mathfrak{G})$ and $\mathfrak{P}(\mathfrak{G}^*)$ coincide (intertwined by Σ),

$$\Sigma:\ \mathrm{ad}_g(X) \to p_X(\mathrm{ad}^*_g(\xi)) \text{ (check).}$$

(iii) The center $\mathfrak{Z}(\mathfrak{G})$ of algebra $\mathfrak{U}(\mathfrak{G})$ (Casimirs) consists of all G-invariant polynomials $\{p_X(\xi)\}$. In case of semisimple algebra $\mathfrak{G}$, the latter can be further reduced.

(iv) (Chevallet): G–invariant polynomials $\{f(\xi)\}$ on $\mathfrak{G}$, when restricted on the Cartan subalgebra $\mathfrak{H} \subset \mathfrak{G}$, become **W**-invariants, where **W** denotes the Weyl group of the pair $\{\mathfrak{G};\mathfrak{H}\}$. Conversely, any **W**-invariant on $\mathfrak{H}$ gives rise to a G–invariant polynomial on $\mathfrak{G}$.

Comments:

The material of §§5.1-6 has become by now fairly standard, and could be found in many textbooks on the representation theory of semisimple Lie groups and algebras ([Ser1]; [We3]; [Hel-2];[Že]). The bulk of results in chapter 5 is a lasting contribution of Hermann Weyl and Eli Cartan. Our presentation was largely influenced and owes a great deal to books [We]; [Že] and [Hel-2] (particularly, in §5.7).

Chapter 6. The Heisenberg group and semidirect products.

§6.1. *Induced representations and Mackey's group extension theory.*

The induction procedure, introduced in §1.2 (chapter 1), and studied in detail in chapter 3 (compact and finite groups), allows one to construct in a canonical way a representation $\{T\}$ of group G, starting from a representation of its subgroup H. In this section we shall extend the results of chapter 3 to infinite semidirect products. The general theory will be illustrated by examples of the Euclidian motion, affine and Poincare groups.

After the standard induction procedure we turn to an important modification, called *holomorphic induction.* The latter casts a different light on the representation theory of classical compact semisimple groups G. (chapters 4-5). It turns out that all irreducible representations of such G are *holomorphically induced,* via the celebrated Borel-Weil-Bott Theorem. Its meaning and contents will be further explored in §6.3.

1.1. **Induction:** We remind the basic constructions and results of §1.2 and §2.2. Induced representations arise in the context of group G acting on a homogeneous G-space $\mathfrak{X} \simeq H\backslash G$, $g: x \to x^g$. Here $H = H_{x_0}$ denotes the stabilizer of a fixed point $x_0 \in \mathfrak{X}$. The G-action on space $\mathfrak{X}$ defines a H-valued cocycle, $h: \mathfrak{X} \times G \to H$,

$$h(x; gu) = h(x; g)h(x^g; u), \text{ for all } x \in \mathfrak{X} \text{ and } g, u \in G. \tag{1.1}$$

Given a representation S of subgroup H in space $\mathcal{V}$, the induced representation $T = \mathrm{ind}(S \mid H; G)$ of G, acts on a suitable space of $\mathcal{V}$-valued functions on $\mathfrak{X}$ (continuous, L^2, etc.) by operators

$$T_g f(x) = S_{h(x,g)}[f(x^g)];\ g \in G,\ f = f(x). \tag{1.2}$$

The L^2-function space on $\mathfrak{X}$ in (1.2) is taken with respect to a G-invariant measure dx (in case it exists). Otherwise, we pick any measure $\mathrm{d}\mu(x)$, whose translates $\{\mathrm{d}\mu(x^g): g \in G\}$ are absolutely continuous with respect to $\mathrm{d}\mu$,

$$\frac{d\mu(x^g)}{d\mu(x)} = \rho(x; g),$$

and modify (1.2) by the factor $\sqrt{\rho(x;g)}$,

$$T_g f = \sqrt{\rho(x; g)} S_{h(x;g)}[f(x^g)]. \tag{1.3}$$

Factor $\sqrt{\rho}$ makes representation T in (1.3) unitary with respect to $L^2(\mathfrak{X};\mathrm{d}\mu) \otimes \mathcal{V}$-inner product. Cocycle h clearly depends on the choice of coset-representatives $\{\gamma(x): x \in \mathfrak{X}\}$ in G, as $\gamma(x)^g = h(x,g)\gamma(x^g)$, but the resulting induced representation T (1.2), (1.3) is independent. Let us also remark that equivalent representations $S^1 \sim S^2$ of H induce equivalent representations of G.

Another description of induced T_g is given in terms of an associated *G-invariant vector bundle,*

$$\mathcal{W} = \bigcup_{x \in \mathfrak{X}} \mathcal{V}_x,$$

over space $\mathfrak{X} = H\backslash G$, with fibers $\mathcal{V}_x \simeq \mathcal{V}$. Group G acts on $\mathcal{W}$ and on base $\mathfrak{X}$, and two actions commute with the natural projection $P: \mathcal{W} \to \mathfrak{X}$. Furthermore, element g takes fiber $\mathcal{V}_x \to \mathcal{V}_{xg}$, and thus defines a linear map $\sigma(x,g): \mathcal{V}_x \to \mathcal{V}_{xg}$; that satisfies a cocycle condition (1.1), a consequence of the group multiplication. Hence, stabilizer subgroup H_x is linearly represented on the fiber-space $\mathcal{V}_x$,

$$h \to \sigma(x;h) = S^x(h).$$

The G-action on the vector-bundle yields the induced action by linear operators on vector-spaces of *cross-sections* $\Gamma(\mathfrak{X};\mathcal{W}) = \{f(x) \in \mathcal{V}_x\}$, equipped with the standard $(L^2, L^\infty$, etc.) norm[1],

$$(T_g f)(x) = \sigma(x,g)^{-1}[f(x^g)];\; f \in \Gamma(\mathfrak{X};\mathcal{W}). \tag{1.4}$$

A third realization of T_g is given in certain subspaces of $\mathcal{V}$-valued functions on G, like $\mathfrak{C}(G) \otimes \mathcal{V}$; $L^2(G) \otimes \mathcal{V}$, etc. Namely, we consider functions, that *transform according to* S, under left translations with H,

$$\mathfrak{C}(G;S) = \{f(g): f(h^{-1}g) = S_h[f(g)],\ \text{all } h \in H,\ g \in G\}. \tag{1.5}$$

Obviously, space $\mathfrak{C}(G;S)$ is invariant under right translates with $g \in G$, so we get a representation

$$T_g f(x) = f(xg);\ \text{all } x, g \in G. \tag{1.6}$$

In other words T is realized as a subrepresentation of certain multiple of the regular representation on G, $T \subset R^G \otimes I_{d(S)}$.

One can easily check that all three definitions of T, (1.2); (1.6) and (1.6), are equivalent by writing an intertwining operator, $W: \mathfrak{C}(G;S) \to \mathfrak{C}(\mathfrak{X};\mathcal{V})$. Indeed, each function $F(g)$ on G, that transforms according to S, is uniquely determined by its values at coset representatives $\{\gamma_x : x \in \mathfrak{X}\}$. So we get $W: F(g) \to f(x) = F(\gamma_x)$. The second construction also demonstrates independence of T of the choice of coset representatives $\{\gamma_x\}$!

1.2. Commutator algebra and the irreducibility test for Ind. Our next goal is to characterize the commutator algebra Com(T), in particular to find conditions for Com(T) to be scalar, i.e. T - irreducible.

[1] All fiber-spaces are assumed to possess a G-invariant metric, and space $\mathfrak{X}$ is assumed to have a G-invariant volume element.

By a general functional-analytic result[2]: *any linear operator W on space $\mathfrak{C}(\mathfrak{X};\mathcal{V})$ is given by a (distributional) operator-valued kernel $A(x,y): \mathfrak{X}\times\mathfrak{X}\to\mathfrak{B}(\mathcal{V})$* --the algebra of linear operators on $\mathcal{V}$,

$$Wf(x) = \int_{\mathfrak{X}} A(x;y)f(y)dy;$$

For each $x\in\mathfrak{X}$ we denote by σ^X a representation of stabilizer H_X. Group G acts on the product $\mathfrak{X}\times\mathfrak{X}$, $g:(x;x)\to(x^g;x^g)$, and splits it into the union of G-orbits $\{\omega\}$: one of them is the main diagonal $\omega_0=\{(x;x)\}$, all others have a marked point $(x_0;y)$, hence $\{\omega=\omega(x_0;y)\}$ are identified with H_0-orbits in $\mathfrak{X}$ (H_0-stabilizer of x_0). For instance, the orbit-space of the orthogonal group on the product of $(n-1)$-spheres is

$$S^{n-1}\times S^{n-1}/SO(n)\simeq S^{n-1}/SO(n-1)=[-1;1].$$

The commutator algebra of the induced representation can be characterized by the Mackey Test (§3.2).

Theorem 1: *The commutator algebra of $T=\operatorname{ind}(\sigma\mid H;G)$ consists of all operator-valued integral kernels $A(x,y)$ on $\mathfrak{X}\times\mathfrak{X}$, satisfying,*

$$\boxed{A(x^g;y^g)=\sigma^{-1}_{h(x;g)}\,A(x;y)\sigma_{h(y;g)}} \tag{1.7}$$

for all (almost all) $x;y\in\mathfrak{X}$, and $g\in G$.

Formula (1.7) implies that any kernel $A(x;y)$ is uniquely determined on orbit $\omega\subset\mathfrak{X}\times\mathfrak{X}$ by its value a single point (e.g. $(x_0;y)\in\omega$). Moreover, the diagonal value $\{A(x;x)\}$ belongs to the commutator $\operatorname{Com}(\sigma^x\mid H_x)\simeq\operatorname{Com}(\sigma\mid H)$, while off-diagonal entries $A(x;y)$ lies in the intertwining space $\operatorname{Int}(\sigma^x\mid H_x\cap H_y;\sigma^y\mid H_x\cap H_y)$ for any pair of points $x;y\in\mathfrak{X}$.

By analogy with the finite case (chapter 3) we can formally decompose the commutator algebra of T into the *direct integral* (see subsection 1.7),

$$ComT=\operatorname{Com}(\sigma\mid H)\oint Int(\sigma\mid H\cap H_\omega;\sigma^\omega\mid H\cap H_\omega)d\mu(\omega), \tag{1.8}$$

integration over the orbit-space $(\mathfrak{X}\times\mathfrak{X})/G\simeq\mathfrak{X}/H$. As the result we obtain the following irreducibility test for induced representations.

Theorem 2: *Representation $T=\operatorname{ind}(\sigma\mid H;G)$ is irreducible iff (i) σ is irreducible, and (ii) all intertwining spaces: $\operatorname{Int}(\sigma^x\mid H_x\cap H_y;\sigma^y\mid H_x\cap H_y)=0$, for any pair*

---

[2]kernel $A(x;y)$ is derived by observing that point evaluations, $f\to f(x)=\langle f\mid\delta_x\rangle$ are bounded linear functionals on space $\mathfrak{C}(\mathfrak{X};...)$. Any such functional is given by a finite (Borel) measure, $d\mu$. Hence, $(Wf)(x)=\langle f\mid W^*[\delta_x]\rangle=\int f(y)d\mu_x(y)$. The density factor,
$A(x;y)=\frac{\partial\mu_x(y)}{\partial y}$, gives the requisite "distributional measure", so $\mu_x=W[\delta_x]$,.

$x \neq y$.

The argument, outlined in §2.2 in the finite/discrete case, extends with some technical modifications to infinite topological spaces (manifolds) $\mathfrak{X}$. We shall mention a special case of the normal subgroup $H \subset G$. Here space $\mathfrak{X} = H\backslash G$ becomes the factor-group, all conjugate stabilizers $\{H_x\}_{x \in \mathfrak{X}}$ coincide with H, G-action on H, $h \to g^{-1}hg$, induces the "dual" action[3] on the dual object $\widehat{H}$,

$$g: \sigma \to \sigma^g = \sigma(g^{-1}...g).$$

Note that elements $h \in H$ transform σ into an equivalent representation: $\sigma^h \sim \sigma$. So σ^g depends only on the class of g in $H\backslash G$, i.e. $\sigma^g = \sigma^x$ $(x = Hg \in \mathfrak{X})$. Formal expansion (1.8) amounts to

$$\mathrm{Com}(T) = \mathrm{Com}(\sigma \mid H) \oint \mathrm{Int}(\sigma \mid H; \sigma^\omega \mid H) d\mu(\omega), \tag{1.9}$$

If σ is irreducible and all $\{\sigma^x : x \neq x_0\}$ belong to different equivalence-classes in $\widehat{H}$, then T is also irreducible! This observation will be essential in the study of semidirect products $G = H \rhd U$.

1.3. Characters of induced representations. Character of a finite-D representation T was defined by $\chi_T(g) = \mathrm{tr}\, T_g$. In chapters 3-5 we studied irreducible characters on compact (Lie) groups, and found them to form a family of nice (differentiable) functions, orthogonal in $L^2(G)$. For ∞-D unitary representations $\mathrm{tr}\, T_g$ does not strictly speaking make sense, so characters should be understood as distributions on G, defined via pairing to suitable test-functions $\{f(x)\}$, $\langle \chi_T \mid f \rangle = tr(T_f)$. Precisely, functions $\{f\}$ that produce a trace-class operator T_f (see Appendix B).

For induced representations $T = \mathrm{Ind}(\sigma \mid H; G)$, the corresponding group-algebra operators $T_f = \int_G f(g) T^s_g dg$, are given by integral kernels

$$K_f(x; y) = \int_{H_x} \sigma(x; h\gamma_y) f(h\gamma_y) dh, \tag{1.10}$$

where H_x denotes a stabilizer of point $x \in \mathfrak{X} \simeq H\backslash G$, and coset representatives $\{\gamma_y\}$ map $x \to y$. Picking a suitable class of functions $\{f\}$ on G (typically smooth, compactly supported), one can show integral kernels, $\{K_f(x,y)\}$ to be also smooth and rapidly decaying. Hence, by the standard functional analysis (Appendix B) operators $\{T^s_f : f \in \mathfrak{C}^\infty_0\}$ belong to the trace-class (or Hilbert-Schmidt class) on space $L^2(\mathfrak{X})$ (as well as spaces of holomorphic functions, like Hardy $\mathcal{H}_n$). Furthermore, for representations in

[3]The same holds for any automorphism $\alpha \in Aut(H)$, $\alpha: \sigma \to \sigma^\alpha_h = \sigma(h^\alpha)$!

L^2-spaces, we have

$$\mathrm{tr}T_f^s = \int K_f^s(x,x)\,dx.$$

So character χ_T can be defined as a distribution on space $\mathfrak{C}_0^\infty(G)$,

$$\langle \chi_T \mid f\rangle \stackrel{def}{=} \mathrm{tr}(T_f).$$

It often happens, however, that such distributions $\{\chi_T\}$ have piecewise continuous (or smooth) densities with respect to the natural coordinates on Lie group G. We shall see it to be the case for semidirect products: Euclidian, affine, Heisenberg groups, as well as noncompact semisimple groups, like SL_2 (chapter 7).

In §3.2 we established a general character formula for induced representations

$$\boxed{\chi_T(g) = \frac{1}{|H|}\sum \chi_\sigma(\gamma(x_j)g\gamma(x_j)^{-1})]}, \tag{1.11}$$

summation over all fixed points $\{x_j\} \subset \mathfrak{X}$ of element $g \in G$, equivalently all elements $\{s_j\}$ of G that conjugate g into H. Formula (1.11) can be extended to Lie groups G. We assume that all $g \in G$ (or a sufficiently large set of them) has only finitely many fixed points $\{x_1;...x_m\} \subset \mathfrak{X}$, all of which are nondegenerate, in the sense that Jacobian $g^*(x_j)$ of smooth map $g: x \to x^g$, at $\{x_j\}$ has no eigenvalue 1. Then for such elements g one has,

$$\boxed{\chi_T(g) = \sum_j \frac{\chi_\sigma(s(x_j)gs(x_j)^{-1})}{|\det(g^*(x_j)^{1/2} - g^*(x_j)^{-1/2})|}} \tag{1.12}$$

the summation extends over all fixed points $\{x_j\}$ of g, and $\{s(x_j)\}$ denote coset representatives of points $\{x_j\} \subset H\backslash G$ in group G. Formula (1.12) is closely connected with the Atiyah-Bott trace-formula for elliptic operators on vector bundles [AB].

To proceed further we shall need a brief diversion into a general subject of decomposition and inversion/Plancherel formula on noncompact groups.

1.4. Direct-integral decomposition and Plancherel formula. A *direct integral* $\oint_\Omega \mathcal{H}_s d\mu(s)$, of Hilbert spaces $\{\mathcal{H}_s\}$, labeled by points of a measure-space $(\Omega; d\mu)$, consists of all $\mathcal{H}_s$-valued functions $\{f(s)\}$ on Ω, with the standard L^2-norm,

$$\int_\Omega \|f(s)\|_{\mathcal{H}_s}^2 d\mu(s);$$

We say that an operator (or representation) T in Hilbert space $\mathcal{H}$ is decomposed into the *direct integral* $\oint_\Omega T^s d\mu(s)$, if there exists a family of Hilbert spaces $\{\mathcal{H}_s\}_{s\in\Omega}$, together with a family of operators $\{T^s \text{ in } \mathcal{H}_s\}_{s\in\Omega}$, and an intertwining map

$$\mathcal{F}:\mathcal{H}\to\oint_{\Omega}\mathcal{H}_s\,d\mu(s),$$

that transforms each vector $f\in\mathcal{H}$ (or a dense subspace $\mathcal{H}_0\subset\mathcal{H}$) into a $\mathcal{H}_s$-valued function $\hat{f}(s)$ on Ω, $\mathcal{F}:f\to\hat{f}(s)$, so that operator T goes into a multiplication with an operator-valued function T^s,

$$\mathcal{F}:T[f]\to(T^s f)(s),\text{ for } s\in\Omega.$$

We can also assume map $\mathcal{F}:\mathcal{H}\to\int\mathcal{H}_s d\mu(s)$, to be unitary, so

$$\|f\|_{\mathcal{H}}^2=\int_{\Omega}\|\hat{f}(s)\|_{\mathcal{H}_s}^2 d\mu(s);$$

This notion clearly generalizes the usual direct sum decomposition, and it turns into the latter for atomic measures $d\mu=\sum a_j\delta_{s_j}$. But in general $\{\mathcal{H}_s\}$ are not subspaces of $\mathcal{H}$! The following examples should clarify the concept.

Example: The regular representation R of group $\mathbf{Z}^n$ on $\mathcal{H}=L^2(\mathbf{Z}^n)$, $R_\alpha f=f_{\nu-\alpha}$ is decomposed into the direct integral of 1-D representations $T^\theta=e^{i\theta\cdot\nu}I$ in spaces $\{\mathcal{H}_\theta\simeq\mathbf{C}\}_{\theta\in\mathbf{T}^n}$,

$$R=\oint_{\mathbf{T}^n}e^{i\theta\cdot\nu}d\theta,\text{ and } L^2(\mathbf{Z}^n)\simeq\oint_{\mathbf{T}^n}\mathcal{H}_\theta\,d\theta.$$

Here map $\mathcal{F}: L^2\to\oint\dots$ is the standard Fourier transform on $\mathbf{Z}^n$,

$$\mathcal{F}:f_\nu\to\hat{f}(\theta)=\sum f_\nu e^{i\theta\cdot\nu},$$

and properties (i-ii) are easily verified. A similar decomposition of R into the direct sum/integral of characters (1-D irreducible representations) by the Fourier transform holds on any locally compact abelian group, e.g. $\mathbf{R}^n$. It was also proven for fairly general classes of noncommutative groups and more general representations, the role of characters being played now by unitary irreducible representation of G.

The general noncommutative *Plancherel formula,* due to Segal and Mautner [Seg] extends the classical commutative result,

$$f(0)=\int_{\hat{G}}\hat{f}(\xi)d\mu(\xi),\tag{1.13}$$

integration over the dual (character) group $\hat{G}$ of an abelian group G with respect to the properly normalized Haar measure dξ on $\hat{G}$. On $\mathbb{R}^n$ dμ is equal to $(2\pi)^{-n}$-multiple of the standard Lebesgue measure. A noncommutative analog of (1.13) yields a similar representation for any function f from a dense subspace of $L^1\cap L^2$ (in our case $C_0^\infty(G)$) in terms of irreducible characters on G. Namely,

$$\boxed{f(e)=\int_{\widehat{G}}\langle\chi_\pi\mid f\rangle d\mu(\pi)};\tag{1.14}$$

where $\langle\chi_\pi\mid f\rangle=tr\widehat{f}(\pi)$, and the generalized Fourier transform $\widehat{f}(\pi)$ is defined via integral

$$\widehat{f}(\pi)=\int_G f(x)\pi_x^{-1}dx.$$

Integration in (1.14) extends over the dual object $\widehat{G}$ of G, (the set of equivalence classes of unitary irreducible representations π), and $d\mu(\pi)$ is called the *Plancherel measure* of group G. From the inversion formula (1.14) one can easily get the Plancherel Theorem for L^2-norms of f and $\hat{f}$. Indeed, applying (1.14) to a convolution $f*f^*$, where $f^*(x)=\overline{f(x^{-1})}$, the integrand becomes

$$tr[\widehat{f}(\pi)\widehat{f}(\pi)^*]=\left\|\widehat{f}(\pi)\right\|_{HS}^2 \text{ (Hilbert-Schmidt norm)}$$

so (1.14) turns into

$$\boxed{\int_G|f(x)|^2dx=\int_{\widehat{G}}\left\|\widehat{f}(\pi)\right\|_{HS}^2 d\mu}\tag{1.15}$$

in other words the L^2-norm of f is equal to the L^2-norm of (generalized) Fourier modes of f. Let us also remark that inversion formula (1.14) at a single group element $g_0=e$ yields values $f(g)$ at all other points $\{g\}$,

$$\boxed{f(g)=\int_{\widehat{G}} tr[\widehat{f}(\pi)\,\pi_g^{-1}]\,d\mu(\pi)}\tag{1.16}$$

Finally, formulae (1.14) and (1.15) lead to a *direct-integral* decomposition of the regular representation on G (or $\mathfrak{G}$), into the sum of *primary components*,

$$\boxed{R_g\simeq\oint_{\widehat{G}}\pi_g\otimes I_{d(\pi)}d\mu(\pi)}\tag{1.17}$$

where multiplicities $\{d(\pi)\}$ could be finite or infinite.

To show the latter we consider the family of Hilbert-spaces $\{\mathcal{H}_\pi=\mathrm{HS}(\mathcal{V}_\pi):\ \pi\in\widehat{G}\}$, where $\{\mathcal{V}_\pi\}$ are the representation-spaces of π's, and define the map $\mathcal{F}:f\to\widehat{f}(\pi)$, from $L^2(G)$ into $\oint\mathcal{H}_\pi d\mu(\pi)$. Group G acts on each space $\mathcal{H}_\pi$,

$$g:\widehat{f}(\pi)\to\pi_g\widehat{f}(g),$$

and the resulting action is a $d(\pi)$-multiple of π.

1.5. Semidirect products. For the sake of presentation we restrict ourselves to semidirect products with a commutative normal subgroup H. We shall restate the principal result of §3.3.

Theorem 3: *Irreducible representations T of semidirect products, $G = H \rhd U$, with commutative H, are parametrized by pairs $\{(\omega;\sigma)\}$, where ω - an orbit of U in the dual group $\hat{H}$, and $\sigma \in \hat{U}_x$- an irreducible representation of stabilizer U_x of the point (character) $x \in \omega$. Furthermore, T is equivalent to induced representation,*

$$T \sim ind(\chi \otimes \sigma \mid H \rhd U_x;G).$$

Proof essentially follows the lines of the finite-D case (§3.3). We take an irreducible T of $H \rhd U$, and decompose $T \mid H$ into the direct integral of characters (e.g. $\lambda_x(a) = e^{ix \cdot a}$, on $\mathbf{R}^n$),

$$T \mid H = \oint \lambda_x(h) \otimes m(x) d\mu(x);\ \lambda_x \in \hat{H},$$

where $m(x)$ denotes the multiplicity (finite or infinite) of λ_x in $T \mid H$, and $d\mu(x)$ – the corresponding spectral measure. Irreducibility and U-invariance imply that $d\mu$ is supported on a single U-orbit $\omega \subset \hat{H}$, and coincides with the unique U-invariant measure $d_\omega(x)$ (provided the latter exists). The representation space $\mathcal{V} = \mathcal{V}(T)$ can then be identified with sections of a U-invariant vector bundle over ω, in each fiber-space $\mathcal{V}_x$ $(x \in \omega)$ one gets a representation σ^x of stabilizer U_x. Hence $T \mid U$ turns into an induced representation $\mathrm{ind}(\sigma \mid U_x;U)$. The same holds for the entire representation T on G,

$$T \simeq ind(\lambda_x \otimes \sigma^x \mid H \rhd U_x;G). \tag{1.18}$$

Conversely, any induced T of type (1.18) is shown to be irreducible by the Schur's Lemma. Indeed, any operator W in space $L^2(\omega;\mathcal{V}')$ (L^2-sections of the vector bundle $\mathcal{V}'$ over ω) that commutes with $T \mid H$ must be a multiplication with a (operator-valued) function,

$$(W\psi)(x) = A(x)[\psi(x)].$$

Then irreducibility of inducing σ and the relation $T_h^{-1} W T_h = W$, for all $h \in H$, implies $W = cI$, to be scalar.

Now we shall illustrate the general results with a few specific examples.

Affine group: $G = \mathbf{R} \rhd \mathbf{R}_+^*$ is made of transformations $\{(a;b): x \to ax + b; a > 0\}$ of $\mathbf{R}$. It has a normal subgroup $H = \mathbf{R}$, whose dual $\hat{H} \simeq \mathbf{R}$ *contains* 3 *orbits*: $\omega_+ = \{\lambda > 0\}$; $\omega_- = \{\lambda < 0\}$ and $\omega_0 = \{0\}$. Hence there are 3 types of irreducible representations of G:

1) a pair of ∞-D unitary representations T^+; T^-, corresponding to orbits $\omega_\pm$;

2) one-parameter family of 1-D representations $\{T^\mu : \mu \in \mathbf{R}\}$ of the commutative quotient $\mathbf{R}_+ \simeq G/\mathbf{R}$.

The former are realized in space $L^2(\mathbf{R}_+;\frac{dx}{x})$ by operators,

$$T^{\pm}_{(a,b)}\psi(x) = e^{\pm ibxa}\psi(xa),$$

and 1-D characters $T^{\mu}_{(a,b)} = a^{i\mu}$. The group-algebra operators $\{T_f\}$ are given by integral kernels,

$$K_f(x;y) = \int_{-\infty}^{\infty} f(\tfrac{y}{x};b)e^{\pm ib/y}db.$$

We leave as an exercise to the reader to compute characters of $T^{\pm}$, and to prove the Plancherel/inversion formula,

$$f(e) = \tfrac{1}{2}\mathrm{tr}(T_f^{+} + T_f^{-}),$$

or

$$\int |f|^2 dg = \tfrac{1}{2}\Big(\|\hat{f}(+)\|^2_{HS} + \|\hat{f}(-)\|^2_{HS}\Big),$$

for a suitable class of test-functions $\{f\}$ on G. Here $\hat{f}(\pm)$ denotes operators $\{T_f^{\pm}\}$. In other words the Plancherel measure is concentrated at two ∞-dimensional "fat" points of the dual object: $\hat{G} = \{+\} \cup \{-\} \cup \mathbb{R}$.

Euclidian motion group: $\mathbb{E}_n = \mathbb{R}^n \rhd SO(n)$, has normal subgroup $H = \mathbb{R}^n$. Subgroup $U = SO(n) \subset \mathbb{E}_n$ acts on $\hat{H}$ by rotations, so orbits are spheres $\omega = \omega_r = \{|x| = r > 0\}$, and point $\{0\}$. Irreducible representations $T^{r,\sigma}$ act on spaces $L^2(S;\mathcal{V})$, where $S = S^{n-1} \simeq \omega$; $\mathcal{V} = \mathcal{V}_\sigma$ - the space of $\sigma \in \hat{U}_x$ (stabilizer $U_x \simeq SO(n-1)$),

$$T_{(a,u)}\psi(x) = e^{irx^u \cdot a}\sigma(x;u)[\psi(x^u)]. \tag{1.19}$$

Here x is a unit vector on S, $u \in SO(n)$, $a \in \mathbb{R}^n$, and $x \to x^u$ denotes the orthogonal transformation u applied to the unit x. So $T^{r,\sigma}$ is induced by the representation $\sigma \otimes e^{irx \cdot a}$ of $\mathbb{R}^n \times SO(n-1)$. The trivial orbit $\{0\} \subset \mathbb{R}^n$ has stabilizer $K = SO(n)$. So the corresponding representations are those of the compact factor-group $SO(n) = \mathbb{E}_n/\mathbb{R}^n$. One can easily compute integral kernels $\{K_f(x;y)\}$ of the group-algebra operators T_f and the irreducible characters.

In the simplest case $\mathbb{E}_2$ we get a one-parameter family of ∞-D (induced) representations $\{T^r : r > 0\}$, acting in space $L^2(S^1)$, and a sequence of 1-D representations of $SO(2) \simeq \mathbb{E}_2/\mathbb{R}^2$. The characters of $\{T^r\}$ are computed to be

$$\chi_r(a;u) = 2\pi\delta(u)J_0(r|a|), \tag{1.20}$$

where J_0 is the Bessel function of order 0 (problem 2).

The Plancherel/Inversion formula on $\mathbb{E}_2$ has the form,

$$\boxed{f(e) = \int_0^{\infty} \mathrm{tr}(T_f^r)\,r\,dr} \tag{1.21}$$

with the Plancherel measure $d\mu(r) = rdr$, supported on an ∞-D "$\mathbb{R}_+$-part" of the dual object $\hat{\mathbb{E}}_2 = \mathbb{R}_+ \bigcup \mathbb{Z}$. Similarly for $\mathbb{E}_n$ we get

$$\boxed{f(e) = \sum_{\sigma} \int_0^{\infty} \mathrm{tr}(T_f^{r;\sigma}) r^{n-1} dr} \tag{1.22}$$

The proof of both formulae is left to the reader (problem 2).

The Poincare group $\mathbb{P}_n$ of special relativity is a semidirect product of the Minkowski space $\mathbb{M}^n$, with indefinite metric (+---), and the Lorentz group $K = SO(1;n-1)$ of isometries of $\mathbb{M}^n$. The K-orbits in $\mathbb{M}^n$ fall into 4 classes:

point: $\{0\}$

two-sheet hyperboloids: $\omega_r = \{x : x \cdot x = {x_0}^2 - \sum {x_i}^2 = r > 0\}$

one-sheet hyperboloids: $\omega_r = \{x : x \cdot x = {x_0}^2 - \sum {x_i}^2 = -r < 0\}$

the light-cone: $\omega_0 = \{x : r = x \cdot x = {x_0}^2 - \sum {x_i}^2 = 0\}$

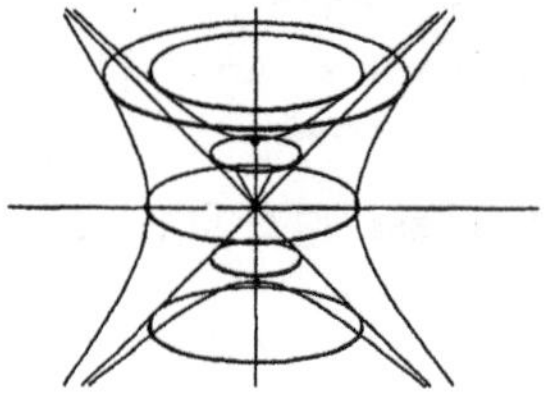

Fig.1: *Orbits of the Poincare group in* $\mathbb{M}^n$, $\{0\}$, *light-cone and 2 kinds of hyperboloids.*

The stabilizer of $\{0\}$ is group $SO(1;n-1)$ itself. Its representation theory, in case n=3, will be developed in the next chapter 7. Two-sheet hyperboloids ω have the compact stabilizer $K_0 \simeq SO(n-1)$, whose representations were studied in chapter 5. They are labeled by highest weights α. The corresponding irreducible representations of $\mathbb{P}_n$ consists of $\{T^{r,\alpha}\}$, $r > 0$ (orbit parameter). They act in space $L^2(\omega) \otimes \mathcal{V}_\alpha$ and are induced by characters $\chi_r(x) = e^{irx_0}$, tensored with α. Such representations were first introduced by E.Wigner [Wig2], and proved to play an important role in the quantum field theory. The one-sheet hyperboloids have non-compact stabilizers, $K_0 = SO(1;n-2)$. Their representations are still induced by characters $\chi_{-r}(x) = e^{irx_1}$, tensored with irreducible representations $\{\sigma\}$ of $K_0 = SO(1;n-2)$. In the simplest case, $n = 3$, stabilizer $K_0 = SO(1;1)$ is $\mathbb{R}$, so $\sigma \in \mathbb{R}$, but in higher dimensions it involves the machinery of chapter 7. The stabilizer of the light-cone is the motion group $K_0 \simeq \mathrm{E}_{n-1} = \mathbb{R}^{n-2} \triangleright SO(n-2)$ (problem 4). Its irreducible representations have form $\sigma = \sigma^{\rho,\alpha}$ $(\rho > 0;\ \alpha \in \widehat{SO}(n-2))$, and the corresponding representations T of $\mathbb{P}_n$ are induced by such σ's.

1.6. Holomorphically induced representations. The standard induced representations $T = \mathrm{ind}(S \mid H; G)$, were realized in several different ways: on sections of G-invariant vector bundles $(\mathfrak{X};\mathcal{V})$, over quotients $\mathfrak{X} = H \backslash G$, where each stabilizer $\{H_x = g_x^{-1} H g_x\}$, acts the fiber $\{\mathcal{V}_x{:}x \in \mathfrak{X}\}$. Alternatively, they were realized in certain subspaces of functions on group G itself (1.7), namely,

$$\mathfrak{C}(G;S) = \{f(hx) = S(h)^{-1} f(x){:}\, h \in H; x \in G\} \subset \mathfrak{C}(\mathrm{G}).$$

For Lie groups G such subspace is characterized in terms of generators $\{X\}$ of the Lie subalgebra $\mathfrak{H}$, acting on $\mathfrak{C}^\infty(G;S)$ by 1-st order differential operators,

$$(\partial_X + S_X) f = 0; \tag{1.23}$$

where ∂_X denotes a left-invariant vector field of element X on G.

The action of H by left translations on G foliates the latter into a union of fibers (H-cosets) and the representation space of T consists of all functions on G, that satisfy certain differential system (1.23) along the H-fibers.

One would like to extend this construction to situations when (1.23) becomes a system of *Cauchy-Riemann equations:* $\partial_{\bar{z}} f = 0$, on certain *complex line bundles* (i.e. manifolds with analytic local structure holomorphic transition functions between fibers). So the representation-space will consists of *holomorphic sections.* More generally, we shall see that homogeneous spaces $\mathfrak{X} = H \backslash G$ of Lie groups can combine both "real" and "complex" parts, so the corresponding functions/sections $\{f(x_1 ... x_k; z_1 ... z_m)\}$ will depend on k real and m complex variables.

We consider a complex Lie algebra $\mathfrak{G}_{\mathbb{C}}$ with a real form $\mathfrak{G}$, and denote by $z \to \bar{z}$, conjugation in $\mathfrak{G}_{\mathbb{C}}$ relative to $\mathfrak{G}_{\mathbb{R}}$. Assume that $\mathfrak{G}_{\mathbb{C}}$ contains a complex subalgebra $\mathfrak{N}$, and a real subalgebra $\mathfrak{H} \subset \mathfrak{G}$, with the following properties:

• $\mathfrak{N} \cap \bar{\mathfrak{N}} = \mathfrak{H}_{\mathbb{C}}$ - a complexification of real $\mathfrak{H}$, and the adjoint action of $\mathfrak{H}$ on $\mathfrak{G}_c$, $\{\mathrm{ad}_X(Y){:}\, X \in \mathfrak{H}\}$ leaves $\mathfrak{N}$ (and $\bar{\mathfrak{N}}$) invariant;

• $\mathfrak{N} + \bar{\mathfrak{N}} = \mathfrak{M}_c$ - complex subalgebra, whose real part $\mathfrak{M} = \mathfrak{M}_r \subset \mathfrak{G}$. Clearly, $\mathfrak{H} \subset \mathfrak{M} \subset \mathfrak{G}$.

Hence we get similar inclusions for the corresponding (real) Lie groups: $H \subset M \subset G$.

Proposition 4: *Homogeneous space $H\backslash G$ has the structure of a fiber bundle with complex fibers $\simeq H\backslash M$ (of complex dim $= m$), over the base $\mathfrak{Y} = M\backslash G$ (of dim $= k$).*

Clearly, quotient $H\backslash G$ is naturally projected onto $M\backslash G = \mathfrak{Y}$,

$$p\text{: "}H\text{-coset"} \to \text{"}M\text{-coset"}.$$

We take a fiber space, $p^{-1}(x) \simeq H\backslash M$, and claim that the latter is isomorphic to a complex quotient $N\backslash M_c$, of two complex subgroups $N; M_c \subset G_c$. It suffices to compare the tangent spaces of both. We observe that $\mathfrak{M}_c = \mathfrak{N} + \mathfrak{M}$, and $\mathfrak{N} \cap \mathfrak{M} = \mathfrak{H}$, hence $\mathfrak{M}/\mathfrak{H} \simeq \text{tangent}(H\backslash M) \simeq \mathfrak{M}_c/\mathfrak{N}$.

Let us illustrate the above constructions for the unitary algebra $\mathfrak{G} = su(n)$. Its complexification $\mathfrak{G}_c = sl(n;\mathbb{C})$ has the (Cartan) conjugation: $X \to -X^*$. We pick a pair of Borel subalgebras $\mathfrak{N}$; $\bar{\mathfrak{N}}$, made of upper/lower triangular matrices:

$$\mathfrak{N} = \left\{ \begin{bmatrix} a & * & * \\ & \ddots & * \\ & & c \end{bmatrix} \right\}; \bar{\mathfrak{N}} = \left\{ \begin{bmatrix} a & & \\ * & \ddots & \\ * & * & c \end{bmatrix} \right\}; \tag{1.24}$$

whose sum $\mathfrak{M}_c = \mathfrak{N} + \bar{\mathfrak{N}}$, comprises entire $\mathfrak{G}_c$ (hence, $\mathfrak{M} = su(n)$!), and whose intersection is the diagonal (Cartan) subalgebra, $\mathfrak{H} = \text{Re}(\mathfrak{N} \cap \bar{\mathfrak{N}}) = \{h = \text{diag}(ia_1;...ia_n)\}$. So subgroup H is the maximal torus $\mathbb{T}^{n-1} \subset G = SU(n)$, and the quotient

$$H\backslash G = H\backslash M = \mathbb{T}^{n-1}\backslash SU(n),$$

acquires the natural complex structure, inherited from space $SL(n)/N$, called *flag manifold* (problem 3).

In the simplest case, $\mathfrak{G} = su(2)$, the complexification $\mathfrak{G}_{\mathbb{C}} = sl(2;\mathbb{C})$ contains the upper and lower-triangular (Borel) subalgebras:

$$\mathfrak{N} = \left\{ \begin{bmatrix} \alpha & \beta \\ & -\alpha \end{bmatrix} \right\}; \bar{\mathfrak{N}} = \left\{ \begin{bmatrix} \alpha & \\ \beta & -\alpha \end{bmatrix} \right\};$$

their sum $\mathfrak{N} + \bar{\mathfrak{N}} = \mathfrak{M}_c = \mathfrak{G}_c$; while the "real part" of the intersection,

$$\text{Re}(\mathfrak{N} \cap \bar{\mathfrak{N}}) = \mathfrak{H} = \left\{ \begin{bmatrix} i\alpha & \\ & -i\alpha \end{bmatrix} \right\}.$$

A similar construction can be given for any simple/semisimple compact algebra $\mathfrak{G}$ (see chapter 5). Here $\mathfrak{N}$; $\bar{\mathfrak{N}}$ are two Borel subalgebras of $\mathfrak{G}_c$, made of all positive/negative root vectors. Their intersection is once again a Cartan subalgebra $\mathfrak{H}$ of G, and the quotient $\mathbb{T}\backslash G$ (of compact Lie group modulo maximal torus), has a natural complex structure (problem 3).

Compact Lie groups, and their quotients $\mathbb{T}\backslash G$ provide one extreme, when the entire homogeneous space turns into a single complex leaf. Another extreme arises when algebra $\mathfrak{N}=\bar{\mathfrak{N}}=\mathfrak{M}_c$, hence $\mathfrak{H}=\mathfrak{N}_r=\mathfrak{M}$, so $H=M\subset G$, and we don't get any complex leaves, the entire quotient being the standard $H\backslash G$!

In the context of Proposition 4 we consider functions/sections $\{f(x;z)\}$ on space $\mathfrak{X}=H\backslash G$, holomorphic in variables $z=(z_j)$, and define induced representations in those. Precisely, a character (representation) S of the real subgroup H, extends to a holomorphic representation $\sigma(X)$ of the complex algebra $\mathfrak{N}$. Then we replace condition (1.24) with

$$[\partial_{\overline{Z}}+\sigma(Z)]f=0;\ \text{for all } Z\in\mathfrak{N}. \tag{1.25}$$

and call the corresponding function-space $\mathfrak{C}(G;\sigma\mid N)$.

Here $\partial_{\overline{Z}}$ indicates the Cauchy-Riemann $\bar{\partial}$-derivative in the complex (fiber) variables. Indeed, in the simplest case: $\mathfrak{G}=\mathbb{R}^2$; $\mathfrak{G}_c=\mathbb{C}^2$; $\mathfrak{N}=\mathrm{Span}(X+iY)$, $\{X;Y\}$-basis in $\mathbb{R}^2$, and $\sigma(X+iY)=\lambda$, space $H\backslash G=\{0\}\backslash\mathbb{R}^2\simeq\mathbb{C}$, and (1.25) turns into the standard $\bar{\partial}$- equation,

$$\bar{\partial}f=\tfrac{1}{2}(\partial_x+i\partial_y)f=-\lambda f,$$

whose solution is

$$f=e^{-\lambda\bar{z}}\times\text{holomorphic function}.$$

The resulting representations of G in spaces $\mathfrak{C}(G;\sigma\mid N)$ are called (partially) *holomorphically-induced*, and denoted by $\mathrm{ind}(S\mid H;\sigma\mid N;G)$.

Next one needs to introduce a G-invariant product in representation spaces $\mathfrak{C}(G;\sigma\mid N)$. In some cases it proves to be possible, for instance, if the complex component of $H\backslash G$ is compact (like $SU(n)$), or is equivalent to a bounded domain in $\mathbb{C}^n$. In such cases, Hilbert space $\mathfrak{L}(G;\sigma\mid N)$ turns into a closed subspace of $L^2(G;\Upsilon_\sigma)$ - the representation space of $\mathrm{ind}(\sigma\mid H;G)$.

We shall not delve deeply into the subject of holomorphically induced representations here, but refer to [Kir1] for further details. Let us mention, however, an important result, due to Borel-Weil-Bott. They proved, that the holomorphic induction on compact Lie groups G, by characters σ of a Borel subgroup N (1.24), yields all irreducible all irreducible representation $\pi \in \hat{G}$.

To give a precise statement, we recall that irreducible $\pi \in \hat{G}$ are labeled by weights (characters) $\{\lambda\}$ of the Cartan subgroup $H \simeq \mathbb{T}^n \subset G$, the highest weight λ is unique, i.e. the corresponding eigenspace $\mathcal{V}_\lambda$ is 1-dimensional. Weight λ, rather eigenspace $\mathcal{V}_\lambda$, gives rise to a *holomorphic line bundle* $\mathcal{E} = \mathcal{E}(\lambda)$ over the complex manifold $\mathfrak{X} = \mathbb{T}^n \backslash G$ with fibers $\simeq \mathcal{V}_\lambda$. Hence, we get a finite-dimensional vector space $\mathcal{L}(\mathfrak{X};\mathcal{E})$ of holomorphic sections of $\mathcal{E}$. It turns out that $\dim \mathcal{L} = \deg(\pi^\lambda)$. Furthermore,

Theorem (Borel-Weil-Bott): *The natural (induced) action of G on $\mathfrak{X}$ and $\mathcal{E}(\lambda)$, considered on space $\mathcal{L}$ of holomorphic sections is equivalent to π^λ.*

Proof exploits the highest-weight theory of chapter 5 and consists of several steps.

1) The Weyl "unitary trick" allows to identify representations π of G with "holomorphic representations" of $G_{\mathbf{C}}$ - its complexification, so that matrix-entries of π on $G_{\mathbf{C}}$ become analytic continuations of entries $\{f(x) = \langle \pi_x \xi \mid \eta \rangle : x \in G\}$.

2) We realize π on by functions (matrix-entries) on $G_{\mathbf{C}}$ (or its real non-compact form $G_{\mathbf{R}}$), that satisfy

$$\begin{cases} f(\zeta x) = f(x); \text{ for all } \zeta \in N_- \\ f(hx) = \lambda(h) f(x); \text{ for all } h \in H \end{cases} \tag{1.26}$$

where H is the Cartan (diagonal) subgroup of $G_{\mathbf{C}}$ ($G_{\mathbf{R}}$), while N_- is made of negative root-vectors (lower-triangular matrices). The embedding of space $\mathcal{V} = \mathcal{V}_\pi$ into functions (1.26) is given by the lowest-weight vector $\eta_0 \in \mathcal{V}$,

$$\xi \to f_\xi(x) = \langle \pi_x(\xi) \mid \eta_0 \rangle.$$

Condition (1.26) can be written in the infinitesimal form as

$$\begin{cases} \partial_Y f = 0, \text{ for all negative root-vectors } Y = Y_\alpha \in \mathfrak{N}_- \\ \partial_H f(x) = \langle \lambda \mid H \rangle f(x); \text{ for } H \text{ in the Cartan subalgebra } \mathfrak{H} \end{cases} \tag{1.27}$$

where ∂_Y; ∂_H denote the right-invariant vector fields on $G_{\mathbf{R}}$ ($G_{\mathbf{C}}$) (generators of left translations).

3) An elegant observation of Borel-Weil-Bott was to note that condition (1.27) has an equivalent form in terms of holomorphic differentiation ∂_Z in the complex domain $G_{\mathbf{C}}$, where $\overline{Z}_\alpha = X_\alpha + iY_\alpha$ (α - root of $\mathfrak{G}$). Namely,

$$\begin{cases} \partial_{\overline{Z}} f = 0; \text{ for all } \overline{Z} = \overline{Z}_\alpha \\ \partial_{iH} f = i\langle \lambda \mid H \rangle; \text{ for all } H \in i\mathfrak{H} \text{ (compact Cartán part)} \end{cases} \tag{1.28}$$

One first observes that space (1.28) is finite-dimensional (it suffices to prove it for the restrictions $\{f\}$ on the compact quotient $\mathbb{T}\backslash G \subset H\backslash G_{\mathbf{C}}$, then apply a unique analytic continuation[4] from the former to the latter). Hence, space (1.28) has the highest weight function $\phi(x)$. Since,

$$\boxed{\partial_{\bar{Z}}\,\phi = 0, \text{ for all } \bar{Z}} \text{ implies } \boxed{\partial_Y \phi = 0, \text{ for all } Y \text{ (negative root-vectors)}}$$

on $G_{\mathbf{R}}$, hence (analytic continuation) on $G_{\mathbf{C}}$, and ϕ is an eigenfunction of Cartan $\mathfrak{H}$, it follows that ϕ is equal to the standard matrix-entry $\phi_0 = \langle \pi_x \xi_0 \mid \eta_0 \rangle$, that couples the highest and the lowest weight λ. Thus we get

$$\text{"space(1.28)"} = \text{"space (1.27)"},$$

and two (induced) representations are equivalent. Finally, it remains to observe that space (1.28), when restricted on compact $G \subset G_{\mathbf{C}}$, defines a holomorphic line-bundle $\mathcal{E} = \mathcal{E}(\lambda)$, and all analytic sections of $\mathcal{E}$, so π^{λ} becomes a holomorphically induced representation in "sections" $\mathcal{L}(\mathcal{E})$, QED.

We shall illustrate the holomorphic induction and the Borel-Weil-Bott Theorem for group $SU(2)$. Here a unitary character $S(\theta) = e^{im\theta}$ extends through a holomorphic representation/character

$$\sigma\left(\exp\begin{bmatrix} \alpha & \beta \\ & -\alpha \end{bmatrix}\right) = e^{im\alpha}. \tag{1.29}$$

It turns out that character $\sigma = \sigma_m$ gives rise to a complex line-bundle $\mathcal{L}_m$ over the quotient-space $\mathfrak{S} = SU(2)/\mathbb{T}$, whose *first Chern-class*[5] is equal to m. The space of holomorphic sections of $\mathcal{L}_m$ has $\dim = m$, and the resulting holomorphically-induced

[4]The simplest prototype of the triple $G_{\mathbf{C}};G_{\mathbf{R}};G$ are multiplicative groups: $\mathbf{C}^*$; $\mathbf{R}_+$ and $\mathbb{T} = \{e^{i\theta}\}$ (unit circle). Representations (characters) $\{\chi_m(\theta) = e^{im\theta}\}$ of $\mathbb{T}$ have a unique extension to holomorphic characters $\{\chi_m(z) = z^m\}$ of real group $\mathbf{R}_+$ and complex $\mathbf{C}^*$.

[5]Chern classes describe the degree of "twisting" of a line (or more general vector) bundle $\mathcal{L}$ over $\mathfrak{S}$. The topological structure of any vector bundle is determined by a family of transition coefficients $\{g_{UV}(x): U,V\text{- a pair of neighborhoods of } x \text{ in } \mathfrak{S}\}$. Coefficients $\{g_{UV}\}$ take on values in a structure group G of $\mathcal{L}$ (GL, SL, SO, SU, etc.), which depends on specific geometric features of $\mathcal{L}$ (Riemannian, Hermitian, holomorphic, etc.). They satisfy the cocycle condition:

$$g_{UV}(x)g_{VW}(x)g_{WU}(x) = 1, \text{ for all triples } U,V,W \ni x; \text{ and } g_{VU} = g_{UV}^{-1}$$

So $\{g_{UV}\}$ defines a G-valued 2-nd cohomology class on $\mathfrak{S}$. But for line bundles ($\dim[\text{fiber}] = 1$), group $G = \mathbf{C}^*$, so multiplicative G-cocycle can be turned into the standard (additive) 2-cocycle:

$$g_{UV} = e^{i\alpha(U;V)},$$

The resulting $\alpha \in H^2(\mathfrak{S};\mathbf{C})$ is the first Chern class of $\mathcal{L}$. In fact, Chern classes always define integral cohomologies, $\alpha \in H^2(\mathfrak{S};\mathbb{Z})$, in other words, differential 2-form $\alpha = \sum \alpha_{ij} dx_i \wedge dx_j$ on $\mathfrak{S}$, integrated over any 2-cycle, (closed 2-D submanifold) $S \subset \mathfrak{S}$, yields integral values, $\oint_S \alpha \in \mathbb{Z}$. The second cohomology group of the sphere $S^2 = SU(2)/\mathbb{T}$ is well known to consist of integers, hence $\alpha = m$.

representation is equivalent to the familiar $\frac{m}{2}$-spin representation $\pi^{m/2}$ of §4.2. The details will be outlined in §6.3 (example 3).

In conclusion let us remark that the induction procedure in any of its modifications: standard, holomorphic, or, more general, partially holomorphic ("combined") seems to provide a universal method of constructing irreducible representations. We shall further explore its meaning and significance in §6.3, based on Kirillov's orbit method and geometric quantization.

Problems and Exercises:

1. Establish formula (1.24) for characters of irreducible representations (Hint: let $g_0 \in G$ and x_0 be its fixed point. Show that the map $(x;h) \to s(x)^{-1}hs(x)$, from $\mathfrak{X} \times \mathrm{H} \to \mathrm{G}$ is a diffeomorphism, that takes invariant measure $dx\,dh$ on $\mathfrak{X} \times \mathrm{H}$, into the measure $d\mu(g)$ on G, whose density relative to the Haar measure dg at the point g_0 is equal to $|\det(1 - g_0^*)|$. Use also formula (1.15) for induced representations and the known relation, that gives Jacobian g_0^*, as the ratio of two modular functions on G and H,
$$\frac{\Delta_G(h)}{\Delta_H(h)} = \det(g_0^*).$$
The modular function on a locally compact group G is a ratio of the left-to-right invariant Haar measures:
$$\Delta(g) = \frac{d_l(g)}{d_r(g)}.$$
On Lie group G, $\Delta(g) = \det(\mathrm{Ad}_g)$.

2. i) Derive the character formulae for representations $\{T^r\}$ of E_2 and $\{T^{r,\sigma}\}$ of E_n.

   ii) Derive the Plancherel-inversion formulae (1.21)-(1.22) for motion groups.

3. Establish the complex structure of the quotient-space $\mathfrak{X} = SU(n)/\mathbb{T}^{n-1}$, using a parametrization by the flag-manifold $SL(n;\mathbb{C})/B_n$.

4. Find stabilizers of orbits of the Lorentz group $SO(1;n-1)$ in $\mathbf{M}^n$. Show that stabilizer K_0 of the light-cone coincides with the motion group E_{n-1} (Hint: write stabilizer algebra $\mathfrak{K}_0$ of point (1;-1;0;...) by block matrices,
$$\begin{bmatrix} 0 & 0 & a_i & \cdots \\ 0 & 0 & -a_i & \\ {}^{T}a_i & {}^{T}a_i & b_{ij} & \\ \vdots & & & \ddots \end{bmatrix}$$
the off-diagonal $2 \times (n-2)$ blocks $\{{a \atop -a}\}$; $\{{}^{T}a; {}^{T}a\}$ forms a $\mathfrak{P}$-component of the Cartan decomposition of $\mathfrak{K}_0$ (§5.7); show that the commutator $[\mathfrak{P};\mathfrak{P}]$ is zero!).

§6.2. The Heisenberg group and the oscillator representation.

The Heisenberg group plays the fundamental role in many areas of harmonic analysis, differential equations, number theory and quantum Physics. It gives a mathematical formulation of the Heisenberg uncertainty principle of quantum mechanics, and reveals close connections to the harmonic oscillator. The latter in turn gives rise to the fundamental creation-annihilation structure of the quantum field theory.

In this section we shall develop the representation theory of the Heisenberg group, based on Stone-von Neumann Theorem. Then we apply it to spectral theory of the harmonic oscillator in $\mathbf{R}^n$, establish connections of the Heisenberg group to symplectic/metaplectic groups and the Weyl algebra, and construct the oscillator representation.

2.1. The Heisenberg group $\mathbb{H}_n$ consists of all triples $\{g=(x,y,t): x,y\in\mathbb{R}^n;\ t\in\mathbb{R}\}$ with multiplication given by

$$g\cdot g' = (x,y;t)(x',y';t') = (x+x';y+y';t+t'+x\cdot y'). \tag{2.1}$$

Its Lie algebra $\mathfrak{H}_n \simeq \mathbb{R}^{2n+1}$ is also made of triples $\{X=(x,y;t)\}$ with the Lie bracket

$$[X;X'] = (0,0;x\cdot y' - y\cdot x').$$

Group $\mathbb{H}_n$ (respectively, algebra $\mathfrak{H}_n$) has a 1-D center $Z=\{(0,0;t)\}\simeq\mathbb{R}$, and the commutative factor-group $G/Z\simeq\mathbb{R}^{2n}$, so $\mathbb{H}_n$ forms a *2-step nilpotent group*: $[G;G]\subset Z$. In fact, $\mathbb{H}_n$ is the simplest among all noncommutative (nilpotent) groups, the *cocycle*[6]

$$\phi(g;g') = x\cdot y', \text{ from } G\times G\to Z, \tag{2.2}$$

measuring a deviation from commutativity. Group $\mathbb{H}_n$ can be viewed as a semidirect product, $G=H\rhd N$, of two commutative subgroups: $H=\{(0,y;t)\}\simeq\mathbb{R}^{n+1}$, and $N=\{(x,0;0)\}\simeq\mathbb{R}^n$, with N acting on H by *unipotent automorpisms,*

$$\alpha_x: (y;t)\to(y;\ t+x\cdot y). \tag{2.3}$$

---

[6]A function $\phi(x;y): G\times G\to\mathbf{R}$, on group G with values in $\mathbf{R}$ (or in more general commutative group Z) is called a 2-*cocycle*, if it satisfies:

$$\phi(a;b)-\phi(a;bc)+\phi(ab;c)-\phi(b;c)=0, \text{ for all } a,b,c\in G.$$

Here we use an additive convention $\{\pm\}$ for the group operation in Z, and multiplicative for G. In other words, the *coboundary*, $\partial\phi(a;b;c)=0$, identically. A 2-cocycle is trivial (*coboundary*), if it can be expressed through a single-variable function $\psi(a)$,

$$\phi(a;b)=\psi(a)-\psi(ab)+\psi(b)=\partial\psi(a;b), \text{ for all } a,b\in G.$$

Any 2-cocycle gives rise to a *central extension* of G by Z, i.e. group H with center $\mathfrak{Z}(H)\simeq Z$, so that $H/Z\simeq G$. The elements of H are pairs $\{(x;t): x\in G; t\in Z\}$, and the multiplication is defined by ϕ,

$$(x;t)\cdot(y;s)=(xy;t+s+\phi(x;y)).$$

The cocycle condition ensures associativity of the group multiplication in H, while trivial cocycle ϕ yields the trivial central extensions, direct product, $H\simeq Z\times G$ (problem 1).

It is often convenient to write $\mathbb{H}_n$ in the complex form: $\mathbb{C}^n \times \mathbb{R}$ with the product

$$(z;t)\cdot(w;\tau) = (z+w; t+\tau+\Im z\cdot\overline{w}).$$

Here cocycle ϕ of (2.2) is replaced by an equivalent (antisymmetric) cocycle (problem 1),

$$\phi'(a;b) = \Im z\cdot\overline{w} = \frac{1}{2i}(z\cdot\overline{w}-\overline{z}\cdot w). \tag{2.4}$$

One might wonder, to what extent the Heisenberg group is unique among all 2-step nilpotent groups with 1-D center. The answer proves to be positive, within the class of *bilinear central extensions*, i.e. extensions defined by bilinear forms, $\phi(a;b) = Qa\cdot b$, $a,b\in\mathbf{R}^n$. Trivial ϕ clearly correspond to symmetric (quadratic) forms Q. So nontrivial possible central extensions, should correspond to the quotient-space of "all bilinear forms $\{Q\}$", modulo "symmetric $\{Q\}$". In other words central (bilinear) extensions of $\mathbf{R}^n$ are labeled by all antisymmetric matrices $\{Q\}$. But any such Q can be brought by the change of basis into the form,

$$Q = \begin{bmatrix} & I & \\ -I & & \\ & & 0 \end{bmatrix},$$

or the direct sum of such matrices. This means that space $\mathbf{R}^n$ is decomposed into the sum, $\mathbf{R}^{2m}\oplus\mathbf{R}^k$, where Q is nondegenerate on the former (like ϕ' of (2.4)), and annihilates the latter. The corresponding Q-central extensions is obviously isomorphic to the sum $\mathbb{H}_m\times\mathbb{R}^k$ (Heisenberg plus abelian).

Sometimes one considers the factor-group $\mathbb{H}_n$, modulo a discrete subgroup $\mathbb{Z} = \{(0,0;k)\}$ of the center Z, and calls it the Heisenberg group. In this form space $\mathbb{H}_n$ is identified with $\mathbb{C}^n\times\mathbb{T}$, and the multiplication is given by

$$(z;t)\cdot(w;s) = (z+w; ts\,e^{i\Im z\cdot\overline{w}});\ t,s \text{ - unit complex numbers.}$$

Finally, let us remark that $\mathbb{H}_1$ can be realized by 3×3 upper triangular matrices

$$\mathbb{H}_1 = \left\{\begin{bmatrix} 1 & x & t \\ & 1 & y \\ & & 1 \end{bmatrix} : x,y,t\in\mathbb{R}\right\},$$

with the standard matrix multiplication. Similarly, $\mathbb{H}_n$ is realized by $(n+2)\times(n+2)$-matrices, with row-vector $\vec{x}$; column-vector $\vec{y}$ and the $n\times n$-identity diagonal block in the middle.

Lie algebra $\mathfrak{H}_n$ has 3 types of generators:

$$p_i = (e_i,0;0);\ q_j = (0,e_j;0);\ Z = (0,0;1),$$

where e_i denotes the i-th basic vector in $\mathbb{R}^n$. They satisfy the *canonical (Heisenberg) commutation relations (CCR)*,

$$[p_i; q_j] = \delta_{ij} Z. \tag{2.5}$$

2.2. Canonical commutation relations. An important example of the Heisenberg-pair is given by operators of multiplication and differentiation,

$$Q: f(x) \to ix\, f(x);\; P: f \to \lambda(\partial f)(x);\; \text{in } \mathbb{R},$$
$$Q_j: f(x) \to ix_j f(x);\; P_k: f \to \lambda(\partial_k f)(x);\; \text{in } \mathbb{R}^n, \tag{2.6}$$

which obey the relations,

$$[P; Q] = i\lambda, \text{ or } [P_j; Q_k] = i\lambda\delta_{jk}. \tag{2.7}$$

We shall see that (2.6), in fact, serves a model example for the Heisenberg relations. Namely, any CCR (2.7) can be realized by a pair of operators (2.6), by a Theorem of Weyl and von Neumann.

Historical Remarks: Heisenberg commutation relations first appeared in the context of Quantum mechanics. The *states* of a quantum system are usually described by Hilbert space vectors $\{\psi \in \mathcal{H}, \text{ e.g. } \mathcal{H} = L^2(\mathbb{R})\}$, while the *observables* are given by certain (typically symmetric, but often unbounded!) operators in $\mathcal{H}$. Examples include the so called *position* and *momentum operators*,

$$Q_i: \psi(x) \to x_i \psi(x);\; P_j: \psi \to i\hbar\partial_j \psi, \text{ on } \mathcal{H} \quad L^2(\mathbb{R}^n),$$

angular momentum operator (see J4.4),

$$M_{ij} = x_i \partial_j - x_j \partial_i,$$

and the most important of all the *energy (Hamiltonian)* operator[7] (e.g. Schrödinger operator with potential V),

$$H = -\frac{\hbar^2}{2}\Delta + V(x) = \tfrac{1}{2}P^2 + V(Q).$$

An expected value of an observable A at a state $\psi \in \mathcal{H}$ is given by the quadratic form

$$\overline{A} = \langle A \rangle_\psi \quad \langle A\psi \mid \psi \rangle,$$

while the "mean quadratic deviation",

$$\epsilon_\psi(A) = \sqrt{\langle (A - \overline{A})^2 \rangle_\psi} = \| (A - \overline{A})\psi \|, \tag{2.8}$$

measures the error in observation. So the precise knowledge of observable A at state ψ (zero-error!) is attainable only for special states: the *eigenvectors* of A. In the physical parlor they are called *bound states* of hamiltonian A. Thus measurability (or observability) of A becomes paramount to its diagonalization.

---

[7] Hamiltonian H completely determines the evolution of quantum system, namely the initial state $\psi_0 \in \mathcal{H}$, evolves at time t into a state $\psi(t) = e^{itH}[\psi_0]$. In other words, quantum evolution is given by a unitary group, generated by H.

Obviously, any pair of commuting observables $\{A;B\}$ can be simultaneously diagonalized, i.e. observed to any degree of precision. However, noncommuting observables, like position $\{Q_i\}$ and momenta $\{P_i\}$, cannot be diagonalized. We shall see that a Heisenberg canonical pair $\{X;Y\}$ has no nontrivial 1-D, even finite-D representations (problem 2)!

It was observed experimentally, that the position and momentum of an electron can not be accurately measured at once, the product of errors always remained greater than the Plank constant $\hbar$. This led Heisenberg to state his famous *Uncertainty principle* of the quantum theory in the form of the commutation relation,

$$[P;Q] = i\hbar. \tag{2.9}$$

Indeed, one can easily check that for any pair of observables $\{A;B\}$ errors (2.8) satisfy,

$$\epsilon(A)\epsilon(B) \geq \Im\langle[A;B]\rangle, \text{ in any state } \psi.$$

So the Heisenberg relation (2.9) provided a mathematical formulation of experimentally observed uncertainty between P and Q:

$$\epsilon(P)\epsilon(Q) \geq \hbar. \tag{2.10}$$

Relation (2.10) has a simple reformulation in terms of the Fourier transform (see §2.1): for any function ψ in the weighted Sobolev space: $\int |x\psi|^2 dx < \infty$; $\int |\xi\hat{\psi}|^2 d\xi < \infty$,

$$\|\psi\|_{L^2} \leq \|x\psi\|\|\xi\hat{\psi}\|.$$

In fact, for any constants a, b we have

$$\boxed{\|\psi\|_{L^2} \leq \|(x-a)\psi\|_{L^2}\|(\xi-b)\hat{\psi}\|_{L^2}}$$

2.3. Representations of $\mathbb{H}_n$; Stone-von Neumann Theorem. Irreducible representations of $\mathbb{H}_n$ fall into two classes: *1-D characters* of the commutative factor group $G/Z \simeq \mathbb{R}^{2n}$,

$$\boxed{\chi^{\alpha\beta}(g) = e^{i(\alpha \cdot x + \beta \cdot y)}}, \text{ for } g = (x, y; t), \tag{2.11}$$

and a one-parameter family of ∞-D representations T^λ ($\lambda \in \mathbb{R}$) realized in Hilbert space $\mathcal{H}$ $L^2(\mathbb{R}^n)$ by operators

$$\boxed{T_g^\lambda \psi(x) = e^{i\lambda(b \cdot x + t)}\psi(x+a)} \tag{2.12}$$

One can check (problem 3), that T^λ are induced by 1-D representations (characters) $\chi^\lambda(b;t) = e^{i\lambda t}$, of subgroup $H = \{(0, b; t)\}$, as explained in §6.1,

$$T^\lambda = \text{ind}\ (G; \chi^\lambda\ H).$$

Let us remark that, the structure and realization of $\{\pi\text{'s}\}$, (2.11), (2.12), in the form of induced representations, can be easily derived from a semidirect product decomposition of $\mathbf{H}_n$, and the Mackey's theory of §6.1 (problem 3).

However, we shall produce a direct argument based on the Stone-von Neumann Theorem. Let us observe that infinitesimal generators of $\{T^\lambda\}$ (representation of the Lie algebra $\mathfrak{H}_n$) consists of multiplications, and differentiations:

$$q_j \to T^\lambda(q_j) = i\lambda x_j = Q_j;\ p_j \to \partial_j = P_j;\ Z \to i\lambda I, \tag{2.13}$$

the center being represented by scalar operator $Z \to i\lambda I$. Conversely, an infinitesimal representation (2.13) of Lie algebra $\mathfrak{H}_n$ integrates through the Lie group-representation (2.12).

Theorem (Stone-von Neunmann): *A pair of antisymmetric (unbounded) operators P and Q in Hilbert space $\mathcal{H}$ satisfying the Heisenberg commutation relations: $[P;Q] = i\lambda I$, can be realized as a multiplication and a differentiation:*

$$Q\psi(x) = i\lambda x\psi(x),\ P\psi(x) = \partial\psi(x),$$

on scalar or vector-valued functions $\psi \in L^2(\mathbb{R})$ (or $L^2 \otimes \mathcal{V}$). In other words there exists a unitary intertwining map $W: \mathcal{H} \to L^2$, that takes Q into $i\lambda x$ and P into ∂_x.

The irreducible action (representation) of P, Q in $\mathcal{H}$ corresponds to $\dim\mathcal{V} = 1$ (scalar functions), so any action is equivalent to a "$\dim\mathcal{V}$-multiple" of an irreducible action.

The proof exploits spectral decomposition of an antisymmetric operator Q (Appendix A). Any such Q can be realized by multiplication in the direct integral space:

$$\mathcal{H} = \oint_{\mathbf{R}} \mathcal{V}_x dx = \{\mathcal{V}_x\text{-valued } L^2\text{-functions } \psi(x) \text{ on } \mathbf{R}\},$$

$$Q: \psi(x) \to ix\,\psi(x).$$

Here we assumed λ 1, without loss of generality. Operator P generates a one-parameter unitary group $U_t = e^{Pt}$. The Heisenberg commutation relations imply

$$U_t Q U_t^{-1} = Q + itI. \tag{2.14}$$

The latter means that conjugation with U_t shifts "spectrum of Q" (spectral subspaces)[8] by the amount $\{t\}$. Consequently, all spaces $\{\mathcal{V}_x : x \in \mathbf{R}\}$, labeling values of $\{\psi(x)\}$, become isomorphic (equidimensional), $\mathcal{V}_x \simeq \mathcal{V}$, and operators $\{U_t\}$ define a family of $\mathcal{V}$-unitary maps (cocycle) $\{\sigma(x;t)\}$, so that

$$(U_t\psi)(x) = \sigma(x;t)[\psi(x+t)].$$

[8]If Q had discrete spectrum $\{\lambda_k\}$, the relation (2.14) would mean that the eigensubspace E_λ is shifted by U_t onto $E_{\lambda+t}$. In fact, continuity of t implies that Q has continuous, indeed absolutely continuous (Lebesgue) spectrum!

Cocycle σ is easily verified to be trivial: $\sigma(x;t) = \sigma(x)^{-1}\sigma(x+t)$, which allows to transform operators $\{U_t\}$ into shifts of $\mathcal{V}$-valued L^2-functions on $\mathbf{R}$,

$$U_t\psi(x) = \psi(x-t),$$

via map $W: \psi(x) \rightarrow \sigma(x)[\psi(x)]$. But the latter is obviously generated by operator $P = \partial_x$, QED.

We have stated Theorem 1 for a single Heisenberg pair $\{P;Q\}$. The result easily extends to n-tuples $\{P_1; \dots P_n; Q_1; \dots Q_n\}$, satisfying CCR (2.5). Here one simultaneously diagonalizes all $\{Q_j\}$, and analyzes the action of n-parameter unitary group

$\{U_t = exp(t_1P_1 + \dots t_nP_n): t = (t_1; \dots t_n) \in \mathbb{R}^n\}$ on joint "spectral subspaces" of Q's. Hilbert space $\mathcal{H}$ then turns into $\{\mathcal{V}$-valued L^2-functions on $\mathbb{R}^n$; with translation-invariant (Lebesgue!) measure$\}$, operators $\{Q_j\}$ become multiplications by independent variables $\{x_j\}$ on $L^2(\mathbb{R}^n)$, while $\{P_j\}$ turn into differentiations $\{\partial_j\}$.

To apply Theorem 1 to irreducible representations of $\mathbb{H}_n$, we observe that any such T restricted on center Z must be scalar, $T \mid Z = e^{i\lambda t}$. If $\lambda = 0$, then T factors through the representation of the commutative quotient-group G/Z, so it becomes a character (2.11). For nonzeros λ generators (Lie algebra) of $\mathbb{H}_n$ obey the Heisenberg CCR. So Theorem 1 applies, and we get subgroups $\{(a;0;0)\}$ and $\{(0;b;t)\}$, acting by translations and multiplications on $L^2(\mathbb{R}^n)$,

$$T_a\psi = \psi(x+a); T_b\psi = e^{i\lambda(b \cdot x + t)}\psi(x),$$

whence follows (2.12).

Figure 1 below illustrates the *dual object* (set of all irreducible representations) of $\mathbb{H}_n$: it consists of a series of infinite-D representations $\{T^\lambda: \lambda \in \mathbb{R}\backslash 0\}$, and the hyperplane of 1-D representations (characters) of $\mathbb{H}_n/Z$ $\{\mu(g) = e^{i\mu \cdot w}; g = (w;t)\}$.

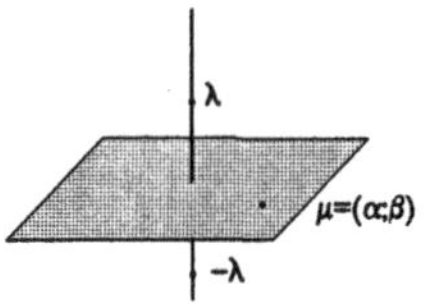

Fig.2: *The dual object of the Heisenberg group is made of 1-parameter family of $\infty - D$ representations $\{T^\lambda\}$, and a two-parameter (or 2n-parameter) family of characters $\{\mu\}$.*

2.4. Characters of T^λ and the Plancherel formula: Representations T^λ extend in the usual way to any of convolution algebras on $G = \mathbb{H}_n$: $\mathcal{C}_0(G)$, $\mathcal{C}_0^r(G)$, $L^1(G)$, etc.,

$$f \to T_f^\lambda = \int_G f(g) T^\lambda(g) dg.$$

Formula (2.12) implies that operators $\{T_f^\lambda\}$ are given by integral kernels on $\mathbb{R}^n$,

$$K_f(x;y) = \tilde{f}\,(y-x;\lambda x;\lambda), \tag{2.15}$$

where $\tilde{f}$ denotes the Fourier transform of $f(a,b;t)$ in the second and third variables,

$$\tilde{f}\,(...;\xi;\lambda) = \int_{\mathbb{R}^{n+1}} f(...;b;t) e^{i(\xi \cdot b + \lambda t)} db dt. \tag{2.16}$$

Evidently smooth compactly supported functions $\{f\}$ on G yield nice (compact; Hilbert-Schmidt; trace-class) kernels K_f (Appendix B). As before we define the character of T^λ as a distribution on G, $\chi^\lambda = tr\, T^\lambda$, via pairing to nice test-functions f,

$$\langle \chi^\lambda \mid f \rangle = tr\, T^\lambda(f).$$

The latter is computed by (2.15)-(2.16),

$$tr K_f = \int K_f(x;x) dx = \int_{\mathbb{R}^n} \tilde{f}\,(0;\lambda x;\lambda) dx = \left(\tfrac{2\pi}{\lambda}\right)^n \hat{f}(0;0;\lambda),$$

where $\hat{f}$ denotes the 1-variable Fourier transform in t:

$$f \to \hat{f}(...;\lambda) = \int f(...;t) e^{i\lambda t} dt.$$

Thus we get the character formula for irreducible representations $\{T^\lambda\}$ of $\mathbb{H}_n$,

$$\boxed{\langle \chi^\lambda \mid f \rangle = \left(\tfrac{2\pi}{\lambda}\right)^n \int f(0;0;t) e^{i\lambda t}\, dt} \tag{2.17}$$

We can interpret (2.17) by saying that distribution χ^λ is supported on a (one-parameter subgroup) center Z, and is equal to "δ-function of Z" $\times$ "G-invariant density $\{(\frac{2\pi}{\lambda})^n e^{i\lambda t}\}$". Integrating (2.17) in λ we immediately derive the Plancherel/inversion formula on $\mathbb{H}_n$,

$$f(e) = \int tr\, T^\lambda(f)\, d\mu(\lambda), \text{ for all } f \in C_0^\infty(G), \tag{2.18}$$

with the Plancherel measure

$$\boxed{d\mu(\lambda) = \frac{\lambda^n}{(2\pi)^{n+1}} d\lambda} \tag{2.19}$$

supported on the set of infinite-D irreducible representations $\{T^\lambda\colon \lambda \in \mathbb{R}\backslash\{0\}\}$.

The inversion formula (2.18) yields as usual the Plancherel formula

$$\boxed{\int_G |f(x)|^2 dx = \int_{\hat{G}} \left\| \hat{f}(\lambda) \right\|_{HS}^2 d\mu(\lambda), \text{ for all } f \in L^2(G)}, \tag{2.20}$$

where $\hat{f}(\lambda) = T^\lambda(f)$ means the noncommutative (operator-valued) "*Fourier transform*" of f, and $\left\| \hat{f}(\lambda) \right\|_{HS}^2 = \mathrm{tr}\,(\hat{f}(\lambda)^* \hat{f}(\lambda))$ - its Hilbert-Schmidt norm.

We remark that all integral-operators $\{K_f\}$ (2.15), with $f \in C_0^\infty(G)$, belong to the Hilbert-Schmidt class for all λ. Indeed,

$$\|K_f\|_{HS}^2 = \int\int \left| \tilde{f}\,(y-x;\lambda x;\lambda)\right|^2 dx\,dy < \infty!$$

2.5. The harmonic oscillator. The quantum-mechanical harmonic oscillator is a Schrödinger operator with quadratic potential. We shall take it in the simplest form (from which more general cases could be easily deduced):

$$H = \tfrac{1}{2}(-\Delta + |x|^2), \text{ in } L^2(\mathbb{R}^n) \text{ or } H = \tfrac{1}{2}(-\partial^2 + x^2), \text{ in 1-D}.$$

One is interested is spectral theory of H: its *eigenvalues* and *eigenfunctions.* The harmonic oscillator happened to belong to a rear and beautiful species, called *solvable models.* In other words one can write down explicitly all eigenvalues and eigenfunctions of H.

Theorem 2: *Spectrum of operator H is purely discrete. In 1-D it consists of a sequence of eigenvalues: $\lambda_k = k + \frac{1}{2}; k = 0;1;...;$ the corresponding eigenfunctions being the classical Hermite functions* on $\mathbb{R}$,

$$\psi_k(x) = h_k(x)e^{-x^2/2}. \tag{2.21}$$

Here $\{h_k\}$ denote the classical Hermite polynomials[9],

$$h_k(x) = \frac{1}{\sqrt{2^k k!}}e^{x^2/2}(e^{-x^2})^{(k)}. \tag{2.22}$$

The multi-D oscillator on $\mathbb{R}^n$ has eigenvalues $\{\lambda_k\}$, labeled by n-tuples of integers, $k = (k_1;...k_n)$, $\lambda_k = (k_1 + ... + k_n) + \frac{n}{2}$; and the corresponding eigenfunctions are products of 1-variable Hermite functions, $\psi_k(x) = \psi_{k_1}(x_1)...\psi_{k_n}(x_n)$.

Normalizing factors $\left\{\frac{1}{\sqrt{2^k k!}}\right\}$ render system $\{\psi_k\}$ orthonormal in $L^2(\mathbb{R})$, $\|\psi_k\|_{L^2} = 1$.

We shall establish Theorem 1, as a simple application of the Heisenberg CCR. But this time we choose a different realization of generators $\{p;q\}$ of $\mathbb{H}_1$, namely by the so called *creation/annihilation pair*[10],

$$a = \frac{1}{\sqrt{2}}(\partial + x);\; a^\dagger = \frac{1}{\sqrt{2}}(-\partial + x), \tag{2.23}$$

or similarly defined creation/annihilation n-tuples

$$a_j = \frac{1}{\sqrt{2}}(\partial_j + x_j);\; a_k^K = \frac{1}{\sqrt{2}}(-\partial_k + x_k);\; 1 \le j;k \le n. \tag{2.24}$$

Clearly, daggered operator $\{a_k^K\}$ are adjoint to $\{a_k\}$ in $L^2(\mathbb{R}^n)$. One can easily verify that

[9]Hermite polynomials form one of three known families of classical orthogonal polynomials, along with Laguerre (chapter 8), and Jacobi (chapter 4) (see [Erd];[Leb]).

pair $\{a;a^{\dagger}\}$ obey the Heisenberg CCR,

$$[a;a^{\dagger}]=\tfrac{1}{2}[\partial+x;\partial-x]=1;$$

or

$$[a_j;a_k^{\dagger}]=\delta_{jk}I.$$

But unlike, the position and momentum, $Q=x;P=i\nabla$, operators a and $a^{\dagger}$ are not self-adjoint in L^2. The harmonic oscillator can be represented in terms of the pair $\{a;a^{\dagger}\}$,

$$H=\tfrac{1}{2}(-\partial^2+x^2)=a^{\dagger}a+\tfrac{1}{2}=aa^{\dagger}-\tfrac{1}{2}. \tag{2.25}$$

Finally, triple $\{a;a^{\dagger};H\}$ obeys the commutation relations,

$$[H;a^{\dagger}]=a^{\dagger};\ [H;a]=-a. \tag{2.26}$$

In this regard $\{a^{\dagger};a\}$ behave, like the raising/lowering elements $\{X;Y\}$ of the Lie algebra $\mathfrak{sl}(2)$ (chapter 4). The oscillator, however, is not a Cartan (diagonal) element of pair $\{a^{\dagger};a\}$, since $[a^{\dagger};a]=I$, rather than $\frac{1}{4}H$, as in $\mathfrak{sl}(2)$!. The nature of the triple $\{a^{\dagger};a;H\}$ in relation to $\mathfrak{sl}(2)$ will be elucidated below. Commutation relations (2.26) readily yield all spectral results of Theorem 2. Indeed, if λ is an eigenvalue of H and ψ - the corresponding eigenvector,

$$H[\psi]=\lambda\psi,$$

then all $\psi_k=(a^{\dagger})^k[\psi]$, and $\psi_{-m}=a^m[\psi]$ are also eigenvectors,

$$H[\psi_{\pm k}]=(\lambda\pm k)\psi_{\pm k}.$$

Since operator H is positive,

$$\langle H\psi\mid\psi\rangle=\tfrac{1}{2}\int(\mid\nabla\psi\mid^2+x^2\mid\psi\mid^2)dx>0,\text{ for all }\psi\in L^2,$$

it has the lowest eigenvalue $\lambda_0\geq 0$, and eigenfunction ψ_0, called the ***ground-state***, and the ***ground-state energy***, i.e. the lowest-energy states of the quantum system. These must be annihilated by the lowering operator,

$$a[\psi_0]=(\partial+x)\psi_0=0;$$

which immediately yields the ground-state

$$\psi_0(x)=e^{-x^2/2}\text{- the Gaussian.}$$

[10]The terminology came from the quantum-field theory, whose main task is to account for the creation-annihilation of quantum particles (or better to say, "particle-states" of quantum-fields) in subatomic interactions. The mathematical structure for the creation-annihilation processes is based on the notion of (multi-particle) Fock-space, discussed below. It typically has a (unique) vacuum-state (vector) ω, and the corresponding space $\mathcal{H}_0=\text{span}\{\omega\}$, as well "1-particle", "2-particle",... spaces: $\mathcal{H}_1$; $\mathcal{H}_2$; ... The creation operators $\{a^{\dagger}\}$ send $\mathcal{H}_0\to\mathcal{H}_1$; $\mathcal{H}_1\to\mathcal{H}_2$; ... ; the particles being "created from the vacuum", while annihilation operators go in the opposite direction, $a:\mathcal{H}_n\to\mathcal{H}_{n-1}\to\ldots\to\mathcal{H}_0$, thus diminishing the particle number. The simplest model that accommodates such features are raising/lowering elements $\{X;Y\}$ of $\mathfrak{sl}(2)$ (chapter 4).

Substitution of ψ_0 in H (2.25) yields the lowest eigenvalue,

$$H[\psi_0] = (aa^\dagger + \tfrac{1}{2})\psi_0 = \tfrac{1}{2}\psi_0 \Rightarrow \boxed{\lambda_0 = \tfrac{1}{2}},$$

hence all other eigenvalues,

$$\lambda_k = \tfrac{1}{2} + k; k = 0; 1; 2; \ldots$$

Let $\psi_k = (a^\dagger)^k[\psi_0]$ denote the k-th eigenfunction of H. To get a Hermite-representation of ψ_k we apply yet another identity,

$$-a^\dagger = \partial - x = e^{x^2/2}\partial[e^{-x^2/2}],$$

in other words the raising operator $a^\dagger$ is conjugated to a derivative-operator ∂, via multiplication with the Gaussian. Hence,

$$(a^\dagger)^k = e^{x^2/2}(-\partial)^k[e^{-x^2/2}].$$

Applying the latter to the ground-state ψ_0 we get the *Rodriguez formula* (2.22) for ψ_k, QED.

2.6. The oscillator representation and the metaplectic group. The Heisenberg Lie algebra $\mathfrak{H}_n$ acts naturally on $\mathbb{R}^n$ by differentiations and multiplications (position-momentum operators) (Theorem 1), and thus gives rise to the *Weyl algebra* $\mathcal{W} = \mathcal{W}_n$ (an associative hull of $\mathfrak{H}_n$), made of all differential operators with polynomial coefficients:

$$A = \sum_{|\alpha+\beta| \leq m} a_{\alpha\beta} x^\beta \partial^\alpha. \tag{2.27}$$

Algebra $\mathcal{W}$ is *graded* according to the total degree m in variables $\{x\}$ and derivatives $\{\partial\}$ of polynomials $A = a(x;\partial)$ in (2.27),

$$\mathcal{W}^0 = \{\text{const}\} \subset \mathcal{W}^1 \subset \mathcal{W}^2 \subset \ldots \subset \mathcal{W}^m = \{A: \deg A \leq m\} \subset \ldots$$

One can easily check (problem 4) the product and the commutation formulae,

$$\mathcal{W}^p\mathcal{W}^q \subset \mathcal{W}^{p+q};\ [\mathcal{W}^p;\mathcal{W}^q] \subset \mathcal{W}^{p+q-2};\ \text{for all } p,q \geq 0. \tag{2.28}$$

The lost of 2 degrees in the commutator $[A;B]$ ($A \in \mathcal{W}^p$, $B \in \mathcal{W}^q$), results from the basic relation $[x;\partial] = \text{const}$, extended to other generators $\{x^\alpha;\partial^\beta\}$ of $\mathcal{W}$ (problem 4). Hence follows

(i) 2-nd degree operators $\{A\}$ ($m = 2$) form a Lie subalgebra $\mathcal{W}^2$ of $\mathcal{W}$, with respect to the natural commutator bracket: $[A;B] = AB - BA$. In fact,

(ii) A subspace of first degree operators $\{A\}$ ($m = 1$) is an ideal of $\mathcal{W}^2$, isomorphic to Heisenberg algebra $\mathfrak{H} = \mathfrak{H}_n$;

(iii) $\mathcal{W}^2$ contains a symplectic subalgebra $\mathfrak{A} \simeq \mathbf{sp}(n)$, spanned by all 2-nd order operators $\{\partial^2_{ij};\ x_i\partial_j;\ x_ix_j\}$.

So Lie algebra $\mathcal{W}^2$ factors into a semidirect product $\mathfrak{H} \triangleright \mathfrak{A}$, with $\mathfrak{A}$ acting by derivation on $\mathfrak{H}$. Consequently, Lie group of $\mathcal{W}^2$ also breaks into a semidirect product: $\mathbb{H}_n \triangleright G$, where G denotes a simply connected cover of the symplectic group $\mathbf{Sp}(n)$. So group $\mathbf{Sp}(n)$, and its cover G, act by automorphisms of Lie algebra $\mathfrak{H}_n$ and group $\mathbb{H}_n$.

The latter could be also verified by the direct computation. Namely any

$$\mathbf{Sp}(n) \ni g = \begin{bmatrix} a & b \\ c & d \end{bmatrix} : (x, y; t) \to (ax+cy; bx+dy; t), \tag{2.29}$$

respects the Heisenberg Lie bracket on triples $\{(x; y; t): x, y \in \mathbf{R}^n; t \in \mathbf{R}\}$ (problem 5).

In the previous subsection we have constructed an irreducible representation of $\mathbb{H}_n$ in space $L^2(\mathbb{R}^n)$, of the form

$$T^\lambda_g f(x) = e^{i\lambda(x\cdot b+c)} f(x+a);\ g = (a, b; c) \in \mathbb{H}_n;$$

whose generators were given by operators,

$$p_j \to \partial_j;\ q_j \to i\lambda x_j;\ Z \to i\lambda. \tag{2.30}$$

Formulae (2.30) extends through the representation of the Weyl algebra, in particular, its 2-nd order (Lie) part $\mathcal{W}^2$. So we get $\mathbf{sp}(n)$-generators acting in $L^2(\mathbb{R}^n)$ by operators,

$$p_ip_j \to \frac{1}{i\lambda}\partial^2_{ij};\ q_ip_j \to x_i\partial_j;\ q_iq_j \to i\lambda x_ix_j. \tag{2.31}$$

The latter can be lifted to a simply connected cover-group $\widetilde{\mathbf{Sp}}(n)$, and yields the celebrated *oscillator-representation* (also known as *spinor, metaplectic, Borel-Shale-Weil*), that appears in many different places and finds numerous applications. Another way to introduce the oscillator representation comes from the action of $\mathbf{Sp}(n)$ by automorphisms of $\mathbb{H}_n$ (2.29). Let us observe that symplectic automorphisms $\{u\}$ preserve the center of $\mathbb{H}_n$, $z^u = z$, for all $z \in Z$. Hence two representations: $\mathrm{T}^\lambda_{g^u}$ and T^λ_g $(g \in \mathbb{H}_n)$, are equivalent, any T^λ being uniquely determined by its value on Z,

$$T^\lambda_z = e^{i\lambda z}.$$

As a consequence we get a family of intertwining operators, $\{T_u : u \in \mathbf{Sp}(n)\}$, determined modulo scalars (like in the Mackey's theory, §6.1). They define a projective representation of $G = \mathbf{Sp}(n)$ in $L^2(\mathbb{R}^n)$,

$$T_{uv} = \alpha(u, v) T_u T_v;\ \text{for all } u, v \in \mathbf{Sp}(n). \tag{2.32}$$

with cocycle α. Any α-projective representation of any group G was shown in §3.2 to correspond to an "honest" representation of a central extension of G, (group G_α with a center $Z \subset \mathbb{T}$, so that $G_\alpha/Z \simeq G$). In fact, Heisenberg representation (2.12) arises that way (problem 6). Let us also remark that a symplectic extension (2.32) of T^λ depends only on sign of λ, $T = T^{\pm}$. It remains to compute cocycle α. We shall see that α is not trivial on $Sp(n)$, but trivializes on a finite cover of $Sp(n)$. For the sake of presentation we consider here only 1-dimensional case $Sp(1) \simeq SL_2(\mathbb{R})$ (see problem 7 for the general case).

Proposition 3: *Cocycle α on SL_2 is 2-valued, $\alpha(u) = \pm I$, so the corresponding central extension of SL_2 forms a 2-fold cover, called the metaplectic group $Mp(1)$.*

Let $V = \begin{bmatrix} & 1 \\ -1 & \end{bmatrix}$ denote the generator of rotations $SO(2) \subset SL_2$. It corresponds to a quadratic element $\frac{1}{2}(p^2 + q^2)$ in the Weyl algebra $\mathcal{W}^2$, which is taken by representation T (2.31) to the harmonic oscillator,

$$A = \tfrac{1}{2}(-\partial^2 + x^2),$$

The analysis of the oscillator in the previous part showed its spectrum to consist of half-integers: $\{\frac{1}{2}; \frac{3}{2}; \frac{5}{2}; \ldots\}$. Hence, a unitary group, generated by A via (2.31), takes orthogonal rotations

$$u(\phi) = \begin{bmatrix} \cos\phi & \sin\phi \\ -\sin\phi & \cos\phi \end{bmatrix} \in SO(2),$$

into unitary operators

$$T_u \sim \begin{bmatrix} \ddots & & \\ & \exp(i\tfrac{m}{2}\phi) & \\ & & \ddots \end{bmatrix}$$

The resulting representation, $\phi \to T_{u(\phi)}$, becomes single-valued on a 2-fold cover of $SO(2)$, isomorphic to $SU(2)$, so cocycle α becomes trivial on $SU(2)$. The corresponding 2-fold cover of SL_2 "resolves" (trivializes) cocycle α on a subgroup $SO(2)$,

$$\alpha(u; v) = \frac{\beta(uv)}{\beta(u)\beta(v)}; \text{ for all } u; v \in SO(2). \tag{2.33}$$

If G denotes the corresponding 2-fold cover of SL_2, then (2.33), along with the Cartan decomposition, $g = uhv$ ($u, v \in K$; $h \in H$) on G, yields the trivial cocycle α on the entire group G, QED.

2.7. Symmetries and spectral multiplicities of the oscillator. The symplectic group and its spinor representation allow to explain spectral multiplicities of the multi-D harmonic oscillator $H = \frac{1}{2}(-\Delta + |x|^2)$ in $\mathbb{R}^n$. The oscillator has an obvious $SO(n)$-symmetry, since both the Laplacian Δ and potential $|x|^2$ commute with rotations. However, the multiplicity of $\lambda_k = k$,

$$\#(\lambda_k)=\{(i_1...i_n):(i_1+\tfrac{1}{2})+...+(i_n+\tfrac{1}{2})=k\}=\binom{k+n}{n},$$

is much higher than could predicted, based on $SO(n)$-symmetry (compare the obvious cases of $\mathbb{R}^2$ and $\mathbb{R}^3$). This suggests that H might possess a larger symmetry-group. This, indeed, proves to be the case.

Theorem 4: *The symmetry group of the oscillator H in $Sp(n)$ coincides with the unitary group $SU(n)$, and the restriction of $SU(n)$ on eigenspaces is irreducible.*

Let us remark that both $SO(n)$ and $SU(n)$ are subgroups of $Sp(n)$, the former is given by all block-diagonal matrices

$$SO(n)=\left\{\begin{bmatrix} u & \\ & u\end{bmatrix}: {}^{\mathrm{T}}u\cdot u=I\right\},$$

while the latter is made of all $2n\times 2n$ orthogonal matrices that commute with $J=\begin{bmatrix} & I\\ -I & \end{bmatrix}$

We associate with any matrix A on the (Heisenberg) phase-space $\{(x;p)\}=\mathbb{R}^{2n}$, a quadratic form

$$f_A(x;p)=\langle Ax\mid p\rangle=\sum a_{ij}x_ip_j,$$

and the corresponding differential operator L_A, given by any possible convention: left $\{x_i\partial_j\}$; right $\{\partial_ix_j\}$ or symmetric (Weyl) $\{\frac{1}{2}(x_i\partial_j+\partial_jx_i)\}$, so

$$L_A=\sum a_{ij}x_i\partial_j \text{ (in the left convention).}$$

The oscillator clearly corresponds to the identity matrix,

$$H=\tfrac{1}{2}(-\Delta+|x|^2)\leftrightarrow\tfrac{1}{2}\begin{bmatrix} I & \\ & I\end{bmatrix}=I. \tag{2.34}$$

Next we observe that the natural linear action of group $GL(2n)$ on $\mathbb{R}^{2n}$ is transformed under the map $A\to L_A$ to conjugation of matrices,

$$g: A\to {}^{\mathrm{T}}gAg.$$

On the other hand the oscillator representation assigns to any symplectic g a unitary operator T_g in $L^2(\mathbb{R}^n)$, so that

$$T_g^{-1}L_AT_g=L_{({}^{\mathrm{T}}gAg)};\ g\in Sp(n);\ A\in gl(2n).$$

So the commutator of H consists precisely of symplectic matrices, that preserve the Euclidian inner-product form (2.34), $\{g: {}^{\mathrm{T}}g\cdot g=I\}$, i.e. group $SO(2n)\cap Sp(n)=SU(n)$ (the maximal compact subgroup of $Sp(n)$, see §5.7). Conversely, the commutator of $SU(n)$ can be shown to coincide with $\{H\}$. Thus $\{SU(n)\}$ and $\{H\}$ form a maximal commuting pair in the sense of §4.4-4.5 (like the Laplacian Δ on S^n and the orthogonal group $SO(n+1)$). Since the metaplectic representation is irreducible in $L^2(\mathbb{R}^n)$ any pair should break it into the sum of "joint irreducible components", $\bigoplus_k \mathcal{E}_k$. Hence, restrictions $H\mid\mathcal{E}_k=\lambda_k$, and $SU(n)\mid\mathcal{E}_k$-irreducible, QED.

Remark: In the differential-operator form (2.31) Lie algebra $su(n)$ can be represented in terms of the creation-annihilation operators, $\{a_i; a_j^\dagger\}$ of (2.24). We set $X_{ij} = a_i a_j^\dagger$, and have the reader verify that $\{X_{ij}\}$ satisfy the commutation relations of the $su(n)$-basis, and commute with H,

$$[X_{ij}; X_{km}] = \delta_{im} X_{kj} + \delta_{jk} X_{im};\ [H; X_{ij}] = 0.$$

After the eigenspaces of H are shown to be irreducible under $SU(n)$ the natural question arises, what are these irreducibles $\{\pi^k\}$, in terms of their weights, as described in §5.3. It would be difficult to derive the weights directly from the $L^2(\mathbb{R}^n)$-realization of metaplectic T, as the latter does not correspond to any natural (regular/induced) action of $SU(n)$ on its quotients. As we shall see the proper way to interpret $T \mid SU(n)$ is in terms of the holomorphic-induction of §6.1. This would require yet another realization of T in the complex domain, that we shall explain.

2.8. The Bargman-Segal representation. We shall conclude this section with yet another realization of irreducible representations $\{T^\lambda\}$ of $\mathbb{H}_n$ and the Weyl algebra in spaces of holomorphic functions on $\mathbb{C}$ and $\mathbb{C}^n$,

$$\mathcal{H} = \mathcal{H}(\mathbb{C}) = \left\{F(z) : \|F\|^2 = \int |F(z)|^2 e^{-|z|^2} dz\right\}. \tag{2.35}$$

Polynomials $\{z^n\}_0^\infty$ are easily verified (using polar coordinates) to form an orthogonal system in $\mathcal{H}$, with norms,

$$\|z^n\|^2 = \pi n!.$$

So we get an orthonormal basis $\left\{\frac{1}{\sqrt{p\, n!}} z^n : n = 0; 1; \ldots\right\}$. Two operators, multiplication $z: F(z) \to zF(z)$, and complex differentiation $\partial: F \to F'(z)$, are adjoint one to the other, relative to the product (2.35), and obey the Heisenberg CCR,

$$[z; \partial_z] = 1.$$

So they behave like the creation/annihilation pair $\{a; a^\dagger\}$ (2.24), while their real/imaginary parts become the position/momentum operators:

$$p = \frac{1}{\sqrt{2}}(z + \partial);\ q = \frac{\lambda}{\sqrt{2i}}(z - \partial);\ z = i\lambda. \tag{2.36}$$

Infinitesimal representation (2.36) of Lie algebra $\mathfrak{H}$ could be lifted (exponentiated) through a unitary representation of Lie group $\mathbb{H}$ in space $\mathcal{H}(\mathbb{C})$,

$$U^\lambda{}_{(w;t)} F(z) = e^{i\lambda(t - z \cdot \overline{w}) - \frac{1}{2}\lambda^2 |w|^2} F(z + w). \tag{2.37}$$

In fact, the equivalence of all generators and representations: T^λ on $L^2(\mathbb{R})$ (2.12) and U^λ on $\mathcal{H}(\mathbb{C})$ (2.37), can be established via an intertwining map $W: L^2 \to \mathcal{H}$,

$$W: f(t) \to F(z) = \int_{-\infty}^{\infty} e^{\lambda(\sqrt{2}zt - \frac{1}{2}z^2 - \frac{1}{2}t^2)} f(t)\, dt. \tag{2.38}$$

We shall leave the details to the reader (see problem 8), and just mention that Bargman-Segal representation and spaces $\{\mathcal{H}(\mathbb{C})\}$ have many remarkable features. One of them is the *reproducing kernel* $\{K(z;w)\}$ on $\mathbb{C}\times\mathbb{C}$, which gives the value of $F\in\mathcal{H}$ at any point z in terms of integrals over $\mathbb{C}$,

$$F(z)=\langle K(z;...)\,|\,F(...)\rangle_{L^2}=\int_{\mathbb{C}} K(z;w)\bar{F}(w)e^{-|w|^2}d^2w.$$

The reproducing kernel on space $\mathcal{H}(\mathbb{C})$ is equal to $K(z;w)=e^{\lambda z\cdot\bar{w}}$.

Finally, we can go back to the decomposition problem for the metaplectic T, restricted on subgroup $SU(n)$. Let us observe that $SU(n)$ acts on $\mathbb{C}^n\simeq\mathbb{R}^{2n}$ in the natural way, by unitary linear maps, $z\to z^u$. Hence, its action on space $\mathcal{H}$ consists of coordinate transformations,

$$T_u\colon F(z)\to F(z^u),\ \text{on polynomials } \{F(z)\}.$$

Space of polynomials $\mathcal{P}$ is made of homogeneous components of various degrees,

$$\mathcal{P}_k=\Big\{f=\sum_{|\alpha|=k} a_\alpha z^\alpha\Big\},$$

each $\mathcal{P}_k$ identified with symmetric tensors, and $T\,|\,\mathcal{P}_k$ is precisely the k-th symmetric tensor power of the natural representation π in $\mathbb{C}^n$. So it has signature (weight) $\alpha=(k;0;...0)$ (see §5.3)!

Further results and details could be found in [How]; [Ta2]; [GS1].

Problems and Exercises:

1. i) Show that any Z-valued 2-cocycle $\phi(a;b)$ on group G defines a central extension H of G by Z, via multiplication formula: $(x;t)\cdot(y;s)=(xy;t+s+\phi(x;y))$, $x,y\in G; t,s\in Z$.

   ii) The extension H is trivial, iff the cocycle ϕ is trivial:
   $$\phi(a;b)=\psi(a)-\psi(ab)+\psi(b)=\partial\psi(a;b), \text{for some } \psi(a);$$

   iii) Two extensions $H_1;H_2$, defined by cocycles ϕ_1; ϕ_2 are equivalent (isomorphic), iff ϕ_1 and ϕ_2 differ by a trivial cocycle $\partial\psi$ (Hint: construct an isomorphism $\sigma:H_1\to H_2$, via map, $(a;0)\to(a;\psi(a))$.

   iv) Apply the above results to the Heisenberg group to show that cocycle (2.2) is nontrivial, and two cocycles: $\phi_1(z;z')=x\cdot y'$; $\phi_2(z;w)=\Im(z\cdot\overline{w})$. Hence two multiplication formulae are equivalent.

2. Show that the Heisenberg algebra has no nontrivial finite-dimensional representations (Hint: commutator $[T_X;T_Y]$ has trace 0, so it cannot be λI with $\lambda\neq 0$!).

3. Apply the Mackey's group-extension theory for semi-direct products to derive all irreducible representations (2.11)-(2.12) of $\mathbf{H}_n$.

4. (i) Check the commutation relation $[\mathcal{W}^p;\mathcal{W}^q]\subset\mathcal{W}^{p+q-2}$, for the Weyl algebra (Do it first for generators $\{x^\alpha;\partial^\beta\}$ of $\mathcal{W}$, starting with the basic relations $[x_i;\partial_j]=\delta_{ij}$).

   (ii) conclude that the first-degree part $\mathcal{W}^1$ is isomorphic to the Heisenberg algebra $\mathfrak{H}_n$, via map, $P=a\cdot\partial+b\cdot x+c\to(a;b;c)\in\mathfrak{H}_n$ $(a;b\in\mathbb{R}^n;\ c\in\mathbb{R})$.

   (ii) Show that the 2-nd order operators of the form:
   $$P=\sum a_{ij}\partial^2_{ij}+\sum b_{ij}x_i\partial_j+\sum c_{ij}x_ix_j=A\partial\cdot\partial+Bx\cdot\partial+Cx\cdot x,$$
   with symmetric matrices $\{A;B\}$ and an arbitrary $C\in\mathbf{gl}(n)$, form a symplectic Lie algebra via identification:
   $$\tfrac{1}{2}P\to\begin{bmatrix} C & B\\ A & -{}^{\mathrm{T}}C\end{bmatrix}\in\mathbf{Sp}(n).$$

5. Show that (2.29) defines an automorphism of $\mathbf{H}_n$, that preserves its center, and any such automorphism is given by an element of $\mathbf{Sp}(n)$.

6. Show that representation T^λ (2.12) of $\mathbf{H}_n$ comes from a projective representation of the commutative group $\mathbb{R}^{2n}\simeq\mathbb{C}^n$, defined by the cocycle
   $\alpha(z;w)=e^{i\lambda\Im(z\cdot\overline{w})}=e^{i\lambda(x\cdot y'-y\cdot x')}$; where $z=x+iy$; $w=x'+iy'$.

7. Show that metaplectic group $\mathbf{Mp}(n)$ is a 2^n-fold cover of $\mathbf{Sp}(n)$, with center
   $$Z\simeq\underbrace{\mathbb{Z}^2\times\ldots\times\mathbb{Z}^2}_{n-times}$$

   Use commuting oscillators: $A_j=\frac{1}{2}(-\partial_j^2+x_j^2)$; $j=1;\ldots n$, in the Weyl algebra $\mathcal{W}_n$.

8. i) Check that map (2.38) intertwines representations T^λ and U^λ of $\mathbf{H}_n$ in spaces $L^2(\mathbb{R}^n)$ and $\mathcal{H}(\mathbb{C}^n)$ (use Lie algebra generators).

   ii) Find the unitary group ($\mathbf{SO}(2)\subset\mathbf{Sp}(1)$), generated by the oscillator $\frac{1}{2}(p^2+q^2)$ in the $\mathcal{H}$-realization of the metaplectic extension of T^λ, and show that
   $$T^{\pm}_{u(\phi)}F(z)=e^{i\phi/2}F(ze^{i\phi}). \tag{2.39}$$

   iii) Show that the standard Fourier transform $\mathcal{F}:L^2(\mathbb{R})\to L^2(\mathbb{R})$ corresponds to $\phi=\frac{\pi}{2}$ in (2.39).

§6.3. The Kirillov orbit method.

Many results of the representation theory of $\mathbf{H}_n$, outlined in the preceding section can be extended to arbitrary nilpotent, solvable, exponential and more general Lie groups [Kir1,2], [Pu]. The key idea, crystallized in the work of Kirillov, Souriau, Kostant et al., was to associate irreducible representations of G to orbits $\mathcal{O}$ of the co-adjoint action of G in the dual space $\mathfrak{G}^*$ to Lie algebra $\mathfrak{G}$. We shall outline the construction of representations $T^{\mathcal{O}}$, then derive the character formulae, and the Plancherel measure, based on the orbit method.

3.1. Construction of representations $T^{\mathcal{O}}$. A co-adjoint orbit $\mathcal{O} \subset \mathfrak{G}^*$ always carries a natural *symplectic structure*, nonsingular skew-symmetric bilinear form on tangent spaces: $B = B_\xi(X;Y)$; $X,Y \in T_\xi$ - tangent vectors at point $\xi \in \mathcal{O}$. Equivalently, there exists a nonsingular *differential 2-form:*

$$\Omega = \Omega_B = \sum b_{ij} d\xi^i \wedge d\xi^j \text{ (in local coordinates } \{\xi_j\} \text{ on } \mathcal{O}). \tag{3.1}$$

The latter defines a *Poisson-Lie bracket* on functions (observables) $\{f\}$ on $\mathcal{O}$ (see chapter 8):

$$\{f;h\} = B(\nabla f; \nabla h) = \sum b_{ij}(\partial_i f)(\partial_j h).$$

So the space of smooth functions $C^\infty(\mathcal{O})$ turns into an ∞-dimensional Lie algebra. The tangent space of $\mathcal{O}$ at point $\{\xi\}$ is identified with the quotient $\mathfrak{G}/\mathfrak{G}_\xi$ - Lie algebra modulo the *stabilizer* of ξ, $\mathfrak{G}_\xi = \{Y: ad^*_Y(\xi) = 0\}$. By definition form B, at a point $\xi \in \mathcal{O}$ is equal to

$$B_\xi(X;Y) = \langle [X;Y] \mid \xi \rangle, \text{ for any pair of elements } X,Y \in \mathfrak{G}. \tag{3.2}$$

Clearly, B_ξ depends only on the images (projections) of vectors $X;Y$ on the tangent space $T_\xi = \mathfrak{G}/\mathfrak{G}_\xi$. We shall list a few basic properties of form B:

- B_ξ annihilates $X,Y \in \mathfrak{G}_\xi$, so it depends only on classes of X,Y in the tangent space T_ξ $\mathfrak{G}/\mathfrak{G}_\xi$
- B_ξ is nonsingular on T_ξ (symplectic)
- 2-form Ω_B (3.1) is *closed*, i.e. differential

$$d\Omega_B = \sum_{i<j<k} (\partial_i b_{jk} - \partial_j b_{ik} + \partial_k b_{ij}) d\xi_i \wedge d\xi_j \wedge d\xi_k = 0. \tag{3.3}$$

The latter follows from the Jacobi identity on Lie algebra $\mathfrak{G}$ (problem 1). We also have

- bilinear form B is G-invariant on $\mathcal{O}$ (with respect to the co-adjoint action).

In particular, all co-orbits in $\mathfrak{G}^*$ have even dimensions, $\dim \mathcal{O} = 2m$, and the product $\Omega \wedge \ldots \wedge \Omega$ (m-times) defines an invariant volume element on $\mathcal{O}$. In the standard terminology symplectic form B turns $\mathcal{O}$ to a *classical mechanical phase-space* (see chapter 8), with a large symmetry group G. Elements of Lie algebra $\mathfrak{G}$ define a family of *classical observables (hamiltonians)* on $\mathcal{O}$, $\phi_X(\xi) = \langle X \mid \xi \rangle$.

In the general setup of *Geometric quantization* one is given a classical mechanical system (symplectic manifold $\mathcal{O}$) with a Hamiltonian h, or a family of hamiltonians $\{h_j\}$, closed with respect to the *Poisson-Lie bracket* on $\mathcal{O}$, in other words a *Poisson-Lie algebra* $\mathfrak{G}$ of *classical observables* (functions on $\mathcal{O}$). Then one is asked to construct a *Quantum* (Hilbert) *phase-space* $\mathcal{H}$, and to assign *quantum observables* (operators in $\mathcal{H}$) to classical observables, maintaining all possible Lie symmetries of the system. In other words one has to construct an irreducible representation of group G, associated to $\mathcal{O}$.

The general construction exploits the notion of an *admissible subalgebra* (or *polarization)* of $\mathfrak{G}$, i.e. a maximal subalgebra $\mathfrak{H}$, that satisfies $\langle \xi \mid [\mathfrak{H};\mathfrak{H}] \rangle = 0$, so ξ defines a 1-D (local) representation of subgroup $H = \exp \mathfrak{H}$,

$$\alpha(exp Y) = e^{i\langle \xi \mid Y \rangle}, \; Y \in \mathfrak{H}. \tag{3.4}$$

Subalgebras $\mathfrak{H}$ satisfying $\langle [\mathfrak{H};\mathfrak{H}] \mid \xi \rangle = 0$, are said to be *subordinated* to ξ. Any subordinated subalgebra $\mathfrak{H}$ has $\dim(\mathfrak{G}/\mathfrak{H}) \geq \frac{1}{2} dim \mathcal{O}$, since $\mathfrak{H}/\mathfrak{G}_\xi \subset T_\xi(\mathcal{O})$ is an isotropic subspace of symplectic form B. In many cases the maximal allowed dimension ($\mathrm{codim}\mathfrak{H} = \frac{1}{2}\dim\mathcal{O}$) is attained by some $\mathfrak{H} \subset \mathfrak{G}$. Moreover, orbit $\mathcal{O}$ contains the null-space $\mathfrak{H}^{\perp} = \{\eta : \langle \eta \mid \mathfrak{H} \rangle = 0\}$ (the so called *Pukanszky condition*), so orbit $\mathcal{O}$ is foliated into subspaces of dimension $= \frac{1}{2}\dim\mathcal{O}$. This condition holds for large classes of semisimple and solvable Lie groups (examples 1,2 below).

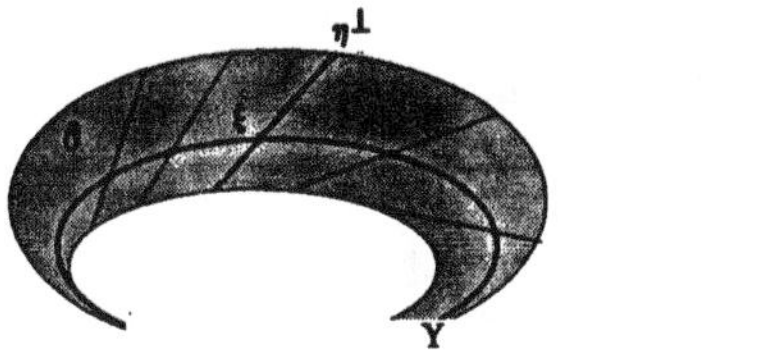

Fig.3 *gives a schematic view of a co-adjoint orbit* $\mathcal{O}$ *foliated into linear subspaces* $\{\mathfrak{H}^{\perp}\}$ *of* $\frac{1}{2}dim\mathcal{O}$ *over the quotient space Y.*

The $\{\mathfrak{H}^{\perp}\}$-foliation of $\mathcal{O}$, or rather its foliation into H-orbits[11], of $\dim = \frac{1}{2}\dim\mathcal{O}$,

[11] Notice that subgroup H does foliate $\mathcal{O}$ into disjoint orbits: for any pair $\xi_1;\xi_2 \in \mathcal{O}$, either $ad(H)\xi_1 = ad(H)\xi_2$, or $ad(H)\xi_1 \bigcap ad(H)\xi_2 = \emptyset$.

allows to introduce local canonically-conjugate coordinates, like the position and momenta $\{q_i;p_j\}$-variables of the previous section. Formula (3.4) then defines a 1-D representation of an admissible Lie algebra $\mathfrak{H}$, or a local representation χ of group H. We don't know, however, whether χ could be extended through the global representation of H, and how many different extensions can be defined by χ. The answer turns out to involve some topological constraints on the orbit, and its quotient-space $\mathcal{Y} = \mathcal{O}/H$ (fig.3).

Theorem 1 (Kirillov): (i) *Local representation* $\chi(expY) = e^{i\langle\xi\,|\,Y\rangle}$, $Y \in \mathfrak{H}$, *extends through the global representation of an admissible subgroup* H, *iff orbit* $\mathcal{O}$ *is integral, i.e. the fundamental form* Ω_B *takes on integral values on any 2-cycle (closed 2-D surface)* $\Sigma \subset \mathcal{O}$),

$$\oint_{\Sigma} \Omega_B = m. \tag{3.5}$$

In the standard terminology, one says that cocycle Ω_B *has integral cohomology,* $H^2(\mathcal{O};\mathbb{Z})$.

(ii) *Different extensions of* χ *are parametrized by characters of the fundamental group* $\Gamma = \pi_1(\mathcal{Y})$, *i.e. by elements of the 1-st cohomology group,* $H^1(\mathcal{Y};\mathbb{T}) = H^1(\mathcal{O};\mathbb{T})$.

Once the character χ is constructed we define $T^{\mathcal{O}}$, as an induced representation of G, $T = ind(\alpha\ H;G)$. More precisely, irreducible representations of group G are labeled by 2 parameters: an integral (quantized) orbit $\mathcal{O} \subset \mathfrak{G}^*$, and an elements β of the first cohomology group $H^1(\mathcal{Y};\mathbb{T})$; so $T = T^{\mathcal{O};\beta}$. Representation T is naturally realized on functions (sections) over the quotient-space $\mathcal{M} = \mathcal{O}/$"foliation $\mathfrak{H}_\xi$", $\dim \mathcal{M} = \frac{1}{2}\dim \mathcal{O}$, and one can show T to be irreducible.

The proof of irreducibility exploits specifics of the Lie group structure of G. For nilpotent Lie groups one applies the dimensional reduction, and the natural "functorial properties" of the correspondence "orbits → representation" with respect to operations of induction and restriction: $S\,|\,H \to T = ind(S\,|\,H;G)$, and $T \to T\,|\,H$, [Kirl,2]. Typically, both operations yield decomposable representations $\{ind\,S$, or $T\,|\,H\}$, and one asks for the direct integral decomposition of both. It turns out [Kir], that

i) an irreducible $T = T^{\mathcal{O}}$ restricted on H, $T^{\mathcal{O}}\,|_H = \oint S^{\omega}$ - direct integral over the set of H-orbits $\omega \subset p(\mathcal{O})$, where p denotes the natural projection $\mathfrak{G}^* \to \mathfrak{H}^*$;

ii) an induced representation $T = ind(S^{\omega}\,|\,H;G)$ is expanded into the direct integral $\oint T^{\mathcal{O}}$ over the set of G-orbits $\mathcal{O}$, intersecting $p^{-1}(\omega) \subset \mathfrak{G}^*$.

The latter shows that irreducibility of *ind*(χ;...) is equivalent to the inverse image of point $\chi \in \mathfrak{H}^*$, $p^{-1}(\chi)$ being contained in a single G-orbit. This explains irreducibility of induced T for admissible subgroups H. We refer to [Kirl] for further details.

A modification of the above construction involves complexified Lie algebras $\mathfrak{G}$ and $\mathfrak{H}$, and "holomorphically induced representations" (§6.1) in spaces of holomorphic functions/sections over $\mathcal{O}$. In such context certain integrability conditions on $\mathcal{O}$ arise, so orbits are "quantized", in the sense that certain parameters describing $\mathcal{O}$ become discretized. The induced (irreducible) representations on $\mathcal{O}$, that arise in this way, are parametrized by characters of the fundamental group $\pi_1(\mathcal{O})$. The foremost example will be the *Borel-Weil-Bott Theorem* for compact Lie groups, which gives irreducible $\{\pi\}$ realized in sections of "holomorphic vector bundles over $\mathcal{O}$".

Let us illustrate the foregoing with 3 examples.

3.2. Heisenberg groups $\mathbb{H}_1$ and $\mathbb{H}_n$ are conveniently represented by $(n+1)\times(n+1)$-matrices

$$\mathbb{H}_n = \left\{ g = g(\boldsymbol{x},\boldsymbol{y};t) = \begin{bmatrix} 1 & \boldsymbol{x} & t \\ & \ddots & \boldsymbol{y} \\ & & 1 \end{bmatrix} \right\},$$

where row $\boldsymbol{x}$ and column $\boldsymbol{y} \in \mathbb{R}^n$, $t \in \mathbb{R}$. Their Lie algebras $\mathfrak{G}$ have the same (upper-triangular) form with zeros on the main diagonal. So the dual space $\mathfrak{G}^*$ can be identified with the lower-triangular matrices

$$\left\{ \psi = \begin{bmatrix} 0 & & \\ \boldsymbol{\xi} & 0 & \\ \alpha & \boldsymbol{\eta} & 0 \end{bmatrix} \right\}$$

via pairing of elements $\{X = X(x,y;t) \in \mathfrak{G}\}$ to functionals $\{\psi = \psi(\xi;\eta;\alpha) \in \mathfrak{G}^*\}$,

$$\langle \psi \mid X \rangle = \mathrm{tr}(\psi X) = \boldsymbol{\xi} \cdot \boldsymbol{x} + \boldsymbol{\eta} \cdot \boldsymbol{y} + \alpha t.$$

The co-adjoint action consists of conjugation of ψ by elements $\{g\}$: $\psi \to g^{-1}\psi g$, and subsequent truncation of the lower-triangular part of the product,

$$ad_g^*(\psi) = [g^{-1}\psi g]_-. \tag{3.6}$$

The reader easily computes (3.6) to be

$$(\boldsymbol{\xi} - \alpha \boldsymbol{b}; \boldsymbol{\eta} + \alpha \boldsymbol{a}; \alpha); \text{ for } g = g(\boldsymbol{a}; \boldsymbol{b}; t).$$

So the co-orbits are hyperplanes $\mathcal{O} = \mathcal{O}_\alpha = \{(\xi,\eta;\alpha) : \alpha = \text{const}\} \simeq \mathbb{R}^{2n}$, and all points $\{(\xi;\eta;0)\}$ in the hyperplane $\alpha = 0$ (see fig.4). No *orbit-quantization* (3.5) required here, as all $\{\mathcal{O}_\alpha\}$ (and all their cohomologies) are trivial. Each orbit has the standard polarization, and corresponds to an irreducible representation T^λ ($\lambda \neq 0$) induced by an admissible subgroup $H = \{g = g(0;b;t)\}$ of G.

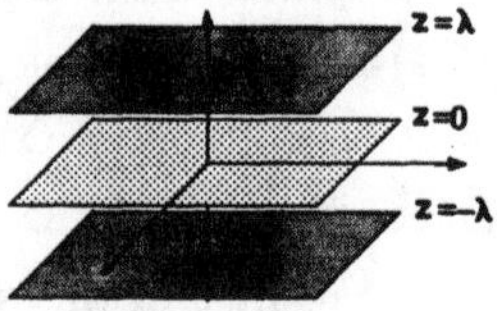

Fig.4. *Co-orbits of Heisenberg group* $\mathbf{H}_n$ *consist of two families: 2n-D hyperplanes:* $z = \lambda \neq 0$, *that correspond to irreducible representations* $\{T^\lambda\}$ *of §6.2. The second family is made of all points (0-D orbits) of the plane* $z = 0$.

3.3. Euclidian motion group $\mathbb{E}_n = \mathbb{R}^n \rhd SO(n)$ is also realized by $(n+1)\times(n+1)$-matrices:

$$g = (u;a) = \begin{bmatrix} u & a \\ & 1 \end{bmatrix};\ \text{column } a \in \mathbb{R}^n;\ u \in SO(n).$$

Hence its Lie algebra consists of matrices,

$$\mathfrak{G} = \left\{ X = \begin{bmatrix} K & x \\ & 0 \end{bmatrix};\ K \in so(n);\ x \in \mathbb{R}^n \right\},$$

with the bracket,

$$\left[\begin{pmatrix} K & x \\ & 0 \end{pmatrix}; \begin{pmatrix} L & y \\ & 0 \end{pmatrix} \right] = \begin{pmatrix} [K;L] & K(y)-L(x) \\ & 0 \end{pmatrix};$$

and the dual space $\mathfrak{G}^*$ could be represented by lower-triangular matrices

$$\left\{ \psi = \begin{bmatrix} P & \\ \xi & 0 \end{bmatrix};\ P \in so(n);\ \text{row } \xi \in \mathbb{R}^n \right\},$$

with the pairing

$$\langle \psi \mid X \rangle = tr(\psi X) = tr(PK) + \xi \cdot x.$$

The adjoint action of G on $\mathfrak{G}$ then becomes

$$ad_g \begin{pmatrix} K & x \\ & 0 \end{pmatrix} = \begin{pmatrix} u^{-1}Ku & u^{-1}(x+Ka-a) \\ & 0 \end{pmatrix};\ g = (u;a);$$

while the co-adjoint action

$$ad_g^* \begin{pmatrix} P & \\ \xi & 0 \end{pmatrix} = \begin{pmatrix} u^{-1}(P - a \wedge \xi)u & \\ \xi u & 0 \end{pmatrix}. \tag{3.7}$$

Here ξu denotes the left multiplication of row-vector ξ by orthogonal matrix u. Symbol $a \wedge \xi$ stands for the antisymmetric part of rank-one matrix ${}^{\mathrm{T}}a\xi = (a_i \xi_j)$, so

$$(a \wedge \xi)_{ij} = a_i \xi_j - \xi_i a_j.$$

Clearly, co-orbit $\mathcal{O} \subset \mathfrak{G}^*$ is fibered over its projection onto the abelian component $\mathbb{R}^n$ of $\mathfrak{G}^*$, a sphere $S_r = \{\eta = \xi u : u \in SO(n)\}$ of radius $r = |\xi|$. We leave it to the reader to verify that fibers $\{\Upsilon_\eta : \eta \in S_r\}$ are $(n-1)$-dimensional linear subspaces of the antisymmetric $so(n)$-component of $\mathfrak{G}^*$.

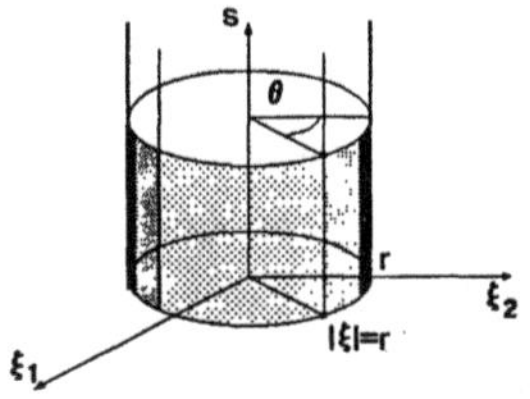

Fig.5. *Co-orbits of the motion group $\mathbb{E}_2$ are cylinders $\{(\xi;s) : |\xi| = r > 0\}$, and a discrete (quantized) set of points (0−D orbits) $\{(0;m)\}$ on the s-axis. The former give a family of ∞-D representations $\{T^r\}$, the latter correspond to characters $\{\chi_m\}$ of the compact quotient $\mathbf{M}_2/\mathbb{R}^2 \simeq SO(2)$.*

We shall elaborate the rest of construction in the simplest case of plane motions $\mathbb{E}_2$. Here co-adjoint elements are identified with pairs $\{(s;\xi)\}$, where $s \leftrightarrow P_s = \begin{bmatrix} & s \\ -s & \end{bmatrix}$ labels the anti-symmetric component. There are 2 classes of co-orbits: cylinders

$$\mathcal{O} = \mathcal{O}(\xi) = \{(s - a \wedge \xi; \xi u) : a \in \mathbb{R}^2;\ u = u_\theta \in SO(2)\},$$

of radius $r > 0$, fibered over the circle $S_r = \{|\xi| = r\}$, and 1-parameter family of degenerate (0-D) orbits $\{(s;0) : s \in \mathbb{R}\}$ (fig.5). The canonical 2-form on cylinders $\mathcal{O}_r$ is

$$\Omega_r = r d\theta \wedge ds.$$

No quantization condition (3.5) arises here, as $\oint_\gamma \Omega_r = 0$, for all 2-cycles $\gamma \subset \mathcal{O}$ (there are no nontrivial 2-cycles!). The reduced space (base) $\mathcal{Y} \simeq \mathbb{T}$ has fundamental group $\Gamma = \pi_1(\mathcal{Y}) = \mathbb{Z}$, whose characters (dual group) form a torus $\hat{\Gamma} = \mathbb{T}$.

We fix a point $\xi = (0;\xi) \in \mathcal{O}$, and find an admissible subalgebra $\mathfrak{H} = \mathbb{R}^2$ of ξ. According to the general theory, character $\chi_\xi(a) = e^{i\xi \cdot a}$ $(a \in \mathfrak{H})$, extends through a representation of the subgroup $H = exp\mathfrak{H}$, in $\hat{\Gamma}$-different ways, labeled by angle $\phi \in [0;2\pi]$. So we get a 2-parameter family of characters $\{\chi_{r;\phi} : r > 0; \phi \in [0;2\pi]\}$ of the admissible subgroup H, that induce irreducible representations: $T^{r;\phi} = ind(\chi_{r;\phi} \mid H;G)$.

The appearance of a second parameter ϕ seems to contradict our previous description of the dual object of $\mathbf{M}_2$ in J6.1. There we have shown $\hat{\mathbf{M}}_2$ to consist of a 1-parameter family $\{T^r\}$, plus a discrete (quantized) set of characters $\{T^m = e^{im\theta}\}$ of the quotient $SO(2)$. The contradiction is resolved by remembering that the general results (Theorem 1) apply to a simply connected cover $\widetilde{\mathbf{M}}_2$, which represents a central extension of $\mathbf{M}_2$ via group $\mathbb{Z}$. The condition $T^{r;\phi} \mid \mathbb{Z} = I$ (trivial) selects a single member T^r of the family $\{T^{r,\phi} : \phi\}$.

The quantization of degenerate (one-point) orbits $\{(s;0): s = 2\pi m\}$ results from compactness of the admissible quotient $H/[H;H] \simeq SO(2)$, $H = \mathbb{M}_2$.

Let us remark that all higher-D motion groups $\mathbb{E}_n$ $(n \geq 3)$ are themselves simply connected, so the quantization rules of Theorem 1 apply directly to them (problem 2). Co-orbits of examples 3.1-3.2 had real polarizations $\{H\}$, so the resulting irreducible representations $\{T^{\mathcal{O}}\}$ were obtained by the usual induction procedure. Our next example, compact group $SU(2)$, lies at the opposite extreme. Its irreducible representations are all finite-dimensional (chapters 4-5), so they can not be induced in the usual sense. However, we shall see all of them realized as "holomorphically induced representations", according to the Borel-Weil-Bott prescription of §6.1.

3.4. Quantization of $SU(2)$ and the Borel-Weil-Bott Theorem. Here $\mathfrak{G} = \mathfrak{G}^* \simeq \mathbb{R}^3$ and co-adjoint orbits are spheres

$$\mathcal{O}_r = \{\mathrm{tr}(X^2) = -r^2\} \simeq SU(2)/U(1).$$

We identify each orbit with a complex projective space $\mathbb{CP}^1$ via the map

$$G \ni \begin{bmatrix} u & w \\ -\overline{w} & \overline{u} \end{bmatrix} \mapsto (u;w)/\{(e^{i\theta}u; e^{i\theta}w)\} \simeq \mathbb{C}\times\mathbb{C}/\mathrm{mod}(\mathbb{C});$$

where matrices $\left\{\begin{bmatrix} \lambda & \\ & \overline{\lambda} \end{bmatrix} : \lambda = e^{i\theta}\right\}$ make up the diagonal subgroup $U(1) \subset SU(2)$.

The projective space $\mathcal{O} \simeq \mathbb{CP}^1$ has standard homogeneous coordinate z in two neighborhoods $\mathcal{O}_{\pm}$, that cover it:

$$z = \tfrac{u}{w} \text{ in } \mathcal{O}_+ = \{(u;w): w \neq 0\}, \text{ and } z = \tfrac{w}{u} \text{ in } \mathcal{O}_- = \{(u;w): u \neq 0\}.$$

Furthermore, space $\mathcal{O}$ is equipped with a family of symplectic structures which come from different orbits $\{\mathcal{O}_r\}$,

$$\Omega_\alpha = \frac{\alpha}{2\pi i}\frac{dz\wedge d\overline{z}}{(1+z\overline{z})^2}. \tag{3.8}$$

One can show that Ω_α is the only G-invariant 2-form on $\mathbb{CP}^1$ (problem 3). The argument exploits the G-action on $\mathbb{CP}^1$ by fractional-linear transformations:

$$g = \begin{bmatrix} a & b \\ -\overline{b} & \overline{a} \end{bmatrix} : z \to \frac{az+b}{-\overline{b}\,z+\overline{a}}. \tag{3.9}$$

Parameter α is simply related to the radius r of $\mathcal{O}$. The quantization condition of Theorem 1 requires α to be an integer m. The corresponding representation space is formed by sections of the holomorphic line bundle $\mathcal{L}$, whose *1-st Chern class* is given by Ω_m. Following the general prescription of geometric quantization we express form Ω (3.9) as the exterior derivative of a 1-form[12] $\omega = \omega_{\pm}$ (on each hemisphere $\mathcal{O}_{\pm}$). So we write $\Omega = d\omega_+ = d\omega_-$. The difference

$$\omega_+ - \omega_- = d\phi = \frac{1}{2\pi i} g^{-1} dg,$$

where $g = g(z)$ is a holomorphic transition function on the overlap $\mathcal{O}_+ \cap \mathcal{O}_- \simeq \mathbb{C}\backslash\{0\}$. But any holomorphic nonvanishing function $g(z)$ in $\mathbb{C}\backslash\{0\}$ can be written as $f_+(z) z^m f_-(\frac{1}{z})$, for some integer m, with holomorphic functions $f_\pm$ on $\mathbb{C}$. This shows that any holomorphic line-bundle over $\mathbb{CP}^1$ is equivalent to the one with transition function $g = z^n$. Taking such $g(z_\pm)$ we calculate 2-form Ω_α on the 2-cycle $\mathcal{O} \simeq S^2$, and find

$$\int_{\mathcal{O}} \Omega = \int_{\mathcal{O}_+} \Omega + \int_{\mathcal{O}_-} \Omega = \int_{\mathcal{O}_+} d\omega_+ + \int_{\mathcal{O}_-} d\omega_- = \oint_{equator} g^{-1} dg = m.$$

So the requirement that $\mathcal{L}$ have a proper line-bundle structure sets the integral over $\mathcal{O}$ (cohomology class) of Ω to an integral value m!

Holomorphic sections of $\mathcal{L}$ consist of functions $\{f_\pm\}$ on $\mathcal{O}_\pm$ related by the transition function g:

$$f_+(z_+) = g(z_+) f_-(z_-) = z^m f_-(z_-).$$

So if one asks both functions be analytic, both must be polynomials in z of degree m. Thus we get $(m+1)$-dimensional space of polynomials. To establish the Borel-Weil-Bott Theorem for $SU(2)$ we only need to compute the $SU(2)$-action on sections of $\mathcal{L}$. Remembering the fractional-linear action of G on $\mathbb{CP}^1$, we get the familiar form irreducible representations of chapter 4 (§4.2),

$$\pi_g^m f(z) = (-\bar{b}\, z + \bar{a}\,)^m f\left(\frac{az+b}{-\bar{b}\, z + a}\right).$$

Remark: The above construction of irreducible representations, based on co-orbits in $\mathfrak{G}^*$, is a special case of the more general *Geometric quantization* procedure on an arbitrary symplectic manifold (classical-mechanical phase-space) $\mathcal{P}$. The latter is typically equipped with a canonical 2-form, and a Poisson-Lie algebra $\mathfrak{G}$ of so-called *primary observables*, functions of $\mathcal{P}$. In some cases, e.g. Lie group $G = exp(\mathfrak{G})$ acting transitively on $\mathcal{P}$, the phase-space can be identified with a co-orbit $\mathcal{P} \simeq \mathcal{O} \subset \mathfrak{G}^*$, via the momentum-map (see 8.1), so the above theory applies here. We refer to [Kirl]; [Kos]; [Woo]; [Hur]; [Sni] for further details.

---

[12] Form ω can be interpreted as a connection form of a line-bundle $\mathcal{L}$ over $\mathbb{CP}^1$. An easy computation shows,

$$\omega_\pm = \frac{m}{2\pi i} \frac{\bar{z}\, dz}{(1 + z\bar{z}\,)}; \; z = z_\pm.$$

3.5. The character formula and the Plancherel measure. We shall conclude this chapter with a universal character- and Plancherel-formula on Lie group G in terms of the orbit structure of $\mathfrak{G}^*$, due to A. Kirillov. These results apply to large classes of groups and representations, including compact Lie groups; $SL_2(\mathbb{R})$; exponential groups[13] G (whose ad_X-map has no purely imaginary eigenvalues, so $exp:\mathfrak{G}\to G$ becomes a diffeomorphism); principal series representations of noncompact semisimple groups, and many more.

Characters $\{\chi_T = \mathrm{tr}(T)\}$ of irreducible representations of G are given by certain conjugate-invariant distributions on G, $\chi(g^{-1}xg) = \chi(x)$, for all $x, g \in G$. We pick a pair of neighborhoods $U \subset G$ and $V \subset \mathfrak{G}$, related by the exponential map, $exp: V \leftrightarrow U$, and define an invariant distribution associated to an orbit $\mathcal{O} \subset \mathfrak{G}^*$,

$$\langle \phi_{\mathcal{O}} \mid f \rangle = \int_{\mathcal{O}} \int_U f(expX) e^{i\langle X \mid \xi \rangle} dX \, d\beta_{\mathcal{O}}(\xi); \tag{3.10}$$

for any test-function f on $\mathfrak{G}$. Here dX denotes the usual Lebesgue measure on $\mathfrak{G}$, while $\beta = \beta_{\mathcal{O}}$ is a $2n$-form (volume element) on $\mathcal{O}$, obtained by taking $n = \frac{1}{2}\dim\mathcal{O}$ wedge-factors of $\Omega = \Omega_{\mathcal{O}}$,

$$\beta = \frac{1}{n!} \underbrace{\Omega \wedge \ldots \wedge \Omega}_{n-times}. \tag{3.11}$$

So distribution $\phi_{\mathcal{O}}$ can be thought of as the Fourier transform $\mathfrak{F}_{\xi \to X}$ of an ad_G-invariant measure $d\beta$ on $\mathcal{O}$, pulled back to the group by the exp-map. Since exp respects adjoint action/conjugation on G and $\mathfrak{G}$,

$$g^{-1}(expX)g = exp(ad_g X),$$

the resulting distribution $\phi_{\mathcal{O}}$ is clearly conjugate-invariant. Distribution $\phi_{\mathcal{O}}$ is closely related to character χ of representation $T^{\mathcal{O}}$.

Proposition: *There exists a conjugate-invariant function* $p = p_{\mathcal{O}}$ *on* G, *equal* 1 *at the identity* $\{e\}$, *and different from zero on* $U \subset G$, *so that*

$$\boxed{\chi_{\mathcal{O}} = \frac{1}{p_{\mathcal{O}}} \phi_{\mathcal{O}}} \tag{3.12}$$

Clearly (3.12) is equivalent to

$$\langle \chi_{\mathcal{O}} \mid f \rangle = tr(T_f) = \int_{\mathcal{O}} \int f(expX) \frac{1}{p(expX)} e^{2\pi i \langle \xi \mid X \rangle} dX \, d\beta. \tag{3.13}$$

The correspondence between characters $\{\chi_{\mathcal{O}}\}$, given by (3.13), and representations $\{T^{\mathcal{O}}\}$, based on co-orbits, has not been proven yet in a complete generality, although the result is believed to be true.

[13]This class includes fairly many nilpotent and solvable groups, affine, Heisenberg, etc.

Let us remark that formula (3.12) reveals the nature of distribution $\chi_{\mathcal{O}}$, that is closely connected with the geometry of orbit $\mathcal{O}$. Thus compactness of $\mathcal{O}$ (e.g. compact G) implies that $\chi_{\mathcal{O}}$ is a regular function (hence $T^{\mathcal{O}}$- finite-dimensional!). If $\mathcal{O}$ is a cylinder (example 2), so $\mathcal{O}$ is made of subspaces $\{\xi_0 + L\}$, then distribution $\chi_{\mathcal{O}}$ contains δ-type factors. A particular case: $L = \mathfrak{H}^{\perp}$- annihilator of a subalgebra $\mathfrak{H} \subset \mathfrak{G}$, implies that $\chi_{\mathcal{O}}$ is supported on a subgroup $H = exp\mathfrak{H} \subset G$.

For orbits of maximal dimension density $p_{\mathcal{O}}$ in the denominator (3.13) proves to be a universal function:

$$\boxed{p(expX) = detF(ad_X);\ \text{where } F(t) = \frac{sinh(t/2)}{t/2} = \sum_0^{\infty} \frac{(t/2)^{2k}}{(2k+1)!}} \tag{3.14}$$

Next we proceed to the Plancherel formula for Lie groups. We remind that the Plancherel measure $d\mu$ on any unimodular[14] group has the form

$$\int_G |f(x)|^2 dx = \int_{\widehat{G}} tr[T_f T_f^*] d\mu(T), \tag{3.15}$$

for a suitable class of test-functions $\{f\}$ on G, e.g. $f \in L^1 \cap L^2(G)$.

One could understand (3.15) as a decomposition of δ-function on G into the the sum/integral of irreducible characters,

$$f(e) = \int_{\widehat{G}} tr(T_f) d\mu(T);\ \text{or } \delta(e) = \int_{\widehat{G}} \chi_T d\mu(T). \tag{3.16}$$

In case when the character-formula (3.12) holds the meaning of (3.16) becomes transparent: it gives a decomposition of the Lebesgue measure on $\mathfrak{G}^*$ into canonical measures on orbits.

Indeed, let us assume that there exists a 1-1 correspondence between orbits and representations, and that the orbits of maximal dimension have the universal density function $p_{\mathcal{O}}$ of (3.14). The Lebesgue measure $d\xi$ on $\mathfrak{G}^*$ is clearly ad^*_G-invariant, hence it can be decomposed (resolved) in the sum/integral of the canonical measures $\{\beta_{\mathcal{O}}\}$ (3.11) on orbits,

$$d\xi = \int_{orbit-space} \beta_{\mathcal{O}} d\mu(\mathcal{O}). \tag{3.17}$$

Fourier-transforming (3.17) we get

[14]Group G is called unimodular, if the right- and left-invariant Haar measures on G are equal. Most examples considered in the book are unimodular: all compact groups; simple and semisimple noncompact groups; nilpotent (Heisenberg) groups; motion groups, etc.

$$\delta(X) = \int_{orbit-space} \phi_{\mathcal{O}}(expX)\, d\mu.$$

Now we apply the character-formula (3.12), remembering that $p_{\mathcal{O}}(e) = 1$, and find

$$\delta(x) = \int_{orbit-space} \chi_{\mathcal{O}}(x)\, d\mu(\mathcal{O}),$$

which yields the requisite decomposition (3.16).

To compute $d\mu$ explicitly we need a set of ad_G-invariant coordinates on $\mathfrak{G}^*$. Let us assume that maximal-D orbits are joint level-sets of the family of functions $\lambda_1;...\lambda_k$ ($k = \mathrm{codim}\mathcal{O}$), so $\{\lambda_j\}$ parametrize the orbit-space. Let us also introduce coordinates $\varphi_1;...\varphi_{2n}$ on each orbit $\mathcal{O}$ ($2n = \dim\mathcal{O}$). The orbit-space $\mathfrak{T} = \mathfrak{G}^*/\mathrm{ad}_G^*$ can then be identified with the level-set: $\varphi_1 = ... = \varphi_{2n} = 0$. To proceed further we need a notion of *Pfaffian* of a skew-symmetric matrix $A = (a_{kj})$. Any such A corresponds in a 1-1 way to an exterior 2-form: $\omega_A = \sum a_{kj} e_k \wedge e_j$. The n-th exterior power of ω_A is a constant multiple of the only highest rank $2n$-form,

$$\omega_A \wedge ... \wedge \omega_A = C\, e_1 \wedge ... \wedge e_{2n}.$$

Constant C, represents a polynomial of degree n in variables $\{a_{kj}\}$, called the Pfaffian of A, and denoted by $\mathrm{Pf}A$. Pfaffian has many interesting properties, for instance, $(\mathrm{Pf}\,A)^2 = detA$ (problem 4).

Returning to the Plancherel measure we can state the following general result. Let $\xi(\lambda)$ be a point on an orbit $\mathcal{O}$ with coordinates $\{\lambda_1;...\lambda_k\}$. We denote by A the matrix of Poisson brackets at $\xi(\lambda)$,

$$a_{ij} = \{\varphi_i;\varphi_j\}(\xi(\lambda)).$$

Theorem 3: *The Plancherel measure on group G is given by the formula*

$$\boxed{d\mu(\lambda) = J(\lambda;0)\, \mathrm{Pf}\, A(\lambda)\, d\lambda_1...d\lambda_k} \tag{3.18}$$

where $J(\lambda;\varphi)$ is the Jacobian of the coordinate change: $\xi \to (\lambda;\varphi)$,

$$J = \left| \frac{\partial(\xi)}{\partial(\lambda;\varphi)} \right|.$$

Formula (3.18) takes on a particularly simple form, when coordinate functions $\{\lambda_i;\varphi_j\}$ are linear, i.e. elements of algebra $\mathfrak{G}$. Then

$$a_{ij}(\lambda) = \langle \xi(\lambda) \,|\, [\varphi_i;\varphi_j] \rangle = \sum c_{ij}^m \lambda_m;$$

where $\{c_{ij}^m\}$ are structure constants of the algebra $\mathfrak{G}$. Now the Plancherel measure turns into

$$\boxed{d\mu = P(\lambda)d\lambda_1...d\lambda_k} \tag{3.19}$$

with a homogeneous polynomial $P(\lambda)$ of degree $n = \frac{1}{2}\dim\mathcal{O}(\lambda)$ in $\{\lambda_j\}$, with coefficients depending on the structure constants of $\mathfrak{G}$.

It is worth to remark that formal application (3.19) yields the correct Plancherel measure for complex semisimple groups (e.g. $SL_2(\mathbb{C})$), and compact Lie groups. In the latter case integration in variables λ's must be replaced by summation over the discrete set of the "properly (Borel-Weil-Bott) quantized" orbits (problem 5)! In the real semisimple case (e.g. $SL_2(\mathbb{R})$) it predicts the correct answer only for the so called *discrete-series* representations, as we shall demonstrate in the next chapter.

Problems and Exercises.

1. Show that closedness of the canonical 2-form Ω_B (3.3) on a co-adjoint orbit $\mathcal{O} \subset \mathfrak{G}^*$ follows from the Jacobi identity for the Lie bracket on $\mathfrak{G}$.

   i) Use the general formula for differential of a 2-form Ω on a manifold $\mathcal{M}$: given any triple of (tangent) vector fields $X;Y;Z$ on $\mathcal{M}$,

   $$d\Omega(X;Y;Z) = X[\Omega(Y;Z)] - Y[\Omega(X;Y)] + Z[\Omega(X;Y)] +$$
   $$+\Omega([X;Y];Z) - \Omega([X;Z];Y) + \Omega([Y;Z];X);$$

   Here $X[...]$ means (Lie) derivative of function (e.g. $\Omega(Y;Z)$) along vector field X, and $[X;Y]$, etc. - Lie commutators of vector fields.

   ii) Consider functions $\{f(\xi)\}$ on $\mathfrak{G}^*$, restricted on orbit $\mathcal{O}$, pick an element $X \in \mathfrak{G}$ (view it as a vector field on $\mathcal{O}$), and show that the Lie derivative,

   $$X[f](\xi) = \langle \xi \mid [X; \nabla f(\xi)] \rangle$$

   Here $\nabla f(\xi)$ denotes the gradient of $f \mid \mathfrak{G}^*$ (an element of $\mathfrak{G}$), and $[...]$ is the Lie-algebra bracket of X and ∇f.

   iii) Apply part (ii) functions $\{f = \Omega(Y;Z) = \langle \xi \mid [Y;Z] \rangle$; etc.$\}$ on $\mathfrak{G}^*$, and fields $\{X;Y;Z \in \mathfrak{G}\}$, and show the differential of Ω_B is expressed in terms of the Jacobi-term,
   $d\Omega = 2\langle \xi \mid Jac(X;Y;Z) \rangle$; where $Jac = [X;[Y;Z]] + [Y;[Z;X]] + [Z;[X;Y]]$.

2. Verify Theorem for Euclidian motion groups $\mathbf{E}_n$ $(n \geq 3)$ (Hint: reduced co-orbit space $\mathcal{Y} \simeq S^{n-1}$, have 2-nd cohomology group $H^2(S^{n-1}) \simeq \mathbf{Z}$).

3. Prove that the only G-invariant 2-form on $\mathbf{CP}^1 \simeq SU(2)/U(1)$ is Ω_α (3.8) (Use fractional-linear action of $G = SU(2)$ on $\mathbf{CP}^1$, take $\Omega = f(z;\bar z)dz \wedge d\bar z$, and compute $f = const\frac{1}{(1+z\bar z)^2}$.

4. Check the identity: $\mathrm{Pf}(A)^2 = \det A$, for any skew-symmetric A (Hint: bring A to a canonical form, where $\omega_A = \sum a_j e_j \wedge e_{n+j}$).

5. Verify the Plancherel formula (3.19) for (i) $SU(2)$, $SU(n)$ and other compact Lie groups; (ii) the Heisenberg group; (iii) Euclidian motion groups.

Additional comments:

The imprimitivity systems and representation theory of group extensions (§6.1) were developed by G. Mackey [Mac]. Our approach to induction, holomorphic induction and Borel-Weil-Bott Theorem follows [Kir1] and [Ta2]. There are many excellent expositions of the ubiquitous Heisenberg group and the oscillator representation (see for instance [How];[Ta2];[Kir1]).

The orbit method appeared first in the work of Kirillov [Kir2] on representation theory of nilpotent groups. It was further developed by Kostant, Auslander, Pukanszky et al ([Kir1];[Pu];[Kos]). Later an important connection between the orbit method and classical mechanics was discovered by B. Kostant, and further developed by Souriau [Sou], Kirillov, Kostant [Kos] and others. Since its inception the orbit method came to play the ever increasing role in many parts of geometry, mechanics, geometric quantization ([How]; [Hur]; [Sn]; [Be]; [How]; [Kos]; [Kir1]; [Dir]), and more recently in connection to the string theory. Further details and examples of quantization will be given in §8.5.

Chapter 7. Representations of SL_2.

Lie group SL_2 consists of 2×2 matrices of determinant 1. It is the first (and most prominent) member in the large family of ***noncompact simple/semisimple Lie groups.*** It appears in many different disguises: conformal $SU(1;1)$; Lorentz $SO(1;2)$; symmetries (fractional-linear maps) of the hyperbolic (Poincare-Lobachevski) geometry in $\mathbf{H}$ (§1.1), and other. The analysis and representation theory of SL_2 reveals many fascinating features, that combine both "compact" (chapters 2;6) and "noncompact" theories (chapters 4-5). They find deep connections to the number theory, automorphic forms, Riemann surfaces, spectral geometry, some of which will be explored in §§7.5-7.6. Most of the chapter (§§7.1-7.6) deals with the real unimodular group $SL_2(\mathbf{R})$. The last section (§7.7) indicates some extensions of the theory to the complex (Lorentz) group $SL_2(\mathbf{C})$, and its higher-dimensional cousins $SO(1;n)$.

§7.1. *Principal, complementary and discrete series.*

Our goal in this section is to construct and analyze unitary irreducible representations of $SL_2(\mathbf{R})$. Those turned out to consists of 3 series: ***principal, complementary*** and ***discrete.*** In all 3 cases the induction procedure of §1.2; §3.2 and §6.1 plays an important role.

1.1. Principal series. Group G acts by fractional-linear transformations $g\colon x\to x^g=\frac{ax+c}{bx+d}$; on $\mathbb{R}$ (or $\mathbb{C}$). We shall show that $\mathbb{R}$ is the homogeneous space of G, as well as the upper/lower half-planes $\mathbb{C}_+=\{\Im z>0\}$; $\mathbb{C}_-=\{\Im z<0\}$. Namely,

Proposition: i) *Real line $\mathbb{R}$ is isomorphic to a quotient-space* of G, *modulo the Borel subgroup of the upper-triangular matrices,*

$$B==h=\begin{bmatrix} a & b \\ & \frac{1}{a} \end{bmatrix};\ a\in\mathbb{R}^\times;\ b\in\mathbb{R}\ .$$

ii) *The Poincare upper half-plane $\mathbb{H}\simeq G/K$, where $K=SO(2)$ is the maximal compact subgroup of G.*

To demonstrate (i) we use a unique factorization of (almost) any $g\in G$ into its upper/lower-triangular parts,

$$g=\begin{bmatrix} a & b \\ c & d \end{bmatrix}=\begin{bmatrix} a & b \\ 0 & d \end{bmatrix}\begin{bmatrix} 1 & 0 \\ c/d & 1 \end{bmatrix}=hx, \tag{1.1}$$

where $h\in B$; $x\in X==\begin{bmatrix} 1 & 0 \\ x & 1 \end{bmatrix}$. So family $\{x\}$ provides coset representatives for the quotient G/B. The G-action on the quotient-space $G/B\simeq\mathbb{R}$ is computed by multiplying a coset representative x by $g\in G$, and factoring out the upper-triangular term,

$$xg = h(x;g)x^g. \tag{1.2}$$

Here $h(x;g)$ denotes a B-valued cocycle ($h: X\times G \to B$), and x^g marks the image of a coset $\{x\}$ under the action of g. Then we get[1]

$$xg = \begin{bmatrix} 1 & \\ x & 1 \end{bmatrix}\begin{bmatrix} a & b \\ c & d \end{bmatrix} = \begin{bmatrix} a & b \\ ax+c & bx+d \end{bmatrix} = \begin{bmatrix} (bx+d)^{-1} & b \\ & bx+d \end{bmatrix}\begin{bmatrix} 1 & 0 \\ \frac{ax+c}{bx+d} & 1 \end{bmatrix} \tag{1.3}$$

The second statement (ii) was explained back in chapter 1 (§1.1), where we realized the Poincare-Lobachevski half-plane $\mathbb{H}$ as $SL_2(\mathbb{R})/SO(2)$. It can also be demonstrated by the so-called *Iwasawa* decomposition (problem 1).

We denote by $\epsilon_{\pm}(a)$ two special characters on the multiplicative group of reals $\mathbb{R}^{\times} = \mathbb{R}_+ \bigcup \mathbb{R}_- \simeq \mathbb{R}_+ \times \mathbb{Z}_2$,

$$\epsilon_+(a) = 1;\ \epsilon_-(a) = sgn\, a,$$

and write any character χ on $\mathbb{R}^{\times}$ as a pair $(s;\epsilon_{\pm})$, or ($s\pm$, for brevity), with $s \in \mathbb{R}$,

$$\chi(a) = |a|^{is}\epsilon_{\pm}(a) = \begin{cases} |a|^{is};\ \text{for } \epsilon_+ \\ sgn(a)\,|a|^{is};\ \text{for } \epsilon_- \end{cases}.$$

The *principal series representations* of $SL_2(\mathbb{R})$ act on space $L^2(\mathbb{R})$, by the formula

$$\boxed{(T_g^{s\pm}f)(x) = |bx+d|^{is-1}\epsilon_{\pm}(bx+d)f\left(\frac{ax+c}{bx+d}\right)} \tag{1.4}$$

Another realization of the principal series representations is obtain from the natural action of $SL_2(\mathbb{R})$ on space $L^2(\mathbb{R}^2) = \mathcal{H}^+ \oplus \mathcal{H}^-$, sum of even and odd functions,

$$T_g^{\pm} f(x;y) = f(ax+cy;bx+dy);\ f \in \mathcal{H}^{\pm}.$$

Each of two spaces can be decomposed into the direct integral of spaces of homogeneous functions, $\mathcal{H}^{\pm s} = \{f \in \mathcal{H}^{\pm}:\ f(tx;ty) = t^s f(x;y)\}$, by writing $f = f(r;\theta)$ in polar coordinates and Mellin-transforming the radial variable,

$$f(r;\theta) \to \int_0^{\infty} f(r;\theta)\, r^{-s-1}dr,\ s = i\rho.$$

The latter can be identified either with functions on $\mathbb{R}$ with the L^2-product,

$$f(x;y) = |y|^s f(x/y;\pm 1) \to f(x;\pm 1),$$

or with L^2-functions on the circle, $f = r^s f(1;\theta) \to f(1;\theta)$. It is easy to see that the restricted operators $T^{\pm}|\mathcal{H}^{\pm s}$ are equivalent to the principal series $T^{s\pm}$, for purely

[1] The factorization formula (1.1) as well as coset-transformations (1.4) are particular cases of a general setup, namely a subgroup $H \subset G$, and a subset $X \subset G$ of coset representatives $X \bigcap H = \{e\}$, with the unique factorization $g = hx$, for almost any $g \in G$. Given such pair $(H;X)$ one can identify X with the homogeneous space $H\backslash G$, and produce an H-valued cocycle $h(x,g)$ by factoring product $xg = h(x;g)x^g$.

imaginary s. Indeed, taking the **R**-realization of $\mathcal{H}^{\pm s}$, we get

$$|y|^{-s} f(ax+cy; bx+dy) = |bx+d|^{s} \epsilon_{\pm}(bx+d) f(\frac{ax+c}{bx+d}).$$

Theorem 1: *Formula (1.4) defines a unitary irreducible representation $T^{s\pm}$ of $SL_2(\mathbb{R})$ on $L^2(\mathbb{R})$, for any $s \neq 0$. Representations $T^{s;+}$ and $T^{-s;-}$ are equivalent for all $s \in \mathbb{R}$, while all other pairs are inequivalent.*

Proof: Let us observe that operators (1.4) form a representation of G, due to the cocycle property of $h(x;g)$ of (1.2). Indeed,

$$(T_g f)(x) = \alpha(x,g) f(x^g),$$

where scalar cocycle $\alpha(x;g)$ consists of a B-character χ composed with h. Unitarity of $T^{s\pm}$ follows easily from the change of variable formula. Indeed, operator T_g transforms $f(x) \to \alpha f \circ \phi(x)$, where $\phi: x \to \frac{ax+c}{bx+d}$. So the L^2-norm of $T_g f$ is preserved,

$$\|T_g f\|^2 = \int |(\alpha \circ \phi^{-1}) f|^2 |\phi'|^{-1} dx = \|f\|^2, \text{ all } f,$$

iff $\alpha(x) = \sqrt{|\phi'(x)|}$. Here ϕ' denotes the Jacobian of the coordinate change ϕ. Calculating ϕ' for the fractional-linear $\phi(x) = s^g$, we find $\phi' = \frac{1}{(bx+d)^2}$, whence comes the normalizing factor $|bx+d|^{-1}$ attached to a unitary character, $\chi(\ldots) = |\ldots|^{is} \epsilon_{\pm}(\ldots)$ in (1.4).

Irreducibility: The reader has probably noticed that representations $\{T^{s\pm}\}$ are induced by one-D characters of the Borel subgroup B. There exist a convenient irreducibility test for induced representations based on the notion of *Mackey's imprimitivity system*: group G, homogeneous space $X \simeq G/B$, a (continuous) family of vector spaces (vector bundle) $\{\mathcal{V}_x : x \in X\}$, and a family of operators (cocycle) $\sigma(x;g): \mathcal{V}_x \to \mathcal{V}_{xg}$. In chapter 3 (J3.2) we described a "finite/discrete version" of Mackey's results. The continuous version goes along the same lines, but some technical modifications are needed due to topological setup. Namely, with each point $y \in X$ we associate its stabilizer, $B_y = \{g: y^g = y\} = y^{-1} B y$, and an inner automorphism $Ad_y: h' \to h = y h' y^{-1}$, that takes B_y onto B. Automorphism Ad_y pulls back any character (representation) χ from B to B_y:

$$\chi \to \chi^y(h') \overset{def}{=} \chi(y h' y^{-1}), \text{ for } h' \in B_y .$$

The Mackey's irreducible test for T^C requires 2 steps,

1) to check irreducibility of χ (obvious for characters);

2) to show that two characters (representations) χ and χ^y restricted on the joint intersection $B \bigcap B_y$ are different (non equivalent).

More generally, one shows that the commutator algebra of $T = Ind(\chi \mid B; G)$ is decomposed into the direct sum (discrete or continuous), of the intertwining subspaces,

$$\mathrm{Com}(T) = \mathrm{Com}(\chi \mid B) \underset{\omega}{\oplus} \mathrm{Int}(\chi \mid B \bigcap B_y; \chi^y \mid B \bigcap B_y),$$

the summation extends over all non-diagonal G-orbits $\omega \subset X \times X$, each of them being labeled by a point (e, y). In the continuous case direct sum $\underset{\omega}{\oplus} \ldots$ is replaced by the direct integral $\oint \ldots d\omega$. To apply the Mackey's test in our case, $G = SL_2(\mathbf{R})$, $B = \left\{\begin{bmatrix} a & b \\ & 1/a \end{bmatrix}\right\}$, and $X = \left\{y = \begin{bmatrix} 1 & \\ y & 1 \end{bmatrix}\right\}$, we compute

$$y^{-1}By = \left\{\begin{bmatrix} a+by & b \\ y(\frac{1}{a}-a)-y^2b & * \end{bmatrix}\right\}.$$

The low off-diagonal entry must be 0 on the intersection-subgroup $B \bigcap B_y$, which implies $b = \frac{1}{y}(a+\frac{1}{a})$, i.e.

$$y^{-1}hy = \begin{bmatrix} 1/a & * \\ & a \end{bmatrix}; \text{ for all } h \in B \bigcap B_y.$$

So the "pull-back" character $\chi^y(a) = \chi(\frac{1}{a})$, whenever the off-diagonal entry of h $b \neq 0$. It follows then that two characters χ and χ^y, restricted on $B \bigcap B_y$, are inverse one to the other, so $\chi^2 = 1$. But the latter is impossible for $\chi(a) = e^{isa}\epsilon_{\pm}(a)$, unless $s = 0$. This proves irreducibility of $T^{s\pm}$ for all $s \neq 0$.

More elementary proof of irreducibility utilizing a simple Fourier analysis, rather than the Mackey's test, is outlined in problem 2.

Equivalence: The above argument, based on Mackey's test, also yields an intertwining space $\mathrm{Int}(T^{\chi_1}; T^{\chi_2})$, for a pair of induced representations. Ignoring the diagonal of $X \times X$ (of measure 0): *there exists a nontrivial intertwining* $Q \in Int(T^{\chi_1}; T^{\chi_2})$, *iff two characters are related by* $\chi_1 = \chi_2^y$, *on the intersection* $B \bigcap B_y$ *for (almost) any* $y \in X$. In our case $\chi_1(a) = \chi_2(\frac{1}{a})$, for all (Cartan) diagonal matrices a, hence $\chi_1\chi_2 = 1$. Rewriting the latter in terms of parameters $(s\pm)$ yields: $T^{s+} \sim T^{-s;-}$, while all other pairs are different (non-equivalent), QED.

1.2. Discrete series representations are realized in holomorphic functions on the Poincare-Lobachevski upper-half plane, $\mathbb{P} = \{z = x + iy,\ y > 0\}$. Group G act on $\mathbb{P}$ by fractional-linear transformations: $z \to \frac{az+c}{bz+d}$. In fact, $\mathbb{P}$ is a homogeneous space $K \backslash G$, where subgroup

$$K = \left\{\begin{bmatrix} \cos\phi & \sin\phi \\ -\sin\phi & \cos\phi \end{bmatrix}\right\} \simeq \mathbb{T},$$

coincides with the stabilizer of point $i \in \mathbb{P}$, and represents a maximal compact subgroup in G. There exists an invariant (Haar) measure on $\mathbb{P}$, computed in §1.1, $d\mu = \frac{dxdy}{y^2}$.

We shall consider vector-spaces $\mathcal{H}_n^{\pm}$ of holomorphic/antiholomorphic functions

$f(z)$ in $\mathbb{P}$ with norm

$$\|f\|^2 = \int_{\mathbb{P}} |f(z)|^2 y^{n-1} dydx,\ n = 1,2,...,$$

and define *discrete-series representations* in $\mathcal{H}_n^{\pm}$ by

$$\boxed{(T_g^{\pm n} f)(z) = (bz+d)^{-n-1} f(\tfrac{az+c}{bz+d})} \tag{1.5}$$

The reader can recognize (1.5) as a *holomorphically induced representation* of §6.1, space $\mathbb{P}$ has clearly an analytic structure, and function $\alpha(z;g) = bz+d$, hence $(bz+d)^{\pm m}$, becomes a holomorphic in variable z cocycle α: $\mathbb{P} \times G \to \mathbb{C}^*$.

Density y^{n-1} in the definition of $\mathcal{H}_n$-norm follows from the unitarity requirement of operators[2]

$$T_g f = \alpha(z;g) f(z^g),\ \text{in}\ L^2(\mathbb{P}; w(z)dA).$$

Another realization of discrete series is obtained in the Poincare-disk $\mathbb{D} = \{z: |z| < 1\}$ - a homogeneous space of the conformal twin $SU(1;1)$ of $SL_2(\mathbb{R})$ (see §1.1, problem 5). We recall that the latter means 2×2 complex matrices $\{g\}$, that preserve an indefinite hermitian (1,1)-form in $\mathbb{C}^2$, $\langle \vec{z}; \vec{w} \rangle = z_1 \bar{w}_1 - z_2 \bar{w}_2$, so

$$\left\{ g = \begin{bmatrix} \alpha & \beta \\ \bar{\beta} & \bar{\alpha} \end{bmatrix},\ |\alpha|^2 - |\beta|^2 = 1 \right\}.$$

Lie algebra $su(1,1)$ consists of all complex matrices $\left\{ X = \begin{bmatrix} is & \zeta \\ \bar{\zeta} & -is \end{bmatrix}; s \in \mathbb{R}; \zeta \in \mathbb{C} \right\}$, and the correspondence between two groups (subgroups of $SL(2;\mathbb{C})$!) is given by the Möbius element

$$\sigma = \frac{1}{\sqrt{2}} \begin{bmatrix} 1 & -i \\ -i & 1 \end{bmatrix} \in SL(2;\mathbb{C}).$$

Namely, conjugation, $\sigma: g \to \sigma^{-1} g \sigma$, maps $SL(2;\mathbb{R}) \to SU(1;1)$. Conversely,

$$\sigma^{-1}: \begin{bmatrix} \alpha & \beta \\ \bar{\beta} & \bar{\alpha} \end{bmatrix} \to \begin{bmatrix} Re(\alpha\ \beta) & Im(\alpha+\beta) \\ -Im(\alpha\ \beta) & Re(\alpha+\beta) \end{bmatrix},$$

sends $SU(1,1) \to SL(2;\mathbb{R})$.

The Möbius element $\sigma: z \to z^\sigma = i\frac{z+i}{z-i} = w$, takes $\mathbb{P}$ into $\mathbb{D}$, and conjugates fractional-linear actions of both groups. Clearly, the $SL_2(\mathbb{R})$-invariant volume-element $\frac{dxdy}{y^2}$ on $\mathbb{P}$ is taken into $SU(1;1)$-invariant element $\frac{dwd\bar{w}}{(1-|w|^2)^2}$ on $\mathbb{D}$.

In the unit disk-version space $\mathcal{H}_n$ consists of holomorphic/antiholomorphic functions in $\mathbb{D}$ with norm

---

[2]Indeed, fixing point $z_0 = i$, and a map $g_z: i \to z$, given by matrix

$$g_z = \begin{bmatrix} \sqrt{y} & \\ x/\sqrt{y} & 1/\sqrt{y} \end{bmatrix}$$

we can easily compute the density, that makes $T^{\pm n}$ unitary,

$$w(x) = |\alpha(z_0; g_z)|^2 |(\text{Jacobian } g_z)(z_0)|^2 = (\sqrt{y})^{2(n+1)} y^{-2} = y^{n-1}.$$

$$\|f\|^2 = \int_{\mathbb{D}} |f(z)|^2(1-|z|^2)^{n-1}dA,$$

and operators $T_g^{\pm n}$ act on $\mathcal{H}_n$ by

$$\boxed{(T_g^n f)(z) = (\beta z+\bar{\alpha})^{-n-1} f(\tfrac{\alpha z+\bar{\beta}}{\beta z+\bar{\alpha}})} \qquad (1.6)$$

As above one can easily check unitarity of operators T^n of (1.6).

Theorem 2: *Representation $T^{\pm n}$ of (1.5)-(1.6) are irreducible and distinct for all $\pm n$.*

To prove irreducibility we shall use the infinitesimal method of §1.4 (see [Lan]), and analyze the generators of Lie algebra $su(1,1)$ acting in space h_n. Lie algebra $su(1,1)$ has a Cartan basis of 3 elements

$$h = \begin{bmatrix} i & \\ & -i \end{bmatrix};\ X = \begin{bmatrix} & 1 \\ 1 & \end{bmatrix};\ Y = \begin{bmatrix} & i \\ -i & \end{bmatrix};$$

and the corresponding one-parameter subgroups are

$$\left\{\begin{bmatrix} e^{i\phi} & \\ & e^{-i\phi} \end{bmatrix}; 0 \le \phi \le 2\pi\right\}; \left\{\begin{bmatrix} ch\,t & sh\,t \\ sh\,t & ch\,t \end{bmatrix}; t \in \mathbb{R}\right\}; \left\{\begin{bmatrix} ch\,t & ish\,t \\ -ish\,t & ch\,t \end{bmatrix}; t \in \mathbb{R}\right\}.$$

We compute the corresponding one—parameter groups of operators,

$$T_{exp(th)} f = e^{i(n+1)t} f(e^{i2t}z);\ T_{exp(tX)} f = (zsh\,t+ch\,t)^{-n-1} f(\tfrac{zch\,t+sh\,t}{zsh\,t+ch\,t});$$

$$T_{exp(tY)} f = (z\,i\,sh\,t + ch\,t)^{-n-1} f(\tfrac{\dots}{\dots});$$

and differentiate the latter in t at t=0 to get the generators. This yields

$$T_h = i[(n+1) + 2z\partial_z];\ T_X = -(n+1)z + (1-z^2)\partial_z;$$
$$T_Y = -i[(n+1)z + (1-z^2)\partial_z].$$

Let us observe that monomials $\{z^k\}_0^r$ form an orthogonal basis of eigenfunctions of operator T_h in h_n,

$$\langle z^k \mid z^j \rangle = \left(2\pi \int_0^1 r^{2k+1}(1-r^2)^{n-1}dr\right)\delta_{kj} = \begin{cases} 0; & \text{if } kWj; \\ \frac{k!(n-1)!}{(n+k)!}; & \text{for } k=j \end{cases}$$

and

$$T_h(z^k) = i(n+1+2k)z^k.$$

Applying operators T_X and T_Y to the basis $\{z^k\}$ we find

$$T_X(z^k) = kz^{k-1} - (n+1+k)z^{k+1};\ T_Y(z^k) = -i[kz^{k-1} + (n+1+k)z^{k+1}].$$

So linear combinations T_{X+iY}, T_{X-iY} represent the raising/lowering operators in the basis $\{z^k\}$, and $1 = z^0$ is the lowest-weight vector of T_h, a situation reminiscent of §4.4 (raising/lowering for Legendre functions and and spherical harmonics). Notice, that operators $T_{X0;Y}$ are not in the Lie algebra $su(1;1)$, but in its complexification $SL(2;C)$! Now we can establish irreducibility in two steps:

(i) first we observe that vector $f_0 = 1$ is cyclic in $\mathcal{H}_n$, since powers $\{T^k_{X+iY}(f_0)\}$ span the entire space;

(ii) if a function f has $\langle f \mid f_0\rangle = f(0) \neq 0$, then the projection

$$P_0 : f \to \langle f \mid f_0\rangle f_0 = \int_K e^{-inu} T_u(f)du,$$

takes f into a nonzeros multiple of f_0, hence $Span\{T_g(f)\} \supset Span\{T_g(f_0)\} = \mathcal{H}$, so f is also cyclic. Finally, any vector $f \in \mathcal{H}_n$ can be shifted by an element $g \in G$, so that

$$(T_g f)(0) = \langle T_g f \mid f_0\rangle \neq 0.$$

This shows that any vector f is cyclic, hence T^n is irreducible, QED.

1.3. Complementary series representations are labeled by the real parameter $\lambda \in [-1;1]$, and look similar to the principal series. Namely,

$$\boxed{(T^\lambda_g f)(x) = |bx+d|^{-\lambda-1} f\left(\frac{ax+c}{bx+d}\right)} \tag{1.7}$$

for a suitable class of functions on $\mathbb{R}$. The class of functions $\{f\}$ is defined by the norm

$$\|f\|^2_\lambda = \frac{1}{\Gamma(\lambda)}\int\int |x-y|^{\lambda-1} f(x)\overline{f(y)}dx\,dy, \tag{1.8}$$

in other words

$$\|f\|_\lambda = \langle R_\lambda f \mid f\rangle_{L^2} = \left\|R_{\lambda/2}(f)\right\|_{L^2};$$

where R_λ denotes the *Riesz potential* $\frac{1}{\Gamma(\lambda)}|x-y|^{\lambda-1}$, a fractional power of the Laplacian $(-\Delta)^{-\lambda/2}$ on $\mathbb{R}$ ($\Delta = \partial^2_x$) (see chapter 2). We leave as an exercise to the reader (problem 4) to show that $\{T^\lambda\}$ are, indeed, unitary irreducible representations of SL_2 with respect to the product (1.8). The details are similar to the previous analysis of the principal and complementary series, so we shall skip them (see problem 4).

Finally, we shall state without proof the general classification Theorem of Bargman [Bar], about irreducible representations of $SL_2(\mathbb{R})$.

Theorem 3: *The principal, discrete and complementary series comprise all unitary irreducible representations of* $SL_2(\mathbb{R})$.

The proof (see [Lan],[Ta2]) involves an infinitesimal analysis of the associated representations of Lie algebra $\mathbf{SL}_2$. The latter are shown to possess the highest (lowest) weights vector, labeled by a complex parameter s. Writing the unitarity condition for the resulting (group) representations T^s and analyzing them one comes up with 3 basic series.

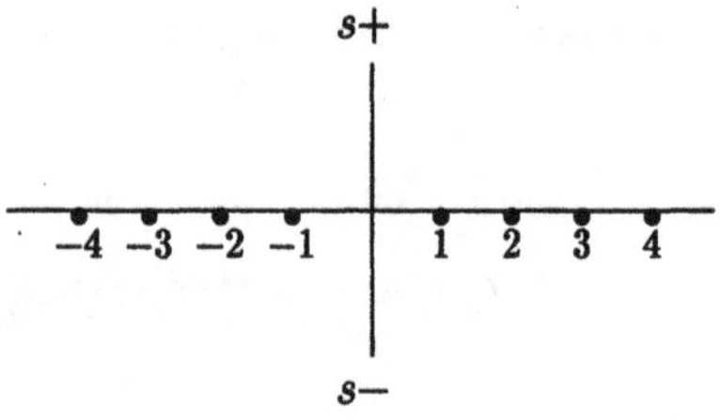

Fig. 1: *Space* $\hat{G}$ *of irreducible representations of* $SL_2(\mathbf{R})$. Here vertical axis represents the *principal series*; interval (-1;1) on the horizontal axis - *complementary series*, while integers $\{\pm m\}$ correspond to the *discrete series*.

Problems and Exercises:

1. Show the Iwasawa decomposition: any matrix $g \in SL_n$ is uniquely factored into the product UN, of a unitary/orthogonal U, and an upper-triangular N with positive entries on the main diagonal (use Gram-Schmidt orthogonalization).

2. Show irreducibility of the principal series representation T^λ $(\lambda = s\pm)$ by taking any $Q \in Com(T)$, and analyzing 3 families of operators that commute with it:
(i) $\left\{T_{\mathring{b}}\colon \mathring{b} = \begin{bmatrix} 1 & \\ b & 1 \end{bmatrix}; b \in \mathbf{R}\right\}$; *(ii)* $\left\{T_{\mathring{a}}\colon \mathring{a} = \begin{bmatrix} a & \\ & a^{-1} \end{bmatrix}; a \in \mathbf{R}^\times\right\}$; *(iii)* $\left\{T_\sigma\colon \sigma = \begin{bmatrix} & 1 \\ -1 & \end{bmatrix}\right\}$.
Step $\mathring{1}$. Show: $T_{\mathring{b}} Q = Q T_{\mathring{b}} \Rightarrow Q = K(\text{x-y})$ is a convolution with a distributional kernel $K(x)$ on $\mathbf{R}$.
$\mathring{2}$. $T_{\mathring{a}} Q = Q T_{\mathring{a}} \Rightarrow$ distribution $K(x) = c_0\delta(x) + c_1 H_+(x) + c_2 H_-(x)$, *where* $H_\pm$ are Heaviside functions on $[0;+\infty)$ and $(-\infty;0]$.
$\mathring{3}$. $T_\sigma Q = Q T_\sigma \Rightarrow K = c\delta(x)$, i.e. $Q = cI$ - scalar!

3. Formulate and prove the continuous version of Mackey's imprimitivity criteria (Do it first for spaces of continuous functions (or cross-sections), $\mathcal{C}(X)$, then use density arguments of chapters 1;2;3).

4. *Complementary series representations:* show
i) inner product (8) is positive-definite, i.e. $\mathcal{H}_\lambda = \{f : \langle R_\lambda f \mid f\rangle < \infty\}$ forms a Hilbert space (use Fourier transform to show positive-definiteness of the convolution operator R_λ);
ii) representation T^λ is unitary in $\mathcal{H}_\lambda$ (note that operators T^λ have the form
$$\phi'^{\frac{\lambda+1}{2}} f \circ \phi,$$
where ϕ denotes the fractional-linear map $x \to x' = \frac{ax+c}{bx+d}$; and check the identity
$$R_\lambda(x-y)[\phi'(x)\phi'(y)]^{\frac{\lambda-1}{2}} = R_\lambda(x'-y'),$$
for Riesz potentials;
iii) establish irreducibility of representations $\{T^\lambda\}$ by analyzing infinitesimal generators, along the lines of Theorem 1.

§7.2. Characters of irreducible representations.

Character of a representation T is given by its trace: $\chi_T(g) = \operatorname{tr} T_g$. In chapters 3-5 we studied characters on compact (Lie) groups, and found them to form a family of nice (differentiable) functions, orthogonal in $L^2(G)$.

For ∞-D unitary representations $\operatorname{tr} T_g$ does not strictly speaking make sense, so characters should be understood as distributions on G, defined via pairing to suitable test-functions $\{f(x)\}$, $\langle \chi_T \mid f\rangle = tr(T_f)$. Precisely, functions $\{f\}$ that produce a trace-class operator T_f (see Appendix B).

In this section we demonstrate that the principal, discrete and complementary series representations have sufficiently many trace-class operators $\{T_f\}$ (all $f \in \mathbb{C}_0^\infty$). Then we compute the corresponding distributions $\{\chi_T\}$ on an open dense set of elliptic and hyperbolic conjugacy classes, and find both of them to be given by nice (analytic) densities on G.

2.1. Characters, as distributions on G. All 3 series (principal, complementary and discrete) were shown to be induced $T = \operatorname{ind}(\chi \mid B; G)$. So the corresponding group-algebra operators $T_f = \int_G f(g) T_g^s dg$, are given by integral kernels

$$K_f(x;y) = \int_{B_x} \chi(x; b\gamma_y) f(b\gamma_y) db, \tag{2.1}$$

where B_x denotes a stabilizer of point $x \in X \simeq B\backslash G$, and coset representatives $\{\gamma_y\}$ map $x \to y$. Picking a suitable class of functions $\{f\}$ on G (typically smooth, compactly supported), one can show integral kernels, $\{K_f(x,y)\}$ to be also smooth and rapidly decaying. Hence, by the standard functional analysis (Appendix B) operators $\{T_f^s : f \in \mathbb{C}_0^\infty\}$ belong to the trace-class (or Hilbert-Schmidt class) on $L^2(X)$, or holomorphic functions, like $\mathcal{H}_n$-spaces. Furthermore, for representations in L^2-spaces[3],

$$\operatorname{tr} T_f^s = \int K_f^s(x,x)\, dx.$$

So character χ_T is defined as a distribution on $\mathbb{C}_0^\infty(G)$,

$$\langle \chi_T \mid f\rangle \stackrel{def}{=} tr(T_f).$$

However, such distributions have often continuous/smooth densities with respect to the natural coordinates on G. Our goal here is to compute these densities for the principal and discrete series representations of $G = SL_2(\mathbb{R})$.

We denote by λ_g, λ_g^{-1} the eigenvalues of matrix $g \in SL_2(\mathbb{R})$,

$$\lambda_g; \lambda_g^{-1} = \frac{tr\, g \pm \sqrt{(tr\, g)^2 - 4}}{2}.$$

[3]The situation is somewhat trickier for operators in holomorphic function-spaces, like $\mathcal{H}_n$, for discrete series representations.

The eigenvalues are real for *hyperbolic* elements g ($tr\, g > 2$), and form a conjugate pair of unit complex numbers $\{e^{\pm i\phi}\}$ for *elliptic* elements g ($tr\, g < 2$). Since characters are conjugate-invariant functions on G, one expects them to depend on $\{\lambda_g^{\pm 1}\}$ only. Indeed,

Theorem 1: *(i) The character χ_s of the principal series representation $T^{s\pm}$ is equal to*

$$\chi_{s\pm}(g) = \begin{cases} \dfrac{|\lambda_g|^{is} + |\lambda_g|^{-is}}{|\lambda_g - \lambda_g^{-1}|}\, \epsilon_\pm(\lambda_g); & \text{for hyperbolic } g \\ 0; & \text{for elliptic } g \end{cases}$$

(ii) *The character of the discrete series representation $T^{\pm n}$ is equal to*

$$\chi_n(g) = \begin{cases} \dfrac{\lambda_g^{-n}}{\lambda_g - \lambda_g^{-1}}; & \textit{hyperbolic } g \\ \dfrac{e^{in\phi}}{e^{-i\phi} - e^{i\phi}}; & \textit{elliptic } g \end{cases} ; \quad \chi_{-n}(g) = \begin{cases} \dfrac{\lambda_g^{-n}}{\lambda_g - \lambda_g^{-1}}; & \textit{hyperbolic } g \\ \dfrac{e^{-in\phi}}{e^{i\phi} - e^{-i\phi}}; & \textit{elliptic } g \end{cases}$$

(iii) *The complementary series character χ_ρ looks like the principal, with the real parameter $\rho \in [-1;1]$ in place of purely imaginary is*, so

$$\chi_\rho(g) = \begin{cases} \dfrac{|\lambda_g|^{\rho} + |\lambda_g|^{-\rho}}{|\lambda_g - \lambda_g^{-1}|}\, \epsilon_\pm(\lambda_g); & \text{for hyperbolic } g \\ 0; & \text{for elliptic } g \end{cases}$$

Proof: (i) The principal series operator $T_g^{s\pm}$ can be thought of as a distributional kernel on $\mathbb{R}$ of the form,

$$K_g(x;y) = |bx+d|^{is-1}\epsilon_\pm(\ldots)\delta(\tfrac{ax+c}{bx+d} - y).$$

So its trace is formally given by

$$\mathrm{tr} T_g^{s\pm} = \int (\alpha^{is-1}\epsilon_\pm)(bx+d)\delta(s^g - x)dx. \tag{2.2}$$

where $\alpha(t)$ denotes $|t|$, $\epsilon_+(t) = 1$, $\epsilon_-(t) = sgn\, t$.

We apply the general change of variable formula for the δ-function to integral (2.2),

$$(\delta \circ \Phi)(x) = \sum_{\{x_j: \Phi(x_j) = 0\}} \frac{1}{|\Phi'(x_j)|}\, \delta(x - x_j).$$

In our case, $\Phi(x) = s^g - x$, so Φ has real zeros (fixed points of $g{:}x \to s^g$) only for hyperbolic g, and these are

$$x_{1,2} = \frac{-d \pm \lambda_g}{b}.$$

Evaluating integrand $\alpha^{is-1}\epsilon = |bx+d|^{\cdots}$, and derivative $\Phi' = (bx+d)^{-2} - 1$ at

$x = x_{1,2}$, we get $(bx+d) = \pm\lambda_g$, and $\Phi'(x) = \lambda_g^{-2} - 1$. Substituting the latter in (2.2) yields the first character formula.

(ii) The proof is more involved for the discrete series, due to the fact that the latter were realized in spaces of holomorphic functions on $\mathbb{P}$ or $\mathbb{D}$ (rather than $L^2(\mathbb{R})$), so trace formulae of type (2.2) are not available here. The relevant formulae involve more complicated reproducing kernels. However, for elliptic g we can easily compute $\chi_{\pm n}$ using the conformal $SU(1,1)$ version of $SL_2(\mathbb{R})$, acting on $\mathcal{H}_n(\mathbb{D})$, and the fact that the operators $T_h^{\pm n}$, $h = \begin{bmatrix} i & \\ & i \end{bmatrix}$, are diagonalized in the basis $\{z^k\}_{k=0}^{\infty}$; with eigenvalues $\{\pm(2k+n+1)\}$. It follows immediately then that

$$\mathrm{tr}\ T^n_{exp\,\phi h} = \sum e^{i(2k+n+1)\phi} = \frac{e^{in\phi}}{e^{-i\phi} - e^{i\phi}};\ \text{and similarly for } T^{-n}.$$

To compute the hyperbolic part of the trace formulae we shall use yet another realization of $T^{\pm n}$ in L^2-spaces on the half-line, $L^2(\mathbb{R}_+; t^{-n}dt)$. Observe that the Fourier/Laplace transform

$$\mathcal{F}: f(t) \to \int_0^\infty f(t)e^{itz}dt = F(z),$$

takes unitarily $L^2(\mathbb{R}_+; t^{-n}dt)$ into holomorphic functions in the upper half-plane $\mathbb{P} = \{Im\, z > 0\}$, square-integrable with weight $y^{n-1}dydx$. Indeed, $F(x+iy) = \mathcal{F}(e^{-ty}f)$, so by the $\mathbb{R}^n$-Plancherel formula,

$$\int_0^\infty |f(t)|^2 e^{-2yt}dt = 2\pi \int_{-\infty}^\infty |F(x+iy)|^2 dx.$$

Multiplying both sides with $y^{n-1}dy$ and integrating from 0 to ∞, we get

$$LHS = \frac{\Gamma(n)}{2^n}\| f \|^2_{L^2(t^{-n}dt)} = RHS = 2\pi \int\int |F(z)|^2 y^{n-1}dydx = 2\pi \| F \|^2_{\mathcal{H}_n}.$$

We shall transform the representation operators $T^{\pm n}$ from spaces $\mathcal{H}_n$ to $L^2(\mathbb{R}_+;...)$ by conjugating with $\mathcal{F}$. Since the inverse transform

$$\mathcal{F}^{-1}: F \to \frac{1}{2\pi} \int_{\Im m\, z = y_0} e^{-izt}F(z)dz,$$

we find that the transformed operators are given by integral kernels

$$K_n(\xi;\eta) = \frac{1}{2\pi}\int e^{i(\eta z^g - \xi z)} \frac{dz}{(bz+d)^{n+1}}. \tag{2.3}$$

It is convenient to change variables: $z \to t = bz+d$. Then we compute: $z = \frac{t-d}{b}$; $dz = \frac{dt}{b}$; $\eta z^g - \xi z = \frac{1}{b}[(a\eta + d\xi) - (\frac{\eta}{t} + \xi t)]$, and after substitution in (2.3), kernel K_n takes the form,

$$K_n(\xi;\eta) = \frac{1}{2\pi b} e^{i\left(\frac{a\eta + d\xi}{b}\right)} \int_{\Im m t = t_0} e^{-i(\eta/t + \xi t)/b} \frac{dt}{t^{n+1}}.$$

Another change of variable, $t \to t^{-1}$, brings the integral over the line $Im t = ...$ into the integral over the circle $C = \{\, | t + \frac{i}{2} | = \frac{1}{2} \}$

$$K_n(\xi;\eta) = \frac{1}{2\pi b} e^{i(...)/b} \oint_C e^{-i(t\eta + \xi/t)/b} t^{n-1} dt,$$

and the latter is more convenient to compute the trace,

$$\text{tr}\, K_n = \int_0^\infty K_n(\xi;\xi) d\xi = \frac{1}{2\pi b} \oint t^{n-1} dt \int_0^\infty e^{i\xi[(a+d)-(t+1/t)]/b} d\xi. \tag{2.4}$$

Remembering, that exponential $\{t^2 - (a+d)t + 1\}$ in (2.4) represents a characteristic polynomial of g, the inner integral in (2.4) yields

$$\frac{b}{i[(t+1/t) - (a+d)]} = \frac{bt}{(t-\lambda_g)(t-\lambda_g^{-1})}.$$

Thus we derive the following trace-formula

$$\boxed{\text{tr}\, K_n = \frac{1}{2\pi i} \oint \frac{t^n dt}{(t-\lambda_g)(t-\lambda_g^{-1})}} \tag{2.5}$$

The contour integral (2.5) is easily evaluated: for hyperbolic g the smallest of two eigenvalues ($\lambda_g^{-1} < 1$) is inside the circle. Taking the appropriate residue in (2.5) we finally get

$$\chi_n(g) = \frac{\lambda_g^{-n}}{\lambda_g - \lambda_g^{-1}},$$

and similar derivation applies for negative n, QED.

§7.3. The Plancherel formula for $SL_2(\mathbf{R})$.

In this section we shall derive the Plancherel/inversion formula for group $G = SL_2(\mathbf{R})$, and get the *direct-integral* decomposition (see §6.1) for the regular representation of G on $L^2(G)$, into the sum of *primary components.*

An interesting feature of SL_2 comes from the fact that its Plancherel formula combines both the *continuous* (principal series), and *discrete* (discrete series) contributions, so SL_2 behaves like compact and noncompact group at once. We find the explicit Plancherel measure on SL_2, including the density of the continuous branch, and multiplicities of discrete components.

3.1. The general noncommutative *Plancherel formula* was discussed in §6.1. It extends the classical Fourier-inversion formula (§2.1),

$$f(0) = \int_{\widehat{G}} \hat{f}(\xi)d\xi, \tag{3.1}$$

$d\xi$ - properly normalized Haar measure on $\hat{G}$, and yields a similar representation for any function $f \in C_0^\infty(G)$, in terms of irreducible characters on G. Namely,

$$\boxed{f(e) = \int_{\widehat{G}} \langle \chi_\pi \mid f\rangle d\mu(\pi)}; \tag{3.2}$$

where $\langle \chi_\pi \mid f\rangle = \mathrm{tr}\hat{f}(\pi)$, and $\hat{f}(\pi) = \int_G f(x)\,\pi_x^{-1}dx$.

Integration in (3.2) extends over the (unitary) dual object $\widehat{G}$ of G, and $d\mu(\pi)$ denotes the *Plancherel measure* of G. From the inversion formula (3.2) one easily derives the Plancherel Theorem for L^2-functions on G. Indeed, applying (3.2) to a convolution $f*f^*$, where $f^*(x) = \overline{f(x^{-1})}$, the integrand becomes

$$tr[\hat{f}(\pi)\hat{f}(\pi)^*] = \left\|\hat{f}(\pi)\right\|_{HS}^2 \text{ (Hilbert-Schmidt norm)}$$

so (3.2) turns into

$$\boxed{\int_G |f(x)|^2 dx = \int_{\widehat{G}} \left\|\hat{f}(\pi)\right\|_{HS}^2 d\mu} \tag{3.3}$$

Let us also remark that inversion formula (3.2) at a particular group element $g_0 = e$ yields values $f(g)$ at all other points of G

$$\boxed{f(g) = \int_{\widehat{G}} tr[\hat{f}(\pi)\,\pi_g^{-1}]\;d\mu(\pi)} \tag{3.4}$$

Finally, from (3.2) and (3.3) we get a decomposition of the regular representation on $L^2(G)$ into the direct integral of irreducibles,

$$\boxed{R_g \simeq \oint_{\widehat{G}} \pi_g \otimes I_{d(\pi)} d\mu(\pi)} \tag{3.5}$$

with multiplicity (finite or infinite) equal to the degree $d(\pi)$.

Our goal here is to prove formula (3.2), and to compute the Plancherel measure $d\mu$ for $SL_2(\mathbb{R})$.

Plancherel Theorem: *Let $\{\chi_{s\pm}\}_{s\in\mathbb{R}}$ and $\{\chi_{\pm n}\}_{n=1}^{\infty}$ denote the principal and discrete series irreducible characters of $SL_2(\mathbb{R})$. Then for any function $f \in C_0^\infty(G)$ one has*

$$\boxed{f(e) = \frac{1}{2\pi}\left\{\sum_{n\neq 0} |n| \langle \chi_n | f\rangle + \frac{1}{2}\int \langle \chi_{s+} | f\rangle\, s \tanh(\tfrac{\pi s}{2}) ds + \frac{1}{2}\int \langle \chi_{s-} | f\rangle\, s \coth(\tfrac{\pi s}{2}) ds\right\}} \tag{3.6}$$

Formula (3.6) shows that the Plancherel measure $d\mu$ is supported on the union of the principle parts $\widehat{G}_{p+} \cup \widehat{G}_{p-}$ ($\widehat{G}_{p\pm} \simeq \mathbb{R}$), and the discrete part $\widehat{G}_d \simeq \mathbb{Z}\backslash\{0\}$, of the dual object $\widehat{G}$, and is equal to

$$\boxed{d\mu = \frac{1}{2\pi}\left\{\sum_{n\in\widehat{G}_d} |n|\, \delta_n + \frac{s}{2}\tanh(\tfrac{\pi s}{2})\, ds\Big|_{\widehat{G}_{p+}} + \frac{s}{2}\coth(\tfrac{\pi s}{2})\, ds\Big|_{\widehat{G}_{p-}}\right\}} \tag{3.7}$$

In the rest of this section we shall outline the proof of the Plancherel formula for $SL_2(\mathbb{R})$.

3.2. Decompositions of G, Conjugacy classes, and the Haar measure. There different ways to decompose G, i.e. choose coordinates in G. Typically they are defined in terms of three special subgroups of G: maximal compact group

$$K = \left\{u = u_\theta = \begin{bmatrix} \cos & \sin \\ -\sin & \cos \end{bmatrix};\ 0 \le \theta < 2\pi\right\};$$

diagonal (Cartan) group

$$A = \left\{a = \pm a_t = \begin{bmatrix} \pm e^t & \\ & \pm e^{-t} \end{bmatrix};\ t \in \mathbb{R}\right\}$$

along with its positive part, semigroup

$$A_+ = \{a_t \colon t > 0\};$$

and upper/lower *unipotent subgroups:*

$$N = N_\pm = \left\{n = \begin{bmatrix} 1 & n \\ & 1 \end{bmatrix} \text{ and } \begin{bmatrix} 1 & \\ n & 1 \end{bmatrix};\ n \in \mathbb{R}\right\}.$$

We shall use two decompositions of SL_2, called *Iwasawa:* $G = KN_-A$, or AN_+K, and *Cartan:* $G = KA_+K$. This means that any (almost any) matrix $g \in G$ can be uniquely factored into the product:

(i) $g = una$, or anu, where $u \in K$, $n \in N_{\pm}$, $a \in A_+$ (Iwasawa)

(ii) $g = uav = u_\phi a_t u_\psi$, $u, v \in K$, $a \in A_+$ (Cartan).

To demonstrate the first decomposition[4] we observe that $K\backslash G \simeq \mathbb{P}$- the Poincare half-plane, and K fixes point $\{i\}$. So any $g: i \to i^g = z = x+iy$, can be decomposed as

$$g = ug_z;\ g_z = \begin{bmatrix} \sqrt{y} & \\ x/\sqrt{y} & 1/\sqrt{y} \end{bmatrix}.$$

But

$$g_z = \begin{bmatrix} 1 & \\ x/y & 1 \end{bmatrix}\begin{bmatrix} \sqrt{y} & \\ & 1/\sqrt{y} \end{bmatrix} = na.$$

The second decomposition is easier to demonstrate in the conformal $SU(1,1)$-set-up. Then

$$K = \left\{\begin{bmatrix} e^{i\theta} & \\ & e^{-i\theta} \end{bmatrix}\right\};\ A = \left\{\begin{bmatrix} ch & sh \\ sh & ch \end{bmatrix}\right\}.$$

The quotient-space $K\backslash G \simeq \mathbb{D}$ (Poincare disk), subgroup K stabilizes $\{0\}$, so each $g = ug_z$, where g_z: $0 \to z = re^{i\theta}$, is given by

$$g_z = \frac{1}{\sqrt{1-r^2}}\begin{bmatrix} e^{i\theta/2} & re^{-i\theta/2} \\ re^{i\theta} & e^{-i\theta/2} \end{bmatrix}.$$

The latter obviously factors into $a_t u_{\theta/2}$ with $\cosh t = \frac{1}{\sqrt{1-r^2}}$. Thus the NA-factorization in the Iwasawa's KNA can be thought of as real-imaginary axis in the half-plane-realization of $K\backslash G$, while A_+K of Cartan's KA_+K correspond to polar coordinates in the disk-realization of $K\backslash G$.

We associate to either of two decompositions an appropriate disintegration of the Haar measure on G.

Proposition 1: *The Haar measure dg on G is decomposed as*

(i) Iwasawa: $g = una \Rightarrow dg = du\, dn \frac{da}{|a|}$;

(ii) Cartan: $g = u_\phi a_t u_\psi \Rightarrow dg = \sinh 2t\, dt \frac{d\phi\, d\psi}{(2\pi)^2}$.

Let us remark that the Cartan's KA_+K-decomposition for $SU(1,1)$ represents a hyperbolic analog of the Euler-angle decomposition in $SU(2)$, with hyperbolic rotations $\begin{bmatrix} ch & sh \\ sh & ch \end{bmatrix}$ in place of ordinary rotations $\begin{bmatrix} cos & sin \\ -sin & cos \end{bmatrix}$.

The similarity extends also to the Haar measure on both groups,

$$\boxed{dg = \sin\theta\, d\theta\, d\phi\, d\psi} \text{ for } SU(2) \text{ vs. } \boxed{dg = \sinh t\, dt\, d\phi\, d\psi} \text{ for } SU(1,1).$$

---

[4]For another derivation of Iwasawa decomposition see problem 1 of §7.1.

The proof of Proposition 1 follows from the general factorization formula of the Haar measure for any pair $G \supset K$ (compact subgroup),

$$d_G(g) = d_K(u) d_{K\backslash G}(z), \text{ for } g = ug_z,$$

and an explicit form of the invariant measure on the quotient $K\backslash G$, both in the half-plane-realization: $\frac{dxdy}{y^2}$ on $NA \simeq \mathbf{P}$, and the disk-realization: $\frac{rdrd\phi}{(1-r^2)^2}$ on $\mathbf{D} \simeq \mathrm{A}_+K$.

Indeed, for $K\backslash G \simeq \mathbf{P}$, $g = ug_z$ with

$$g_z = \begin{bmatrix} 1 & \\ x/y & 1 \end{bmatrix} \begin{bmatrix} \sqrt{y} & \\ & 1/\sqrt{y} \end{bmatrix} = na.$$

Hence follows,

$$dn \frac{da}{|a|} = \frac{dxdy}{2y^2},$$

which proves (i). Similarly one can verify (ii) for $K\backslash G \simeq \mathbf{D}$: given any $g = u_\phi g_z$, where

$$g_z = \frac{1}{\sqrt{1-r^2}} \begin{bmatrix} e^{i\psi/2} & re^{-i\psi/2} \\ re^{i\psi/2} & e^{-i\psi/2} \end{bmatrix}; \text{ for } z = re^{i\psi} \in \mathbf{D}.$$

Writing polar radius r in terms of the hyperbolic radius t, $r = \tanh t$, we get

$$d(vol) = \frac{rdrd\psi}{(1-r^2)^2} = \sinh 2t \, dt d\psi, \text{ QED.}$$

Corollary: The discrete series representations $\{T^{\pm n}\}$ are subrepresentations of the regular representation[5] R in $L^2(G)$.

An irreducible representation T can be embedded in the regular R, if its matrix-entries, $t(g) = \langle T_g \phi \mid \psi \rangle$, are L^2-functions on G. Due to irreducibility of T it suffices to compute the L^2-norm of a single matrix-entry, $f(g) = \langle T_g 1 \mid 1 \rangle$, where 1 means a constant function in the disk-realization of space $\mathcal{H}_n$. Using the Cartan decomposition $g = u_\phi a_t u_\psi$, and the transformation properties of $1 \in \mathcal{H}$ under the K-action, $T_{u_\phi}(1) = e^{in\phi} 1$, we see that f depends on variable t only, and

$$f(t) = \int_{\mathbf{D}} (z \sinh t + \cosh t)^{-n-1} (1 - |z|^2)^{n-1} dz d\bar{z} =$$

$$= \cosh^{-(n+1)} t \int\int (1 + re^{i\psi} \tanh t)^{-n-1} r(1-r^2) dr d\psi.$$

But the later integral is evaluated in a straightforward way, so one gets $f(t) = \frac{\pi}{n} \cosh^{-(n+1)} t$, and

$$\| f \|^2_{L^2(G)} = \int | f(t) |^2 \sinh 2t \, dt < \infty.$$

3.3. Conjugacy classes and the Harish-Chandra transform. One of the interesting features of the $SL_2(\mathbb{R})$-representation theory is the presence of different types of conjugacy classes in $G = G_+ \cup G_- \cup G_0$, (all three open subsets in G). These are

[5] So they could be thought of as "discrete components" of the regular representation, whence the terminology!

(i) *hyperbolic* classes $G_{\pm} = \bigcup_{t>0} \{g^{-1}(\pm a_t)g\colon\ g \in G\}$. We shall parametrize them by variable $t \in \mathbb{R}_{\times}$

$$C_{\pm t} \simeq A\backslash G = N_+K.$$

(ii) *elliptic* classes $G_0 = \bigcup_{0<\theta<\pi} \{g^{-1}u_\theta g\colon g \in G\}$, labeled by θ $(0 \le \theta \le \pi)$,

$$C_\theta \simeq K\backslash G = N_-A \text{ (or } A_+K).$$

The Haar measure on G can be disintegrated into invariant measures on classes C_t (or C_θ), and certain density on the sets of classes: two half-lines $\pm(0;\infty)$ and a semicircle $(0;\pi)$. These are given by the formulae of the type we derived in §5.6 for compact Lie groups:

$$\begin{aligned} dg \mid_{G_\pm} &= (e^t - e^{-t})^2 dt\, d_{C_t}(nu) = 4\, sinh^2 t\, dt\, d_{N_+K}(\ldots) \\ dg \mid_{G_0} &= \left|e^{i\theta} - e^{-i\theta}\right|^2 d\theta\, d_{C_\theta}(z) = 4\, sin^2\theta\, d\theta\, d_{\mathbf{P}}(z) \end{aligned} \tag{3.8}$$

Now we are ready to proceed to the Plancherel Theorem. The proof will be given in 3 steps. In step 1° we shall introduce and compute the so-called *Harish-Chandra (conjugacy-class), or H-transforms*, of C_0^∞-functions f on G, which essentially amount to averaging f over the elliptic, or hyperbolic conjugacy class. Then (step 2°) we shall link 2 H-transforms (hyperbolic and elliptic) at points, where two sets of conjugacy classes meet. Finally, in step 3° we shall express irreducible characters in terms of H-transforms, which would eventually lead us to the Plancherel formula on G.

Step 1°: Given any function f on G, we shall define its *elliptic* and *hyperbolic H-transforms*, by averaging f over the corresponding (elliptic/hyperbolic) conjugacy classes. Precisely,

$$\text{(elliptic) } f \to F_f^K = F^K(\theta) = (e^{i\theta} - e^{-i\theta}) \int_{K\backslash G} f(g^{-1}u_\theta g)dg, \text{ on } G_0;$$

$$\text{(hyperbolic)} \quad f \to F_f^A = F^{A\pm}(t) = F^A(\pm a_t) = \mid e^t - e^{-t} \mid \int_{A\backslash G} f(g^{-1}a_t g)dg, \quad \text{on } G_\pm. \tag{3.9}$$

In other words, "averaged function" f is multiplied by the "square-root" of the Weyl-type density (3.8). From the disintegration formula (3.8) it follows,

$$\int_G f(g)dg = \int_{K_+} F^K(\theta)(e^{-i\theta} - e^{i\theta})d\theta + \int_{A_+} F^A(a_t)\left|e^t - e^{-t}\right|dt + \int_{A_+} F^A(-a_t)\left|e^t - e^{-t}\right|dt.$$

Densities that appear in the above integrals represent the omitted "half-powers" of Weyl-factors (3.8). We can also express any character, or more generally any class-function (distribution) χ on G in terms of its H-transforms,

$$\langle\chi \mid f\rangle = \int_{K_+} \chi(\theta)F^K(\theta)2i\,sin\theta\,d\theta + \int_{A_+} \chi(a_t)F^A(a_t)2\,|\,sinh\,t\,|\,dt + \int_{A_+} \chi(-a_t)... \quad (3.10)$$

In the proof of the Plancherel Theorem we shall need the H-transforms of K-*invariant* functions on the group,

$$f(u^{-1}gu) = f(g),\ u \in K.$$

Let us remark that the Plancherel formula (3.6) is sufficient to establish for K-invariant functions $\{f\}$. Indeed, any function-space $\mathcal{L}$ on G, (like C_0^∞), can be splits into the direct sum of m-isotropic components with respect to the K-action on $\mathcal{L}$ by the conjugation. Precisely,

$$f = \sum_{-\infty}^{\infty} f_m;$$

where each term,

$$f_m(g) = \int_K e^{-im\theta} f(u_\theta^{-1} g u_\theta) d\theta,$$

satisfies the equation $f(u_\phi^{-1} g u_\phi) = e^{im\phi} f(g)$.

Since averaging over G-conjugacy classes includes integration over K, it follows immediately, that all nonzero Fourier-components $\{f_m : m \neq 0\}$, will be H-transformed to 0. Consequently, any character $\langle\chi \mid f_m\rangle = 0$, unless $m = 0$, so $\langle\chi \mid f\rangle = \langle\chi \mid f_o\rangle$, for all characters χ and any function f on G.

Proposition 2: *Two H-transforms of a K-invariant function: $f(u^{-1}gu) = f(g)$, for all $u \in K$, are given by*

$$\text{(i) } F^K(\theta) = 2\pi i\,sin\theta \int_0^\infty f\left(\begin{bmatrix} cos\theta & e^{2r}sin\theta \\ -e^{-2r}sin\theta & cos\theta \end{bmatrix}\right)(e^{2r} - e^{-2r})dr;$$

$$\text{(ii) } F^A(t) = 2\pi \int_{-\infty}^{\infty} f\left(\begin{bmatrix} e^t & x \\ & e^{-t} \end{bmatrix}\right)dx. \quad (3.11)$$

The derivation of (3.11) is fairly straightforward. For the elliptic F^K we evaluate integral (3.9), using the Cartan KA_+K-decomposition, $g = ua_rv$. This yields the first relation (3.11-i),

$$F^K(\theta) = 2\pi(e^{i\theta} - e^{-i\theta}) \int_0^\infty f(a_r^{-1} u_\theta a_r)\, sinh\, 2r\, dr,$$

To get hyperbolic F^A we evaluate second equation (3.9) using the Iwasawa AN_+K-decomposition, i.e. $g = anu$, with $n = \begin{bmatrix} 1 & \xi \\ & 1 \end{bmatrix}$. Then

$$F^A(t) = |\,e^t - e^{-t}\,| \int_{NK} f(u^{-1}n^{-1}a_t nu)dn\,du.$$

The inner conjugation, $n^{-1}a_tn$, in the variable of f yields matrix $\begin{bmatrix} e^t & \xi(e^t - e^{-t}) \\ & e^{-t} \end{bmatrix}$

while the outer conjugation, $u^{-1}(...)u$, leaves f invariant, and thus only contributes factor 2π to the integral. So

$$F^A(t) = 4\pi \sinh t \int_{-\infty}^{\infty} f\left(\begin{bmatrix} e^t & 2\xi \sinh t \\ & e^{-t} \end{bmatrix}\right) d\xi,$$

whence combining $2\xi \sinh t$ into a single variable x, we get the second relation (3.11-ii). Next we shall recast the F^K-transform in a more convenient form by making the change of variables: $s = \sin\theta$, $\xi = e^{2r}\sin\theta$, $\eta = e^{-2r}\sin\theta$, and introducing a new function

$$f \to \tilde{f}(\xi,\eta) = f\left(\begin{bmatrix} \sqrt{1-\xi\eta} & \xi \\ -\eta & \sqrt{1-\xi\eta} \end{bmatrix}\right), \tag{3.12}$$

defined in the region $\{(\xi,\eta): |\xi\eta| \le 1\}$. Then (3.11) becomes,

$$\boxed{F^K(s) = 2\pi i \left\{ \int_s^{\infty} \tilde{f}(x; \tfrac{s^2}{x}) dx - \int_0^s \tilde{f}(\tfrac{s^2}{x}; x) dx \right\};\ F^A(\mathrm{t}) = 2\pi \int_{-\infty}^{\infty} f\left(\begin{bmatrix} e^t & x \\ & e^{-t} \end{bmatrix}\right) dx} \tag{3.13}$$

Step 2. Our goal now will be to relate two types of H-transforms at the points of G, where elliptic and hyperbolic conjugacy classes meet, namely elements $\{\pm I\} \in G$, or points $\theta = \{0;\pi\}$ in K, respectively $\{a_0; -a_0\}$ in $\pm A$ (fig.2). To accomplish this objective we need some properties of functions F^K and F^A defined by (3.13).

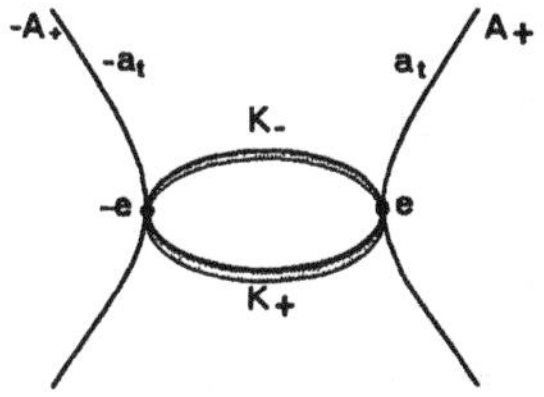

Fig.2. *gives a schematic view of the two types of conjugacy classes: elliptic (represented by the circle) and hyperbolic (infinite lines $A_{\pm}$). Two classes meet at the points $\{\pm e\}$. The Plancherel formula results from matching elliptic and hyperbolic Harish-Chandra transforms at $\{\pm e\}$.*

Lemma 3: *The elliptic H-transform, $F^K(\theta)$, has a jump-discontinuity at $\theta = 0$, that is equal to $2iF^A(a_0)$-the value of the hyperbolic F^A at $a_0 = \mathrm{I}$, while the right and left derivatives of F^K* at $\theta = 0$ *are equal. Similarly at $\theta = \pi$, the jump-discontinuity of F^K is equal to $2iF^A(-a_0)$ - the value of the hyperbolic F^A at $-a_0 = -I$, while derivatives are equal. Precisely,*

$$\text{(a) } F^K(0_+) = -F^K(0_-) = 2\pi i \int_0^{\infty} f\left(\begin{bmatrix} 1 & x \\ & 1 \end{bmatrix}\right) dx = iF^A(0);$$
$$\text{(b) } \tfrac{d}{ds}F^K(0_+) = \tfrac{d}{ds}F^K(0_-) = -2\pi i \tilde{f}(0;0) = -\pi i f(e), \text{ at } \theta = 0, \tag{3.14}$$

and similar relations hold at $\theta = \pi$.

The Lemma's conclusions follow immediately from equations (3.14).

The first of them (3.14a) is obtained by the substitution $s = 0_+$ into integrals representing F^K in (3.12), and the obvious relation $\tilde f(x;0) = f\left(\begin{bmatrix} 1 & x \\ & 1 \end{bmatrix}\right)$. To treat F^K at negative values of s, we observe that the integrand $\tilde f(\xi;\eta)$ is an even function on $\mathbb{R}^2$, due to K-invariance of the original function f on G, which implies in particular,

$$f\left(\begin{bmatrix} a & -b \\ -c & d \end{bmatrix}\right) = f\left(\begin{bmatrix} a & b \\ c & d \end{bmatrix}\right), \text{ for all } g = \begin{bmatrix} a & b \\ c & d \end{bmatrix}.$$

So changing variables from s to $-s$ in (3.13) results in the change of sign of two integrals

$$F^K(-s) = -2\pi i\left\{\int_s^\infty \tilde f(x;\tfrac{s^2}{x})dx - \int_0^s \dots\right\} \to -2\pi i\int_0^\infty \tilde f(x;0)dx = -iF^A(a_0); \text{ as } s \to 0.$$

Next we differentiate function F^K of (3.13) in variable s, and evaluate at $s = 0$ (respectively $\theta = 0$),

$$\tfrac{d}{ds}F^K = 2\pi i\left\{-2\tilde f(s;s) + 2s\left[\int_s^\infty \dots - \int_0^s \dots\right]\right\} \to -4\pi i\tilde f(0;0) = -4\pi i f(e).$$

As a corollary we get the requisite relations between F^K and F^A at two limit-points $\pm I$, as well as the representation of $f(\pm e)$ in terms of derivatives $\frac{d}{d\theta}F^K$,

$$\boxed{2\pi i F^A(a_0) = F^K(0_+) - F^K(0_-);\ -4\pi i f(e) = \tfrac{d}{d\theta}F^K(0_\pm)} \text{ at } \theta = 0$$

$$\boxed{2\pi i F^A(-a_0) = F^K(\pi_+) - F^K(\pi_-);\ -4\pi f(-e) = \tfrac{d}{d\theta}F^K(\pi_\pm)} \text{ at } \theta = \pi. \quad (3.15)$$

Step 3. Now we shall evaluate the principal series characters for K-invariant functions f utilizing their conjugacy-class H-transforms. We recall the character formulae, obtained in §7.2,

$$\chi_{s\pm}(a_t) = \frac{e^{ist} + e^{-ist}}{|e^t - e^{-t}|}\epsilon_\pm;\ \text{on } G_\pm;\ \text{and } \chi_{s\pm}(u) = 0,\ \text{on } G_0.$$

Then we can write by (3.10),

$$\langle \chi_{s\pm} \mid f\rangle = \int_0^\infty \chi_{s\pm}(a_t)F^A(a_t)\,|e^t - e^{-t}|\,dt + \int^\infty [\textit{same integrand at } (-a_t)].$$

Both integrals reduce to

$$\int (e^{ist} + e^{-ist})F^A(\pm a_t)dt.$$

Remembering that characters are even functions of t on $\pm A$, the latter become simple "Fourier transforms" of functions $F^A(\pm a_t)$. Let us also remember that

$$\chi_{s+}(a_t) = \chi_{s+}(-a_t),$$

while χ_{s-} changes sign, when $a_t \to -a_t$. So their sum $\chi_{s+} + \chi_{s-}$ cancels out on $-A$ and equals to $2\chi_s$ on A, while their difference, $\chi_{s+} - \chi_{s-}$, cancels out on A and is equal to $2\chi_s$ on $-A$. Thus we get

$$\boxed{\langle \chi_{s+} \mid f\rangle + \langle \chi_{s-} \mid f\rangle = \mathcal{F}\{F^A\mid_A\};\ \langle \chi_{s+} \mid f\rangle - \langle \chi_{s-} \mid f\rangle = \mathcal{F}\{F^A\mid_{-A}\}} \tag{3.16}$$

Next we evaluate the discrete series characters, $\langle \chi_{\pm n} \mid f\rangle$, where

$$\chi_{\pm n}(\pm a_t) = \frac{e^{-nt}}{e^t - e^{-t}}\epsilon_{\pm};\ \text{and}\ \chi_{\pm n}(u_\theta) = \frac{e^{in\theta}}{e^{i\theta} - e^{-i\theta}}.$$

For our purpose it will be more convenient to evaluate the sum of two $\chi_n + \chi_{-n}$. We find

$$\begin{aligned}\langle \chi_n \mid f\rangle &= \int_0^\infty \chi_n(t) F^A(t) \mid e^t - e^{-t} \mid dt + \int_0^{2\pi} \chi_n(\theta) F^K(\theta)(e^{i\theta} - e^{-i\theta}) d\theta = \\ &= \int_0^\infty e^{-nt}\, F^A(t)\, dt + \int_0^{2\pi} F^K(\theta) e^{in\theta} d\theta.\end{aligned} \tag{3.17}$$

The latter equation yields Fourier coefficients $\{b_n\}$ of function F^K (3.16) in terms of the discrete series characters and the hyperbolic F^A,

$$b_n = \langle \chi_n \mid f\rangle - \int_0^\infty e^{-|n|t} F^A(t) dt.$$

Differentiating the Fourier series $F^K = \sum b_n e^{in\theta}$, we get

$$\frac{d}{d\theta} F^K = i \sum_{n \neq 0} n \left\{ \langle \chi_n \mid f\rangle - \int_0^\infty \dots \right\} e^{in\theta} + \left(F^A(0_+) - F^A(0_-)\right)\delta_0 + \left(\dots\right)\delta_\pi.$$

Next we rewrite the $(-n \int \dots)$- term of the n-th coefficient of $(F^K)'$, as

$$-\int_0^\infty e^{-tn} \frac{d}{dt} F^A(\pm a_t),$$

and taking into account the parity (even/odd) of function $F^A(\pm a_t)$, and that of $F^K(\theta)$, we represent $\frac{d}{d\theta} F^K$ by the following expansions

$$i \sum_1^\infty n \langle \chi_n \mid f\rangle + \frac{i}{2} \sum_1^\infty \int_{-\infty}^\infty e^{-|t|n} sgn\, t \frac{d}{dt} F^A(a_t) dt + \frac{i}{2} \sum_1^\infty (-1)^n \int_{-\infty}^\infty e^{-|t|n} sgn\, t \frac{d}{dt} F^A(-a_t) dt \tag{3.18}$$

The rest of argument involves some basic Fourier analysis on $\mathbb{R}$. First we shall apply the Plancherel formula on $\mathbb{R}$ for all integrals of (3.18),

$$\int f(t) g(t) dt = 2\pi \int \widehat{f}(s)\, \widehat{g}(s) ds.$$

Then use a simple Fourier-transform relation

$$\mathcal{F}\colon e^{-|t|n} sgn\, t \to \frac{-2is}{s^2 + n^2}.$$

and find by (3.16) the $\mathfrak{F}$-transformed derivatives of F^A in terms of characters,

$$\mathfrak{F}: \tfrac{d}{dt}F^A(a_t) \to is\,[\langle\chi_{s+}\mid f\rangle + \langle\chi_{s-}\mid f\rangle];\ \text{and}\ \tfrac{d}{dt}F^A(-a_t) \to is\,[\langle\ldots\rangle - \langle\ldots\rangle].$$

These formulae allow us to rearrange terms in (3.18) into those of χ_{s+}, and those of χ_{s-}; namely,

$$\chi_{s-}\sum_{n=even}\left(\frac{s^2}{s^2+n^2}\right) + \chi_{s+}\sum_{n=odd}\left(\frac{s^2}{s^2+n^2}\right).$$

Two series are easily summable by the Poisson summation formula (§2.1 of chapter 2),

$$\sum_{even}\left(\frac{s^2}{s^2+n^2}\right) = \tfrac{s}{2}\sum e^{-\pi s\,|k|} = \tfrac{s}{2}\coth\tfrac{\pi s}{2};\quad \sum_{odd}\left(\frac{s^2}{s^2+n^2}\right) = \ldots = \tfrac{s}{2}\tanh\tfrac{\pi s}{2}.$$

Thus we get the final result,

$$2\pi i f(e) = \tfrac{d}{d\theta}F^K(0) = i\sum_{n\neq 0}|n|\,\langle\chi_n\mid f\rangle + \tfrac{i}{2}\int_{-\infty}^{\infty}\langle\chi_{s-}\mid f\rangle s\coth\tfrac{\pi s}{2}\,ds + \tfrac{i}{2}\int_{-\infty}^{\infty}\langle\chi_{s+}\mid f\rangle s\tanh\tfrac{\pi s}{2}\,ds,$$

which completes the proof.

Corollary: *The regular representation of $SL_2(\mathbb{R})$ on $L^2(G)$ is decomposed into the sum of the continuous and discrete parts,*

$$R_g \simeq \underbrace{\left\{\oint (T_g^{s+}\otimes I)(s\tanh s)ds\right\}}_{continuous\ +} \oplus \underbrace{\left\{\oint (T_g^{s-}\otimes I)(s\coth s)ds\right\}}_{continuous\ -} \oplus \underbrace{\left\{\sum_{n\neq 0}(T^n\otimes I)\,|n|\,\delta_n\right\}}_{discrete}.$$

The L^2-Plancherel formula,

$$\boxed{\|f\|^2 = \int \|\hat f(s;+)\|_{HS}^2 (s\tanh s)ds + \int \|\hat f(s;-)\|_{HS}^2 (s\coth s)ds + \sum_{n\neq 0} n\,\|\hat f(n)\|_{HS}^2}$$

Here $\hat f$ denotes the usual noncommutative Fourier transform of f, i.e. $\hat f(\pi) = \pi_f$, for the principal and discrete series irreducible representations $\{\pi\}$.

Problems and Exercises:

1. Consider the natural action of $G = SL_2(\mathbb{R})$ on $L^2(\mathbb{R}^2)$,

$$(T_g f)(x,y) = f(ax+cy; bx+dy),$$

and show that representation T is the direct integral of the principal series representations,

$$T = \oint T^{s+} ds \oplus \oint T^{s-} ds.$$

Hint: i) observe that $\mathbb{R}^2 \backslash \{0\}$ can be identified with the homogeneous space $N \backslash G$, where

$$N = \left\{ \begin{bmatrix} 1 & b \\ & 1 \end{bmatrix} : b \in \mathbb{R} \right\}$$

consists of upper-triangular unipotent matrices, so T is the regular representation of G on $L^2(N \backslash G)$.

ii) Note, that representation spaces $\mathcal{H}_s^{\pm}$ can be identified with even/odd function $f(x,y)$ on $\mathbb{R}^2$ homogeneous of degree $(is-1)$,

$$\mathcal{H}_s^{\pm} = \{f(tx; ty) = t^{is-1} f(x,y)\}.$$

Any such f is uniquely determined by its values on the line $\{(x;1)\}$, and can set

$$\| f \|_s^2 = \int | f(x;1) |^2 dx.$$

iii) Show that $T \mid \mathcal{H}_s^{\pm}$ is equivalent to the principal series $T^{s\pm}$ (but $\mathcal{H}_s$ are not subspaces $\mathcal{H}$!).

iv) Establish the direct-integral decomposition by explicitly writing map

$$\mathcal{F}: L^2(\mathbb{R}^2) \rightarrow \oint \mathcal{H}_s ds,$$

with properties (i-ii), as *Mellin transform* of f (§2.1) in the radial directions,

$$\mathcal{F}: f \rightarrow \tilde{f}(x,y;s) = \int_0^\infty f(tx,ty)\, t^{-is} dt. \tag{3.19}$$

v) Use the Mellin inversion/Plancherel formula:

$$u(t) = \frac{1}{2\pi} \int_{-\infty}^{\infty} \tilde{u}(s) t^{is-1} ds; \text{ and } \int_0^\infty | u(t) |^2 t dt = 2\pi \int_{-\infty}^{\infty} | \tilde{u}(s) |^2 ds.$$

to resolve any function f on $\mathbb{R}^2$ into the integral

$$f(x,y) = \frac{1}{2\pi} \int \tilde{f}(x,y;s) ds = \int \tilde{f}(s) ds, \text{ so that } T_g f = \int T_g^s[\tilde{f}(s)] ds,$$

and to show

$$\| f \|_{L^2}^2 = \int \| \tilde{f}(s) \|_{\mathcal{H}_s}^2 ds.$$

Change variables $x = t\kappa$, $y = t$ ($t > 0$; $\kappa \in \mathbb{R}$), and write the L^2-norm as

$$\| f \|^2 = \int \int | f(t\kappa, t) |^2 t dt\, d\kappa;$$

Then apply Mellin-Plancherel formula, to integral

$$2\pi \int \int | \tilde{f}(\kappa, 1; s) |^2 d\kappa ds,$$

to get the latter in the form $\| \tilde{f}(...;s) \|_{\mathcal{H}_s}^2$, QED.

§7.4. Infinitesimal representations of SL_2; spherical functions and characters.

Infinitesimal method of §1.4 reduces representations of Lie group to those of its Lie algebra, and greatly facilitates their study. Further reduction for SL_2 is achieved by passing to a single generator L of the center of its enveloping algebra $\mathfrak{A}$ (called Casimir). Any irreducible T^s must be scalar on L, thus we get an infinitesimal character $\chi(s)$. Character χ uniquely identifies all three irreducible series, and determines the eigenvalues of the Laplacian on G and the hyperbolic plane $\mathbf{H} = K\backslash G$.

Another reduction arises from the compact subgroup K of G. The latter gives rise to the algebra of spherical functions and spherical transform. Once again irreducible $\{T^s\}$ are completely determined by "spherical reduction", and spherical functions give generalized eigenfunctions of the hyperbolic Laplacian.

We shall keep the above convention for the principal, discrete and complementary series: $T^{is\pm}$ $(-\infty < s < \infty)$; T^s $(-1 < s < 1)$; and T^n $(n = \pm 1; \pm 2; ...)$.

4.1. K-invariants and spherical functions. We shall show that the even part of the principal series T^{is+}; as well as the complementary series T^s possess a unique K-invariant vector $\phi_0 \in \mathcal{H}_s$:

$$T_u\phi_0 = \phi_0; \text{ for all } u \in K.$$

Such vector is easy to exhibit in any of realizations of the principal and complementary T, constructed §7.1. In the $\mathcal{H}_s$-realization (a subspace of $L^2(\mathbb{R}^2)$), we find

$$\phi_0 = (x^2 + y^2)^{\frac{is-1}{2}} \text{ (for principal } T\text{)}, \ \phi_0 = (x^2 + y^2)^{\frac{s-1}{2}} \text{ (for complementary } T\text{)}. \quad (4.1)$$

A K-invariant allows to embed an irreducible representation T into the regular representation in bounded functions on homogeneous space $\mathcal{M} = K\backslash G$, by assigning each $\phi \in \mathcal{H}$ - a representation space of T, a function on G equal to its matrix entry,

$$\phi \to f_\phi = f(g) = \langle T_g\phi \mid \phi_0 \rangle. \quad (4.2)$$

Clearly, f depends only on the class of g (mod K), $f = f(z)$, $z \in K\backslash G$. Thus we get a map $\mathcal{W}: \mathcal{H} \to L^\infty(\mathcal{M})$, that intertwines T and the regular representation R in $L^\infty(\mathcal{M})$.

A particular matrix entry $f_0(g) = \langle T_g\phi_0 \mid \phi_0 \rangle$ is double-invariant (under both left and right multiplication with K),

$$f_0(ugv) = f_0(g); \text{ for all } u; v \in K.$$

Such functions $\{f\}$ on G are called *zonal spherical functions*, by analogy with spherical harmonics on S^2 (chapter 4). One can show that spherical functions $\{f \in L^1(G)\}$ form a commutative (convolution) subalgebra of $L^1(G)$, which could be reduced to a function-algebra on the half-line $[0;\infty)$, via the Cartan decomposition (problem 1).

Let us remark that embedding (4.2) does not make T a subrepresentations (discrete component) of the regular representation R on $\mathcal{M}$, which was the case with compact groups. Indeed, $\mathcal{W}$ embeds $\mathcal{H}$ into $L^\infty(\mathcal{M})$, rather than a subspace of $L^2(\mathcal{M})$. Irreducible decomposition of R for non-compact groups and homogeneous spaces are typically of the continuous (direct integral) type,

$$R \simeq \oint T^{is+} d\mu(s).$$

In the next section we shall compute the corresponding Plancherel measure[6] $d\mu$ for the regular representation of SL_2 on the Poincare half-plane $\mathbb{H} \simeq G/K$.

4.2. Representations $T\,|\,K$. In chapter 3 (§3.1) we have shown that any representation T of a compact group K is decomposed into the direct sum of *primary components* (multiples of irreducibles),

$$T \simeq \bigoplus_{\pi \in \widehat{K}} \pi \otimes d(\pi);\ d(\pi) \text{ - multiplicities (finite } d = 0;1;\dots \text{ or } \infty).$$

In particular, an irreducible representation T^s of $G = SL_2$ ($s = is\pm$; s; or n), restricted on $K = SO(2)$, admits such decomposition,

$$T^s\,|\,K = \bigoplus_m e^{im\theta} \otimes d_m; \text{ for } u = u_\theta = \begin{bmatrix} \cos\theta & \sin\theta \\ -\sin\theta & \cos\theta \end{bmatrix} \in K, \tag{4.3}$$

direct sum of characters. We shall establish the exact form of decomposition (4.3) for all 3 series.

Theorem 1: *For any irreducible representation T^s of SL_2 all multiplicities $\{d_m\}$ in (4.3) are 1, the set of characters $\{\chi_m = e^{im\theta}\}$ (spectrum of $T^s\,|\,K$), coincides with an arithmetic sequence of the difference 2: $m = m_0; m_0 \pm 2; m_0 \pm 4; \dots$; (or half-sequence, positive or negative). Precisely,*

(i) principal series:

$$\boxed{T^{is+}|K = \bigoplus_{m-even} \chi_m};\ \boxed{T^{is-}|K = \bigoplus_{m-odd} \chi_m}$$

(ii) *discrete series:*

$$\boxed{T^{+n}|K = \bigoplus_{m \geq n} \chi_m};\ \boxed{T^{-n}|K = \bigoplus_{m \leq -n} \chi_m}$$

[6] Complementary series $\{T^s\}$ do not enter regular representation R on $\mathcal{M}$, as they don't enter a larger regular representation R on G, of which $R^{\mathcal{M}}$ can be shown to form a subrepresentation, $R^{\mathcal{M}} \subset R^G$.

(iii) *complementary series has the same expansion as an even-principal series part.*

Remark: Theorem 1 rephrased in terms of generator $H = \begin{bmatrix} & 1 \\ -1 & \end{bmatrix}$ of rotations states: *operator T_H is diagonalized with a discrete spectrum of eigenvalues, $\{m_0 \pm 2k\}$. The principal-series spectra contain all even/odd integers, while the discrete-series has either the lowest weight vector ($m_0 = n$, for positive discrete series), or the highest weight vector ($m_0 = -n$, for the negative discrete series).*

The reader may compare this result to weight structure of irreducible representations of $SU(2)$, and other classical compact Lie groups in chapters 4-5.

Proof: The discrete-series case was already established in §7.1. For the principal series we shall use yet another realization of $T^{is\pm}$, in even/odd functions on the circle K. This form of $T^{is\pm}$ results either from Iwasawa decomposition $G = BK$ (remembering that principal series T are induced by characters of B), or from the $\mathcal{H}_s$-form realized as homogeneous functions in $\mathbf{R}^2$, and restricting those on a unit circle $\mathbf{T} = \{x^2 + y^2 = 1\} \simeq K$. In either case, space $\mathcal{H}_{s\pm}$ is identified with even/odd parts of $L^2(K)$, where K acts by translations. So $T^s \mid K$ turns into the regular representation on K, and the Theorem easily follows.

4.3. Infinitesimal generators and Laplacians on G and $\mathbf{H}$. We consider a basis of 3 elements in the Lie algebra $\mathfrak{G} = sl_2(\mathbb{R})$:

$$\left\{H = \begin{bmatrix} & 1 \\ -1 & \end{bmatrix}; V = \begin{bmatrix} & 1 \\ 1 & \end{bmatrix}; W = \begin{bmatrix} 1 & \\ & -1 \end{bmatrix}\right\} \tag{4.4}$$

which generate one-parameter subgroups:

$K = \{\exp tH : 0 \le t \le 2\pi\}$ - rotations;
$M = \{\exp tV : -\infty < t < \infty\}$ - hyperbolic rotations;
$A = \{\exp tW : -\infty < t < \infty\}$ - diagonal (Cartan) subgroup

Here W denotes the Cartan element, while V, H are combinations of raising/lowering elements (chapter 4),

$$V = X + Y;\ H = X - Y.$$

Any element $X \in \mathfrak{G}$ gives rise to a left-invariant vector field (or 1-st order differential operator) ∂_X. Vector fields $\{\partial_X\}$ generate the algebra of all left-invariant differential operators on G:

$$\mathfrak{D}(\mathfrak{G}) = \left\{\Omega = \sum a_{i_1 \dots i_m} \partial_{X_{i_1}} \partial_{X_{i_2}} \dots \partial_{X_{i_m}} : X_{i_1}; \dots; X_{i_m} \in \mathfrak{G}\right\},$$

which coincides with the *associative hull* (universal enveloping algebra) $\mathfrak{A} = \mathfrak{A}(\mathfrak{G})$.

Algebra $\mathfrak{D}(\mathfrak{G})$ contains a central (*Casimir*) element

$$L = \Delta_G = \tfrac{1}{4}(W^2 + V^2 - H^2) = \tfrac{1}{4}[W^2 - 2(XY+YX)], \tag{4.5}$$

(problem 2). In fact, one can show that L generates the center of $\mathfrak{A}$, so any central element is a "polynomial of L". The proof however is more involved. The corresponding differential operator (still written as L, or Δ_G, to indicate manifold G) commutes with all left and right translations,

$$LR_g f = R_g Lf, \text{ for all } g \in G, \text{ and smooth functions } f.$$

So we get an invariant Laplacian on G.

The reader should not confuse Δ with the Laplace-Beltrami operator on G, with respect to some left (or right) invariant Riemannian metric. The natural *bi-invariant* (left and right) metric on tangent spaces $\{T_x(G)\}$ is given by the Killing form (§5.1) at the point $\{e\}$:

$$\langle X \mid Y\rangle = \mathrm{tr}\,(\mathrm{ad}_X \mathrm{ad}_Y), \text{ for } X,Y \in \mathfrak{G} = T_e(\mathfrak{G}),$$

shifted then to other tangent spaces $\{T_x{:}x \in G\}$ by left/right translations (see problem 3). For noncompact semi-simple Lie groups the Killing metric has typically *indefinite* type[7], so the corresponding Laplacians (Casimir elements, §5.7) represent non-commutative generalizations of the wave operator: $\Box = \partial_0^2 - \sum_1^n \partial_j^2$, on $\mathbb{R}^{n+1}$.

The situation is different on symmetric spaces $\mathcal{M} = K\backslash G$, as we know from ch.5 (§5.7), like the Poincare-Lobachevski half-plane $\mathbb{H}$. Let us remark that the G-action, $\{g{:}x \to \phi_g(x)\}$ on any homogeneous space $\mathcal{M} = H\backslash G$, gives rise to a Lie algebra of vector fields (1-st order differential operators), representing $\mathfrak{G}$,

$$(\partial_X f)(x) = \tfrac{d}{dt}\,|_{\,t=0} f(\phi_{g(t)}(x));\ g(t) = \exp tX.$$

The latter extends to the associative hull $\mathfrak{A}(\mathfrak{G}) \ni p(X) \to p(\partial_X)$, whose elements turn into higher order differential operators on $\mathcal{M}$. The reduced Killing form on the tangent space $\{T_e\}$ (hence, all other $\{T_x\}$) becomes positive-definite (as will be explained in the next chapter), so Δ becomes the standard positive Laplacian on $\mathcal{M}$ (see §5.7). Let us compute the invariant Laplacian for the Poincare-Lobachevski half-plane $\mathbb{H}$. It is convenient to write functions on $\mathbb{H}$, as $f(z;\bar z)$. Given a one-parameter group $t \to \phi(t;z)$ of holomorphic transformations of $\mathbb{H}$, generated by a vector field B, the corresponding 1-st order operator,

$$\partial_B f = \tfrac{d}{dt}\Big|_0 f(\phi(t;z);\phi(t;\bar z)) = [\phi_t(0;z)\partial_z + \bar\phi_t(0;z)\partial_{\bar z}](f). \tag{4.6}$$

[7]The Killing form is positive-definite (Riemannian) iff group G is compact (see §5.7)!

We apply (4.6) to one-parameter subgroups generated by 3 basic elements (4.4),

$$\exp tH: z \to \frac{z\cos t - \sin t}{z\sin t + \cos t};\ \exp tV: z \to \frac{z\,\mathrm{ch}\,t + \mathrm{sh}\,t}{z\,\mathrm{sh}\,t + \mathrm{ch}\,t};\ \exp tW: z \to e^{2t}z;$$

and find

$$\partial_H = -[(z^2+1)\partial_z + (\bar z^2+1)\partial_{\bar z}];\ \partial_V = (1-z^2)\partial_z + (1-\bar z^2)\partial_{\bar z}];\ \partial_W = 2z\partial_z + 2\bar z\,\partial_{\bar z}.$$

Substituting in (4.5) we get the familiar expression of invariant Laplacians on $\mathbb{H}$ and $\mathbb{D}$,

$$\Delta_{\mathbb{H}} = -y^2(\partial_x^2 + \partial_y^2) = \left(\frac{z-\bar z}{2i}\right)^2\partial_{z\bar z}^2;\ \Delta_{\mathbb{D}} = (1-|z|^2)(\partial_x^2+\partial_y^2). \tag{4.7}$$

Let us also remark that Laplacians (4.7) represent the Laplace-Beltrami operators Δ_g on Riemannian manifolds $\mathbb{H}$ and $\mathbb{D}$, with respect to the natural hyperbolic metrics: $ds^2 = \frac{1}{y^2}(dx^2+dy^2)$ on $\mathbb{H}$, and $\frac{1}{1-r^2}(dx^2+dy^2)$ on $\mathbb{D}$. Indeed, given a metric $g = g_{ij}dx^i dx^j$ on manifold $\mathcal{M}$, the standard Laplacian is defined by

$$\Delta_g = \frac{1}{\sqrt{\det(g_{ij})}}\sum \partial_i g^{ij}\sqrt{\det(g_{ij})}\,\partial_j;$$

and (4.7) follows by a straightforward computation.

Since the invariant Laplacian commutes with the group action (belongs to the center) its restriction on any irreducible subspace[8] $\{\mathcal{H}_s\}$, embedded in $L^\infty(\mathbb{H})$, must be scalar,

$$T^s(\Delta) = \chi(s)I.$$

Number $\chi = \chi(s)$ is called the *infinitesimal character* of T^s, (since for higher rank groups, where $\dim\mathfrak{Z}(\mathfrak{A}) > 1$), χ turns into a character on $\mathfrak{Z}$. For SL_2 infinitesimal characters could be computed explicitly.

Proposition 3: *Infinitesimal characters $\{\chi(s)\}$ of the principal, complementary and discrete series are given by*

$$\boxed{\chi(s) = \frac{1-s^2}{4}} \tag{4.8}$$

where parameter s takes on imaginary values ($s \in i\mathbb{R}$) for the principal series; $s \in [-1;1]$, for complementary series; and s - integer $n = \pm 1; \pm 2; ...;$for discrete series. Thus

$$\chi(s) = \begin{cases} \frac{1+s^2}{4};\ (principal) \\ \frac{1-s^2}{4};\ (complementary) \\ \frac{1-n^2}{4};\ (discrete) \end{cases}$$

[8]For a ∞-D representation T in space $\mathcal{H}$, all generators $\{T_X : X \in \mathfrak{G}\}$, hence $T(\Delta)$ are unbounded operators, which should be considered on a dense core $\mathcal{H}_0$ of smooth vectors (§1.4).

The derivation of (4.8) is fairly straightforward. It suffices to compute $\chi(s)$ on any fixed vector $\phi_0 \in \mathcal{H}_s$. For the principal and complementary series $\{T^s\}$ we can choose K-invariants $\{\phi_0\}$ of (4.1), for the discrete series $\{T^{\pm n}\}$ the highest/lowest vectors $\{\phi_n\}$ (problem 5).

Problems and Exercises:

1. (a) Show directly that spherical functions $\{f(ugv) = f(g);\ \text{all}\ u, v \in K\}$ on groups $G = SL_2;\ K = SO(2)$, or compact $G = SU(2); K = U(1) \simeq \mathbf{T}^1$, form commutative subalgebras of $L^1(G)$, by using Cartan decomposition $G = KA_+K$ to reduce spherical functions $\{f(g)\}$ on G to single-variable functions $\{f(t): t > 0\}$ for SL_2, or $\{f(\theta): 0 \le \theta < \pi\}$, for $SU(2)$, then writing the convolution formulae for reduced functions (The general argument based on Cartan automorphism θ for pairs G, K was outlined in problem 1 of §5.7).

2. Show that element Δ (4.5) commutes with all $B \in \mathfrak{G}$, hence all of $\mathfrak{A}(\mathfrak{G})$ (check it for the Cartan basis $\{W; X; Y\}$). Furthermore, the center of $\mathfrak{A}$ consists of polynomials in Δ: $f(\Delta) = \sum a_k \Delta^k$.
(The proof is somewhat involved: each element $A \in \mathfrak{A}$ can be uniquely represented by a polynomial $p = \sum a_{kmn} W^k X^m Y^n$ in 3 (noncommuting!) variables $\{W; X; Y\}$ (in the given order). One has to write the commutation relations: $\mathrm{ad}_W(p) = 0$; $\mathrm{ad}_X(p) = 0; \mathrm{ad}_Y(p) = 0$

3. i) Show that a bi-invariant metric $\langle \xi \mid \eta \rangle_g$ on Lie group G, restricted on the tangent space at $\{e\}$, $T_e(G) \simeq \mathfrak{G}$, is invariant under the adjoint action,

$$\langle \mathrm{Ad}_g \xi \mid \mathrm{Ad}_g \eta \rangle = \langle \xi \mid \eta \rangle, \text{ for all } \xi, \eta \in \mathfrak{G},\ g \in G.$$

Conversely, any Ad_g-invariant metric on $\mathfrak{G} \simeq T_e(G)$ extends to a bi-invariant metric on G.

ii) All Ad-invariant products can be easily described for semisimple Lie algebras. Killing form, $\langle X \mid Y \rangle_0$, is one of them; for simple $\mathfrak{G}$ Killing is the only one; if $\mathfrak{G} = \oplus \mathfrak{G}_p$- direct sum of simple ideals, and $\langle\,|\,\rangle_p$ denotes the Killing form on $\mathfrak{G}_p$, then any Ad-invariant form on $\mathfrak{G}$, is given by $\sum a_p \langle\,|\,\rangle_p$ (Hint: any bilinear form is obtained from a nondegenerate Killing form by an operator B, $\langle X \mid Y \rangle = \langle B(X) \mid Y \rangle$; form $\langle\,|\,\rangle$ is Ad-invariant iff operator B commutes with Ad_g!).

4. Verify directly that Laplacian (4.7) on $\mathbf{H}$ is invariant under all fractional-linear transformations $g: z \to \frac{az+c}{bz+d}$, in SL_2. What happens to Δ under the GL_2-action?

5. Compute the infinitesimal characters of Proposition 3.

§7.5. Selberg trace formula.

In this section we shall study the regular and induced representations R of group $G = SL_2$ on compact quotient-spaces $\mathcal{M} = G/\Gamma$, where Γ is a discrete subgroup of G. We shall show that representation R breaks into the discrete sum of irreducibles, each entering R with a finite multiplicity. We prove a reciprocity Theorem (similar to Frobenius reciprocity of chapter 3), and establish the *Trace-formula* for operators R_f ($f \in C_0^\infty$) - a noncommutative analog of the Poisson summation. Then we shall proceed to evaluate the contribution of various "parts of Γ" to the trace-formula. This eventually lead us to the celebrated *Selberg trace formula* on compact quotient-spaces of SL_2. After proving the general result we shall outline its ramifications and special cases, and find interesting connections to spectral theory of Laplacians on hyperbolic Riemann surfaces $\mathbf{H}/\Gamma$, Poincare half-plane modulo a discrete (Kleinian) subgroup $\Gamma \subset SL_2$. We also give an application of the representation theory of SL_2 to geodesic flows on negatively curved Riemann surfaces.

5.1. Induced representations on G/Γ: reciprocity and the general trace-formula. Let Γ be a discrete subgroup of G, α be a representation of Γ. We denote by $R = R^\alpha = \mathrm{ind}(\alpha \mid \Gamma;G)$ - an induced representation of G. It is an interesting and challenging problem to find a decomposition of R^α into irreducible components of G. A simple commutative prototype of such setup consists of a lattice $\Gamma \subset \mathbb{R}^n$ ($\Gamma \simeq \mathbb{Z}^n$), and a representation/character $\alpha(\gamma) = e^{i\alpha \cdot \gamma}$, $\gamma \in \Gamma$. The induced representation R^α acts on scalar/vector functions $\{f(x)$ on $\mathbb{R}^n\}$, that satisfy the α-Floquet condition (a generalization of the periodic condition),

$$f(x+\gamma) = e^{i\alpha \cdot \gamma} f(x);\ \text{for all } x \in \mathbb{R}^n;\ \gamma \in \Gamma.$$

Representation R^α is easily seen to decompose into the direct sum of characters,

$$R_x^\alpha \simeq \bigoplus_{m \in \Gamma'} e^{i(\alpha+m)\cdot x};\ x \in \mathbb{R}^n, \tag{5.1}$$

sum over the dual lattice Γ', in a special periodic case

$$R_x \simeq \bigoplus_{m \in \Gamma'} e^{im\cdot x};\ x \in \mathbb{R}^n.$$

Both results are simple corollaries of the Poisson summation formula of §2.1. In particular we get spectrum of the Laplacian Δ_α (or any other constant-coefficient differential operator) on $\mathbb{R}^n/\Gamma \simeq \mathbb{T}^n$, with periodic/Floquet boundary conditions,

$$\mathrm{Spec}(\Delta_\alpha) = \{\lambda_m = (m+\alpha)^2 : m \in \Gamma'\}. \tag{5.2}$$

Our main goal in this section is to derive a noncommutative version of (5.1) for quotients SL_2/Γ, and then to link such decomposition to spectral theory of Laplacians on spaces $\mathbb{H}/\Gamma$. Throughout this section quotient-spaces $\mathcal{M} = G/\Gamma$ will be assumed compact. The corresponding Γ are called *co-compact* (or *uniform lattices*), and SL_2 can be shown to have a plenty of those. Uniform lattices are characterized by the following properties (see [GGP], chapter 1):

i) Γ has finitely many generators $\{\gamma_1;...\gamma_m\}$, and a finite number of relations among them;

ii) the G-conjugacy class $\{g^{-1}\gamma g : g \in G\}$ of any $\gamma \in \Gamma$ is closed in G;

iii) Γ contains only *elliptic* and *hyperbolic* elements.

A hyperbolic element γ is called *primitive,* if $\gamma \neq {\gamma_0}^k$, for some $\gamma_0 \in \Gamma$, and $k > 1$. Each elliptic element γ has a finite order $\gamma^m = e$.

Let us remark that (iii) easily follows from (ii), since a conjugacy class $C(\gamma) = \{g^{-1}\gamma g\}$ of any parabolic element γ contains limit points $\pm I$, not in $C(\gamma)$. So parabolic C can not be a closed subset.

First we shall state the *general trace formula* for an arbitrary pair $\Gamma \subset G$, of a (locally compact) group G, and discrete subgroup Γ, with compact quotient G/Γ. Let R^α be induced by a finite-D representation α of Γ. Operators $\{R^\alpha_f\}$ are given by integral kernels,

$$K_f(x;y) = \sum_{\gamma \in \Gamma} f(g_x^{-1}\gamma g_y)\alpha(\gamma);$$

where $g_x;g_y \in G$ are coset representatives of points $\{x;y\}$ in $\mathcal{M}$. Clearly, compactly supported (or rapidly convergent) functions $\{f\}$ give rise to compact operators $\{R_f\}$. In fact, operators $\{R_f\}$ belong in the trace class (see Appendix B), and

$$\operatorname{tr} R_f = \langle \chi_R \mid f \rangle = \int_X K_f(x;x)dx = \sum_{\gamma \in \Gamma} tr\,\alpha(\gamma) \int_X f(g_x^{-1}\gamma g_x). \tag{5.3}$$

On the other hand, any representation T, which contains a compact operator $\{T_f\}$ has "discrete spectrum", in the sense that T is decomposed into the direct sum (vs. direct integral) of irreducible components:

$$T \simeq \bigoplus_s T^s \otimes m(s), \tag{5.4}$$

each one entering T with a finite multiplicity $\{m(s)\}$ (problem 1). If χ_s denotes the character of T^s, then combining (5.4) with (5.3) we get the general *trace-formula,*

$$\boxed{\sum m(s)\langle \chi_s \mid f \rangle = \sum_{\gamma \in \Gamma} tr\,\alpha(\gamma) \int_X f(g_x^{-1}\gamma g_x)} \tag{5.5}$$

for a suitable class functions f on G (e.g. C_0^∞). In what follows, however, it would be more convenient to write the RHS of (5.5) as

$$\sum_{\gamma \in \Gamma} tr\, \alpha(\gamma)\, vol(\Gamma_\gamma \backslash G_\gamma) \int_{G_\gamma \backslash G} f(g^{-1}\gamma g)\, dg, \tag{5.6}$$

where Γ_γ; G_γ are stabilizers (centralizers) of an element γ in Γ and G, respectively, and dg denotes the invariant (Haar) measure on the quotient space $G_\gamma \backslash G$.

5.2. Trace-formula on SL_2. Our main goal is to compute (5.6) explicitly for group $SL_2(\mathbb{R})$, then to deduce all possible information about the LHS of (5.5): irreducible components $\{T^s\}$ of R^α, as well as their multiplicities. To state the general result we remind the reader 3 character formulae for irreducible representations of SL_2, obtained in §7.3. All three express $\{\chi\}$ in terms of the eigenvalue $\{\lambda = \lambda(g)\}$ of matrix g: real $\lambda > 1$, for hyperbolic $g \sim \begin{bmatrix} \lambda & \\ & \lambda^{-1} \end{bmatrix}$; and complex, $\lambda = e^{i\theta}$, for elliptic $g \sim \begin{bmatrix} \cos\theta & \sin\theta \\ -\sin\theta & \cos\theta \end{bmatrix}$;

$$\text{\underline{principal}:} \begin{cases} \chi_{is+}(g) = \dfrac{|\lambda|^{is} + |\lambda|^{-is}}{|\lambda - \lambda^{-1}|}; \\ \chi_{is-}(g) = \dfrac{|\lambda|^{is} + |\lambda|^{-is}}{|\lambda - \lambda^{-1}|} \operatorname{sgn}\lambda; \end{cases} \text{for hyperbolic } g;$$

$$\text{\underline{complementary}: } \chi_s(g) = \frac{|\lambda|^{s} + |\lambda|^{-s}}{|\lambda - \lambda^{-1}|}; \tag{5.7}$$

$$\text{\underline{discrete}: } \chi_{+m}(g) = \begin{cases} \dfrac{\lambda^{-m}}{\lambda - \lambda^{-1}}; \text{ hyperbolic } g \\ \dfrac{e^{-im\theta}}{e^{i\theta} - e^{-i\theta}}; \text{ elliptic } g \end{cases}; \ \chi_{-m}(g) = \begin{cases} \dfrac{\lambda^{m}}{\lambda - \lambda^{-1}}; \text{ hyperbolic } g \\ \dfrac{e^{im\theta}}{e^{i\theta} - e^{-i\theta}}; \text{ elliptic } g \end{cases}.$$

Given a function $f \in C_0^\infty(G)$, we introduced its *generalized Fourier transform/ coefficients* $\{\hat{f}(s\pm); \hat{f}(s); \hat{f}(\pm n)\}$, by integration of f against irreducible characters,

$$\hat{f}(s) = \int f(g)\, \overline{\chi_s(g)} dg = tr\,(T_f^s);\ s = is\pm; s; \pm n, \tag{5.8}$$

so

$$\hat{f}(is+) = \int f(g) \overline{\chi_{is+}(g)}\, dg, \text{ etc.}$$

Following [GGP] we have to compute the contribution of different conjugacy classes of Γ to $\operatorname{tr}(R_f^\alpha)$. Those are conveniently divided into 4 groups: hyperbolic classes, elliptic classes, and 2 special elements $\{e\}$ and $\{-e\}$.

• **Contribution of hyperbolic elements.** We shall split all hyperbolic conjugacy classes $\tilde{\gamma} = \{g^{-1}\gamma g\} \subset \Gamma$, into primitive classes $\mathrm{Pr}_h = \{\tilde{\gamma} : \gamma \neq \gamma_0^k\}$, and their iterates $\{\tilde{\gamma}:\ \gamma = \gamma_0^k;\ \gamma_0\text{-primitive}\}$. We have to evaluate each term of (5.6) for a hyperbolic element $\gamma \sim \begin{bmatrix} \lambda & \\ & \lambda^{-1} \end{bmatrix}$. The RHS of (5.6) involves the conjugacy-class H-transform of f, which was essentially evaluated in §7.4. Namely,

$$\int_{D\backslash G} f(g^{-1}\gamma g)dg = F^A(\gamma) = \frac{1}{4\pi\,|\,\lambda - \lambda^{-1}\,|} \int_{-\infty}^{\infty} \left\{ \widehat{f}(is+) + \widehat{f}(is-)\, sgn\,\lambda \right\} |\,\lambda\,|^{is} ds, \quad (5.9)$$

in other words the hyperbolic H-transform of f is equal to the inverse Mellin transform (§2.3) of the even and odd principal series ***generalized Fourier transforms*** of f.

Next we compute volume of the quotient-space

$$vol(\Gamma_\gamma \backslash G_\gamma) = \int_1^{\lambda} \frac{d\lambda}{\lambda} = ln\,\lambda, \quad (5.10)$$

since $G_\gamma \simeq \mathbb{R}^\times$ (multiplicative group of reals), while $\Gamma_\gamma = \{\gamma^m : m \in \mathbb{Z}\}$ forms a discrete lattice in G_γ. Combining (5.10) and (5.9) we come up with the total hyperbolic contribution to the trace,

$$\sum_{\tilde{\gamma}\,\in Pr_h} \sum_{k=1}^{\infty} \frac{ln\,\lambda(\gamma)\, tr\,\alpha(\gamma)}{2\pi\,|\,\lambda^k - \lambda^{-k}|} \left\{ \int_{-\infty}^{\infty} \left\{ (\tfrac{1+\epsilon}{2})\widehat{f}(is+) + (\tfrac{1-\epsilon}{2})\widehat{f}(is-) \right\} \lambda^{isk} ds \right\} \quad (5.11)$$

Here the outer summation extends over all hyperbolic primitive classes $\{\tilde{\gamma}\}$, while the inner consists of their iterates $\{\tilde{\gamma}^{\,k}\}$; symbol

$$\epsilon_\alpha = \begin{cases} 1; & if\ \alpha(-e) = I \\ -1; & if\ \alpha(-e) = -I \end{cases}.$$

Note that hyperbolic terms involve only the principal-series characters.

• **Contribution of elliptic elements.** We consider the set of primitive elliptic elements, and notice that each of them has a *finite even order*[9]: $\gamma^{2m} = e$. We need to evaluate an elliptic H-transform of f in the RHS of (5.6). For any elliptic element $u = u_\theta \sim \begin{bmatrix} cos\theta & sin\theta \\ -sin\theta & cos\theta \end{bmatrix}$;

$$F^K(u) = \frac{1}{4\pi i\, sin\theta} \left\{ \frac{\widehat{f}(+0) - \widehat{f}(-0)}{2} + \sum_1^{\infty} \left(\widehat{f}(+n)e^{in\theta} - \widehat{f}(-n)e^{-in\theta} \right) \right\} + \quad (5.12)$$

$$+ \frac{1}{16\pi\, sin\theta} \int_{-\infty}^{\infty} \left\{ \widehat{f}(is+) \frac{ch\,(\theta - \frac{\pi s}{2})}{ch\frac{\pi s}{2}} - \widehat{f}(is-) \frac{sh\,(\theta - \frac{\pi s}{2})}{sh\frac{\pi s}{2}} \right\} ds.$$

The first half of formula (5.12) (sum) consists of the discrete series "generalized Fourier coefficients" of f, while the second half (integral) is made of two principal series "generalized Fourier transforms". The derivation essentially follows arguments of §7.4, where (5.12) was established at $\theta = 0_{\pm}$ (problem 2 provides further details).

---

[9] Indeed, if γ had an odd order p, then $-\gamma$ would have an even order $2p$, so $\gamma = (-\gamma)^{p+1}$ could not be primitive. Hence each elliptic γ is equivalent to a rotation by angle $\theta = \frac{\pi}{p}$.

The volume of $\Gamma_\gamma \backslash G_\gamma$ is easily found to be $\frac{\pi}{m}$, where $\gamma^{2m}=e$ (γ-primitive). Indeed, $G_\gamma = SO(2) \simeq [0;2\pi]$, while Γ_γ makes a finite subgroup of order $2m$ in G_γ. Combining contribution of all elliptic classes, we get

$$\sum_{\widetilde{\gamma}}\sum_{k=1}^{m-1}\frac{tr\alpha(\gamma^k)}{4\pi i\, sin\frac{\pi k}{m}}\left\{(1-\epsilon)\frac{\widehat{f}(+0)-\widehat{f}(-0)}{2}+\sum_1^\infty(1-(-1)^n\epsilon)\left(\widehat{f}(+n)e^{in\frac{\pi k}{m}}-\widehat{f}(-n)e^{-in\frac{\pi k}{m}}\right)\right\}+$$

$$+\frac{1}{16m\, sin\frac{\pi k}{m}}\int_{-\infty}^{\infty}\left\{\widehat{f}(is+)\frac{ch\,(\frac{k}{m}-\frac{1}{2})\pi s}{ch\frac{\pi s}{2}}-\widehat{f}(is-)\frac{sh\,(\frac{k}{m}-\frac{1}{2})\pi s}{sh\frac{\pi s}{2}}\right\}ds, \tag{5.13}$$

The outer summation extends over all conjugacy classes of primitive elliptic elements $\{\widetilde{\gamma}\}$, the inner over all iterates $\{\gamma^k\colon 1\le k\le m-1\}$, from k=1 through $\frac{1}{2}\times$order of γ.

• **Contribution of elements $\{e\}$ and $\{-e\}$.** Clearly, terms of (5.5), which correspond to $\{\pm e\}$ are

$$d(\alpha)\, vol(\Gamma\backslash G)\{f(e)+\epsilon f(-e)\},$$

where symbol

$$\epsilon=\begin{cases}1; \text{if } \alpha(\text{-}e)=I\\ -1; \text{if } \alpha(\text{-}e)=-I\end{cases}$$

It remains to express $f(\pm e)$ (or δ-functions at $\{\pm e\}$), through irreducible characters (generalized $\mathfrak{F}$-transforms) on G. But those are precisely the Plancherel formulae obtained in the previous section

$$f(e)=\frac{1}{\pi^2}\left\{\sum_{n=1}^\infty n\left(\widehat{f}(+n)+\widehat{f}(-n)\right)+\frac{1}{4}\int_{-\infty}^{\infty}\left(\widehat{f}(is+)\,s\,th\frac{\pi s}{2}+\widehat{f}(is-)\,s\,cth\frac{\pi s}{2}\right)ds\right\}.$$

A similar relation holds for class {-e}

$$f(-e)=\frac{1}{\pi^2}\left\{\sum_{n=1}^\infty(-1)^{n-1}n\left(\widehat{f}(+n)+\widehat{f}(-n)\right)+\right.$$
$$\left.+\frac{1}{4}\int_{-\infty}^{\infty}(1+\epsilon)\hat{f}(is+)\,s\,th\frac{\pi s}{2}+(1-\epsilon)\hat{f}(is-)\,s\,cth\frac{\pi s}{2}\,ds\right\}.$$

We shall now summarize all 4 contributions into the final result.

Selberg trace formula: *Let Γ be a uniform lattice in $G=SL_2$, $\alpha(\gamma)$ - a finite-D representation of Γ of degree $d(\alpha)$, and $R=R^\alpha$ - the induced representation $ind(\alpha\,|\,\Gamma;G)$ of G. Representation R is decomposed into the direct sum of irreducibles (principal, complementary and discrete series), each entering R with a finite multiplicity,*

$$R\simeq\underbrace{\bigoplus_k T^{is_k\pm}\otimes N^{\pm}_{is_k}}_{principal}\underbrace{\bigoplus_j T^{s_j}\otimes N_{s_j}}_{complementary}\underbrace{\oplus\, T^{\pm n}\otimes N^{\pm}_n}_{discrete}. \tag{5.14}$$

Operators $\{R_f\colon f\in C_0^\infty(G)\}$ belong to the trace-class, and trace of R_f can be *expanded in two different ways, according to* (5.5), *the LHS being*

$$\sum_k \widehat{f}(is_k \pm) N^{\pm}_{is_k} + \sum_j \widehat{f}(s_j) N_{s_j} + \sum_n \widehat{f}(\pm n) N^{\pm}_n; \tag{5.15}$$

while the RHS is made of contributions of various conjugacy-classes of G,

$$\sum_{\widetilde{\gamma}} \sum_{k=1}^{\infty} \frac{ln\lambda(\gamma)\, tr\alpha(\gamma)}{2\pi \mid \lambda^k - \lambda^{-k}|} \left\{ \int_{-\infty}^{\infty} \left\{ (\tfrac{1+\epsilon}{2}) \widehat{f}(is+) + (\tfrac{1-\epsilon}{2}) \widehat{f}(is-) \right\} \lambda^{isk} ds \right\} +$$

$$- \sum_{\widetilde{\gamma}} \sum_{k=1}^{m-1} \frac{tr\alpha(\gamma^k)}{4\pi i\, sin\frac{\pi k}{m}} \left\{ (1-\epsilon) \frac{\widehat{f}(+0) - \widehat{f}(-0)}{2} + \sum_{n=1}^{\infty} (1-(-1)^n \epsilon) \left(\widehat{f}(+n) e^{in\frac{\pi k}{m}} - \widehat{f}(-n) e^{-in\frac{\pi k}{m}} \right) \right\} +$$

$$+ \frac{tr\alpha(\gamma^k)}{16m\, sin\frac{\pi k}{m}} \int_{-\infty}^{\infty} \left\{ \widehat{f}(is+) \frac{ch\,(\frac{k}{m} - \frac{1}{2})\pi s}{ch\frac{\pi s}{2}} - \widehat{f}(is-) \frac{sh\,(\frac{k}{m} - \frac{1}{2})\pi s}{sh\frac{\pi s}{2}} \right\} ds + \tag{5.16}$$

$$+ \frac{d(\alpha) vol(\Gamma \backslash G)}{\pi^2} \left\{ \sum_{n=1}^{\infty} (-1)^{n-1} n \left(\widehat{f}(+n) + \widehat{f}(-n) \right) + \right.$$

$$\left. + \frac{1}{4} \int_{-\infty}^{\infty} (1+\epsilon) \widehat{f}(is+)\, s\, th\frac{\pi s}{2} + (1-\epsilon) \widehat{f}(is-)\, s\, cth\frac{\pi s}{2} ds \right\}.$$

The first line of (5.16) contains the hyperbolic contributions ($\widetilde{\gamma} \in \mathrm{Conj}_h(G)$), lines 2-3 elliptic part ($\widetilde{\gamma} \in \mathrm{Conj}_{ell}(G)$), lines 4-5 come from the classes $\{\pm e\}$.

Looking at a cumbersome 5-line expression one can't help wondering about its meaning and utility? Here we shall give some partial answers to this daunting question, derive a few Corollaries of (5.16), and then mention some interesting connections of the Selberg-trace formula to spectral theory of Laplacians on hyperbolic surfaces.

5.3. Discrete series multiplicities. The first simple consequence of Selberg is the formula for multiplicities of irreducible components of R^α. For discrete series $\{T^{\pm n}\}$ those are precisely the coefficients $N^{\pm}_n$ of $\{\widehat{f}(\pm n)\}$ in (5.15). Hence,

$$N^+_n = \left(1+(-1)^{n-1}\epsilon\right) \left\{ \frac{d(\alpha) vol(\Gamma \backslash G)}{\pi^2} n - \sum_{\widetilde{\gamma}} \sum_{k=1}^{m-1} \frac{\mathrm{tr}\alpha(\gamma^k)}{4ki\, sin\frac{\pi k}{m}} e^{in\frac{\pi k}{m}} \right\}, \tag{5.17}$$

summation over elliptic classes $\{\widetilde{\gamma}\}$, and $\{\pm e\}$, as only those carry the discrete-series components. A similar relation holds for N^-_n. In special cases, when discrete subgroup Γ has no elliptic elements, formula (5.17) simplifies to

$$N^+_n = N^-_n = \left(1+(-1)^{n-1}\epsilon\right) \frac{d(\alpha) \mathrm{vol}(\Gamma \backslash G)}{\pi^2} n.$$

The reason for the splitting of the discrete-series components in both sides of the trace formula (5.15)-(5.16) is fairly general and simple. It has to do with orthogonality properties of the discrete-series matrix entries and characters. Roughly speaking they behave like the characters on compact groups (see chapter 3). The latter, as we already know, have a unique decomposition for any conjugate-invariant function ϕ on compact G,

$$\phi = \sum a_s \chi_s = \sum b_s \chi_s \Rightarrow \boxed{a_s = b_s}; \text{ for all irreducible } \{s\},$$

due to their completeness and orthogonality in $L^2(G)$. For non-compact G, like SL_2, characters (even discrete-series) are no more L^2-functions. However, discrete-series matrix-entries $f(g) = \langle T^n_g \psi \mid \psi \rangle$ do belong in L^2, $\{T^n\}$ being embedded in the regular representation $R \mid L^2(G)$. Moreover, one can show that entry f can be chosen in space $L^1(G)$, e.g. the highest/lowest entry $f_0 = \langle T^n_g \psi_0 \mid \psi_0 \rangle$ (problem 3). Such f can then be paired with any (principal/complementary) matrix entry $f_1(g)$, the result being

$$\langle f \mid f_1 \rangle = 0, \text{ for all } f_1 \in \mathrm{Span}\{\langle T^s \eta \mid \eta \rangle\}.$$

Hence, operators $T^s_f = 0$, for all principal/complementary series $\{T^s\}$, as well as their traces,

$$\mathrm{tr}\, T^s_f = \int f(g)\overline{\chi_s(g)}dg = 0.$$

The latter provides the requisite ***orthogonality relation*** between the discrete and other series matrix entries and characters. It remains now to insert such f in both sides (5.15)-(5.16) of the Selberg trace-formula, and observe that all non-discrete terms vanish.

Along with splitting into the "discrete" and "continuous" parts trace-formula allows yet another splitting into the positive (even) and negative (odd) components[10]. "Odd and even" functions, representations, etc., arise on SL_2 due to the element $\{-e\}$,

$$f(-x) = \pm f(x);\ T_{-e} = \pm I; \text{ etc.}$$

Clearly, principal $\{is+\}$, as well as complementary $\{s\}$-series, hence their characters, are even, while the principal $\{is-\}$-series is odd. Separating the even and odd terms in both sides of the trace formula we get a complete splitting,

$$\text{Discrete: } \sum_{n=0}^{\infty} \widehat{f}(\pm n) N^{\pm}_n = \frac{d(\alpha) vol(\Gamma\backslash G)}{\pi^2} \left\{ \sum_{n=1}^{\infty} (-1)^{n-1} n \left(\widehat{f}(+n) + \widehat{f}(-n) \right) + \right.$$

[10] No simple orthogonality relation would help this time, as principal/complementary characters form an "over-determined system" on G. Indeed, the trace-formula underpin the over-determinancy of $\{\chi_s\}$, as its the LHS (5.15) exhibits a discrete sum of characters,

$$\sum N^{\pm}_{is} \chi_{is\pm} + \sum N_s \chi_s$$

while the other side (5.16) is represented by various combinations of "discrete sums" (over $\{\widetilde{\gamma}\}$) and "continuous integrals" of that same set of characters,

$$\sum_{\widetilde{\gamma}} \sum_{k=1}^{\infty} \frac{ln\lambda(\gamma)\, tr\alpha(\gamma)}{2\pi \mid \lambda^k - \lambda^{-k} \mid} \left\{ \int_{-\infty}^{\infty} \left\{ \left(\frac{1+\epsilon}{2}\right)\chi_{is+} + \left(\frac{1-\epsilon}{2}\right)\chi_{is-} \right\} \lambda^{isk} ds \right\} + \dots \text{ etc.}$$

Notice, that the RHS of Selberg formula (5.16) contains only the principal series characters!

$$-\sum_{\tilde\gamma}\sum_{k=1}^{m-1}\frac{tr\alpha(\gamma^k)}{4\pi i\, sin\frac{\pi k}{m}}\left\{(1-\epsilon)\frac{\hat f(+0)-\hat f(-0)}{2}+\sum_{n=1}^{\infty}(1-(-1)^n\epsilon)\left(\hat f(+n)e^{\frac{in\pi k}{m}}-\hat f(-n)e^{-\frac{in\pi k}{m}}\right)\right\}$$

Even continuous:

$$\sum_k \hat f(is_k+)N^+_{is_k}+\sum_j \hat f(s_j)N_{s_j}= \tag{5.18}$$

$$\sum_{\tilde\gamma}\sum_{k=1}^{\infty}\frac{ln\lambda(\gamma)\,tr\alpha(\gamma)}{2\pi\,|\,\lambda^k-\lambda^{-k}|}\int_{-\infty}^{\infty}(\tfrac{1+\epsilon}{2})\hat f(is+)\lambda^{isk}ds+\sum_{\tilde\gamma}\sum_{k=1}^{m-1}\frac{tr\alpha(\gamma^k)}{16m\, sin\frac{\pi k}{m}}\int_{-\infty}^{\infty}\hat f(is+)\frac{ch\,(\frac{k}{m}-\frac12)\pi s}{ch\frac{\pi s}{2}}+$$

$$+\frac{d(\alpha)vol(\Gamma\backslash G)}{4\pi^2}\int_{-\infty}^{\infty}(1+\epsilon)\hat f(is+)\,s\,th\tfrac{\pi s}{2};$$

Odd continuous:

$$\sum_k \hat f(is_k-)N^-_{is_k}=$$

$$\sum_{\tilde\gamma}\sum_{k=1}^{\infty}\frac{ln\lambda(\gamma)\,tr\alpha(\gamma)}{2\pi\,|\,\lambda^k-\lambda^{-k}|}\left\{\int_{-\infty}^{\infty}(\tfrac{1-\epsilon}{2})\hat f(is-)\,\lambda^{isk}ds\right\}-\frac{tr\alpha(\gamma^k)}{16m\, sin\frac{\pi k}{m}}\int_{-\infty}^{\infty}\left\{\hat f(is-)\frac{sh\,(\frac{k}{m}-\frac12)\pi s}{sh\frac{\pi s}{2}}\right\}ds+$$

$$+\frac{d(\alpha)vol(\Gamma\backslash G)}{4\pi^2}\int_{-\infty}^{\infty}(1-\epsilon)\hat f(is-)\,s\,cth\tfrac{\pi s}{2}\,ds\Big\}.$$

5.4. Recipricity Theorem. Next we turn to the principal/complementary part of *spectrum* of R^α. Our goal is to find all irreducibles $\{is_k\pm\}$ and $\{s_j\}$, as well as their multiplicities $\{N\}$ in (5.15). There are two possible approaches to the problem. One of them, based on a *Frobenius-type reciprocity principle,* allows to reduce a decomposition of induced $R=\mathrm{ind}(\alpha\,|\,\Gamma;G)$, to the study of certain Γ-invariants of irreducible representations $\{T^s\}$ of G.

We remind the reader the classical Frobenius reciprocity Theorem for compact G (§3.2): *irreducible* $\pi\in\widehat G$ *enters an induced representation* $R=\mathrm{ind}(\alpha\,|\,\Gamma;G)$, *iff the restriction* $\pi\,|\,\Gamma$ *contains a copy of* α, $\pi\,|\,\Gamma\supset\alpha\otimes N_\alpha$. *Furthermore multiplicity* $N_\pi(R)$ *(of* π *in* R*) is equal to multiplicity* $N_\alpha(\pi)$ *of* α *in* $\pi\,|\,\Gamma$. For trivial $\alpha=I$, it says: $\pi\subset R$, *iff* $R\,|\,\Gamma$ *contains* Γ*-invariants, and multiplicity* $N_\pi(R)$ *is equal to dimension of* Γ*-invariants in* π.

The main difficulty with non-compact G has to do with the fact, that unitary representations in Hilbert (L^2-type) spaces, like $\mathrm{h}_s\subset L^2(\mathrm{H})$, have typically no Γ-invariants. The latter appear in larger (L^∞-type) spaces, associated to $\{T^s\}$, which have a dense (core) intersection with L^2. Although topologically, representations in L^2- and L^∞-type spaces may look different, their joint core makes them essentially equivalent (they are also equivalent in the sense of characters, and/or infinitesimal characters!).

For group SL_2 such "expanded" spaces $\{\mathrm{h}_s\}$ could be realized[11] in $L^2(H)$. In §7.5

we described them, as eigenspaces of the invariant Laplacian Δ on $\mathbb{H}$,

$$\tilde{\mathcal{H}}_s = \{f(z): \Delta f = \tfrac{1-s^2}{4} f\}.$$

We look for (finite-D) subspaces $\mathcal{A}_s \subset \tilde{\mathcal{H}}_s$, which "transform according to α" under the action of Γ. In other words, there exists a basis of functions $\{f_1(z);...f_d(z)\}$ in $\mathcal{A}_s$ $(d = d(\alpha))$, so that

$$T^s_\gamma \begin{bmatrix} f_1 \\ ... \\ f_d \end{bmatrix}(z) = (\alpha_{jk}(\gamma)) \begin{bmatrix} f_1(z) \\ ... \\ f_d(z) \end{bmatrix}; \gamma \in \Gamma;\ j;k = 1;...d, \tag{5.19}$$

where z^g means the usual fractional-linear action, $\frac{az+b}{cz+d}$, of SL_2 on $\mathbb{H}$. In the case of trivial $\alpha = I$, we look for Γ-invariant eigenfunctions of the Laplacian,

$$\Delta f = \tfrac{1-s^2}{4} f; \text{ and } f(z^\gamma) = f(z); \text{ all } \gamma \in \Gamma,$$

the so called *automorphic forms* on $\mathbb{H}$. Functions (5.19) could be thought of as *generalized (vector-valued) automorphic forms* of *weight* α. The main reciprocity result for pairs $(SL_2;\Gamma)$ can now be stated as follows.

Reciprocity Theorem: *Irreducible representation T^s of the principle/complementary series enters an induced representation $R^\alpha = \mathrm{ind}(\alpha \mid \Gamma;G)$ with multiplicity N_s, iff the expanded space $\tilde{\mathcal{H}}_s$ contains a subspace $\mathcal{A}_\alpha$ of automorphic forms of weight α. Furthermore, multiplicity N_α is equal to $\frac{dim \mathcal{A}_\alpha}{d(\alpha)}$ (i.e. multiplicity of α in $T^s \mid \Gamma$).*

For 1-D representations α (characters of Γ), reciprocity Theorem states that N_α is equal to the number (dimension) of automorphic form of weight α. The proof of reciprocity can be obtained by a careful analysis of automorphic forms (eigenfunctions of the Laplacian), for all 3 series of SL_2, as in [GGP] (chapter 1). But there is also a fairly general argument (due to Piatetski-Shapiro), valid for all semi-simple Lie groups G. We shall skip further details and refer the reader to [GGR].

Reciprocity Theorem reduces our "spectral problem" for R^α to the study of automorphic forms, a difficult and largely unresolved question by itself. So its utility in this regard is somewhat limited (as compared to, say, Frobenius reciprocity for Laplacians on spheres and other compact symmetric spaces).

Another approach to the *Spectral decomposition problem* for R^α is closely related (in fact, includes as a special case) *Spectral theory of Laplacians* on Riemannian surfaces

---

[11] We have done it in §7.1 for the discrete series, and for principal/complementary series it follows from the results of §7.5: existence of K-invariants in $\{T^s\}$.

$\mathcal{M} = \mathbb{H}/\Gamma$. Such spectral problems (if one aims at a complete solution) are notoriously difficult, if solvable at all. So one often sets less ambitious goal of finding some *asymptotic spectral data,* which involve irreducibles $\{is_k \pm ; s_j\}$ along with their multiplicities.

5.5. We shall conclude this section with the *Weyl-type* formula for spec(R^α). Let $N(\lambda)$ count the number of the principal/complementary series terms of parameter $s_k \leq \lambda$. We assume that $\epsilon = 1$ (i.e. $\alpha(-e) = I$), so R^α does not contain the odd-principal series. Then one has

$$\boxed{N(\lambda) \sim \frac{d(\alpha)}{4\pi^2} vol(\Gamma\backslash G)\,\lambda^2, \text{ as } \lambda \to \infty} \tag{5.20}$$

In other words the number of irreducibles $\{T^s\}$ (counted with multiplicities) in the parameter-range $[0;\lambda]$ increases as Const $\times\, \lambda^2$, with constant proportional to volume of the quotient space $\mathcal{M} = \Gamma\backslash G$.

To establish (5.20) we formally apply the "even-continuous part" of the Selberg formula (5.18) to a function f on G, with the step-like generalized Fourier transform $\{\widehat{f}(s)\colon s = is+;\ \text{or } s\}$,

$$\widehat{f}(s) = \begin{cases} 1; & s \leq \lambda \\ 0; & s \geq \lambda \end{cases}.$$

Then the LHS of (5.18) becomes exactly $N(\lambda)$, while the leading asymptotic in the RHS is given by the last integral-term of (5.18),

$$\frac{d(\alpha)\, vol(\Gamma\backslash G)}{4\pi^2} \int_0^\lambda 2\, s\, \mathrm{th}(\tfrac{\pi s}{2})\, ds \approx \frac{d(\alpha)\, vol}{4\pi^2} \lambda^2.$$

Strictly speaking functions $\{f\}$ with the Heaviside-type "Fourier transform" are not allowed in the Selberg-trace formula, they can not be C_0^∞, and the corresponding operators $\{T_f^s\}$ do not belong to the trace-class. However, any such $\{f\}$ could be approximated by a family of nice (regular) functions $\{f_\epsilon\}$, which suffices for the proof of (5.20) (see [GGP]).

Formula (5.20) means, in particular, that the number of complementary series constituents of R^α is always finite, all of them being located on interval $[0;1]$, whereas no accumulation of $\{s_j +\}$ and $\{s_k\}$ is allowed at any finite point λ_0 of $[0;\infty)$.

We shall indicate the connection of representation R^α to Laplacians on Riemann surfaces $\Gamma\backslash\mathbb{H}$. To simplify matters let us take trivial $\alpha = I$. The representation space $L^2(\Gamma\backslash G)$ contains a subspace $\mathcal{V}$ of K-invariants (made of all K-invariants of the even-principal and complementary components of R^α). Space $\mathcal{V}$ can be identified with $L^2(\Gamma\backslash\mathbb{H}) = L^2(\Gamma\backslash G/K)$. Furthermore, the Casimir (central) element

$$\Delta_G = \tfrac{1}{4}\{W^2 - 2(XY+YX)\} \in \mathfrak{Z}(\mathfrak{A})$$

turns into an invariant Laplacian on $\mathbb{H}$. Taking quotient $\mathcal{M} = \Gamma\backslash\mathbb{H}$ reduces $\Delta_{\mathbb{H}}$ to the natural Laplace-Beltrami operator on the Riemann surface $\mathcal{M}$ (group Γ plays the role of "periodic boundary conditions" for $\Delta_{\mathbb{H}}$, the same way as discrete lattice $\mathbb{Z}^n \subset \mathbb{R}^n$ yields the "periodic" torus Laplacian $\Delta_{\mathbb{T}^n}$ from the $\mathbb{R}^n$-Laplacian).

We observe that the Casimir becomes scalar on each *primary subspace* $\mathcal{L}_s \subset L^2(\Gamma\backslash G)$ ($\mathcal{L}_s \simeq \mathcal{H}_s \otimes N_s$ - a multiple of irreducible T^s, contained in R),

$$R_\Delta \mid \mathcal{L}_s = \frac{1-s^2}{4} I.$$

Furthermore, each primary (spectral) subspace $\mathcal{L}_s$ contains exactly N_s "K-invariants" (since each irreducible $\mathcal{H}_s$ has a unique K-invariant, described in §7.5). We denote by $\mathcal{V}_s$ a subspace of K-invariants in $\mathcal{L}_s$. Then the entire space $\mathcal{V}$ of K-invariants is decomposed into the direct sum of eigenspaces of $R_\Delta \mid \mathcal{V} \simeq \Delta_{\mathcal{M}}$,

$$\mathcal{V} = \bigoplus_s \mathcal{V}_s;$$

the eigenvalue: $\lambda_s = \frac{1-s^2}{4}$, on $\mathcal{V}_s$ has multiplicity $N_s = \dim \mathcal{V}_s$. Here $\{s\}$ varies over the spectrum (principal; complementary; discrete) of the induced representation R.

Hence, solution to the spectral problem for R yields a complete solution of the eigenvalue problem for the Laplacians $\Delta_{\mathcal{M}}$ on Riemann surfaces $\Gamma\backslash\mathbb{H}$, and vice versa. Both problems, however, are equally difficult to solve exactly. In the next section we shall explore some aspects of asymptotic spectral analysis of $\Delta_{\mathcal{M}}$ and establish interesting links to geometry of $\mathcal{M}$, particularly the relation between eigenvalues of $\Delta_{\mathcal{M}}$ and the length of closed (periodic) geodesics on $\mathcal{M}$.

5.6. Geodesic flows on negatively curved Riemann surfaces. Finally, we shall give a simple application of spectral decomposition $L^2(\mathcal{M}) = \bigoplus_s \mathcal{V}_s$, to a geodesic flow on $\mathcal{M} = \Gamma\backslash\mathbb{H}$. The geodesic flow (Appendix C) is a 1-parameter family of transformations on the unit cotangent bundle

$$S^*(\mathcal{M}) = \{(x;\xi): x \in \mathcal{M}; \|\xi\| = 1\} = \Omega,$$

generated by the vector field $\Xi = \partial_x a \cdot \partial_\xi - \partial_\xi a \cdot \partial_x$, where $a(x;\xi) = \sqrt{g^{ij}(x)\xi_i\xi_j}$ - square root of the metric tensor form on covectors ξ.

Theorem 3: *Geodesic flow on any compact Riemann surface $\mathcal{M} = \Gamma\backslash G$ is ergodic, in the sense that any function $f \in L^2(\Omega)$, invariant under the flow, is constant.*

Indeed, vector field Ξ on $\mathcal{M}$ is the image of geodesic generator on the Poincare-Lobachevski plane (or disk) $\mathbb{D}$. But $S^*(\mathbb{D})$ is naturally identified with the group $G = SU(1;1)$ itself, and since field Ξ on $S^*(\mathbb{D})$ commutes with the G-action (by isometries) it must be a left-invariant field, given by a Lie algebra element $X \in su(1;1)$. To find X we just take a geodesics through $\{0\}$ in the direction of x-axis, and find the 1-parameter group of X to be,

$$\begin{bmatrix} cht & sht \\ sht & cht \end{bmatrix}$$

hence $X = \begin{bmatrix} & 1 \\ 1 & \end{bmatrix}$. Any (geodesically) invariant function f on Ω must be a null-vector of an operator $\{\pi_X^s\}$, for some s in the spectrum of $R^{\mathcal{M}}$. But we know all generators $\{\pi_X\}$ for (principal, discrete, complimentary series π), neither one has a nontrivial kernel for $s \neq 0$. Hence, kernel$(R_X) = 0$, and the flow $\exp(t\Xi)$ is ergodic.

Problems and Exercises:

1. Consider a unitary group representation T with sufficiently many compact operators $\{T_f\}$, (for a suitable class of test-functions $\{f\}$ in the group-algebra, and show that such T has always a discrete (direct sum) decomposition into irreducible components: $T \simeq \bigoplus_p T^p \otimes m(p)$, of finite multiplicity, $m(p) < \infty$. Steps:

   i) take a symmetric element $f = f^*$ on G, and the corresponding self-adjoint compact operator T_f; show that a non-zero eigensubspace $\mathcal{E}_\lambda$ $(\lambda \neq 0)$ of T_f intersects G-invariant subspaces $\{\mathcal{H}\}$ of T; choose minimal among such $\mathcal{H}$ (always exists, due to fin of $\mathcal{E}_\lambda$!); and show that $T \mid \mathcal{H}$ is irreducible.

   ii) derive the direct sum decomposition of T, and show that all multiplicities must be finite.

2. Derive formula (5.12) for the H-transform of an elliptic element $u = u_\theta$. Steps:

   i) Show: $\int_{-\pi}^{\pi} F^K(u_\theta)\phi(u_\theta)d\theta = \int_{G_{el}} f(g)\phi(g) \mid e^{it} - e^{-it}\mid^{-2} dg,$

   integration over all elliptic conjugacy classes in G, where $\{e^{\pm it}\}$ denote eigenvalues of matrix $g \in G_{el}$; ϕ - an arbitrary test-function, constant on conjugacy classes.

   ii) Apply (i) with $\phi = e^{-int}(e^{-it} - e^{it})$ to get
   $$\int_{-\pi}^{\pi} F^K(u_\theta)e^{-in\theta}(e^{-i\theta} - e^{i\theta})\, d\theta = \int_{G_{el}} f(g)\frac{e^{-int}}{e^{it} - e^{-it}}dg;$$
   and compare it with formulae (5.16) for characters and the ensuing transforms of f, $\{\hat{f}(s) = \langle f \mid \chi_s\rangle$: $s = \pm is$; s; or $\pm n\}$, to show
   $$\begin{aligned}\hat{f}(n+) &= \int_{G_{el}} f(g)\frac{e^{-int}}{e^{it} - e^{-it}}dg + \int_{G_{hyp}} f(g)\frac{\lambda^{-n}}{\lambda - \lambda^{-1}}dg;\\ \hat{f}(n-) &= -\int_{G_{el}} f(g)\frac{e^{int}}{e^{it} - e^{-it}}dg + \int_{G_{hyp}} f(g)\frac{\lambda^{-n}}{\lambda - \lambda^{-1}}dg;\end{aligned} \tag{5.21}$$
   where first integrals are taken over elliptic and second over hyperbolic conjugacy classes.

   iii) Deduce from (5.21) the formula for the elliptic Harish-Chandra transform of f
   $$F^K(u_\theta) = \frac{1}{2\pi(e^{-i\theta} - e^{i\theta})}\left\{\frac{\hat{f}(0+) - \hat{f}(0-)}{2} + \sum_1^\infty(\hat{f}(n+)e^{in\theta} - \hat{f}(n-)e^{-in\theta}) - \underbrace{- \sum_1^\infty(e^{in\theta} - e^{-in\theta})\int_{G_{hyp}} f(g)\frac{\lambda^{-n}}{\lambda - \lambda^{-1}}dg}_{F_h}\right\}. \tag{5.22}$$

   iv) It remains to express the last sum F_h in (5.22) through the continuous series characters and transforms $\{\hat{f}(is\pm)\}$. Use the Fourier expansion of the conjugate Poisson kernel in $\mathbb{D}$,
   $$\sum_1^\infty \lambda^{-n}(e^{in\theta} - e^{-in\theta}) = \frac{\lambda(e^{i\theta} - e^{-i\theta})}{1 - 2\lambda\cos\theta + \lambda^2};\ \text{to get}$$
   $$F_h(\theta) = \int_{G_{hyp}} f(g)\frac{\lambda}{(\lambda - \lambda^{-1})(1 - 2\lambda\cos\theta + \lambda^2)}dg.$$

   v) Use the formulae for the hyperbolic H-transform (cf §7.4),
   $$F_h(\theta) = \frac{1}{8\pi^2}\int_1^\infty\int_{-\infty}^\infty(\hat{f}(is+) + \hat{f}(is-))\frac{\lambda^{is}}{1 - 2\lambda\cos\theta + \lambda^2}ds\,d\lambda +$$
   $$+ \frac{1}{8\pi^2}\int_1^\infty\int_{-\infty}^\infty(\hat{f}(is+) - \hat{f}(is-))\frac{\lambda^{is}}{1 + 2\lambda\cos\theta + \lambda^2}ds\,d\lambda.$$

vi) Formula (v) simplifies, using the relations $\widehat{f}(-is\pm)=\widehat{f}(is\pm)$, and the well known Fourier integrals:

$$\int_0^\infty \frac{\lambda^{is}}{1+2\lambda\cos\theta+\lambda^2}d\lambda=\frac{\pi\sinh s\theta}{\sin\theta\sinh\pi s};\ -\pi<\theta<\pi$$

$$\int_0^\infty \frac{\lambda^{is}}{1-2\lambda\cos\theta+\lambda^2}d\lambda=\frac{\pi\sinh s(\theta-\pi)}{\sin\theta\sinh\pi s};\ 0<\theta<\pi;$$

$$\int_0^\infty \frac{\lambda^{is}}{1-2\lambda\cos\theta+\lambda^2}d\lambda=-\frac{\pi\sinh s(\theta+\pi)}{\sin\theta\sinh\pi s};\ -\pi<\theta<0.$$

vii) Complete the derivation.

3. Take a discrete-series representation T^n, pick its highest/lowest weight-vector ψ_0 ($T^n_u\psi_0=e^{in\theta}$; for $u=u_\theta\in K$), and show
i) matrix entry $f_0(g)=\langle T^n_g\psi_0\mid\psi_0\rangle$ is in $L^1(G)$ (use Iwasawa decomposition);

ii) show that operators: $T^s_{f_0}=0$, for all principal and complementary-series representations $\{T^s\}$ (Pair f_0 to a spherical function $\phi_0(g)=\langle T^s\eta_0\mid\eta_0\rangle$ ($T^s_u\eta_0=\eta_0$; for $u=u_\theta\in K$); check that $\langle f_0\mid\phi_0\rangle=0$, by comparing K-actions on $f_0;\phi_0$; then use irreducibility of T^s!).

§7.6. Laplacians on hyperbolic surfaces $\mathbb{H}/\Gamma$.

Section §7.7 develops the Spectral theory of hyperbolic Laplacians $\Delta_{\mathcal{M}}$ on Riemann surfaces $\mathcal{M} = \mathbb{H}/\Gamma$, from a different prospective. It does not directly involve the representation theory of the preceding sections (§7.1-7.6), but exploits the heat-kernel method, following McKean's approach [Mc].

The hyperbolic space can be realized either as Poincare-Lobachevski half-plane $\mathbb{H}$, or disk $\mathbb{D}$. The distance between two points $x = x_1 + ix_2$ and $y = y_1 + iy_2$ in the half plane $\mathbb{H}$ is given by

$$d(x;y) = ch^{-1}(1 + \frac{|x-y|^2}{2x_2y_2}), \tag{6.1}$$

while in the disk realization (problem 4) it corresponds to

$$d(z;w) = \frac{1}{2}log\left(\frac{|1-z\overline{w}| + |z-w|}{|1-z\overline{w}| - |z-w|}\right).$$

The geodesics in $\mathbb{D}$ are all circles perpendicular to the boundary, while in $\mathbb{H}$ they also include all lines parallel to the y-axis (see fig.3 and Appndix C).

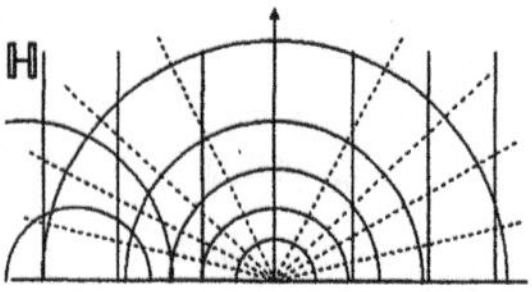

D

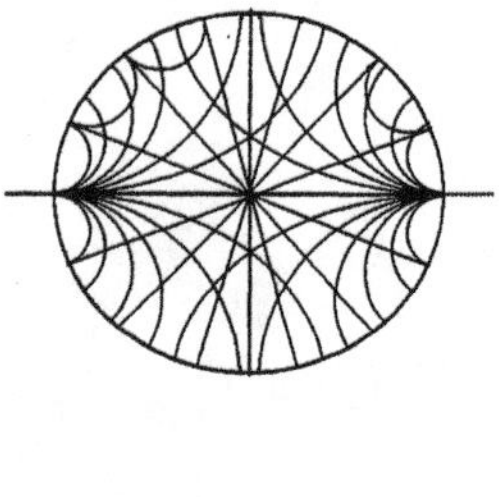

Fig.3: *illustrates geodesics in the hyperbolic geometry of the half-plane (left) and the disc (right). We have shown 2 families of geodesics. One consists of concentric circles, centered at {0} in $\mathbb{H}$, which are Möbius transformed into the family of vertical circles, centered on the horizontal axis of $\mathbb{D}$ (the left and right poles in $\mathbb{D}$ correspond to points {0} and {∞} in $\mathbb{H}$). The dashed circles connecting two poles of $\mathbb{D}$ represent radial (dashed) rays in $\mathbb{H}$. The second family of geodesics is made of parallel lines in $\mathbb{H}$, that get transformed into a family of circles converging to the right pole {∞} in $\mathbb{D}$. The hyperbolic geometry clearly violates the Euclid parallel lines axiom.*

6.1. Fundamental regions in $\mathbb{H}$. A discrete subgroup Γ of SL_2 divides space $\mathbb{H}$ into a union of fundamental regions: $\bigcup_{\gamma\in\Gamma} \mathcal{M}_\gamma$, non-overlapping geodesic polygons $\{\mathcal{M}_\gamma = \gamma(\mathcal{M}):\ \gamma \in \Gamma\}$, each bounded by a finite number of geodesic arcs $\{C_j\}$. Such tessellations of $\mathbb{H}$ are similar to partition of the Euclidian plane (or $\mathbb{R}^n$-space) into the union of fundamental rectangles by a discrete lattice $\Gamma \subset \mathbb{R}^2$. Each fundamental region $B \subset \mathbb{R}^2$ is then mapped onto the quotient space, $B \to \mathbb{R}^2/\Gamma \simeq \mathbb{T}^2$ (2-torus).

Throughout this section we shall assume that group Γ has no fixed points inside

$\mathbb{H}$, i.e. no $\gamma \in \Gamma$ fixes a point z_0 of $\Im z_0 > 0$. This means Γ has no elliptic elements (fixed points in $\mathbb{H}$ turn into "sharp vertices" of the quotient-space, where $\mathcal{M}$ loses its smooth structure, a situation reminiscent to quotients of sphere S^2, modulo any of discrete "Platonic" symmetries, chapter 1). Compactness of $\mathbb{H}/\Gamma$ (hence, of G/Γ) also implies the absence of parabolic elements. Thus Γ is made entirely of hyperbolic elements.

To construct Γ-fundamental polygons in $\mathbb{H}$ we pick any point $z_0 \in \mathbb{H}$, $\gamma \in \Gamma$, and consider a subset

$$\mathcal{M} = \{z \in \mathbb{H}\colon \mathrm{d}(z;z_0) \leq \mathrm{d}(z;\gamma_j(z_0));\ j = 1;2;...\}.$$

Such $\mathcal{M}$ forms a geodesic $4N$-gone in $\mathbb{H}$, the boundary arcs $\{C_j\}$ being transformed one into the other by elements $\{\gamma\}$. Boundary arcs $\{C_j\}$ are naturally divided into opposite pairs, and under proper identification we get a smooth (analytic) Riemann surface $\mathcal{M} = \mathbb{H}/\Gamma$ of genus N. Figures 5 below illustrates a deformation of octagon ($N = 2$), into a surface of genus 2.

Group Γ then turns into the *fundamental group*[12] of $\mathcal{M}$, in other words each element $\gamma \in \Gamma$ corresponds to a class of *equivalent* closed loops $\{\omega \sim \gamma\}$. Furthermore, any fundamental class $\{\omega\}$ of $\Gamma \simeq \pi_1(\mathcal{M})$, contains a *minimal length path* γ_0, which coincides with the closed geodesics of the given homotopy type. We shall illustrate the foregoing with a few examples of fundamental regions.

Examples of fundamental regions $\mathbb{H}/\Gamma$:

1. The modular group $\Gamma = SL_2(\mathbb{Z})$ is generated by 2 elements: translation by 1 in the horizontal direction $h: z \to z+1$, and reflection $\sigma: z \to -1/z$, given by matrices

$$h = \begin{bmatrix} 1 & 1 \\ & 1 \end{bmatrix}; \sigma = \begin{bmatrix} & 1 \\ -1 & \end{bmatrix}.$$

Two types of fundamental regions of Γ are shown on fig.4:

[12]Two path γ_1; γ_2 on a manifold $\mathcal{M}$ are called *(homotopy) equivalent,* if γ_1 can be continuously deformed into γ_2 inside $\mathcal{M}$. Any pair $\{\gamma_1;\gamma_2\}$ of closed path (loops) passing through a fixed point $x_0 \in \mathcal{M}$ can be formally *multiplied* by combining them into a single path $\gamma_1 \circ \gamma_2$ ("γ_1" followed by "γ_2"). Such multiplication is easily verified to respect the (homotopy) equivalence on space $\Omega(x_0)$ of closed path, through $\{x_0\}$.

So the set of equivalence classes of $\{\gamma \in \Omega(x_0)\}$ acquires the group structure: the identity consisting of all path contractible to $\{x_0\}$, while the inverse $\{\gamma^{-1}\}$, being given by γ, traversed in the opposite direction). The resulting group is called the *fundamental (or 1-st homotopy) group* of $\mathcal{M}$, and denoted by $\pi_1(\mathcal{M})$. It carries some important topological information about $\mathcal{M}$. For instance, the fundamental group $\Gamma = \pi_1(\mathcal{M})$ of a 2-D surface $\mathcal{M}$, when quotient modulo its commutator $\Gamma' = [\Gamma;\Gamma]$, yields the so-called homology group $H_1(\mathcal{M}) = \Gamma/\Gamma'$. The latter forms a commutative group, isomorphic to $\mathbb{Z}^N \times G$ (direct sum of the "free and torsion components": $\mathbb{Z}^N$and G), and dimension N of the free component is precisely the genus of $\mathcal{M}$!

(I) vertical strips (light shading) are shifts of the basic strip $\{-\frac{1}{2} \leq \Re z \leq \frac{1}{2}; |z| > 1\}$;

(II) Fundamental triangles (dark shading), if fact, quadrilateral, since point $i \in C_3$ should be considered a vertex, obtained by shifting the basic triangle $\{0;\ \pm\frac{1}{2}+i\frac{\sqrt{3}}{2}\}$. The latter is bounded by geodesic arcs: $C_{1,2}=\{|z \pm \frac{1}{2}|=1\}$; and $C_3 = \{|z| = 1\}$.

The first type regions are obtained by fixing point $\{2i\}$, shifting it by $h: 2i \to 2i \pm 1$, and inverting by $\sigma: 2i \to \frac{1}{2}i$, then writing the corresponding geodesic bisectors:

$$d(z;2i) = d(z;\ \pm 1+2i) \text{ (vertical lines); and } d(z;2i) = d(z;\tfrac{1}{2}i) \text{ (circle } |z|=1).$$

Obviously, the *fundamental triangle* is obtained by σ-inverting the *fundamental strip*. Let us remark that the fundamental region $\mathcal{M}$ of the modular group $SL_2(\mathbb{Z})$ is not compact, it has *cusp* $\{0\}$ at $\{\infty\}$ (i.e. at the "infinite boundary" of $\mathbf{H}$, $\partial\mathbf{H} = \mathbf{R}\bigcup\{\infty\}$). But $\mathcal{M}$ has finite volume (problem 2). Topologically manifold $\mathcal{M}$ looks like a sphere with an (infinitely) long spike (Fig.5).

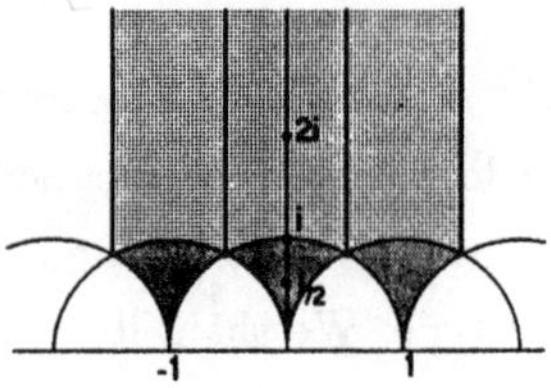

Fig.4: *Two types of fundamental regions of the modular group* $\Gamma = SL_2(\mathbb{Z})$.

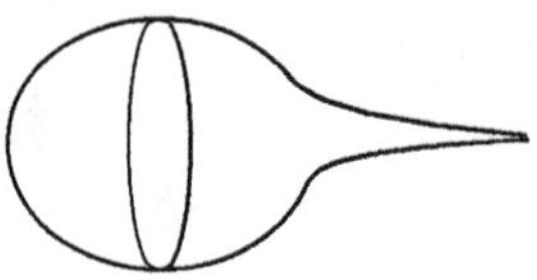

Fig.5. *Riemann surface* $\mathbf{H}/SL_2(\mathbb{Z})$ with a cusp (spike) at $\{\infty\}$.

2. Our next example (Fig.6) demonstrates a typical fundamental polygon in $\mathbf{H}$, whose vertices are (finite order) fixed points of certain elliptic elements $\{\gamma_j \in \Gamma\}$.

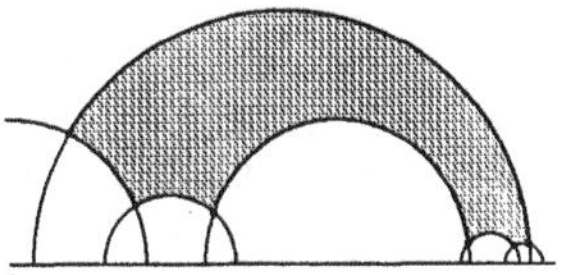

Fig.6: *A typical fundamental polygon in* $\mathbb{H}$.

Topologically quotients $\mathcal{M} = \mathbb{D}/\Gamma$, drawn in fig.6, turn into Riemann surfaces of higher genus g. Fig. 7 below demonstrates a transformation of an octagon with properly identified sides into a genus 2 Riemann surface. Let us notice that images of the fundamental region $\mathcal{M}$ under all elements $\{\gamma \in \Gamma\}$ form a regular tessellation of $\mathbb{D}$ into $4g$-gons, an impossible fit in the Euclidian flat-land, for any $g \geq 1$. A remarkable feature of the hyperbolic plane is the existence of regular tessellations of any order, hence of higher genus hyperbolic surfaces $\mathcal{M}/\Gamma$ of constant negative curvature!

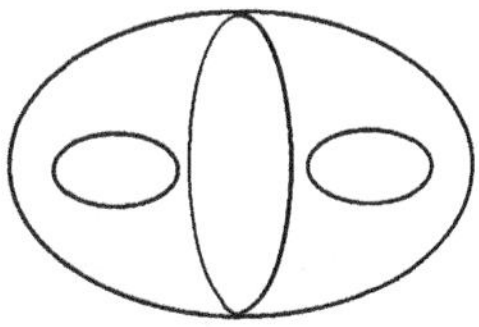

Fig.7. (i) *A Riemann surface of genus 2 (A) is cut into 2 truncated tori (B);*

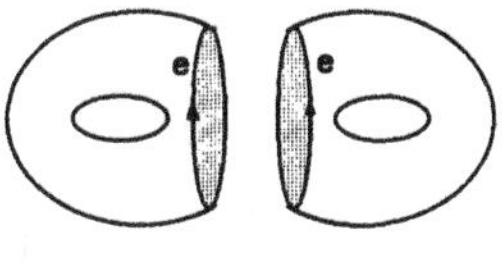

(ii) *The tori are then unwrapped into squares (with holes in the middle, which correspond to the original cut) via 2 standard fundamental cuts;*

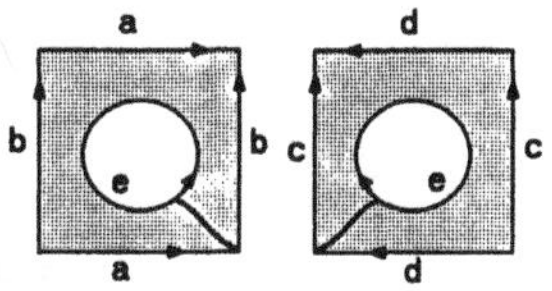

(iii) *The inner circle is stretched through a vertex to turn each square into a pentagon (D),*

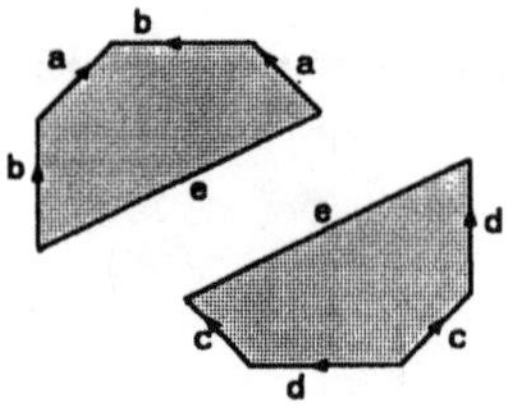

(iv) *Finally two pentagons are glued together along the diagonal (the original cut) to form an octagon (E). finally two pentagons are glued together along the diagonal (the original cut) to form an octagon (E). The fundamental group Γ of genus-2 surface has 4 generators* $\{a;b;c;d\}$, *and a single relation:*

$$aba^{-1}b^{-1}cdc^{-1}d^{-1}=e,$$

which corresponds to traversing the octagon clockwise.

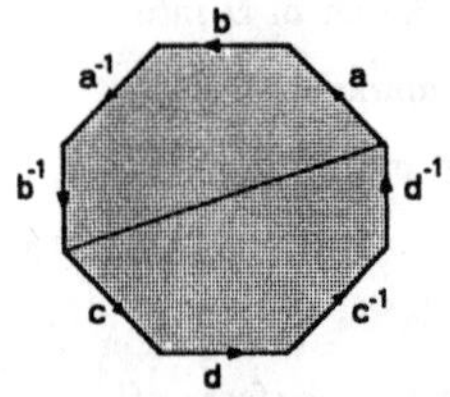

(v) *The resulting fundamental group Γ of genus-2 surface has 4 generators* $\{a;b;c;d\}$, *and a single relation:* $aba^{-1}b^{-1}cdc^{-1}d^{-1}=e$, *which corresponds to traversing the octagon clockwise.*

6.2. Kernels, traces and heat-invariants. We are interested in eigenvalues $\{\lambda_k\}$ of the Laplacian Δ on hyperbolic surfaces $\mathcal{M}=\mathbb{H}/\Gamma$. These are usually impossible to compute exactly (in any closed form), so one would like to get some approximate or asymptotic expansions of $\{\lambda_k\}$. The latter often involves the study of certain "means" of $\{\lambda\text{'s}\}$, (cf. chapter 2), and the related *transforms* (Laplace, Fourier, Stieltjes, etc.) of the counting (spectral) function $N(\lambda)=\#\{k:\lambda_k\le\lambda\}$. We shall mention a few of them

Heat-kernel (Laplace): $\Theta(t)=\sum e^{-t\lambda_k}=\int_0^\infty e^{-t\lambda}dN(\lambda)=\operatorname{tr}(e^{t\Delta})$;

Zeta-function: $Z(s)=\sum\lambda_k^{-s}=\int_0^\infty\lambda^{-s}dN(\lambda)=\operatorname{tr}(\Delta^{-s})$;

Resolvent (Cauchy-Stieltjes): $R(\zeta)=\sum\frac{1}{\zeta-\lambda_k}=\int_0^\infty\frac{dN(\lambda)}{\zeta-\lambda}=\operatorname{tr}(\zeta-\Delta)^{-1}$; or

$$\sum\frac{1}{(\zeta-\lambda_k)^m}=\int_0^\infty\frac{dN(\lambda)}{(\zeta-\lambda)^m}=\operatorname{tr}(\zeta-\Delta)^{-m};$$

Wave-trace (Fourier): $\chi(t)=\sum e^{it\sqrt{\lambda_k}}=\int e^{it\sqrt{\lambda}}dN(\lambda)=\operatorname{tr}(e^{it\sqrt{\Delta}})$.

The first two are called *theta* and *zeta*-functions of Δ, by analogy with the classical (Jacobi, Riemann) theta and zeta functions, which correspond to an "integral spectrum" $\{\lambda_k=k\}$. All transforms represent traces (regular or generalized) of various Green-functions of Δ: theta gives the trace of the *heat-kernel*, $\operatorname{tr}(e^{-t\Delta})$, (the fundamental solution of the heat problem: $u_t = \Delta u$), Cauchy-Stieltjes transforms R corresponds to *resolvent* of Δ (and its powers), while distribution[13] $\chi(t)$ on $\mathbb{R}$ yields the *wave-trace*, $\mathrm{tr}(e^{it\sqrt{\Delta}})$ (Green's function of the wave equation: $u_{tt} - \Delta u = 0$) (see §2.4). All transforms are related one to the other, for instance, *zeta* is obtained from *theta* via Γ-integral,

$$Z(s) = \frac{1}{\Gamma(s)}\int_0^\infty \Theta(t)t^{s-1}dt,$$

a consequence of the obvious relation

$$\lambda^{-s} = \frac{1}{\Gamma(s)}\int_0^\infty e^{-t\lambda}t^{s-1}dt.$$

This often allows to link large-k asymptotics of spectrum $\{\lambda_k\}$, to asymptotics of zeta, theta, etc. functions, via the so called *Abelian/Tauberian Theorems.* If for instance, sequence $\lambda_k \sim k^p$, as $k \to \infty$, for some $p > 0$, then its theta-function

$$\Theta(t) = \sum e^{-t\lambda_k} \sim \int^{+\infty} e^{-t\lambda}\lambda^p d\lambda,$$

is easily seen to obey

$$\Theta(t) \sim \Gamma(p+1)t^{-p-1}; \text{ at small } t,$$

and similar relations could be derived for other "kernels of Δ".

The converse, however, is not true, typically means, like Θ, Z, etc., behave much better, than irregularly distributed sequence of eigenvalues. For instance, the heat-trace admits an asymptotic expansion, due to Munakshisundaram-Plejel,

$$\mathrm{tr}\, e^{t\Delta} \sim t^{-n/2}\{b_0 + b_1 t + b_2 t^2 + \ldots\}, \tag{6.2}$$

whose coefficients $\{b_k\}$, called *heat-invariants,* carry some important geometric-topological information about manifold m. For instance, b_0 is proportional to $vol(m)$, with Const, depending on n; $b_1 = \frac{1}{6}\int Kdx$ - integral of scalar curvature[14], $K(x)$, over the natural (Riemannian) volume element on m. Higher invariants involve certain universal polynomial expressions in curvature and its covariant derivatives, integrated over m. All of them could be computed in principle, but the formulae quickly become unmanageable [Gil]. So to get a better insight into the structure of spec(Δ) one need to

[13]It is no difficult to show that heat-semigroup $e^{-t\Delta}$, and resolvent $(\zeta - \Delta)^{-1}$, of Laplacians on compact Riemannian manifolds m, are given by nice integral kernels: $K_t(x;y)$ and $R_\zeta(x;y)$ (cf. chapter 2). Hence both are compact, and belong to the trace-class. But the wave-kernel (made of unitary operators $\{W(t) = exp(it\sqrt{\Delta})\}$) could be "traced" only in a generalized sense, as a distribution on R (see chapters 1-2).

[14]Scalar (Gauss) curvature $K(x_0)$ at a point x_0 on a surface e measures the deviation of surface area of the geodesic sphere: $S_\epsilon = \{d(x;x_0) < 0\}$, from the Euclidian (flat) area: $|S_\epsilon| = \pi\epsilon^2 - K\epsilon^4 + \ldots$. For an n-D manifold we take all geodesic 2-planes $\{e_\alpha\}$ through x_0 (images under the *exp*-map of all planes in the tangent space T_{x_0}), and average the Gauss curvature of e_α at x_0 over all $\{e_\alpha\}$. The formal definition of K involves the trace of the Ricci tensor.

look for a different set of "geometric spectral data". One such class consists of the length of all closed path (geodesics) in $\mathcal{M}$.

In this section we shall compute the heat-trace (theta-function) for Laplacians on $\mathcal{M}$, and will link it to closed geodesics in $\mathcal{M}$. A simple prototype of the main result and the basic approach below is the flat (torus) case $\mathbb{T} \simeq \mathbb{R}^n/\Gamma$, discussed in chapter 2.

Any lattice (discrete subgroup) $\Gamma \subset \mathbb{R}^n$ can identified with the image of the standard (unit-cell) lattice $\mathbb{Z}^n$ under a linear map $A: \mathbb{R}^n \to \mathbb{R}^n$. So the Laplacian on $\mathbb{R}^n/\Gamma$, pulled back to the standard n-cell $\mathbb{T}_0 = [0;1] \times \ldots \times [0;1]$, turns into a 2-nd order constant-coefficient elliptic operator: $L = B\partial \cdot \partial = \sum b_{ij}\partial^2_{ij}$, where matrix $B = ({}^tAA)^{-1}$.

The eigenvalues of L in $\mathbb{T}_0$ (respectively, Δ in $\mathbb{T}$) are readily computed

$$\lambda_k = 4\pi^2 Bk \cdot k;\ k = (k_1;\ldots;k_n) \in \mathbb{Z}^n.$$

while closed path in $\mathbb{T}$, labeled by n-tuples $m = (m_1;\ldots m_n)$, have length

$$\ell_m = |Am|^2 = B^{-1}m \cdot m.$$

To get the heat-kernel L on $\mathbb{T}_0$, we take the free-space Gaussian, e^{-tL},

$$G_t(z) = \sqrt{\frac{detB}{(2\pi t)^n}}\, exp\{-\tfrac{1}{2t}Bz \cdot z\};\ (z = x - y),$$

averaged it over the lattice Γ,

$$G(z) \to \sum_{m \in \Gamma} G(z+m),$$

then applies the Poisson summation formula. This yields

$$\sum_k e^{-t(4\pi^2 Bk \cdot k)} = \frac{\sqrt{detB}}{(4\pi t)^{n/2}} \sum_m e^{-\frac{1}{4t}B^{-1}m \cdot m}, \tag{6.3}$$

whose LHS gives the theta-function, $\sum e^{-t\lambda_k}$, while the RHS involves exponentials of length of all closed path,

$$t^{-n/2} \sum e^{-\frac{1}{4t}\ell_m^2}.$$

In the hyperbolic setup the role of lattice Γ is played by a discrete subgroup of SL_2, and "tori" become fundamental regions $\mathbb{H}/\Gamma$. We shall take as above the free space heat-kernel K_t on $\mathbb{H}$, which depends only on the distance $d(x;y)$, between $\{x\}$ and $\{y\}$[15], $K = K(d(x;y))$, average it via the Γ-action, and analyze all terms of the resulting

[15] The latter follows from the symmetry properties of the Laplacian Δ on $\mathbf{H}$: Laplacian is invariant under the SL_2-action, hence any "function of Δ", e.g. $K_t = e^{-t\Delta}$, is also G-invariant. But a G-invariant integral kernel $F(x;y)$ must depend on the distance $d(x;y)$ only, due to a double-transitive action of SL_2 on $\mathbf{H}$: any pair $\{x;y\}$ is taken to any other equidistant pair $\{x';y'\}$ by an element $g \in SL_2$ (problem 3). This is a general feature of so called rank-one symmetric spaces (see J5.7).

series.

The analysis could be carried out for any *radial* function $K(r)$, $r = d(x;y)$, decaying sufficiently fast at $\{\infty\}$, to assure convergence of the Γ-series below. Any such K defines an SL_2-invariant integral kernel $K(x;y) = K(d(x;y))$ on $\mathbb{H}$, which can then be reduced (quotient) to the manifold $\mathcal{M}$. The resulting kernel

$$K_{\mathcal{M}}(x;y) = \sum_{g \in \Gamma} K(x^g;y) \text{ on } \mathcal{M}, \tag{6.4}$$

belongs to the trace class.

The trace formula below gives $\mathrm{tr} K_{\mathcal{M}}$ in terms of *length of primitive inconjugate elements* $\{p \in \Gamma\}$. A hyperbolic element γ is called *primitive*, if γ is not an iterate of another element, i.e. $\gamma \neq \gamma_0^m$, for $m \geq 2$, and $\gamma_0 \in \Gamma$. We shall introduce the *length* of $q \in G$, as

$$\ell(q) = ch^{-1}(\tfrac{1}{2}\sqrt{\mathrm{tr}\, qq^* + 2}). \tag{6.5}$$

One can check (problem4) that ℓ does, indeed, represent the length of the smallest closed path (geodesics) in the fundamental class of $q \in \Gamma$. For hyperbolic $q \sim \begin{bmatrix} m & \\ & m^{-1} \end{bmatrix}$ one finds

$$\ell(q) = \int_1^{m^2} \frac{dy}{y} = log(m^2), \text{ hence } \ell(q^n) = 2\,log(m^n). \tag{6.6}$$

Now we can state the main result of the section.

6.3. Trace formula: *Given a radial function $K(r)$, $r = ch\, d(x;y)$, on $\mathbb{H}$, the reduced kernel $K_{\mathcal{M}}$ (6.4) on a compact quotient space $\mathcal{M} = \mathbb{H}/\Gamma$ belongs to the trace-class. Its trace, $trK_{\mathcal{M}} = \int_{\mathcal{M}} K_{\mathcal{M}}(x;x)dx$, can be expanded in a double series, over integers $n = 1;2;...$ and over all inconjugate, primitive elements $\{p \in \Gamma\}$,*

$$\boxed{\mathrm{tr} K_{\mathcal{M}} = |\mathcal{M}|\, K(1) + \sum_{n=1}^{\infty} \sum_{\substack{inconj \\ prim\{p\}}} \frac{\ell(p)}{\sqrt{2}\, sh\frac{1}{2}\ell(p^n)} \int_{ch\,\ell(p^n)}^{\infty} \frac{K(u)\, du}{\sqrt{u - ch\,\ell(p^n)}}} \tag{6.7}$$

Here p^n denotes the n-th iterate of the primitive path p, and $\ell(...)$ - the length of the path.

The proof involves several steps, some of them similar to the derivation of the trace formulae in the preceding section. Namely,

1) Group Γ is split into the union of conjugacy classes of powers of inconjugate primitive elements:

$$\Gamma = \{e\} \bigcup \{\gamma^{-1} p^n \gamma \colon \gamma \in \Gamma \setminus \Gamma_p\},$$

where Γ_p denotes the centralizer (commutator) of element p in Γ.

2) We observe that integrals $\int_{\mathcal{M}} K(x^g;x)\,dx$, over a fundamental region $\mathcal{M}$, are equal, for all conjugate elements $\{g=\gamma^{-1}p^n\gamma:\gamma\in\Gamma\}$. So

$$\mathrm{tr}K_{\mathcal{M}}=\int_{\mathcal{M}} K(x;x)+\sum_n\sum_p[\Gamma:\Gamma_p]\int_{\mathcal{M}} K(x^{p^n};x),$$

where $[\Gamma:\Gamma_p]$ denotes the index of subgroup $\Gamma_p\subset\Gamma$. The first integral corresponds to the trivial class $\{e\}$, while the remaining terms give nontrivial classes $\{p^n\}$. Notice that index $[\Gamma:\Gamma_p]=\#\{\text{cosets in } \Gamma\ \Gamma_p\}$ is always finite .

3) Let also remark that

$$[\Gamma:\Gamma_p]\int_{\mathcal{M}} K(x^{p^n};x)=\sum_{g\in class\{p^n\}}\int_{\mathcal{M}} K(x^g;x)dx,$$

is identified with an integral of K over yet larger fundamental region $\mathcal{M}_p\simeq\mathbb{H}/\Gamma_p$, of subgroup $\Gamma_p\subset\Gamma$, $\mathcal{M}_p$ is made of $[\Gamma:\Gamma_p]$ non overlapping copies of the original $\mathcal{M}$!

4) Next we note that $\Gamma_p=\{p^k:k\in\mathbb{Z}\}$ coincides with the subgroup generated by $\{p\}$. A hyperbolic primitive element $p\in\Gamma$ is conjugate equivalent to a diagonal matrix $q=\begin{bmatrix}m&\\&m^{-1}\end{bmatrix}$. Such q acts on $\mathbb{H}$ by dilations with m^2, so a fundamental region of q can chosen to consist of two strips $\{1\le|\Im x|\le m^2\}$ (fig.8). By the distance formula (6.1) the argument of K becomes

$$\mathrm{ch}d(x^{q^n};x)=1+\frac{(1-m^{2n})^2\,|x|^2}{2m^{2n}{x_2}^2}.$$

Then integration over the fundamental region $\mathcal{M}_q$ yields,

$$\int_{\mathcal{M}_q}K=\int_1^{m^2}\frac{dx_2}{{x_2}^2}\int_{-\infty}^{\infty}dx_1K\Big(1+\tfrac{1}{2}(m^n-m^{-n})^2(1+\Downarrow\tfrac{x_1}{x_2})^2)\Big).$$

Changing variables: $x\to\Downarrow x_2;t=\frac{x_1}{x_2})$, we get

$$log\,m^2\int_{-\infty}^{\infty}K\Big(1+\tfrac{1}{2}(m^n-m^{-n})^2(1+t^2)\Big)dt. \tag{6.8}$$

Remembering that $\ell(q)=log\,m^2$, $\ell(q^n)=log\,m^{2n}$, the argument of (6.8) turns into

$$u=+2sh^2\frac{n\ell(q)}{2}\bigcap 1+t^2).$$

We call the new variable u, and make yet another change, $t\to u$, to brings (6.8) to the form

$$\frac{\ell}{2\,sh\frac{n}{2}\ell}\int_{ch(n\ell)}^{\infty}\frac{K(u)\,du}{\sqrt{u-ch(n\ell)}};\ell=\ell(p)$$

which completes the proof of (6.7).

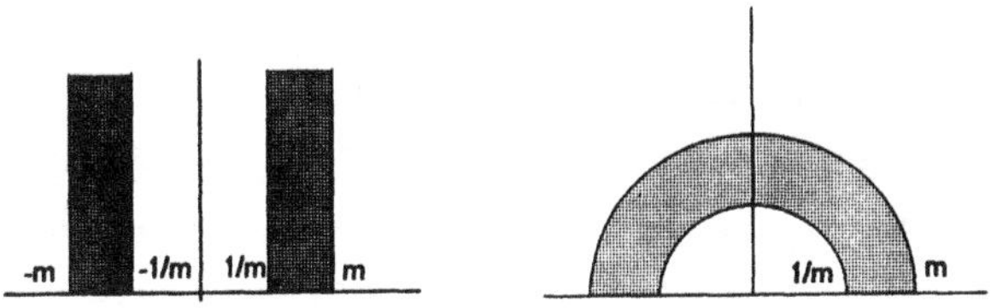

Fig.8. *Fundamental region of a hyperbolic element h can be chosen either as a strip* $\{\frac{1}{m} \le |\Re z| \le m\}$ *(A), or annulus* $\{\frac{1}{m} \le |z| \le m\}$ *(B).*

The heat-trace. Now it remains to specify (6.8) to the heat-kernel $K_t(x;y)$ on $\mathcal{M}$. But first let us find the heat-kernel on $\mathbb{H}$. We have already observed that K depends only on the hyperbolic distance $r = d(x;y)$ between points $x,y \in \mathbb{H}$. There is a well-known formula for K_t in terms of $ch\, r$ (see problem 5),

$$K_t(ch\, r) = \frac{\sqrt{2}\, e^{-t/4}}{(4\pi t)^{3/2}} \int_r^\infty \frac{\tau e^{-\tau^2/4t}}{\sqrt{ch\, \tau - ch r}} d\tau. \tag{6.9}$$

Next we shall compute the contributions of various conjugacy classes in (6.7),

$$K(1) = \frac{\sqrt{2}\, e^{-t/4}}{(4\pi t)^{3/2}} \int_0^\infty \frac{\tau e^{-\tau^2/4t}}{\sqrt{ch\, \tau - 1}} d\tau,$$

while the p^n-term contributes

$$\int_{ch\,\ell}^\infty \frac{K(\tau)\, d\tau}{\sqrt{\tau - ch\,\ell}} = \frac{\sqrt{2}\, e^{-t/4}}{(4\pi t)^{3/2}} \int_{ch\,\ell}^\infty d\tau \int_{ch^{-1}(\tau)}^\infty \frac{\lambda e^{-\lambda^2/4t}}{\sqrt{(\tau - ch\,\ell)(ch\,\lambda - \tau)}} d\lambda =$$

$$= \frac{\sqrt{2}\, e^{-t/4}}{(4\pi t)^{3/2}} \int_\ell^\infty \lambda e^{-\lambda^2/4t} d\lambda \int_{ch\,\ell}^{ch\,\lambda} \frac{d\tau}{\sqrt{(\tau - ch\,\ell)(ch\,\lambda - \tau)}}.$$

The inner integral gives the standard Beta-factor $B(\frac{1}{2};\frac{1}{2}) = \pi$, while the outer integration results in

$$\frac{\sqrt{2}\, e^{-t/4}}{(4\pi t)^{1/2}} e^{-\ell^2/4t}.$$

Substitution in (6.9) yields the following version of the *Selberg trace formula* for the heat-kernel $\Theta(t) = tr\left(e^{t\Delta}\right) = \sum e^{-t\lambda_k}$, of the hyperbolic Laplacian $\Delta = \Delta_{\mathcal{M}}$

$$\boxed{\Theta(t) = \frac{e^{-t/4}}{(4\pi t)^{3/2}} \left\{ C(t)\, |\mathcal{M}| + \pi t \sum_{n=1}^\infty \sum_{\substack{inconj \\ prim\, \{p\}}} \frac{\ell(p)}{sh \frac{n}{2} \ell(p)} e^{-n\ell(p)^2/4t} \right\}} \tag{6.10}$$

with coefficient

$$C(t) = \int_0^\infty \frac{\tau e^{-\tau^2/4t}}{\sqrt{ch\tau - 1}} d\tau.$$

Remarks: Formula (6.10) has many interesting interpretations and applications in Spectral theory of differential operators, and in quantum mechanics. It links directly the "Laplace-transformed" eigenvalues $\{\lambda_k\}$ of $\Delta_{\mathcal{M}}$ on the one hand, and the *length (period) spectrum* of manifold $\mathcal{M}$ $\{\ell(\gamma)$: over all closed geodesics $\gamma \subset \mathcal{M}\}$. In the picturesque language of Mark Kac [Ka]: *One can hear the length spectrum of* $\mathcal{M}$. This allows to address the Inverse spectral problem: *Can one hear (uniquely determine) the geometry (metric) of* $\mathcal{M}$ *from* $spec\Delta_{\mathcal{M}}$?

The length spectrum reduces the Inverse spectral question to a Problem in geometry/dynamics on manifolds: *given two metrics on* $\mathcal{M}$, *whose geodesic flows have identical period/length spectra can one conclude that metrics are isometrically equivalent?* In the classical case of hyperbolic surfaces $\mathcal{M} = \mathbf{H}/\Gamma$ of constant negative curvature, McKean [Mc] showed that there are at most finitely many isospectral surfaces $\{\mathcal{M}\}$. He utilized the Selberg trace formula (6.10), and the Klein-Fricke double and triple traces $\{\mathrm{tr}(\gamma_k\gamma_m);\ \mathrm{tr}(\gamma_i\gamma_k\gamma_m)$: $\gamma_k \in \Gamma\}$. The latter served to identify elements $\{\gamma_k\}$ up to Γ-conjugacy, hence the length of closed geodesics $\{\ell(\gamma)\}$, and their orientations.

The work of McKean [Mc] still left a possibility of a *unique solution* of the Inverse spectral Problem, until Vigneras [Vi] found in 1980 finite (but arbitrary large!) families of distinct isospectral surfaces of any genus. These examples, however turned out to be rather exceptional. Wolpert [Wo] showed that a generic hyperbolic surface is uniquely determined by spec(Δ)! For more general hyperbolic manifolds of non-constant curvature Guillemin and Kazhdan [GK] established "*infinitesimal rigidity*". They also used a reduction to a dynamical Problem.

In the study of dynamical systems it was found that *uniqueness* usually requires an additional "discrete set of data", for instance association of homotopy classes to the periods $\{L_m\}$, [KB]. The latter was previously used by McKean [Mc] in the constant curvature case. Under such hypotheses, the geometric Problem was shown to have a unique solution [KB]. Though the length spectrum brings one tantalizingly close to settling the inverse spectral question (at least on hyperbolic surfaces), there remains the frustrating "labeling problem", whose spectral content is unclear.

In the quantum-mechanical context Laplacian $\Delta_{\mathcal{M}}$ represents the *hamiltonian (energy-operator)*, which describes the evolution of a quantum system (particle), constraint to move on a surface/manifold $\mathcal{M}$. The underlying *classical mechanical system* coincides with the geodesic flow on the ***phase-space*** $T^*(\mathcal{M})$ - cotangent bundle of $\mathcal{M}$, given by the corresponding classical hamiltonian/energy function: $h(x;\xi) = \sum g_{ij}(x)\xi_i\xi_j$ - metric tensor, as a function of phase-variables: $x \in \mathcal{M}$, $\xi \in T_x{}^*$.

Formula (6.10) gives then a correspondence between *"eigenvalues of Δ"* (quantum energy levels), and the *"length spectrum"* (energies/actions) of classical trajectories (geodesics). Such relations exemplify the *Correspondence principle* of Quantum theory, which asserts close links between the classical and quantum systems. Usually, the correspondence holds only approximately/asymptotically at "large energies" (or for small Plank parameter), where the quantum dynamics becomes *quasiclassical.* The remarkable feature of the flat ($\mathbb{T}^n$) and hyperbolic Laplacians is that the quasiclassical expansions (6.3) and (6.10) turns out to be exact!

In the next chapter we shall explore in the greater detail the structure of classical hamiltonian systems, the quantization procedure and the role of group-symmetries in the classical and quantum mechanics.

Problems and Exercises:

1. Find the distance between two points: $x = x_1 + ix_2$; and $y = y_1 + iy_2$; in the Poincare half-plane $\mathbf{H}$ (6.1) (hint: it is easy to compute the distance between purely imaginary points $d(ia;ib) = \ln\frac{b}{a}$. Use it and find a fractional linear map $g:\mathbf{H}\to\mathbf{H}$, that takes $x\to i$; $y\to ib$, and compute b).
2. Find the hyperbolic volume of the fundamental region $\mathcal{M} = \mathbf{H}/\Gamma$; $\Gamma = SL_2(\mathbb{Z})$ (fig.4).
3. Show that any SL_2-invariant integral kernel $F(x;y)$ depends only on the distance $d(x;y)$. Hint: $F(x^g;y^g) = F(x;y)$, for all $x,y\in\mathbf{H}$, $g\in G$; move $x\to i$, and use the fact that stabilizer of $\{i\}, K = SO(2)$, acts transitively on all spheres $S_r(i) = \{y: d(i;y) = r\}$, centered at $\{i\}$, of radius r.
4. Show that the hyperbolic distance between points $\{z\}$ and $\{z^q\}$ $(q\in SL_2)$ is given by (6.6). Use the disk realization of $\mathbf{H}$, and show that
$d(0;z) = \frac{1}{2}\ln(\frac{1+|z|}{1-|z|})$; and $d(0;0^q) = \ln(|\alpha| + |\beta|)$, where $q = \begin{bmatrix}\frac{\alpha}{\beta} & \frac{\beta}{\alpha}\end{bmatrix}\in SU(1;1)$!
5. Derive the heat-kernel $K_t(ch\, r)$ (6.9) on the Poincare half-plane, in terms of hyperbolic radius $r = d(x;y)$. Steps: (i) the Laplacian in hyperbolic polar coordinates $(r;\theta)$ of $\mathbf{D}$ is given by
$$\Delta = \partial_r^2 + cth\, r\partial_r + sh^{-2} r\partial_\theta^2.$$
Changing $r\to x = ch\, r$, bring it into the Legendre-type differential operator,
$$L = \partial(x^2-1)\partial + \frac{1}{x^2-1}\partial_\theta^2, \text{ on the half-line } [1;\infty)\times\mathbf{T};$$
ii) Laplace-transform the heat-kernel
$$K(t;...)\to Q(r;\lambda) = \int_0^\infty e^{-\lambda t}K_t(r)dt,$$
and check the resulting function $Q(r;\lambda)$ to solve the equation:
$$(L+\lambda)[Q] = \delta(r), \tag{6.11}$$
i.e. $Q(r;\lambda)$ gives the resolvent-kernel of L. But for radial function $Q = Q(r)$ equation (6.11) reduces to the Legendre ODE:
$$[(x^2-1)Q']' + \lambda Q = 0,$$
with a suitable "source condition" at $x=0$ (cf. J2.3, chapter 2). Its solution is known to be a Legendre function of 2-nd kind and order ν, $Q = Q_\nu(x)$, where $\lambda = \nu^2 + \frac{1}{4}$.
iii) The Legendre function $Q_\nu(r)$ has an integral representation ([Erd];[Leb]),
$$Q_\nu(r) = \int_r^\infty \frac{e^{-(\nu+\frac{1}{2})z}}{\sqrt{2(ch\, z - ch\, r)}}dz.$$
iv) Use this representation of Q, and compare it with the Laplace transform of the integrand of (6.9),
$$\mathcal{L}[t^{-3/2}exp\{-t/4-\tau^2/4t\}].$$
The later yields a modified Bessel (Kelvin) function $K_{1/2}(r\sqrt{\lambda+1/4})$, via an integral representation of Kelvin functions,
$$\int_0^\infty exp\{-\tfrac{1}{2}(at+\tfrac{b}{t})\}t^{-s-1}dt = 2(\tfrac{a}{b})^{s/2}K_s(\sqrt{ab}).$$
But in our case order $s = \frac{1}{2}$, the corresponding Kelvin function becomes classical,
$$K_{1/2}(x) = K_{-1/2}(x) = \sqrt{\frac{2\pi}{z}}e^{-z}!$$
Complete the derivation!

§7.7. $SL_2(\mathbb{C})$ and Lorentz groups.

In the last section of ch.7 we shall briefly review the representation theory of the complex group $SL_2(\mathbb{C})$, the related Lorentz group of special relativity, and also higher-dimensional (pseudo-orthogonal) Lorentz groups $SO(1;n)$. Most results will be stated without proof.

7.1. The Lorentz group $SO(1;3)$ of special relativity preserves the Minkowski (1;3)-form in $\mathbb{M}^4$. Its 2-fold cover is the complex unimodular group $SL(2;\mathbb{C})$.

To show the correspondence $SL_2 \to SO(1;3)$ we note that the real 4-space $\mathbb{M}^4 = \{(x_0;x_1;x_2;x_3)\}$ can be identified with the space of complex hermitian matrices,

$$\mathfrak{P} = \left\{ X = \begin{bmatrix} x_0 - x_1 & w \\ \bar{w} & x_0 + x_1 \end{bmatrix}; w = x_2 + ix_3 \right\},$$

where the determinant-form $det(X) = \langle X \mid X \rangle = x_0^2 - \sum x_j^2$, defines the Minkowski product on $\mathfrak{P}$, and conjugations,

$$g: X \to g^* X g,$$

preserve hermitian symmetry, as well as the Minkowski norm, $det(g^*Xg) = det(X)$. So we get a map (isomorphism) of $SL_2/\{\pm I\}$ onto the Lorentz group $SO(1;3)$.

Irreducible representations of consist of 2 series: principal and complementary. **Principal series** representations are induced from the Borel subgroup B of all upper-triangular matrices,

$$B = \left\{ h = \begin{bmatrix} a & b \\ & \frac{1}{a} \end{bmatrix}; a \in \mathbb{C}^*; b \in \mathbb{C} \right\},$$

here $\mathbb{C}^*$ denotes the multiplicative group of complex numbers, $\mathbb{C}^* \simeq \mathbb{T} \times \mathbb{R}^*$. The characters of B are labeled by pairs $\{(s;m): s \in \mathbb{R}; m \in \mathbb{Z}\}$,

$$\pi^{s,m}(z) = |z|^{is}\left(\frac{z}{|z|}\right)^m. \tag{7.1}$$

The quotient-space $\mathfrak{S} = B \backslash G$ can be identified with the complex plane $\mathbb{C}$ (via the Gauss decomposition $g = n_- h n_+$), and G acts on $\mathbb{C}$ by fractional-linear transformations, as in §7.1,

$$g: z \to \frac{az+c}{bz+d}.$$

So principal series representations $\{T^{s,m}\}$ can be realized in space $L^2(\mathbb{C}; d^2z)$, by operators,

$$(T_g^{s,m} f)(z) = |bz+d|^{is-2}\left(\frac{bz+d}{|bz+d|}\right)^m f\left(\frac{az+c}{bz+d}\right) \tag{7.2}$$

Here d^2z denotes the standard area element $\frac{1}{2i} dz \wedge d\bar{z}$ in $\mathbb{C}$, factor $|bz+d|^{-2}$ serves to unitarize $T^{s,m}$, and the complex argument $\frac{bz+d}{|bz+d|}$ plays the role of "sgn$(bx+d)$" in the real case (§7.1). So representation $T^{s,m}$ is induced by the character

$\pi^{s,m}$ of B. An alternative realization of $T^{s,m}$ on the 2-sphere will be explained below in the context of the general Lorentz group $SO(1;n)$. We remark that the Riemann sphere S^2 gives a 1-point compactification of the complex plane, $S^2 = \mathbb{C} \bigcup \{\infty\}$.

Irreducibility of $T^{s,m}$ can be now established either by the Mackey's test (Theorem 1 of §7.1), or directly (see problem 2 of §7.1). The principal series representations have characters, given by

$$\chi_{s,m}(g) = \frac{|\lambda_g|^{is+m}\lambda_g^{-m} + |\lambda_g|^{-(is+m)}\lambda_g^{m}}{|\lambda_g - \lambda_g^{-1}|}; \tag{7.3}$$

where $\{\lambda_g;\lambda_g^{-1}\}$ denotes the eigenvalue of matrix $g \in SL_2(\mathbb{C})$. Representations $T^{s,m}$ and $T^{-s,-m}$ are equivalent, which can be verified either directly (by constructing intertwining operator W), or as a consequence of (7.3).

Complementary series correspond to the imaginary $s = i\sigma$, $-2 < \sigma < 2$, and $m = 0$ in (7.3). As in the real case §7.1 they can be realized in the Hilbert space, determined by the Riesz potential in $\mathbb{C}$,

$$\|f\|^2 = \iint |z-w|^{\sigma-2} f(z)\overline{f(w)}d^2z d^2w.$$

Their characters are analytically continued in s functions $\{\chi_{s,m}\}$ of (7.3) (for more details we refer to [GGV], chapter 2).

Group $SL_2(\mathbb{C})$ has no discrete series as will be explained below. The discrete series of semisimple Lie groups were directly linked by Harish-Chandra [Har 5] to the existence of compact the maximal abelian (Cartan) subalgebras, and $SL_2(\mathbb{C})$, as any other complex G, has none.

7.2. The Plancherel-inversion formula for $SL_2(\mathbb{C})$ has the standard formulation: for any smooth rapidly decaying function f on G,

$$\boxed{f(e) = \sum_m \int_{\mathbb{R}^+} \langle \chi_{s,m} | f \rangle (s^2+m^2) ds}$$

where $\{\chi_{s,m}\}$ denote the principal-series characters. So the Plancherel measure $d\mu$ is supported on principal-series part (a countable union of half-lines), and has density,

$$d\mu(s,m) = (s^2+m^2)ds.$$

Gelfand, Graev, Piatetski-Shapiro [GGP] derive a general Plancherel formula on $SL_2(\mathbb{F})$ over any locally compact field $\mathbb{F}$ (reals, complex, p-adics). Their result for complex $\mathbb{C}$ states,

$$\delta(g) = \int_{\widehat{G}_{princ}} \mathrm{tr}(T_g^\pi)\mu(\pi)d\pi,$$

where the Plancherel density,

$$\mu(\pi) = \mathrm{Const}\int_{\mathbb{C}} \frac{\pi(z)}{|1-z|^2} d^2z; \tag{7.4}$$

and $\{\pi = \pi^{s,m}\}$ varies over all characters (7.2) of B. Singular integral (7.4) is understood in the regularized sense, whence follows,

$$\mu(\pi) = \mathrm{Const}(s^2+m^2), \text{ with Const} = \frac{1}{32\pi^4}.$$

Selberg-trace formula for compact quotients $\Gamma\backslash G$ of $SL_2(\mathbb{C})$ looks similar to §7.5, but many simplifications arise due to the absence of the discrete series and the "elliptic conjugacy classes" (any g in $SL_2(\mathbb{C})$ is conjugate to a diagonal matrix!). Chapter I of [GGP] provides the complete details.

7.3. Lorentz groups $SO(1;n)$ preserves the Minkowski product, $x_0y_0 - \sum_1^n x_jy_j$ in $\mathbb{M}^{n+1}$. The preceding discussion of SL_2 has prepared the reader for the introduction to representations of more general semisimple groups. Here we shall briefly discuss one such class higher-dimensional Lorentz groups. The key to the analysis lies in two decompositions, Cartan and Iwasawa (see §§5.7 and 7.1). Group G contains a maximal compact subgroup $K = SO(n)$, and an abelian subgroup of hyperbolic rotations,

$$A = \left\{\begin{bmatrix} ch\,t & sh\,t \\ sh\,t & ch\,t \end{bmatrix}; t \in \mathbf{R}\right\},$$

generated by a (relativistic-boost) operator $H = \begin{bmatrix} & 1 \\ 1 & \end{bmatrix}$. The commutator of H in G is he product $A\cdot M$, where compact subgroup $M = SO(n-1) \subset K$ gives the commutator (centralizer) of A in K (M consists of orthogonal rotations in the span$\{x_2;...x_{n+1}\}$!). Element H (its adjoint map ad_H) splits Lie algebra $\mathfrak{G} = so(1;n)$ into the direct sum of eigenspaces:

$$\mathfrak{G}_{-1} \oplus \mathfrak{G}_0 \oplus \mathfrak{G}_1 \tag{7.5}$$

of eigenvalues $\{-1;0;1\}$. Here subalgebra $\mathfrak{G}_0 = \mathfrak{A} \oplus \mathfrak{M}$ - (Lie algebras of A and M), while subspaces $\mathfrak{G}_{\pm 1}$ are made of matrices of the type,

$$X = \begin{bmatrix} 0 & 0 & v_2 & \cdots v_{n+1} \\ 0 & 0 & u_2 & \cdots u_{n+1} \\ v_2 & -u_2 & 0 & \vdots \\ v_{n+1} & -u_{n+1} & \cdots & 0 \end{bmatrix};$$

where $(n-1)$-vectors $\boldsymbol{v} = (v_2;...v_{n+1})$; $\boldsymbol{u} = (u_2;...u_{n+1})$ satisfy the relations $\boldsymbol{v} = \boldsymbol{u}$ for $\mathfrak{G}_1$, $\boldsymbol{v} = -\boldsymbol{u}$ for $\mathfrak{G}_{-1}$. Indeed, element H acts by left multiplications on row-matrices $\left[\begin{smallmatrix} v \\ u \end{smallmatrix}\right]$.

Decomposition (7.5) clearly obeys the Lie-bracket relations,

$$[\mathfrak{G}_\mu;\mathfrak{G}_\lambda] \subset \mathfrak{G}_{\mu+\lambda};\ \text{for } \mu,\lambda = 0; \pm 1,$$

so $\mathfrak{G}_{\pm 1}$ form two commutative subalgebras $\mathfrak{N}\pm$ of $\mathfrak{G}$, and the sum $\mathfrak{A} \oplus \mathfrak{N}_+$ becomes a 2-step solvable subalgebra of $\mathfrak{G}$ (§1.4). Now the entire space $\mathfrak{G}$ can be split into the direct sum of subalgebras,

$$\mathfrak{G} = \mathfrak{K} \oplus \mathfrak{A} \oplus \mathfrak{N}.$$

With some additional effort (cf. [Hel]) one can show that group G can be similarly factored into the *Iwasawa product,*

$$G = KAN.$$

We define the *Borel subalgebra* of $\mathfrak{G}$, as $\mathfrak{B} = \mathfrak{M} \oplus \mathfrak{A} \oplus \mathfrak{N}$, using the fact that the centralizer $\mathfrak{M}$ of $\mathfrak{A}$ leaves subalgebra $\mathfrak{N}$ invariant. The corresponding *Borel subgroup* is also factored in the product,

$$B = MAN.$$

The quotient-space $B\backslash G$ can be identified either with the n-sphere $S^n = M\backslash K$, or with a "closure" of N, using a different (Gauss) factorization $G = B \cdot N$,

Principal series representations of $SO(1;n)$ are induced by irreducible representations of the Borel subgroup B, so $T = T^{s,m}$ are labeled by the pairs[16] $\{(s,m): s \in \mathbb{R}; m \in \widehat{M}\}$. Let $\mathcal{V} = \mathcal{V}_m$ denote the space of m, and $\pi = \pi^{s,m} = \chi_s \otimes \pi^m$ the corresponding (inducing) representation of B,

$$\pi_b^{s,m} = e^{isa}\pi_u^m;\ b = u \cdot a \cdot h,\ (u \in M,\ a \in A,\ h \in N).$$

Then $T^{s,m} = \mathrm{ind}(\pi^{s,m} \mid B; G)$ acts in the usual way on the product-space $L^2(S^n) \otimes \mathcal{V}_m$, and can be shown to be irreducible (problem 1). A different realization of $T^{s,m}$ could be given in spaces $L^2(N;...)$ using a Gauss-type decomposition $G = B \cdot N$. Let us compare these constructions with special cases of $SO(1;2)$ (or $SL_2(\mathbb{R})$) and $SO(1;3) = SL_2(\mathbb{C})/\mathbb{Z}_2$.

Examples.

- $SL_2(\mathbb{R})$. Here the abelian subgroup,

$$A = \left\{\begin{bmatrix} \lambda & \\ & \lambda^{-1}\end{bmatrix} : \lambda > 0\right\} \text{(positive diagonal)};\ K = \left\{\begin{bmatrix} \cos\phi & \sin\phi \\ -\sin\phi & \cos\phi \end{bmatrix}\right\} \simeq \mathbb{T};$$

[16] The dual object of the orthogonal group $SO(n-1)$ was described in J5.3. It is made of all ordered p-tuples $\{m_1 \geq m_2 \geq ... \geq \pm m_p;\ p = [\frac{n}{2}]\}$.

$$M = \{\pm I\};\ N = \left\{\begin{bmatrix} 1 & b \\ & 1 \end{bmatrix}\right\} \text{ - upper-triangular.}$$

Borel B made of all upper-triangular matrices. The principal series $\{T^{s,\pm}\}$ were realized in $L^2(\mathbb{R})$, where $\mathbb{R} \simeq B\backslash G$, or $L^2(\mathbb{T})$, another (projective) form of $B\backslash G$.

• $SL_2(\mathbb{C})$ has A - positive real diagonal matrices; $K = SU(2)$, or $SO(3)$ for the Lorentz $SO(1;3)$;

M - unitary diagonal $\left\{\begin{bmatrix} e^{i\theta} & \\ & e^{-i\theta} \end{bmatrix}\right\}$ (maximal torus in K);

N, B consist of upper-triangular matrices;

the quotient-space $B\backslash G = M\backslash K \simeq S^2$ (from the Iwasawa factorization), or $\mathbb{C} = \mathbb{R}^2$ (from the Gauss $G = B \cdot N$). In (7.2) we were able to write down explicitly operators $\{T_g^{s,m}\}$, using the fractional-linear (conformal) action of $SL_2(\mathbb{C})$ on the quotient $B\backslash G \simeq \mathbb{C}$.

It is interesting to note that the action of higher Lorentz groups on the homogeneous space $S^n \simeq B\backslash G$ is also conformal. The conformal action could be implemented by a stereographic map from the hyperboloid $\mathbb{H}^n = \{x_0^2 - |x|^2 = 1\} \simeq K\backslash G$ (where G acts isometrically), onto the sphere $S^n = \{x_0^2 + |x|^2 = 1\}$. So right translations of G on the quotient $B\backslash G$ are equivalent to its conformal action on S^n (problem 2).

As their low "SL_2-cousins" the higher Lorentz groups have complementary series. But the discrete series appear in only half of them $\{SO(1;2n)\}$ with even space-like part (like $SL_2(\mathbb{R})$), and are absent in the odd case $\{SO(1;2n+1)$, e.g. $SL_2(\mathbb{C})\}$. The Plancherel formula includes only the principal and discrete series (when present). The detailed analysis of $SO(1;n)$ is contained in the last chapter of [Wal,73], and the original papers [Hir].

Problems and Exercises:

1. Establish irreducibility of the principal series representations of $SO(1;n)$ by the Mackey's method.

2. We define the stereographic map (fig.9) from the unit sphere $S^n = \{x_0^2 + |\boldsymbol{x}|^2 = 1\}$ to the hyperboloid $\mathbf{H}^n = \{x_0^2 - |\boldsymbol{x}|^2 = 1\}$, parametrizing both surfaces by a hyperplane $\mathbf{R}^n = \{x = (0;\boldsymbol{x})\}$ in $\mathbf{R}^{n+1}$. Here maps $\Phi: S^n \to \mathbf{R}^n$ and $\Psi: \mathbf{H}^n \to \mathbf{R}^n$ are given by

$$\Phi: y = (y_0;\boldsymbol{y}) \to \boldsymbol{x} = \frac{\boldsymbol{y}}{1-y_0}; \text{ and } \Psi: z = (z_0;\boldsymbol{z}) \to \boldsymbol{x} = \frac{\boldsymbol{z}}{1-z_0},$$

where $y_0 = \pm\sqrt{1-|\boldsymbol{y}|^2}$ denotes a point in S^n, and $z_0 = \sqrt{1+|\boldsymbol{z}|^2}$ - a point in $\mathbf{H}^n$.

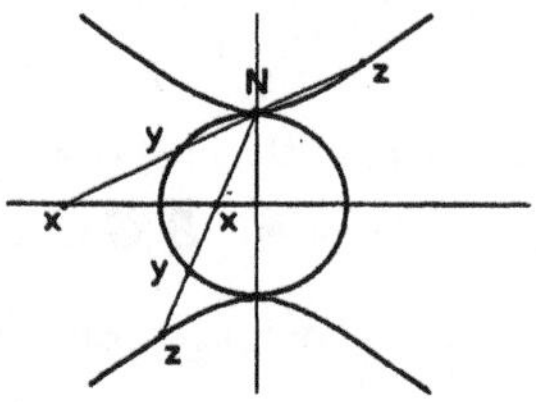

Fig.9. *Two stereographic maps Φ, Ψ take the unit sphere $\{y\}$ and the hyperboloid $\{z\}$ onto* $\mathbf{R}^n$, hence define a conformal map $\Phi^{-1} \circ \Psi$ from $\mathbf{H}^n$ into S^n.

(i) Show that both maps are conformal from S^n, $\mathbf{H}^n$ into $\mathbf{R}^n$. Hint: use the polar coordinates on all three surfaces: $\{(r;\theta)\}$ in $\mathbf{R}^n$, $\{(\phi,\theta): y_0 = \sin\phi\}$ in S^n, and $\{(t;\theta): z_0 = \sinh t\}$ in $\mathbf{H}^n$. Note that Φ and Ψ transform only the radial variable

$$r = \cot\frac{\phi}{2} = \coth\frac{t}{2}.$$

(ii) Combine Φ and Ψ we get a conformal map $\Phi^{-1} \circ \Psi$ from $\mathbf{H}^n$ into S^n,

$$y \to z: \frac{\boldsymbol{y}}{1 \mp \sqrt{1-|\boldsymbol{y}|^2}} = \frac{\boldsymbol{z}}{1 \mp \sqrt{1+|\boldsymbol{z}|^2}}.$$

(iii) Since $G = SO(1;n)$ acts by isometries on the symmetric space $\mathbf{H}^n = K\backslash G$ (K = stabilizer of the pole $N = (1;0)$), the stereographic map $\Phi^{-1} \circ \Psi$ takes it into the conformal group of the sphere S^n. One can show ([Tay], chapter 10), that $SO(1;n)$ comprises the entire conformal group of S^n.

(iv) Compute the conformal factor $\rho(x;g)$ for any $g \in SO(1;n)$ on $x \in S^n$.

Additional results and historical comments.

A. Kirillov [Kir1] traces the onset of the representation theory to the late XIX century, and divides its history into 3 periods. First connected with the names of Frobenius, Schur, Burnside dealt mostly with algebraic aspects: finite group and algebras, characters, projective representations, as was outlined in chapters 1-3. Second period brought in compact groups with the contributions of Haar and von Neumann (invariant integration), Peter-Weyl (completeness of finite-D representations). At the same time E. Cartan and H. Weyl [We3] (1939) unveiled the structure and built the representations theory of simple and semisimple Lie algebras (chapters 4-5). These results not only strike with their profound inner beauty, but find deep applications in a wide range of mathematical and physical subjects: geometry, differential equations, analysis on symmetric spaces (the orginal Cartan's motivation), quantum mechanics and particle Physics. Soon the need came to study noncompact groups and their infinite-D representations. The first result along these lines, the celebrated Stone-von Neumann Theorem, essentially amounted to classification of unitary irreducible representations of the Heisenberg group. In 1939 E. Wigner [Wig2] made the first attempt to build the theory of elementary particles, based on the infinite-D representations. However, more systematic study (third period) began in late 40's, when first classification Theorems for the classical complex Lie groups: $SL(n;\mathbb{C})$; $SO(n;\mathbb{C})$; $Sp(n;\mathbb{C})$, were obtained by Gelfand-Naimark [GN], and Bargman [Bar]. Since then the development of theory went at an ever increasing pace. For general surveys on the role of the harmonic analysis and group representations we recommend surveys [Mac2,5] and [Gr].

Semisimple groups: The basic features of the representation theory of complex semisimple groups were unveiled in the monograph [GN]. It was shown that the principal part here is played by the so-called *parabolic subgroups:* $P \subset G$. These are characterized by the property, that P contains the maximal solvable (Borel) subgroup (e.g. all upper-triangular matrices in SL_n), equivalently, the quotient-space $P\backslash G$ - compact. Želobenko and Naimark [Že] proved that all irreducible (even nonunitary) representations of G are *elementary*[17], i.e. induced by 1-D representations of the parabolic subgroup.

Monograph [GN] derived characters of elementary representations, and established the Plancherel formula for classical complex groups. It was also found that semisimple groups have complementary series, that do not "belong" to the regular representation R^G. Then Harish-Chandra [Har1] extended the "classical results" to arbitrary complex semisimple groups. He also found the general Plancherel formula for semisimple groups in terms of his celebrated c-function ([Har2]),

$$d\mu(\lambda) = \mathrm{Const}\, |\, c(\lambda)\, |^{-2} d\lambda.$$

This yields an explicit expression of $d\mu$ on an arbitrary group G, or a symmetric space $X = G/K$, in terms of the (principal series) weight λ, and the root system $\Sigma = \{\alpha\}$ of G, or the restricted root system of space X (§5.7). Namely ([Har2];[Hel2]),

$$d\mu(\lambda) = \mathrm{Const} \prod_{\alpha \in \Sigma_+} \phi_\alpha\left(\frac{\langle \lambda \mid \alpha\rangle}{\langle \alpha \mid \alpha\rangle}\right);$$

where function ϕ_α depends on multiplicity m_α of root α, and is given by

---

[17] The study of elementary representations is often facilitated by the fact that the double conjugacy classes $P_1\backslash G/P_2$ are finite for any pair of parabolics $P_1; P_2$ (this result is known as Bruhat's Lemma). In particular, for the Borel (maximal solvable) subgroup B, they coincide with the elements of the Weyl group $\mathbf{W}$ of G!

$$\phi_\alpha(\xi) = \begin{cases} \xi^2 \prod(\xi^2+k^2);\ 0<k<\frac{1}{2}m_\alpha \\ \xi\tanh(\pi\xi)\prod(\xi^2+k^2);\ 0<k(\text{odd integer})<\frac{1}{2}m_\alpha \\ \xi\tanh(\frac{\pi\xi}{2})\prod(\xi^2+k^2)^2\prod(\xi^2+l^2);\ 0<k(\text{odd})<\frac{1}{2}m_\alpha;\ \frac{1}{2}m_\alpha<l<\frac{1}{2}m_\alpha+m_{2\alpha} \\ \frac{\xi^3}{tanh(\frac{\pi\xi}{2})}\prod(\xi^2+k^2)^2;\ 0<k(\text{even integer})<\frac{1}{2}m_\alpha \end{cases}$$

All 4 cases are determined by the relative values of multiplicities m_α and $m_{2\alpha}$, according to the table,

$$\begin{cases} m=2,4,6,8\ldots \text{ and } m_{2\alpha}=0 \\ m=1,3,5,7\ldots \text{ and } m_{2\alpha}=0 \\ m=4,8,12,\ldots \text{ and } m_{2\alpha}=1,3,7 \\ m=2,6,10,14,\ldots \text{ and } m_{2\alpha}=1 \end{cases}$$

The classification of all nonunitary (Banach-space) irreducible representations of semisimple groups was given by Berezin [Ber] (1962), based on his study of Laplacians (Casimirs operators) on G. But the problem of selecting unitary representations among all elementary ones turned out to be technically difficult, and has not been yet completely solved. Significant progress has been made in the past decades, that has brought about a complete classification for a growing list of groups: $SO(2;3)$; $SU(2;2)$; $Sp(1;n)$; $SL_n(\mathsf{F})$, for any $\mathsf{F}=\mathbf{R};\mathbf{C};\mathbf{Q}$. For surveys of this important work we refer to [Kn];[KS];[Sp];[Vo].

The representation theory of real semisimple groups added further difficulties. Here elementary representations proved insufficient to build a complete system, even if one allows the holomorphic induction (§6.1). First example of non-elementary "strange series" appeared in the work of Gelfand-Graev for group $SU(1;2)$. But only after the Langlands' work [JL], it became clear that "strange series" can not be realized in "function-spaces" (0-forms), but require higher-rank differential forms. Langlands conjectured that all of them could be realized in higher "L^2-cohomologies", by a combination of the regular and holomorphic induction (see §6.1). The most significant contribution to representations of real semisimple groups came from the works of Harish-Chandra. His papers [Har] gave, in particular, a complete classification of so called discrete series, i.e. $\{T\}$ with square-integrable matrix-entries, equivalently $\{T\}$, embedded in the regular representation R^G. Harish-Chandra associated discrete series to compact Cartan subalgebras in $\mathfrak{G}$. So certain semisimple algebras have them, like $\mathfrak{so}(1;n)$, while others, e.g. $\mathfrak{sl}(n), n\geq 3$, or any complex $\mathfrak{G}$), do not. Harish-Chandra characterized discrete series by means of their characters. This left an open problem to explicitly construct such $\{T\}$. The latter was accomplished by Parthasarathy [Par]; Schmid [Sch] and Atiyah-Schmid [AS].

Our main sources in §7.1-7.5; 7.7 were books [GGP];[GGV];[GMS];[Lan];[Ta].

Chapter 8. Lie groups and Hamiltonian mechanics.

We shall outlines some interesting applications of Lie groups and symmetries to hamiltonian mechanics. Our main emphasis will be on integrable systems and systems that possess large symmetry groups. Although most of the discussion does not directly involve the representation theory of chapters 1-7 (save for the last section, §8.5), the Lie structure theory of groups and algebras will enter many times.

To make our presentation self-contained we included in the first section, §8.1, some basics of the hamiltonian mechanics: Lagrangian formulation, Minimal action principle; Euler-Lagrange equations; Canonical formalism; symplectic/Poisson structure; conserved integrals, integrability and Darbeaux Theorem.

§8.1. *Minimal action principle; Euler-Lagrange equation; Canonical formalism.*

1.1. Hamilton's Minimal Action principle. The state of a classical mechanical system of n degrees of freedom is described by its *position vector:* $q = q(t) = (q_1; \dots q_n)$, varying over $\mathbb{R}^n$, or more general Riemannian manifold $\mathcal{M}$, called *configuration space,* and *velocity (tangent) vector:* $\dot{q} = (\dot{q}_1; \dots \dot{q}_n)$. Its dynamical evolution is determined by the *action functional,*

$$S = \int_{t_0}^{t_1} \mathcal{L}(q; \dot{q}) dt,$$

whose integrand $\mathcal{L}(q; \dot{q})$, called *Lagrangian,* depends on position and velocity, and has physical dimensionality of energy. In many cases of interest Lagrangian represents the difference of *Kinetic* and *Potential* energies, $\mathcal{L} = K - P$, where

$$K = \tfrac{1}{2}\dot{q}^2;$$

or more generally, is given by a Riemannian metric-tensor $\{g_{ij}(q)\}$ on $\mathcal{M}$,

$$K = \tfrac{1}{2}\sum g_{ij}\dot{q}_i\dot{q}_j;$$

while $P = V(q)$ - a potential function.

According to the *Hamilton's principle of Minimal Action:* a trajectory (evolution) of the classical system must minimize (or give a stationary path) of the action-functional. So it satisfies the *Euler-Lagrange equation:*

$$\frac{\delta S}{\delta q(t)} = \mathcal{L}_q - \frac{d}{dt}(\mathcal{L}_{\dot{q}} \Leftrightarrow = 0. \tag{1.1}$$

Equation (1.1) represents a 2-nd order OD system in n variables. In the classical case of $\mathcal{L} =$ "kinetic" – "potential", (1.1) turns into the *Newton's equation,*

$$\ddot{q} = -\partial V(q) = F\text{- force.}$$

The canonical formalism reduces the 2-nd order Euler-Lagrange equations to a $1-$st order system of size $2n$. We introduce a new set of variables:

$$p_i = \partial_{\dot{q}_i}\mathcal{L}(q; \dot{q}) \text{ - } \textit{conjugate momenta}, \tag{1.2}$$

and

$$H(q;p) = p \cdot \dot{q} - \mathcal{L}, \text{ *hamiltonian/energy function.*}$$

Solving a system of equations (1.2) for $\dot{q}$-variables[1], we get $\dot{q} = \dot{q}(q;p)$, and these are substituted in the hamiltonian $H(q;p)$. Then the Euler-Lagrange equations (1.1) are shown to be equivalent to a *hamiltonian system*

$$\tfrac{d}{dt}q = \partial_p H; \ \tfrac{d}{dt}p = -\partial_q H. \tag{1.3}$$

We shall first demonstrate the canonical formalism in the case of *N-particle systems.* The corresponding Lagrangians are

$$\mathcal{L} = \tfrac{m}{2}\dot{q}^2 - V(q), \text{ or } \tfrac{1}{2}\sum_1^N m_j \dot{q}_j^2 - V,$$

where $\{m_j\}$ denotes masses of particles. Then the conjugate momenta: $p_j = m_j \dot{q}_j$, and the hamiltonian,

$$H = \sum \tfrac{1}{2m_j} p_j^2 + V(q); \text{ “kinetic” + “potential”},$$

a familiar expression from elementary calculus/mechanics. For more general (“kinetic − potential”) Lagrangians on manifolds $\mathcal{M}$, the Euler-Lagrange (1.1) takes the form

$$\tfrac{d}{dt}(\sum g_{ij}\dot{q}_j) - \partial_{q_i} V = 0,$$

while the *canonical variables:*

$$p_i = \sum g_{ij}\dot{q}_j; \ H = \tfrac{1}{2}\sum g^{ij} p_i p_j + V(q),$$

(g^{ij}) denotes the inverse matrix (tensor) to (g_{ij}), and the hamiltonian system becomes

$$\begin{cases} \dot{q}_i = \sum_j g^{ij} p_j; \\ \dot{p}_i = -\partial_i V. \end{cases}$$

So geometrically, momentum variables $\{p\}$ can be identified with *cotangent vectors* on $\mathcal{M}$, and the *(position-momentum) phase space* becomes a *cotangent bundle* $T^*(\mathcal{M})$. The change of variables $(q;\dot{q};\mathcal{L}) \to (q;p;H)$, called the *Legendre transform,* can be interpreted as a map from the *velocity phase space*, tangent bundle $T(\mathcal{M}) = \{(q;\dot{q})\}$, to the *momentum phase space*, cotangent bundle $T^*(\mathcal{M}) = \{(q;p)\}$, that takes solutions (trajectories) $\{q(t);\dot{q}(t)\}$ of the Euler-Lagrange equations (1.1) to those of Hamilton system (1.3) (problem 2).

[1] provided it could be solved, i.e. the Hessian of $\mathcal{L}$ in $\dot{q}$-variables is nonsingular,

$$\det\left(\frac{\partial^2 \mathcal{L}}{\partial \dot{q}_i \partial \dot{q}_j}\right) \neq 0.$$

The latter is always the case with the classical Lagrangians, $\mathcal{L}$ = “kinetic” − “potential”, on $\mathcal{M}$, whose Hessian turns into the metric tensor $\{g_{jk}(q)\}$.

1.2. Symplectic structure and Poisson bracket. Phase space $\mathcal{P} = T^*(\mathcal{M})$ is equipped with the natural *symplectic/Poisson structure,* given by a *differential (canonical) 2-form*[2] on $\mathcal{P}$:

$$\Omega = \sum_i dp_i \wedge dq_i, \tag{1.4}$$

in standard (dual) local coordinates $\{q_1 ... q_n; p_1 ... p_n\}$. In other words, we take a basis $\{\partial_{q_1}; ... \partial_{q_n}\}$ in the tangent space $T_q(\mathcal{M})$ and the dual basis $\{dq_1; ... dq_n\}$ in the cotangent space $T^*_q(\mathcal{M})$, so each point $x = (q; \xi) \in T^*(\mathcal{M})$, $q \in \mathcal{M}$, $\xi = \sum p_j dx_j \in T^*_q$, can be represented by a $2n$-tuple $\{q_i; p_j\}$.

Symplectic structure on a general manifold $\mathcal{P}$, with local coordinates $\{x_1; ... x_m\}$, is given by a *non-degenerate closed 2-form,*

$$\Omega = \sum a_{jk} dx_j \wedge dx_k, \tag{1.5}$$

with the usual proviso that Ω be independent of a particular choice of $\{x_1; ... x_m\}$ (i.e. coefficients $\{a_{jk}\}$ transform as a tensor on under coordinate changes on $\mathcal{P}$). Clearly, non-degeneracy of Ω constraints $\dim \mathcal{P}$ to be even. In addition, one requires *closedness* of Ω, in the sense that its *differential*

$$d\Omega = \sum \epsilon_{ijk} \partial_i (a_{jk}) dx_i \wedge dx_j \wedge dx_k = 0.$$

Here ϵ_{ijk} denotes a completely antisymmetric symbol (tensor) in 3 indices, normalized by $\epsilon_{123} = 1$. An alternative way to describe a symplectic structure on $\mathcal{P}$ is in terms of a skew-symmetric bilinear form $\mathfrak{J}_x$ on tangent spaces $\{T_x(\mathcal{P}) : x \in \mathcal{P}\}$,

$$\langle \mathfrak{J}\xi | \eta \rangle = \sum b_{jk} \xi_j \eta_k; \; \xi, \eta \in T_x.$$

Two structures are related one to the other by $\Omega = \mathfrak{J}^{-1} dx \wedge dx$, in other words matrix $(b_{jk}) = {}^{\mathrm{T}}(a_{jk})^{-1}$. Clearly, the standard 2-form Ω yields the standard symplectic matrix,

$$\mathfrak{J} = \begin{bmatrix} & I \\ -I & \end{bmatrix}. \tag{1.6}$$

Examples of symplectic manifolds include:

1) The standard (flat) phase spaces: $\mathbb{R}^2 = \{(x; p)\}$ with $\Omega = dx \wedge dp$, and $\mathbb{R}^{2n}$ with $\Omega = \sum dx_j \wedge dp_j$. Those are often convenient to write in the complex form: $z = x + ip \in \mathbb{C}^n$, then $\Omega = \frac{1}{2i} dz \wedge d\bar{z}$.

2) Cotangent bundles $T^*(\mathcal{M})$ over manifolds $\mathcal{M}$, with $\Omega = \sum dq_j \wedge dp_j$;

[2] Let us remark that (1.6) is independent of the choice of local coordinates on $\mathcal{M}$. Indeed, if $f: q \to q'$ denotes a coordinate change, then differentials $\{dq_j\}$ are transformed by the Jacobian map $f' = (\frac{\partial q'}{\partial q})$, while differentials of co-vectors by inverse transpose of f'. So the product $\sum dp_j \wedge dq_j$ remains invariant.

3) The 2-sphere $S^2 = \{(\phi;\theta)\}$ with $\Omega = \sin\phi d\phi \wedge d\theta$.

4) Co-adjoint orbits of Lie groups: $\mathcal{O} \subset \mathfrak{g}^*$ (chapter 4). The tangent space T_x ($x \in \mathcal{O}$) is identified with the quotient $\mathfrak{G}/\mathfrak{G}_x$ - Lie algebra modulo stabilizer subalgebra of x $\mathfrak{G}_x = \{\xi : ad^*_\xi(x) = 0\}$. As a bilinear form on tangent spaces,

$$\Omega(\xi;\eta) = \langle x \mid [\xi;\eta] \rangle = \langle ad^*_\xi(x) \mid \eta \rangle; \ \xi,\eta \in \mathfrak{G}.$$

5) Finally, we shall mention the so called *Kähler manifolds*, complex manifolds $\mathcal{M}$ with a hermitian metric-form:

$$ds^2 = \sum b_{\mu\nu} dz_\mu d\bar{z}_\nu; \ \bar{b}_{\mu\nu} = b_{\nu\mu};$$

whose imaginary part is a closed 2-form,

$$\Omega = \frac{1}{2i} \sum b_{\mu\nu} dz_\mu \wedge d\bar{z}_\nu. \tag{1.7}$$

To check skew-symmetry of Ω one notes that in real coordinates $\{x_\mu = \Re z_\mu; y_\mu = \Im z_\mu\}$,

$$\Omega = A dx \wedge dx + 2B dx \wedge dy + C dy \wedge dy,$$

where $A = C$ and B are real (antisymmetric and symmetric) matrices $\Im(b_{\mu\nu})$ and $\Re(b_{\mu\nu})$, i.e. $(b_{\mu\nu}) = B + iA$. Hence Ω defines a symplectic structure on $\mathcal{M}$.

Symplectic structure on any phase-space $\mathfrak{P}$ (cotangent bundle, co-adjoint orbit, etc.), allows to assign certain *vector fields,* called *hamiltonian fields,* to functions (observables) F on $\mathfrak{P}$,

$$F \rightarrow \Xi_F = \mathfrak{J}(\partial F).$$

In other words we take a gradient vector field of F and "twist" it by a skew-symmetric linear map $\mathfrak{J}$ on tangent spaces $\{T_x(\mathfrak{P})\}$. The standard symplectic structure $\mathfrak{J}$ (1.7) on phase-space $\mathbb{R}^n \times \mathbb{R}^n$, yields hamiltonian vector field,

$$\Xi_F = \partial_p F \cdot \partial_q - \partial_q F \cdot \partial_p.$$

Each vector hamiltonian field generates a *hamiltonian flow,* $\{exp(t\Xi_F)\}$-a fundamental solution of ODS,

$$\dot{x} = \mathfrak{J}(\partial F(x)), \tag{1.8}$$

which generalizes the canonical system (1.3). Hamiltonian flows possess many special features, for instance, all of them preserve the *canonical (Liouville) phase-volume,* $d^n q \wedge d^n p$ on $\mathfrak{P}$, since all respect the canonical/symplectic form $\mathfrak{J}$, or Ω (problem 3), and

$$\text{"Liouville volume"} = \underbrace{\Omega \wedge \ldots \wedge \Omega}_{n-times}.$$

Symplectic structure also defines a *Lie-Poisson bracket* on the vector space of observables $\{F(x)\}$ on $\mathfrak{P}$, namely

$$\{F;G\} = \langle \mathfrak{J}(\partial F) \mid \partial G \rangle,$$

which in special cases, $\mathbb{R}^{2n}$ or $T^*(\mathcal{M})$, turns into,

$$\{F;G\} = \partial_p F \cdot \partial_q G - \partial_q F \cdot \partial_p F.$$

The reader can verify directly all properties of Lie bracket (skew-symmetry and Jacobi identity) for $\{F;G\}^3$. In fact, the Poisson bracket of any two observables corresponds to the standard Lie bracket of their hamiltonian vector fields,

$$\{F;G\} \to \Xi_{\{F;G\}} = [\Xi_F;\Xi_G] = \Xi_F\Xi_G - \Xi_G\Xi_F.$$

Thus the space of observables $\{F(x)\}$ on $\mathfrak{P}$ acquires a structure of an ∞-D Lie algebra, a subalgebra of all vector fields $\mathfrak{D}(\mathfrak{P})$. The corresponding Lie group consists of all *canonical transformations* on $\mathfrak{P}$, i.e. diffeomorphisms $\{\phi\}$ that preserve the symplectic structure,

$$^{\mathrm{T}}\phi' \mathfrak{J} \phi = \mathfrak{J};$$

equivalently coordinate changes, $y = \phi(x)$, that preserve the canonical 2-form,

$$\Omega^{\phi} = \sum_{km}\left(\sum_{ij} a_{ij}(y(x))\frac{\partial y_i}{\partial x_k}\frac{\partial y_j}{\partial x_m}\right)dx_k \wedge dx_m = \Omega.$$

We shall list a few examples of canonical transformations:

i) symplectic matrices $\{A \in Sp(n)\}$ in $\mathbb{R}^{2n} = \{(q;p)\}$; they clearly preserve the standard 2-form: $\Omega = \sum dp_j \wedge dq_j$.

ii) any coordinate change (diffeomorphism) $y = \Phi(x)$, from manifold $\mathcal{M}$ to $\mathcal{N}$, induces a canonical map,

$$\phi:(x;\xi) \to (\Phi(x);{}^{\mathrm{T}}A^{-1}(\xi));$$

where A denotes the Jacobian matrix of Φ at $\{x\}$, $A = \Phi'_x$ (problem 4).

The reader has probably noticed some coincidence in terminology: *symplectic Lie groups/algebras* on the one hand, and *symplectic structure/geometries* on the other. The relation between two becomes apparent now: *Jacobian matrices of canonical (symplectic) maps are symplectic matrices*[4]*!*

1.3. Conserved integrals; action-angle variables and the harmonic oscillator. The hamiltonian evolution (1.8) of the position and momenta variables $\{q;p\}$ gives rise to

[3]The Jacobi identity on general symplectic manifolds results from closeness of the canonical 2-form Ω.

[4]In this regard symplectic groups plays the same role in the symplectic geometry, as orthogonal groups in the Riemannian geometry. There exists, however, a striking difference between two kinds of geometries. The Riemannian geometry is fairly rigid in the sense that isometries of m (even in the best case of symmetric spaces) form only a finite-dimensional Lie group, whereas "symplectic isometries" (canonical maps) are always infinite-dimensional!

evolution of any other observable (function) F on $T^*(\mathcal{M})$,

$$\dot{F} = \{F; H\}.$$

Functions that remain constant along trajectories of the hamiltonian flow, $\{F; H\} = 0$, are called ***first integrals*** of (1.8). Indeed, the Legendre "back-transform" $(q; p) \to (q; \dot{q})$, takes such F into a function $F(q; p(q; \dot{q}))$, constant along trajectories (solutions) of the E-L equation (1.1) i.e. gives a 1-st integral of (1.1). Hamiltonian H is itself an integral, as $\{H; H\} = 0$. In dimension n=1, it is the only integral. So 2-nd order E-L equation, $\mathcal{L}_q - \frac{d}{dt}\mathcal{L}_{\dot{q}} = 0$, is reduced to a 1-st order ODE:

$$H(q; p(...; \dot{q})) = E - const.$$

In the classical (Newton) case this yields

$$\tfrac{1}{2}p^2 + V = \tfrac{1}{2}\dot{q}^2 + V = E,$$

whence we get an implicit solution in the form of integral

$$\int \frac{dq}{\sqrt{2(E-V)}} = t - t_0. \tag{1.9}$$

More generally, each Poisson integral F of the hamiltonian system (1.3) reduces its total order $2n$ by 2. The proof is based on the following general result.

Darbeaux Theorem: *Any set of functions (observables) $\{F_1; ... F_k; G_1; ... G_m\}$ on a phase space (symplectic manifold) Ω that satisfy the canonical commutation (Poisson bracket) relations $\{F_j; G_i\} = \delta_{ij}$, can be locally extended to a canonical coordinate system on Ω, $\{p_1; ... p_n; q_1; ... q_n\}$, $p_i = F_i$ (i=1,...k) and q_j=G_j (j=1;...m).*

In particular, a Poisson integral F can be made the 1-st momentum variable p_1 of the new coordinates. Then hamiltonian $H(p; q)$ becomes independent on the corresponding canonically conjugate position q_1, and p_1 is constant along the flow,

$$\dot{p}_1 = \frac{\partial H}{\partial q_1} = \{p_1; H\} = 0.$$

Setting p_1= E_1-const we reduce H to a hamiltonian $H(E_1; p_2; ... p_n; q_2; ... q_n)$ in (n−1) degrees of freedom, so the system becomes:

$$\begin{cases} \dot{p}_i = \partial_{q_i} H; \\ \dot{q}_i = -\partial_{p_i} H; \end{cases} i\text{=2;...}n; \text{ and } p_1 = E_1;\ q_1 = q_1(0) + \int_0^t H(...)ds.$$

A hamiltonian system (1.3) in n degrees of freedom is called ***completely integrable*** (in the Liouville sense), if it has n functionally independent, Poisson commuting integrals $\{F_1; F_2; ... F_n\}$,

$$\{H; F_i\} = 0; \text{ and } \{F_i; F_j\} = 0, \text{ for all } i, j. \tag{1.10}$$

Functional independence means gradient-vectors $\{\nabla F_i\}$ are linearly independent at each point $x=(q;p)$. In the standard terminology, commuting integrals (1.10) are said to be *in involution.* Let us remark that any set of n functionally independent integrals $\{F_1;...F_n\}$ allows to reduce the total order of the system by n, i.e. to bring (1.3) to a first order system in n variables:

$$\begin{cases} F_1(q;\dot{q}) = E_1 \\ F_n(q;\dot{q}) = E_n \end{cases}.$$

Complete integrability yields, however, more than a mere reduction of order by n, as each subsequent integral F_k respects the joint level sets and evolutions of all preceding variables in the canonical reduction. Hence, the system could (in principle) be reduced to order 0, i.e. solved completely! The procedure is best illustrated by the harmonic oscillator

$$H=\tfrac{1}{2}\sum p_i^2+\omega_i^2 q_i^2 \text{ on } \mathbb{R}^{2n}.$$

As first integrals of H one could take all coordinate oscillators,

$$\{H_i=\tfrac{1}{2}(p_i^2+\omega_i^2 q_i^2)\colon 1\le i\le n\}.$$

Obviously, $H=\sum H_i$. The joint level sets of $\{H_i=E_i\}$ form a n-parameter family of invariant tori in $\mathbb{R}^{2n}$, and the hamiltonian dynamics consists in a uniform motion in some direction along the torus. Introducing polar coordinates $(r;\theta)$ in the i-th phase plane $\{q_i;p_i\}$, after rescaling, $q_i \to \omega_i q_i$, the j-th hamiltonian becomes

$$H_i=\tfrac{1}{2}(p_i^2+q_i^2)=\tfrac{1}{2}r_i^2.$$

In polar coordinates the flow of each H_i is given by an OD system

$$\begin{cases} \dot{r}=0 \\ \dot{\theta}=r \end{cases}, \text{ i.e. } r=E-const,\ \theta=\theta_0+Et.$$

The Poisson bracket $\{r;\theta\}=\frac{1}{r}$, hence $\{\frac{1}{2}r^2;\theta\}=1$, and the entire set of variables $\{H_1;...H_n;\ \theta_1;...\theta_n\}$ satisfies the canonical commutation relations:

$$\{H_i;\theta_j\}=\delta_{ij}.$$

Returning to a general completely integrable hamiltonian system with n commuting integrals $\{F_1;...F_n\}$ (called *actions)*, the Darbeaux Theorem implies the existence of a canonically conjugate set of *angle variables*: $\{\theta_1;...\theta_n\}$, that satisfy the *canonical relations*

$$\{F_j;\theta_i\}=\delta_{ij}.$$

The hamiltonian H in new coordinates $\{F;\theta\}$ becomes a function of actions only, $H=f(F_1;...F_n)$, since $\frac{\partial H}{\partial\theta_j}=\dot{F}_j=0$. The joint level sets $\{F_i=E_i\}$ form a foliation of the phase-space into invariant tori $\mathbb{T}^n$ (or products $\mathbb{T}^k\times\mathbb{R}^{n-k}$), and the dynamics resembles the oscillator case. Precisely, if Φ denotes the *canonical coordinate change:* $(p;q)\to(F;\theta)$, and

$$F_i = E_i;\ \theta_i(t) = \theta_i(0) + t\frac{\partial H}{\partial F_i}(E_1;...E_n),$$

the hamiltonian flow in the action-angle coordinates, then solution in the original coordinates becomes

$$(p;q)(t) = \Phi^{-1}(...E_i;...\ \theta_i(t)...). \tag{1.11}$$

Let us remark, that solution (1.11), although written explicitly, may be of limited utility unless one is able to compute the canonical map Φ.

Our next goal is to establish integrability, find 1-st integrals and, if possible, explicit solutions of different hamiltonian systems. We shall review a number of the classical models, and also discuss some newly discovered examples. Our principal tools, in the study of conserved integrals, will be symmetries of the problem, and the Noether Theorem, to which we turn in now.

Problems and Exercises.

1. Check the equivalence of the Euler-Lagrange equations (1.1), and the hamiltonian system (1.3).

2. (i) Demonstrate that the "square-root-kinetic-energy" functional of a Riemannian metric $\{g_{ij}(x)\}$, $\mathcal{L}[x] = (\sum g_{ij}(x)\dot{x}_i\dot{x}_j)^{1/2}$, yields geodesics, as extremal curves. So the Euler-Lagrange equation for $\mathcal{L}$ is precisely the equation of geodesics (see Appendix C). The geometric meaning of this result is quite transparent: the Lagrangian density $\mathcal{L}dt$ represents the arc-length element of the metric g!

   (ii) Show that the Legendre transform kinetic energy functional $K = \frac{1}{2}\sum g_{ij}(x)\dot{x}_i\dot{x}_j$ of metric g, becomes the hamiltonian $H = \frac{1}{2}\sum g^{ij}(x)p_ip_j$, of the dual metric $(g^{ij}) = (g_{ij})^{-1}$, on the cotangent space.

3. Check that the Jacobian map $\left[\frac{\partial(a;b)}{\partial(p;q)}\right]$ of a hamiltonian vector field
$$\Xi_F = (a(p;q);b(p;q));\ a = \partial_pF;\ b = -\partial_qF;$$
has determinant 1, hence the hamiltonian flow of F preserves the Liouville volume $d^np \wedge d^nq$.

4. i) Any coordinate change, $q \to \phi(q)$, on configuration space $\mathcal{M}$ defines a canonical transformation,
$$A_\phi: (q;p) \to (\phi(q); {}^T\phi'^{-1}(p));$$
on the phase-space $T^*(\mathcal{M})$ (Check that A_ϕ preserves the canonical 2-form $dq \wedge dp$).

   ii) Apply part (i) to transform the standard canonical $(q;p)$-variables in the phase-space $\mathbb{R}^2 \times \mathbb{R}^2$ to polar $\{(r;\theta)\}$, and spherical $\{(r;\phi;\theta)\}$-coordinates. Compute the hamiltonian $H = p^2 + V(q)$ on $\mathbb{R}^2$ in polar and spherical coordinates, and show
$$H_{polar} = p_r^2 + \frac{1}{r^2}p_\theta{}^2 + V;\ H_{sher} = p_r^2 + \frac{1}{r^2}(p_\phi{}^2 + sin^2\phi\dot{\theta}^2) + V;$$

   iii) Do the same for *elliptical coordinates*:
$$\xi = \frac{r_1+r_2}{2};\ \eta = \frac{r_1-r_2}{2};$$
where $\{r_1;r_2\}$ are distances from $q = (x;y)$ to a pair of focal points: $\{(\pm a;0)\}$ (see fig.1). Show
$$H_{elliptic} = \frac{1}{\xi^2-\eta^2}\{(\xi^2-a^2)p_\xi^2 + (a^2-\eta^2)p_\eta^2\} + V.$$

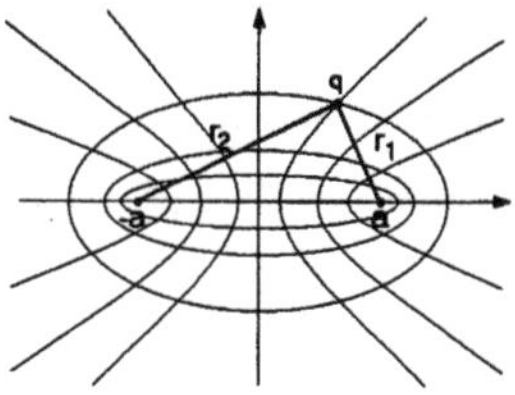

Fig.1: *Elliptical coordinates in* $\mathbb{R}^2$ *are made of confocal ellipsi:*
$$\xi = r_1 + r_2 = \text{Const};$$
and hyperbolae:
$$\eta = r_1 - r_2 = \text{Const}.$$
Here $x = a\,ch\,t\cos\theta$; $y = a\,sh\,t\sin\theta$; *and parameters*
$$\xi = 2a\,ch\,t;\ \eta = 2a\cos\theta.$$
Elliptical coordinate change can also be viewed as a conformal (analytic) map,
$$w = t + i\theta \to z = x + iy,$$
given by
$$z = \frac{a}{2}(e^w + e^{-w}).$$

§8.2. Noether Theorem, conservation laws and Marsden-Weinstein reduction.

Conserved integrals of hamiltonian systems are often derived from the one-parameter groups of symmetries via the celebrated Noether Theorem. In particular all basic conservation laws of Physics: energy, momentum, angular momentum, arise this way from the translational and rotational symmetries of the system. Going further in this direction, we consider systems with "large" Lie groups, or algebras of symmetries (not necessarily commuting). Such symmetries allow to reduce the system to fewer variables, via the Weinstein-Marsden reduction process.

Given a configuration space of a classical mechanical system with Lagrangian $\mathcal{L} = \mathcal{L}(q;\dot{q})$, we consider a one-parameter group of transformations $\Phi_\epsilon:(q;t) \to (q^*;t^*)$ on $\mathcal{M}\times[0;\infty)$ (point transformations along with time reparametrizations),

$$\begin{cases} q^* = q + \epsilon\psi(q;t) + \ldots = \psi_\epsilon(q;t) \\ t^* = t + \epsilon\phi(q;t) + \ldots = \phi_\epsilon(q;t) \end{cases} \tag{2.1}$$

In other words vector field $\psi = \delta q$ and scalar field $\phi = \delta t$, represent infinitesimal generators of the family Φ_ϵ. Obviously any group of transformations (2.1) of $\mathcal{M}$ generates transformations on the *path-space* of $\mathcal{M}$ (trajectories of the system):

$$(q(t);t) \to (q^*(t^*);t^*) = (q+\delta q; t+\delta t) = \psi_\epsilon(q\circ\phi_\epsilon). \tag{2.2}$$

2.1. Noether Theorem: *Any one-parameter group of transformations (2.2) with generators $(\psi = \delta q; \phi = \delta t)$, that leaves invariant the action-functional $S = \int \mathcal{L}(t;q;\dot{q})dt$, gives rise to a conserved integral*

$$\boxed{J = p\cdot\delta q - H\,\delta t = p\cdot\psi - H\phi \quad \text{Const}}$$

Here $p = \partial_{\dot{q}}\mathcal{L}$, and $H = p\cdot\dot{q} - \mathcal{L}$ are the canonical variables (conjugate momenta and the hamiltonian) of $\mathcal{L}$.

The proof follows from the general *variational formula* for a functional $\mathcal{L}$ (problem 1), when we allow free motion of the end points as well as all reparametrizations of the time-variable: $t \to t^* = t+\delta t$,

$$\delta S = (p\cdot\delta q - H\,\delta t)\Big|_{t_0}^{t_1} + \int_{t_0}^{t_1}\left(\mathcal{L}_q - \frac{d}{dt}\mathcal{L}_{\dot{q}}\right)(\delta q - \dot{q}\,\delta t)dt. \tag{2.3}$$

The first factor in the integral represent the E-L equation that vanishes along any critical path. If furthermore functional S is invariant under (2.2), then $\delta S = 0$, and we get

$$J(t_1) = (p\cdot\delta q - H\,\delta t)\,|_{t_1} = J(t_0) \text{ - constant along any critical path! QED.}$$

Remark: Conservation of Noether integral $J = p\cdot\psi - H\phi$ can be recast in the hamiltonian formulation. Indeed, function $J = J(q,p;t)$, considered on the phase space of variables $(q;p)$ clearly satisfies the equation,

$$J_t + \{J;H\} = 0.$$

Thus each symmetry of the Lagrangian $\mathcal{L}$ produces a Poisson commuting integral (symmetry) of the corresponding hamiltonian H. So any Lie group/algebra of symmetries of the Lagrangian gets represented by the Lie group/algebra of hamiltonians.

2.2. Conservation laws. As an application of Noether's Theorem we shall derive the basic conservation laws of classical mechanics.

• **Energy conservation.** If Lagrangian is time independent, $\mathcal{L} \quad \mathcal{L}(q;\dot{q})$, then time shifts: $t \to t+\epsilon$, form a translational symmetry group of $\mathcal{L}$, whose generators: $\psi \;\, 0$, $\phi=1$. Hence the Noether integral becomes hamiltonian/energy function,

$$\boxed{J = H(q;;p) = \mathrm{Const}}.$$

• **Momentum conservation.** We assume now that the space shifts $q \to q+\epsilon u$ in certain directions u leave Lagrangian invariant. Then the generators $\phi = 0$; $\psi = u$ yield

$$J_u = p \cdot u,$$

the u-component of the momentum remains constant. The standard example is a N-particle system in $\mathbb{R}^3$ with pair interactions,

$$\mathcal{L} = \tfrac{1}{2}\sum \dot{q}_i^2 - V,$$

where potential $V = \sum v(q_i - q_j)$. Obviously, simultaneous shifts,

$$(q_1 \dots q_N) \to (q_1+u; \dots q_N+u), \; u \in \mathbb{R}^3,$$

do not change V, hence $\mathcal{L}$. Thus we get conservation of the total momentum of the system,

$$\boxed{P = \sum p_i = \mathrm{Const}}$$

• **Angular momentum conservation.** The source of the angular momentum conservation are rotational symmetries of the Lagrangian, like the central potential (Kepler) problem: $V = V(\,|\,q\,|\,)$ in $\mathcal{L} = \frac{1}{2}\dot{q}^2 - V$. Let us assume that $\mathcal{L}$ is invariant under rotations in the ij-th coordinate plane. The corresponding symmetry generator is a linear vector field

$$\psi \quad \psi_{ij} = \begin{bmatrix} 0 & 1 \\ -1 & 0 \end{bmatrix} \begin{bmatrix} q_i \\ q_j \end{bmatrix}$$

and the Noether integral becomes

$$J_{ij} = p_i q_j - p_j q_i \tag{2.4}$$

- the ij-component of the angular momentum. In the N-body system with potential $V = \sum v(|q_i - q_j|)$, only simultaneous rotations of $(q_1...q_N)$ by $u \in SO(3)$ leave $\mathcal{L}$ invariant. So the total angular momentum is conserved,

$$\boxed{J = \sum p_i \times q_i = \text{Const}}.$$

Remark. A relativistic particle Lagrangian has the form

$$\mathcal{L} \quad \sqrt{\dot{q}_0^2 - \sum \dot{q}_i^2};$$

so the corresponding symmetry group consists of Lorentz transformations $SO(1;3)$, whose generators include rotations in spatial directions,

$$J_{ij} = \begin{bmatrix} 0 & 1 \\ -1 & 0 \end{bmatrix}; \ 1 \leq i; j \leq 3,$$

i.e. Lie algebra $\mathfrak{R} = so(3) \subset so(1;3)$, as well as *relativistic boosts:*

$$J_{0i} = \begin{bmatrix} 0 & 1 & 0 \\ 1 & \ddots & \\ 0 & & 0 \end{bmatrix};$$

generators of hyperbolic rotations, $\mathfrak{P} \subset so(1;3)$. The corresponding Noether integrals are

$$J_{0i} = p_0 q_i + p_i q_0.$$

2.3. The momentum map and Weinstein-Marsden reduction. Let $\mathfrak{G} = \text{span}\{F_1;...F_m\}$ be a Lie algebra of hamiltonians on the phase space $\mathcal{M}$. Then Lie group G of $\mathfrak{G}$ acts on $\mathcal{M}$ by canonical transformations $\{exp(t\Xi_F): F \in \mathfrak{G}\}$. Any such algebra $\mathfrak{G}$ defines a map J from $\mathcal{M}$ into the dual space $\mathfrak{G}^*$, by evaluating hamiltonians $F \in \mathfrak{G}$ at a point $x \in \mathcal{M}$ ("point evaluations" are clearly linear functionals on $\mathfrak{C}(\mathcal{M})$!),

$$\langle J_x \mid F \rangle = F(x), \text{ for all } F \in \mathfrak{G}.$$

The resulting momentum map, intertwines two actions of group G: canonical action on $\mathcal{M}$, and the co-adjoint action on $\mathfrak{G}^*$: if $g = g(t) = exp(tF)$, denote denote a Lie group element (of generator $F \in \mathfrak{G}$), and $\phi_g = exp(t\Xi_F)$ - the corresponding canonical transformation, then

$$J(\phi_g(x)) = Ad_g^*(J(x)), \text{ for all } g \in G; \ x \in \mathcal{M}. \tag{2.5}$$

Furthermore, map J preserves the Poisson structures on $\mathcal{M}$ and $\mathfrak{G}^*$.

We remind the reader that the dual space of a Lie algebra has a natural Poisson bracket (chapter 6 and §8.1), defined in terms of the Lie bracket on $\mathfrak{G}$. Namely, for any pair of functions $F_1(x); F_2(x)$ on $\mathfrak{G}^*$, gradients $\{\partial F_j\}$ belong to the Lie algebra itself, so one can set

$$\{F_1; F_2\}(x) = \langle [\partial F_1; \partial F_2] \mid x \rangle. \tag{2.6}$$

Poisson-Lie algebras $\mathfrak{G}$ often arise as symmetries of hamiltonians H on $\mathcal{M}$. The momentum and angular momentum operators are obvious examples of the momentum map,

$$P_u = u \cdot p \text{ (u-fixed direction), and } J = \{J_{jk} = (p \wedge x)_{jk} = p_j x_k - p_k x_j\}.$$

The former corresponds to translational symmetries (in the u-th direction) $\mathfrak{G} \simeq \mathbb{R}^m$, the latter to rotational symmetries, $\mathfrak{G} \simeq \mathbf{so}(m)$. We shall see that symmetries of hamiltonians $\{H\}$ play a double role. On the one hand they apply to reduce the number of variables (degrees of freedom). On the other hand some, seemingly complicated, systems could be "lifted" to larger, but "simpler" systems, typically based on Lie groups and symmetric spaces. We shall analyze a few important examples of this sort in the next section.

But here we shall concentrate on the *"reduction part"*. Given a Lie algebra $\mathfrak{G}$ of symmetries of hamiltonian H, we consider a *joint level set* of all observables $\{F \in \mathfrak{G}\}$ (it suffices to pick a basis $F_1; F_2; ...F_m$ in $\mathfrak{G}$), at a "level $\xi_0 \in \mathfrak{G}^*$",

$$\mathcal{M}_0 = J^{-1}(\xi_0).$$

Subset $\mathcal{M}_0$ is invariant under the flow of $exp(tH)$. Furthermore, $\mathcal{M}_0$ is invariant under the stabilizer of ξ_0, subgroup $G_0 = \{g \in G : Ad_g(\xi_0) = 0\}$, an obvious consequence of (2.5). The action of group G_0 on $\mathcal{M}_0$ splits it into the union of orbits (a fiber bundle), and the *reduced space* we are interested in is the quotient of $\mathcal{M}_0$ modulo G_0, the *orbit space*[5] Ω.

In our setup space $\Omega_0 \subset \Omega = \mathcal{M}_0/G_0$ also has a natural symplectic structure: the Poisson bracket on $\mathcal{M}_0$ restricted to G_0-invariant functions. This Poisson bracket turned out to be *non-degenerate*, hence yields a symplectic structure on Ω_0 (a Theorem due to Weinstein and Marsden). We shall illustrate the reduction procedure by 3 examples.

2.4. Examples.

• *Radial (spherically symmetric) hamiltonian* $H = H(|q|; |p|)$ in $\mathbb{R}^3 \times \mathbb{R}^3$, a generalization of the classical Kepler central-force problem: $H = \frac{1}{2}p^2 + V(|q|)$. Any such H has an $SO(3)$ (angular momentum) symmetry, the momentum map being,

$$J : (q; p) \to q \times p \in \mathbf{so}(3).$$

Here we identify 3-vectors $\xi = (a; b; c)$ with antisymmetric matrices,

[5] It should be mentioned here that orbit-spaces are typically non-smooth manifolds with corners and edges (take, for instance, $\mathbb{R}^n$, modulo $SO(n)$). But there is always a dense open set Ω_0 of "generic (maximal-D) orbits" in Ω, that possesses a smooth (differentiable) structure!

$$M_\xi = \begin{bmatrix} & c & -b \\ -c & & a \\ b & -a & \end{bmatrix};$$

so $q \times p$ becomes a element of $\mathfrak{so}(3)$. We fix a joint level set

$$\mathcal{M}_0 = \{(q;p): p \times q = J_0\},$$

of the (constant) angular momentum (Noether's Theorem). Without loss of generality vector J_0 could be taken, as $(0;0;j_0)$, where $j_0 = |J_0|$. Then vectors $(q;p)$ belong in the plane orthogonal to the (constant) angular momentum, which reflects the well-known property of central potential forces: planar motion!

A 3-D manifold $\mathcal{M}_0 = \{(q;p) \in \mathbb{R}^2 \times \mathbb{R}^2;\ q \wedge p = j_0\}$ is foliated into orbits of the stabilizer subgroup $G_0 \simeq SO(2)$ of $J_0 \in \mathbb{R}^3 \simeq \mathfrak{so}(3)$. Clearly, any $u \in SO(2)$ takes

$$u: (q;p) \to (q^u;p^u),$$

and preserves the cross (wedge) product $q \wedge p$. So the reduced phase-space $\mathcal{N} = \mathcal{M}_0/SO(2)$ is 2-dimensional! It has a natural set of coordinates: polar radius $\{r = |q|\}$, and its conjugate momentum variable p_r. In polar coordinates $\{r;\theta\}$, we have (problem 4 of §8.1):

$$\boxed{|p|^2 = p_r^2 + \frac{1}{r^2}p_\theta^2;\ p_\theta = j_0}$$

Hence, the reduced dynamics in the phase-space $\mathcal{N} \simeq \mathbb{R}^2$ is given by the hamiltonian,

$$\boxed{H|\mathcal{N} = H\left(r; \sqrt{p_r^2 + \frac{j_0^2}{r^2}}\right)}$$

where $j_0 = |J_0|$ is the total (fixed) angular momentum. Thus we have reduced he number of variables to 1, which shows complete integrability of rotationally symmetric hamiltonians (problem 2).

• **Example:** *Integrable hamiltonians H.* A commuting family of integrals $\{J_k\}$ maps phase-space $\mathcal{P}$ into the Lie algebra $\mathbb{R}^n$. Here group $G \simeq \mathbb{T}^n$ (or $\mathbb{R}^m \times \mathbb{T}^{n-m}$) is generated by the flow $\{exp(t_1\Xi_1 + ... + t_n\Xi_n)\}$ of hamiltonian fields $\{\Xi_k\}$ of $\{J_k\}$; the co-orbits $\{\xi\}$ are points in $\mathbb{R}^n$, stabilizers $G_\xi = \mathbb{T}^n$, and inverse images $J^{-1}(\xi)$ coincide with invariant tori in $\mathcal{P}$. So the reduced space $J^{-1}(\xi)/\mathbb{T}^n$ is trivial.

• **Example:** *Left-invariant metrics on Lie groups.* Tangent/cotangent spaces of Lie group G at each $x \in G$ can be obtained from a single space (Lie algebra), $\mathfrak{G} \simeq T_e$ and its dual $\mathfrak{G}^* \simeq T_e^*$, by left translations,

$$q: \xi \in T_e \to q \cdot \xi \in T_q, \text{ and } p \in T_e^* \to q^{*-1} \cdot p \in T_q^*;\ q \in G.$$

For the sake of presentation we shall consider the matrix group and algebra G and $\mathfrak{G}$. Then $q \cdot \xi$ will be the matrix multiplication. The left-invariant fields on G are of the form: $\xi(q) = q \cdot \xi$ ($\xi \in \mathfrak{G}$), and any left-invariant metric is uniquely determined by its restriction at $\{e\}$, a symmetric bilinear form B on $\mathfrak{G}$,

$$\langle Bq^{-1} \cdot \xi \mid q^{-1} \cdot \eta \rangle; \ \xi; \eta \in T_q.$$

Thus we get a Lagrangian $\mathcal{L} = \frac{1}{2}\langle B ... \mid ... \rangle$, and the corresponding hamiltonian $H = \frac{1}{2}\langle B^{-1} ... \mid ... \rangle$ on phase-space $T^*(G)$, which possess a large symmetry group G of right translations $\{x \to x \cdot q; \ x; q \in G\}$. Those clearly commute with left-invariant vector fields $\{\xi = \xi(q) = q \cdot \xi\}$. The corresponding family of right-invariant hamiltonians (integrals of H) consists of

$$\{J_\xi(q; p') = \langle \xi \cdot q \mid p' \rangle : \xi \in \mathfrak{G}; p' \in T_q^*\}, \text{ on } T^*(G).$$

Identifying co-vector p' at point $\{q\}$ with a co-vector $p = q^{-1} \cdot p' \in \mathfrak{G}^*$ (via left shift with q^{-1}), we can write such functions J's on $G \times \mathfrak{G}^* \simeq T^*(G)$ as,

$$J_\xi(q; p) = \langle q^{-1} \xi q \mid p \rangle = \langle ad_q(\xi) \mid p \rangle = \langle \xi \mid ad_q^*(p) \rangle.$$

So the momentum map becomes

$$\boxed{J(q; p) = ad_q^*(p)}$$

To get the reduced space we take the inverse image of $p_0 \in \mathfrak{G}^*$,

$$J^{-1}(p_0) = \{(q; p) : ad_q^*(p) = p_0\} = \{(q; ad^*_{q^{-1}}(p_0)) : q \in G\}, \tag{2.7}$$

and note that $J^{-1}(p_0)$ projects onto a co-adjoint orbit $\mathcal{O} = \mathcal{O}(p_0)$ (2-nd component in the RHS (2.7)). In fact, space $J^{-1}(p_0)$ is foliated over $\mathcal{O}$ with fibers, isomorphic to G_0 and its conjugates $\{G_q = q^{-1} G q\}$. So the quotient-space $J^{-1}(p_0)/G_0$ can be identified with orbit $\mathcal{O}$, and the reduced hamiltonian becomes,

$$H = \tfrac{1}{2}\langle B^{-1}\xi | \xi \rangle, \text{ restricted to } \mathcal{O}.$$

In the next section we shall illustrate the foregoing discussion of symmetries, conserved integrals, integrability and reduction with a few classical examples, including Kepler one- and two-center gravity problem and the Euler rigid body motion.

Problems and Exercises.

1. Prove the variational formula (2.7). Steps,

i) Consider all possible *variations* of path $q(t)$, $q \to q + \delta q$, along with all reparametrizations, $t \to t^* = t + \delta t$. Write the variation of the functional,

$$\delta S = \int \mathcal{L}(t^*; q^*; \frac{dq^*}{dt^*}) dt^* - \int \mathcal{L}(t; q; \frac{dq}{dt}) dt,$$

ii) Change new parameter t^* back to the old t, and show that the 1-st integrand becomes

$$\mathcal{L}(t+\delta t; q+\delta q; \frac{d(q+\delta q)}{d(t+\delta t)})(1+\delta \dot{t}) = \mathcal{L}(...;...; \frac{\dot{q}+\delta\dot{q}}{1+\delta\dot{t}})(1+...).$$

The dot above any variable/function ($q; \delta q$; etc.) indicates its time derivative $\frac{d}{dt}$.

iii) Expand the latter in the Taylor series to the 1-st order in variations

$$\delta S = \int \mathcal{L}_t \delta t + \mathcal{L}_q \cdot \delta q + \mathcal{L}_{\dot{q}} \cdot (\delta\dot{q} - \dot{q}\,\delta\dot{t}) + \mathcal{L}\,\delta\dot{t},$$

iv) Integrate by parts of all terms containing $\delta\dot{q}$; $\delta\dot{t}$ and get,

$$(\mathcal{L}\delta t + p\cdot\delta q - p\cdot\dot{q}\,\delta t)\Big|_{t_0}^{t_1} + \int (\mathcal{L}_q - \frac{d}{dt}\mathcal{L}_{\dot{q}})\cdot\delta q + (\mathcal{L}_t - \frac{d}{dt}\mathcal{L} + \frac{d}{dt}(\mathcal{L}_{\dot{q}}\cdot\dot{q}))\,\delta t.$$

v) Observe that the off-integral terms above combine to

$$p\cdot\delta q - H\delta t,$$

while the δt-factor inside the integral, after term-wise differentiation and cancellations reduces to $(\mathcal{L}_q - \frac{d}{dt}\mathcal{L}_{\dot{q}})\cdot\dot{q}\delta t$, and completes the proof.

2. Check directly that all 3 components of angular momentum

$$J_{ij} = q_i p_j - q_j p_i;$$

Poisson commute with $H = \frac{1}{2}p^2 + V$, for any radial V,

$$\{J_{ij}; H\} = 0.$$

Show that $\{J_{ij}\}_{i<j}$ satisfy the commutation relations of generators of the Lie algebra $so(n)$ of antisymmetric matrices, $E_{ij} = \begin{pmatrix} 0 & 1 \\ -1 & 0 \end{pmatrix}$ - rotation in the ij-th coordinate plane.

§8.3. Classical examples.

This section will illustrates the concepts and methods of the preceding parts (§§8.1-8.2), particularly the role of symmetries, integrability and reduction, by a few classical examples: spherical pendulum, Kepler problem, Euler rigid body and 2-center gravity problem.

3.1. Spherical pendulum. The configuration space here is the sphere

$$S^2 = \{x^2+y^2+z^2 = 1\} \text{ in } \mathbb{R}^3;$$

and Lagrangian: $\mathcal{L} = \frac{1}{2}(\dot{x}^2+\dot{y}^2+\dot{z}^2) - V(z)$. In spherical coordinates $(\phi;\theta)$,

$$\mathcal{L} = \tfrac{1}{2}(\dot{\phi}^2 + sin^2\phi\,\dot{\theta}^2) - V(cos\phi),$$

and the corresponding hamiltonian:

$$H = \tfrac{1}{2}(p_\phi^2 + sin^{-2}\phi\, p_\theta^2) + V.$$

The system possesses a rotational symmetry about z-axis, whose generator by Noether's Theorem gives a conserved integral (z-component of the angular momentum)

$$p_\theta = \dot{\theta} = g - const.$$

One could verify directly the commutation relation $\{H;p_\theta\} = 0$! As the result the system is reduced to a 1-st order ODE in ϕ-variable,

$$\tfrac{1}{2}(\dot{\phi}^2 + \frac{g^2}{sin^2\phi} + V(cos\phi) = E;\ \dot{\theta} = g$$

which is then integrated as in §8.1,

$$\int \frac{d\phi}{\sqrt{2(E - ...)}} = t - t_0;\ \theta - \theta_0 = gt.$$

3.2. Kepler 2-body Problem. The motion of a 2-body system about the joint center of mass is described by the Lagrangian, $\mathcal{L} = \frac{m}{2}\dot{q}^2 - V$, with potential $V = -\frac{\gamma}{|q|}$, the corresponding hamiltonian $H = \frac{1}{2m}p^2 + V$. In spherical coordinates $(r;\phi;\theta)$,

$$\mathcal{L} \quad \tfrac{1}{2}(\dot{r}^2 + r^2(\dot{\phi}^2 + sin^2\phi\,\dot{\theta}^2)) - V(r),$$

and

$$H = \tfrac{1}{2}(p_r^2 + r^{-2}(p_\phi^2 + sin^{-2}\, p_\theta^2)) + V(r). \tag{3.1}$$

The $SO(3)$-rotational symmetry of central potential V yields conservation of the angular momentum $J = q \times p$, by Noether's Theorem. In particular, the direction of J remains constant, hence 2-body motion is always planar (in the plane orthogonal to L). The Kepler problem is thus reduced to a plane motion, where

$$\mathcal{L} = \tfrac{1}{2}(\dot{r}^2 + r^2\dot{\theta}^2) - V(r);\ H = \tfrac{1}{2}(p_r^2 + r^{-2}p_\theta^2) + V(r).$$

But this $\mathcal{L}$ also possesses a rotational symmetry, $\theta \to \theta + \epsilon$. Hence the Noether integral $p_\theta = g$ – const, i.e. $\{p_\theta; H\} = 0$. Thus the 2-D system is completely (Liouville) integrable. Moreover, two integrals reduce it to a 1-st order ODE, as above

$$\tfrac{1}{2}(\dot{r}^2 + \frac{g^2}{r^2}) + V(r) = E, \Rightarrow \int \frac{dr}{\sqrt{2(E - ...)}} = t;\ \theta = \theta_0 + gt.$$

So we get a family of solutions $\begin{Bmatrix} r(t; E, g) \\ \theta(t; E, g) \end{Bmatrix}$ depending on 2 parameters (constants of integration) E, g. To show complete (Liouville) integrability of the 3-D central potential problem does not require, however, the 2-D (plane) reduction. Indeed, 3 commuting integrals can be exhibited directly:

$$H = \tfrac{1}{2}p^2 + V(r);\ J_z = J_{12} \text{ and } J^2 = J_x^2 + J_y^2 + J_z^2.$$

The commutation relation: $\{J_z; J^2\} = 0$, follows from the $\mathbf{so}(3)$ Lie bracket relations. The reader could easily recognize J^2, as Casimir (central) element of the enveloping algebra of $\mathbf{so}(3)$.

3.3. Euler rigid body problem. The rigid body in $\mathbb{R}^3$ has a density distribution $\{\rho(x)\}$. In the absence of external force its motion consists of the free (linear) motion of the *center of mass*, $\bar{x} = \int x\rho dx$,

$$\bar{x} = vt + b,$$

(momentum conservation), and rotations about $\bar{x}$. We shall restrict our attention to nontrivial rotational components of the dynamics, i.e. study it in an *inertial coordinate system*[6] with origin at the center of mass O.

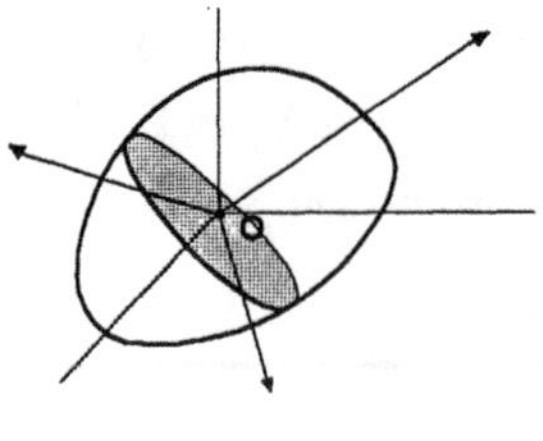

Fig.2: *Rigid body with 2 coordinate frames, the fixed frame (dashed) and the moving frame (solid).*

Two coordinate frames can be assigned to a rotating body: the fixed *rest frame* (of the ambient space), and moving *body frame*, both centered at O (fig.2). We shall

[6] i.e. system whose axis move at a constant velocity v.

consider all kinematic and dynamic parameters of the body in both frames, and use lower case letters for the former and Capitals for the latter. The states of the system are given by orthogonal matrices $\{u = u_t\}$, that rotate a fixed coordinate body-frame at time 0, into its position at t.

	Rest frame	Moving Frame
position:	$q = u_t Q$	Q
velocity:	$v = \dot{q} = \omega \times q = u_t(V)$	$V = \Omega \times Q$
angular velocity:	$\omega = \dot{u}u^{-1} = u_t(\Omega)$	$\Omega = u^{-1}\dot{u}$
angular momentum	$j = \int v \times q d\rho = \int (\omega \times q) \times q d\rho$	$J = \int V \times Q = \int (\Omega \times V) \times V d\rho$
	$j = u_t(J)$	$J = B\Omega$
acceleration:	$a = \ddot{q} = \dot{\omega} \times q + \omega \times (\omega \times q)$	$A = u_t^{-1}(a) = \dot{\Omega} \times Q + \Omega \times \Omega \times Q.$

The angular velocities $\{\omega;\Omega\}$ here could be understood in two possible ways, as antisymmetric matrices (elements of Lie algebra $\mathfrak{so}(3)$), or 3-vectors: $\omega, \Omega \in \mathbb{R}^3 \simeq \mathfrak{so}(3)$. So operation $Q \to \Omega \times Q$, coincides with the adjoint/co-adjoint action, $ad_\Omega(Q)$, under the standard identification of $\mathbb{R}^3$ and $\mathfrak{so}(3)$,

$$Q = (x; y; z) \leftrightarrow \begin{bmatrix} & y & z \\ -y & & x \\ -z & -x & \end{bmatrix}.$$

Similarly, multiplication with $u \in SO(3)$ corresponds to the adjoint action: $\mathrm{ad}_u(Q) = u^{-1}Qu$. A symmetric matrix

$$B = \int (ad_Q^2) d\rho = \int (Q^2 I - {}^{\mathrm{T}}Q\,Q) d\rho$$

represents the *inertia tensor* of the body (in the fixed body-frame),

$$B = \int \begin{bmatrix} y^2+z^2 & -xy & -xz \\ -yx & x^2+z^2 & -yz \\ -zx & -zy & x^2+y^2 \end{bmatrix} d\rho.$$

The inertia tensor relates angular momentum to angular velocity,

$$\boxed{J = B\Omega},$$

the same way as mass/metric tensor relates momentum to velocity in the standard,

$$\boxed{p = mv}$$

Now the basic Newton's law: $\frac{d}{dt}p = \boldsymbol{F}$-force, takes on a form:

$$\frac{d}{dt}j = T = \int F \times q d\rho\text{-torque.}$$

We can write the latter as

$$\frac{d}{dt}j = u_t(\dot{J} + J \times \Omega) = ...$$

In the absence of external force the RHS (torque) is 0. So we get a system of equations for unknown (body frame) parameters: $J(t); \Omega(t)$,

$$\dot{J} + J \times \Omega = 0;$$
$$J = B\Omega$$

or

$$\boxed{B\dot{\Omega} + B\Omega \times \Omega = 0} \tag{3.2}$$

To simplify equations (3.2) and write them down explicitly we introduce a special *inertial body frame*, where matrix/tensor B is diagonalized,

$$B = \begin{bmatrix} I_x & & \\ & I_y & \\ & & I_z \end{bmatrix};$$

the eigenvalues of B are called *principal moments of inertia.* If $\Omega = (\omega_1; \omega_2; \omega_3)$, then (3.2) turns into a system of 1-st order equations for $\{\omega_j\}$ with quadratic nonlinearities,

$$\begin{cases} I_x\dot{\omega}_1 + (I_y - I_z)\omega_2\omega_3 = 0 \\ I_y\dot{\omega}_2 + (I_z - I_x)\omega_1\omega_3 = 0 \\ I_z\dot{\omega}_3 + (I_x - I_y)\omega_1\omega_2 = 0 \end{cases};$$

whose solutions are given in terms of *Jacobi cnoidal (elliptic) functions* [Erd].

Let us remark that equations (3.2) arise in the usual way from the minimal-action principle with (kinetic) Lagrangian on the phase-space $T(G)$ or $T^*(G)$, of Lie group[7] $G = SO(3)$,

$$\mathcal{L} = K = \tfrac{1}{2}\int |\omega \times q|^2 d\rho = \tfrac{1}{2}\int |\Omega \times Q|^2 d\rho = \tfrac{1}{2}\langle B\Omega \mid \Omega\rangle = \tfrac{1}{2}\langle B^{-1}J \mid J\rangle. \tag{3.3}$$

Lagrangian $\mathcal{L}$ has 2 conserved integrals: hamiltonian, $H = \mathcal{L} = \frac{1}{2}\langle B^{-1}J \mid J\rangle$; and total angular momentum: $j^2 = J^2 = |B\Omega|^2$. Third integral can be chosen as any (e.g. z) component of the angular momentum. But more systematic reduction-procedure should exploit the momentum map. Let us notice that $\mathcal{L}$ is a special case of left-invariant Lagrangians on group $SO(3)$, considered in example 3, the metric being given by the

[7]It may be somewhat confusing to see Lagrangian (3.3) without velocity (time-derivatives) terms. The reader should keep in mind, however, that Ω itself represents the "velocity" (tangent) vector on G. So the 1-st order E-L equations (3.2) are, in fact, 2-nd order in the position-variable $\{u(t) \in G\}$. However, (3.2) can solved directly for $\Omega(t)$, whence $u(t)$ is recovered by integration of $u^{-1}\dot{u} = \Omega$.

inertia tensor B,

$$\mathcal{L} = \tfrac{1}{2}tr(\xi B\xi);\ \xi \in \mathfrak{G} = \mathbf{so}(3);\ H = \tfrac{1}{2}tr(pB^{-1}p);\ p \in \mathfrak{G}^*.$$

Here Lie algebra $\mathfrak{G}$ and its dual $\mathfrak{G}^*$ are identified via the ad-invariant product

$$\langle \xi | \eta \rangle = tr(\xi\eta).$$

The momentum-map reduces H to a co-adjoint orbit $\mathcal{O} = \mathcal{O}(p_0)$ - a sphere in $\mathfrak{G} \simeq \mathbb{R}^3$, with the natural (invariant) symplectic structure $\Omega = sin\phi d\phi \wedge d\theta$, and the reduced hamiltonian becomes, $\tfrac{1}{2}\langle B^{-1}p \mid p\rangle$, restricted on $\mathcal{O}$. So we see once again the integrability to result from the symmetry-reduction!

3.4. Two-center gravity problem. We place two centers at points $\{-a; a\}$ on the z-axis. The corresponding Lagrangian is given by

$$\mathcal{L} = \tfrac{1}{2}\dot{q}^2 - V;\ V(q) = -\frac{\gamma_1}{|q-a|} - \frac{\gamma_2}{|q+a|}.$$

It possesses an obvious rotational symmetry about z-axis, so $J_z = p_\theta$ is an integral. The existence of another commuting integral is not so obvious. It results from a hidden symmetry of the *Coulomb potential:* $V = \frac{\gamma}{|q|}$, the so called *Lagrange-Laplace-Runge-Lenz vector*,

$$\boxed{L = p \times J - \gamma\frac{q}{|q|}}$$

where $J = p \times q$ is the angular momentum. One can verify the commutation relation $\{L_i; H\} = 0$, and show that a system of hamiltonians

$$\{J_x;\ J_y;\ J_z;\ L_x;\ L_y;\ L_y;\ L_z\}$$

satisfy the Lie algebra brackets of $\mathbf{so}(4)$ or $\mathbf{so}(1;3)$ depending on sign of γ, with the angular momentum part represented by $\mathbf{so}(3)$-matrices, while the Runge-Lenz occupying the $\mathfrak{P}$-component,

$$J = \begin{bmatrix} 0 & 0 & 0 & 0 \\ 0 & 0 & \gamma & \beta \\ 0 & -\gamma & 0 & \alpha \\ 0 & -\beta & -\alpha & 0 \end{bmatrix};\ \text{and } L = \begin{bmatrix} 0 & a & b & c \\ \pm a & 0 & 0 & 0 \\ \pm b & 0 & 0 & 0 \\ \pm c & 0 & 0 & 0 \end{bmatrix}; \tag{3.4}$$

$J = \alpha J_x + \beta J_y + \gamma J_z$ and $L = aL_x + bL_y + cL_z$ (problem 2).

In particular, hamiltonians J_z and L_z commute. Furthermore, the Coulomb potential is the only one (among other central potentials) to possess the Runge-Lenz type symmetry (problem 1).

For the two-center gravity problem we modify the definition of the *Runge-Lenz vector* to be

$$L = p\times J - \gamma_1 \frac{q-a}{|q-a|} - \gamma_2 \frac{q+a}{|q+a|}.$$

The new L is no more constant under the hamiltonian flow, but one has

Proposition: *Hamiltonians H and L satisfy the Poisson bracket relation*

$$\{J;H\} = a\times\left\{\left(\frac{\gamma_1}{r_1^3}q_1 + \frac{\gamma_2}{r_2^3}q_2\right)\times p\right\},$$

where $q_1 = q-a$; $q_2 = q+a$; $r_{1,2} = |q_1 - q_2|$.

The proof involves standard calculations with Poisson brackets

$$\{L;H\} = \{p\times L - \frac{\gamma_1 q_1}{r_1} - ...; \frac{p^2}{2} + \frac{\gamma_1}{r_1} + ...\} = \{p\times L; \frac{\gamma_1}{r_1} + ...\} - \{\frac{\gamma_1 q_1}{r_1} + ...; \frac{p^2}{2}\}. \quad (3.5)$$

The first bracket

$$\{p\times L; r_1^{-1}\} = -\frac{q_1}{r_1^3}\times L + p\times\left(\frac{q_1}{r_1^3}\times q\right);$$

while the second

$$\{\frac{q_1}{r_1}; \frac{p^2}{2}\} = r_1^{-3}(pq_1^2 - p\cdot q_1 q_1),$$

and similar relations hold for q_2; r_2 -terms. Combining the r_1-terms of (3.5) we get

$$\frac{\gamma_1}{r_1^3}[p\times(q_1\times q) - q_1\times(p\times q) + q_1\times(p\times q_1)]. \quad (3.6)$$

Remembering that $q = q_1 + a$, and interpreting $\times$ as adjoint action of $\mathbf{so}(3)$,

$$a\times b \leftrightarrow [a;b] = ad_a(b),$$

we get in (3.6)

$$ad_{q_1}ad_a(p) - ad_{[q_1;a]}(p) = ad_a ad_{q_1}(p) = a\times(q_1\times p), \text{ QED.}$$

It follows from the Proposition, that z-component of L (in the direction of vector a), Poisson commutes with H, $\{L_z;H\} = 0$. Also J_z and L_z commute, as in the one-gravity-center case (check!). Thus we get a complete set of commuting integrals: $H; J_z; L_z$. In cylindrical coordinates $(r;\theta;z)$ they become

$$\begin{aligned} &H = \tfrac{1}{2}(p_r^2 + p_z^2 + r^{-2}p_\theta^2) - \frac{\gamma_1}{\sqrt{r^2+(z-a)^2}} - \frac{\gamma_2}{\sqrt{r^2+(z+a)^2}} = E; \\ &J_z = p_\theta = g; \\ &L_z = rp_r p_z - z(p_r^2 + r^{-2}p_\theta^2) - \frac{\gamma_1(z-a)}{\sqrt{r^2+...}} - \frac{\gamma_2(z+a)}{\sqrt{r^2+...}} = F. \end{aligned} \quad (3.7)$$

Substituting $p_r = \dot r$; $p_z = \dot z$; $p_\theta = r^2\dot\theta = g$ in (3.7), it is reduced to a 1-st order system for $\dot r$; $\dot z$,

$$\begin{cases} \dot r^2 + \dot z^2 = 2\left(E + \frac{\gamma_1}{\sqrt{...}} + \frac{\gamma_2}{\sqrt{...}}\right) - \frac{g^2}{r^2} \\ r\dot r\dot z - z(\dot r^2 + g^2/r^2) = F + \frac{\gamma_1(z-a)}{\sqrt{...}} + \frac{\gamma_2(z+a)}{\sqrt{...}} \end{cases}$$

from which $\dot r$ and $\dot z$ can be found in terms of other constants and variables,

$$\begin{cases} \dot{r} = f(r;\theta;E,F,g) \\ \dot{z} = g(r;\theta;...) \\ \dot{\theta} = g/r^2 \end{cases}.$$

The latter could in principle be solved exactly.

We shall demonstrate it in the case of the planar motion using elliptical coordinates (see problem 4 of §8.1), and the Liouville's method of separating variables. The 2-center planar problem still possesses a Runge-Lenz symmetry,

$$L_y = (J^2 - a^2 p_y^2) + ax(\frac{\gamma_1}{r_1} - \frac{\gamma_2}{r_2}).$$

Here we use $(x;y)$-coordinates in the plane with centers placed along the y-axis (see fig.3).

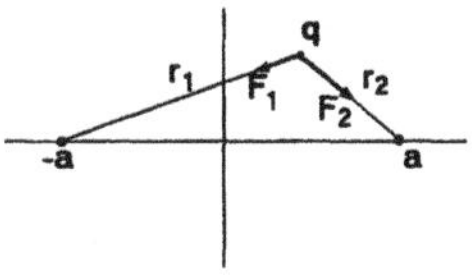

Fig.3. *A two gravity-center system with masses at* $\{\pm a\}$.

Vectors J is orthogonal to the plane, while L lies in the plane. Passing to elliptical coordinates:

$$\xi = \frac{r_1+r_2}{2};\ \eta = \frac{r_1-r_2}{2},$$

where $\{r_1;r_2\}$ are distances the from the moving point to the gravity centers, we get

$$ax = \xi\eta;\ ay = \sqrt{(\xi^2-a^2)(a^2-\eta^2)};\ r^2 = \xi^2 + \eta^2 - a^2.$$

The hamiltonian in elliptical coordinates takes the form

$$H = \frac{1}{2(\xi^2-\eta^2)}\left\{(\xi^2-a^2)p_\xi^2 + (a^2-\eta^2)p_\eta^2\right\} - \frac{\gamma\xi+\gamma'\eta}{\xi^2-\eta^2};\ \gamma = \gamma_1+\gamma_2;\ \gamma' = \gamma_1-\gamma_2.$$

The resulting H belongs to a general class of hamiltonian systems, studied by Liouville (1849), that allow separation of variables. Liouville hamiltonians H have the kinetic and potential-energy terms of the form,

$$K = \frac{1}{2B}\sum_{j=1}^{n} a_j p_j^2;\ \text{and}\ V = \frac{1}{B}\sum_{j=1}^{n} V_j; \tag{3.8}$$

where $B = \sum B_j$, and all functions $\{a_j;\ B_j;\ V_j\}$ depend on a single variable q_j. It is easy to check that (3.8) has n Poisson-commuting integrals

$$F_j = \tfrac{1}{2}a_j p_j^2 + V_j - Hb_j; \tag{3.9}$$

which also commute with H (problem 3). Integrals $\{F_j\}$ are not independent, since their sum,

$$\sum_{j=1}^{n} F_j = 0.$$

But H along with any $(n-1)$ of $\{F_j\}$ form a system of n commuting integrals, so (3.8) is completely integrable. Furthermore, fixing the energy-level: $H = E$, and the values of conserved integrals: $F_j = \alpha_j$, (3.9) is reduced to a system of uncoupled ODE's,

$$\frac{B}{2a_j}\dot{q}_j^2 + V_j - Eb_j = \alpha_j;\ j = 1;2;...$$

The latter could be solve explicitly (in quadratures) by writing

$$\frac{dq_j}{\sqrt{2a_j(\alpha_j + Eb_j - V_j)}} = \frac{dt}{B(q_1...q_n)};\ j = 1;2;...\ n. \tag{3.10}$$

One first introduces "local time"

$$d\tau = \frac{dt}{B(q_1;...q_n)}, \tag{3.11}$$

then solves ODE's (3.10) with the RHS $= d\tau$, to get solutions $\{q_j = q_j(\tau)\}$, and substitutes them in (3.11) to produce "physical time" t in terms of the "local time" τ,

$$t = \int^{\tau} B(q_1;...q_n)d\tau.$$

So any Liouville system is *integrable completely and exactly* (in quadratures)! Returning to the 2-center problem we find

$$B = \xi^2 - \eta^2;\ a_1(\xi) = \xi^2 - a^2;\ a_2(\eta) = a^2 - \eta^2;\ V_1(\xi) = -\gamma\xi;\ V_2(\eta) = -\gamma'\eta.$$

Hence system (3.10)-(3.11) becomes

$$\frac{d\xi}{\sqrt{R(\xi)}} = \frac{d\eta}{\sqrt{Q(\eta)}} = d\tau;\ dt = (\xi^2 - \eta^2)d\tau; \tag{3.12}$$

with quatric polynomials

$$R = 2(\xi^2 - a^2)(E\xi^2 + \gamma\xi + \alpha);\ Q = 2(\eta^2 - a^2)(E\eta^2 - \gamma'\eta - \alpha).$$

So $\xi(\tau)$, $\eta(\tau)$ can be expressed through the Jacobi elliptical functions ([Erd]).

Problems and Exercises:

1. Show, if a "central-force" hamiltonian $H = \frac{1}{2}p^2 + V(|q|)$ Poisson commutes with
$$L = p \times J - f(|q|)\frac{q}{|q|},$$
then $f = \gamma$ - const, and $V = \frac{\gamma}{|q|}$ - Coulomb potential.
2. Verify the Lie bracket relations (3.4) of $so(4)$, or $so(1;3)$ for a combined "angular momentum-Runge-Lenz algebra" $\{J; L\}$.
3. Verify that Liouville integrals $\{F_j\}$ (3.9) are in involution and Poisson-commute with the hamiltonian H of (3.8).

§8.4. Integrable systems related to classical Lie algebras.

Many interesting examples of integrable hamiltonians arise as interacting n-particle systems on $\mathbf{R}$, where $H = \frac{1}{2}\sum p_j^2 + V$, and potential V is made of all two body interactions: $V = \sum v_{ij}$, $v_{ij} = v(q_i - q_j)$, or nearest neighbor interactions. The foremost of those are Calogero-Moser systems: $v = 1/x^2$ (and other special v), and the Toda lattice: a chain of nonlinearly coupled mass-spring systems: $v = exp(q_i - q_{i+1})$. Both types turned out to be closely related to some classical Lie algebras, e.g. $sl(n)$.

In this section we shall develop the basic Lax-pair formalism for such systems, then outline the so called projection method following Olshanetski-Perelomov (see [Pe]). The idea is to view such systems as reductions (projections) of the geodesic flow on "large" symmetry groups, or symmetric spaces $\{\mathcal{H}\}$. Although the number of variables increases, in passing from $\mathbf{R}^n$ to $\mathcal{H}$, we gain in simplicity of the ensuing flows. While the Lax-pair formalism proves integrability of the Calogero-Moser and Toda hamiltonians, the projection method yields an effective computational procedure for conserved integrals.

4.1. N-body hamiltonians. We shall study hamiltonians of the form $H = \frac{1}{2}\sum p_j^2 + V$, where potential

$$V = \sum_{1 \le i < j \le n} v(q_i - q_j); \tag{4.1}$$

- sum of all 2-body interactions, or

$$V = \sum_i v(q_i - q_{i+1}) \tag{4.2}$$

- nearest neighbor interactions. The former are often called *Calogero-Moser systems* (for special v), the latter are exemplified by the well known *Toda lattice.* It turns out that for special classes of 2-body potentials $v(q)$ the resulting n-body hamiltonians (4.1) and (4.2) are completely integrable. For (4.1) these functions are

(I) $v = \frac{1}{q^2}$;

(II) $v = \frac{a^2}{sinh^2 aq}$; or $-\frac{a^2}{cosh^2 aq}$;

(III) $v = \frac{a^2}{sin^2 aq}$; or $-\frac{a^2}{cos^2 aq}$;

(IV) $v = a^2\wp(aq)$ - *Weierstrass $\wp$-function*

(V) $v = q^{-2} + \omega^2 q^2$

(VI) *Toda lattice:* $V = \sum_i b_i^2 e^{-a(q_i - q_{i+1})}$.

Systems (I-VI) are closely related to certain classical Lie groups and their homogeneous spaces G/K. As we shall see they represent "projections" (via symmetry reductions) of the geodesic flow (free motion) on such spaces.

The latter can be illustrated by 1-body potentials of the type (I-VI). Indeed, free motion on the Euclidian 2-plane with hamiltonian

$$H = \tfrac{1}{2}(p_x^2 + p_y^2),$$

becomes upon the radial reduction with fixed angular momentum $p_\theta = g$,

(I) $H = \frac{1}{2}p_r^2 + \frac{g^2}{r^2}$.

Similarly the $\mathbb{R}^2$ harmonic oscillator $H = \frac{1}{2}p^2 + \omega^2(x^2+y^2)$, reduces to

(V) $H = \frac{1}{2}p_r^2 + \frac{g^2}{r^2} + \omega^2 r^2$.

Hamiltonians of type (II)-(III) arise in the radial reduction of the free (geodesic) motion on the hyperboloid

$$\{x \mid x \cdot x = x_0^2 - x_1^2 - x_2^2 = 1\},$$

embedded in the Minkowski 3-space with the indefinite metric:

$$ds^2 = dx_1^2 + dx_2^2 - dx_0^2,$$

and the standard Euclidian 2-sphere

$$\{x_0^2 + x_1^2 + x_2^2 = 1\}$$

respectively. Let us also remark that potentials (I-III) are special cases of the Weierstrass function (IV).

4.2. The Lax-pair formalism. Most known examples of integrable systems (both in finite and infinite-D) can be recast in the form of so called *Lax pair formalism.* Namely, each point in the phase-space $z = (q; p)$ is assigned a pair of linear operators (matrices):

$$L = L(z) \text{ (typically self-adjoint), and } M = M(z),$$

in such a way that the Hamiltonian evolution is equivalent to the *Lax equation*

$$\boxed{i\dot{L} = [L; M] = LM - ML} \tag{4.3}$$

We shall see that evolution equation (4.3) preserves eigenvalues of operator $L(t)$, so we immediately get a family of conserved integrals: eigenvalues $\{\lambda_k(q;p): 1 \le k \le n\}$, or some functions of $\{\lambda_k\}$, e.g. characteristic coefficients of L.

Proposition 1: *For any operator function $M = M(t)$, the eigenvalues of $L(t)$, that solves equation (4.3), remain constant $\{\lambda_k(t) = \lambda_k(0)\}$.*

If M were constant, then $L(t) = e^{-itM} L(0) e^{itM}$ would be given by conjugating $L(0)$ with one-parameter subgroup, generated by M, so the result would obviously hold. In general, the role of the group $\{e^{itM}\}$ is played by the *propagator (fundamental solution)* $U(t)$ of the ODS:

$$\begin{cases} iU' = M(t)U \\ U(0) = I \end{cases}.$$

Then $L(t) = U(t)^{-1}L(0)U(t)$ is still isospectral to $L(0)$. Another argument is based on traces of powers of L. Indeed it suffices to check that all traces

$$\mathrm{tr}(L^m) = \sum \lambda_k^m \text{ - const, for } m{=}1;...n.$$

But derivative, $\frac{d}{dt}(L^m) = \sum L^{m-1-j}\dot{L}L^j$, so differentiating the trace of L^m, we get

$$\mathrm{tr}(\dot{L^m}) = m\,tr(L^{m-1}\dot{L}),$$

and the latter is equal to

$$\mathrm{tr}(L^{m-1}[L;M]) = 0, \text{by (4.3), QED.}$$

Our goal is to construct Lax pairs for the Calogero-Moser and Toda systems (I-VI) and to show that the resulting integrals: eigenvalues $\{\lambda_k(L)\}$, traces $\{\chi_m = tr(L^m)\}$, or the characteristic coefficients $\{b_m(L)\}$ (coefficients of the characteristic polynomial $p(\lambda) = \det(\lambda - L)$, are in involution.

4.3. Construction of Lax pairs. We use the following Ansatz for matrices L and M:

$$L = P + iX, \tag{4.4}$$

where the diagonal part $P = diag(p_1;...p_n)$ depends on the momentum variables, while the off-diagonal part $X = (x_{ij})$ is determined by an odd function $x(q)$ of positions only,

$$x_{ij} = x(q_i - q_j).$$

Similarly, matrix

$$M = Z + Y,$$

with the diagonal part $Z = diag(z_i)$, made of functions

$$z_i = \sum_j{}' z(q_i - q_j),$$

and the off-diagonal part $Y = (y_{ij})$, determined by an even function $y(q)$,

$$y_{ij} = y(q_i - q_j).$$

The Lax equation then takes the form

$$i\dot{P} - \dot{X} = [P;Y] + i[X;Z] + i[X;Z]. \tag{4.5}$$

Writing down matrix entries of (4.5) yields a system of equations for diagonal entries

$$i\dot{p}_j = i\sum{}'(x_{jk}y_{kj} - y_{jk}x_{kj}) = 2i\sum{}' x_{jk}y_{jk}; \tag{4.6}$$

(we took into account the parity of functions x, y). The off-diagonal entries satisfy

$$\dot{x}_{jk} = x'_{jk}(\dot{q}_j - \dot{q}_k) = (p_j - p_k)y_{jk} + i(z_j - z_k)x_{jk} - i\sum_{m \neq j,k}(x_{jm}y_{mk} - y_{jm}x_{mk}), \tag{4.7}$$

Comparing the RHS of (4.6) and (4.7) with the Hamiltonian system of potential

$$V = \sum v_{jk},\ v_{jk} = v(q_j - q_k)$$

we find

$$v'(q) = \tfrac{1}{2}x(q)\,y(q),$$

while the real and imaginary parts of (4.7) yield

$$Re\colon\ x'_{jk} = y_{jk};\ \text{so}\ \boxed{y(q) = x'(q)}$$

$$Im\colon\ x_{jk}(z_j - z_k) = \sum x_{jm}x'_{km} - x_{mk}x'_{jm}. \tag{4.8}$$

Introducing variables: $\xi = q_j - q_m$; $\eta = q_m - q_k$, and remembering the form of matrix entries $\{z_j\}$ of Z, formulae (4.8) yield a functional-differential equation for an odd function x and and even z,

$$\boxed{x(\xi+\eta)(z(\xi) - z(\eta)) = x(\xi)\,x'(\eta) - x(\eta)\,x'(\xi)} \tag{4.9}$$

The analysis of (4.9) is outlined in problem 3 (see [Ca];[Pe]). One can show that

$$\boxed{z(\xi) = \frac{x'(\xi)}{2\,x(\xi)}},$$

and function takes on one of the following expressions,

$$x(\xi) = \begin{cases} \xi^{-1}; \\ a\,coth(a\xi);\ a\,sinh^{-1}(a\xi) \\ a\,cot(a\xi);\ a\,sin^{-1}(a\xi) \\ a\,\frac{cn}{sn}(a\xi);\ a\,\frac{dn}{sn}(a\xi);\ \frac{a}{sn(a\xi)} \end{cases}$$

The last line gives χ in terms of *Jacobi elliptic functions: sn, cn, dn*. The corresponding potentials $v = x^2 + C$ are then found to be those of the list (I-IV).

Potentials of type (V), $v = \omega^2 q^2 + \frac{g^2}{q^2}$, require a slight modification of the basic Lax construction. One takes the evolution equation

$$i\dot{L} = [L; M] \pm \omega L;\ L = L^{\pm}. \tag{4.10}$$

Evolution (4.10) does not preserve eigenvalues of L, or traces $\chi_m = tr(L^m)$, but the latter can be shown to satisfy, $\chi_m(t) = \chi_m(0)\,exp(\pm i\omega t)$. Now the integrals of motion can be determined from a pair of auxiliary operators

$$N_1 = L^+L^-;\ N_2 = L^-L^+.$$

The latter satisfy the usual Lax equation,

$$i\dot{N}_i = [N_i; M],$$

so their powers and eigenvalues provide conserved quantities. Specifically, we take a pair of matrices

$$L^{\pm} = L \pm i\omega Q,$$

with the above L and diagonal $Q = diag(q_1; \dots q_n)$. Then a simple identity

$$[Q;M] = X,$$

with $x_{jk} = \frac{1}{q_j - q_k}$, along with the Lax equation: $i\dot{L} = [L;M]$, leads directly to the modified equation (4.10).

4.4. Complete integrability of Hamiltonian (I-IV). Integrability of systems (I-VI) will be established in two different ways.

1-st argument is fairly general and simple, but it applies to the *repulsive, short-range* potentials $v(q)$. The former means that derivative $v'(q) \leq 0$, so the Newton forces: $F_{ij} = -\partial v(q_i - q_j)$, between the i-th and j-th particles, have always the direction of $q_i - q_j$, i.e. particles repel each other. "Short range" refers to the effective particle interaction at large distances: *if potential $v(q)$ decays sufficiently fast*[8] *at $\{\infty\}$, then hamiltonian trajectories $\{q(t); p(t)\}$ become asymptotic to the free motion* (lines in the phase-space),

$$q_i(t) \sim q_{0i} + p_i t, \text{ as } t \to \infty.$$

So particles do not "effectively feel each other" at large distances. We observe that conserved integrals, $\chi_m = tr(L^m)$ with matrix L of (4.4) represent polynomials in the momentum variables $\{p_1; \dots p_n\}$, whose coefficients depending on $\{x(q_i - q_j)\}_{ij}$. One can show that for any repulsive, short-range potential v, and any hamiltonian path $\{q(t); p(t)\}$, the distances between pairs increase,

$$q_i(t) - q_j(t) \to \infty, \text{ as } t \to \infty,$$

hence their motion becomes asymptotically free. Therefore, large-time limits of $\{\chi_m\}$ turn asymptotically into functions of the momenta variables only. Precisely, if

$$\phi_t: (q_0; p_0) \to (q(t); p(t)),$$

denotes the hamiltonian evolution of H in $\mathbb{R}^{2n}$, then

$$\chi_m(q(t); p(t)) \to \chi_m(p) = \sum p_j^m \text{ - } m^{th} \text{ Newton symmetric polynomial, as } t \to \infty.$$

Obviously, the infinite-time limits of $\{\chi_m\}$ Poisson commute. But the Hamiltonian flow preserves Poisson brackets,

$$\{F \circ \phi_t; G \circ \phi_t\} = \{F; G\} \circ \phi_t,$$

for any pair of observables F, G on $\mathbb{R}^{2n}$. Since a commutator of two integrals $\{\chi_j; \chi_k\}$ is also an integral, and,

$$\lim_{t \to \infty} \{\chi_j; \chi_k\} \circ \phi_t = 0,$$

it follows that $\chi_j; \chi_m$ commute for all time t, including t=0, which proves complete

[8] It is often sufficient to require $\int_0^\infty | v(q) | \, dq < \infty$.

integrability (see problem 4)!

4.2. Remark: Let us briefly discuss the *scattering process* for type-I hamiltonians. Scattering refers to hamiltonian dynamics with certain asymptotic behavior at large time, typically linear (free) motion:

$$(q(t);p(t)) \sim (q^{\pm} + tp^{\pm};p^{\pm}), \text{ as } t \to \pm\infty.$$

So the "$-\infty$" asymptotic state $(q^-;p^-)$ is transformed into "$+\infty$" state $(q^+;p^+)$, by a canonical (symplectic) map $\mathcal{S}$, called *scattering map.* Ordering particles according to their relative position on $\mathbf{R}$ (from left to right), we can write

$$p_1^+ < \dots < p_n^+;\ p_1^- > \dots > p_n^-.$$

Comparing conserved integrals at both extremes, $\chi_m(p^+) = \chi_m(p^-)$, we immediately conclude that, $p_1^- = p_n^+;\ p_2^- = p_{n-1}^+;\ \dots$ A similar relation holds for relative position vectors, $q_1^- = q_n^+;\ q_2^- = q_{n-1}^+;\ \dots$ So the scattering transform amounts to reordering particles.

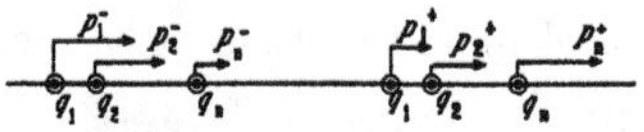

Fig.4 *illustrates the scattering process for a train of particles traveling along* $\mathbf{R}$ *asymptotically free at* $\{\pm\infty\}$. *At* $t = -\infty$ *the first particle* q_1 *has the highest momentum, and the last* q_n *the lowest. They interact via pair potentials* $\{v(q_i - q_j)\}$ *and exchange momenta. As the result the left particle* (q_1) *becomes the slowest, while the right one* (q_n) *gets the highest momentum, with the rest of them occupying intermediate states.*

The above argument does not apply, however, to the *"long-range"*, or *attractive, potentials* (II-VI). So we shall give another (direct) argument to verify vanishing of the Poisson-bracket for eigenvalues $\{\lambda_k(q;p)\}$ of L, which applies to all cases.

2-nd argument. Given a pair of eigenvalues $\{\lambda;\mu\}$ and eigenvectors $\phi = (\phi_1;\dots\phi_n)$; $\psi = (\psi_1;\dots\psi_n)$ of L,

$$L\phi = (P + X)\phi = \lambda\phi;\ L\psi = \mu\psi; \tag{4.11}$$

we differentiate (4.11) in variables q and p,

$$\frac{\partial\lambda}{\partial p_k} = \langle\phi|\frac{\partial L}{\partial p_k}\phi\rangle = \bar\phi_k\phi_k;\ \frac{\partial\lambda}{\partial q_k} = \langle\phi|\frac{\partial L}{\partial q_k}\phi\rangle = i\sum x'_{kj}(\bar\phi_k\phi_j - \bar\phi_k\phi_j).$$

Hence,

$$\{\lambda;\mu\} = i\sum_{kj} x'_{kj}(\bar\phi_k\bar\psi_k R_{kj} - \phi_k\psi_k\bar R_{kj});$$

where $R_{kj} = \phi_k\psi_j - \phi_j\psi_k$. Next we make use of the relation

$$\phi_k\psi_m = \frac{i}{\lambda - \mu}\sum x_{kj}R_{kj};$$

that results from (4.11), and the functional equation (4.9), to bring the bracket $\{\lambda;\mu\}$ into the form

$$\frac{i\lambda}{\mu-\lambda}\sum_{k\neq j}(\bar\psi_k\bar\phi_j R_{kj}+\psi_k\phi_j\bar R_{kj})z_{jk}-\frac{i\mu}{\mu-\lambda}\sum_{k\neq j}(\bar\phi_k\bar\psi_j R_{kj}+\phi_k\psi_j\bar R_{kj})z_{jk}.$$

Since the expressions inside the first and second sum are antisymmetric in (jk), each sum vanishes. Thus we establish integrability of systems (I-VI).

4.5. Integration of the equation of motion. The Projection Method. We have shown complete integrability of hamiltonians (I-VI) by exhibiting a family of Poisson commuting integrals $\{\lambda_k(L)\}$ or $\{\chi_m(L)=\frac{1}{m}\mathrm{tr}(L^m)\}$. Integrals $\{\lambda_k\}$, however, do not yield an explicit solution of the hamiltonian system. The latter will be achieved via the so called Projection method due to Olshanetsky-Perelomov [OP]; [Per]. The idea is to lift the dynamics from n degrees of freedom $(q_1;...q_n)$ to a larger space of dimension $N=n^2-1$, related to some Lie group, where the motion becomes free (geodesic).

Namely, we take the space $\mathcal{H}$ of $n\times n$ hermitian (complex) matrices of trace 0, and consider a free motion: $x(t)=at+b$, i.e. $H=\frac{1}{2}p^2$. The reduction (projection) from $\mathcal{H}$ to $\mathbb{R}^n$ consists of diagonalizing matrix x, $x\to uQu^{-1}$, where $Q=diag(q_1;...\,q_n)$, is made of eigenvalues $\{q_j(x)\}$, and u is unitary.In case n=2,

$$\mathcal{H}=\left\{x=\begin{bmatrix}\alpha & w\\ \bar w & -\alpha\end{bmatrix};\ \alpha\in\mathbb{R};\ w\in\mathbb{C}\right\};$$

is 3-dimensional, $Q=\begin{bmatrix}q & 0\\ 0 & -q\end{bmatrix}$ and the projection consists of radial reduction on the free motion in $\mathbb{R}^3$ for a fixed value of the *total angular momentum*,

$$L^2=|J|^2=p_\phi^2+sin^{-2}\phi\,p_\theta^2;\ J=p\times q.$$

Indeed, in spherical coordinates hamiltonian $H=\frac{1}{2}(p_r^2+\frac{L^2}{r^2})$, turns into the type-I two-body problem upon reduction of the center of mass coordinate[9] $q=\frac{1}{2}(q_1-q_2)$.

Let us return to the general case. To find the evolution of $\{q_j\}$ and of the conjugate momenta $\{p_j=\dot q_j\}$ we differentiate the relation,

$$uQu^{-1}=at+b,$$

$$u\bigcap\dot Q-u^{-1}\dot uQ-Qu^{-1}\dot u\Leftrightarrow u^{-1}=a$$

and rewrite the latter as

[9]The center of mass coordinate $\bar q=\frac{1}{M}\sum m_jq_j$ $(M=\sum m_j$-total mass) moves rectilinearly (at a constant speed), due to the momentum conservation (§8.2):

$$\frac{d}{dt}\bar q=\frac{1}{M}P\text{-const.}$$

Hence, the dynamics of the n-body system can always be reduced (constrained) by the number of degrees of freedom of $\bar q$.

$$\dot{x} = u(t)L(t)u^{-1}(t) = a. \tag{4.12}$$

Introducing operators

$$P = \dot{Q},\ M = -iu^{-1}\dot{u},$$

and

$$L = P + i[M;Q]; \tag{4.13}$$

and differentiating (4.12) once more yields the Lax equation for the pair L, M

$$\dot{L} + i[M;L] = 0.$$

The eigenvalues of L (or its traces) are the familiar conserved integrals. But we we need to compute matrix-function $Q(t)$ explicitly. Let us observe that the type-I Lax pair $L = P + iX$, and $M = Z + Y$ of the previous section ($v = q^{-2}$), does satisfy (4.13). But the initial data a, b can not be chosen arbitrary in $\mathcal{H}$ to be able to project to a type-I trajectory in the q-space. Such data must be constraint by the angular momentum operator[10],

$$J = i[x;\dot{x}] = i[a;b] = iu[Q;L]u^{-1}.$$

Operator J can be computed for L and Q of the previous section, and is found to be a matrix with all off-diagonal entries equal to constant g and the diagonal 0,

$$J = g(\xi \otimes \xi' - I); \xi = (1;1;...1).$$

Now we can write down the solution-curves of the equations of motion. Initial position variable $b = x(0) = Q_0$ can be assumed (without loss of generality) diagonal $(q_1;...q_n)$. Initial momentum a is taken to be

$$a = L_0 = P_0 + iX_0,$$

where P_0 denotes the initial momenta, and matrix X has entries

$$x_{ij}(0) = \frac{1}{q_i - q_j};$$

(so that $J = [a;b]$ is of the right type!). Then $x(t) = Q_0 + tL_0$ describes the free evolution in the extended space $\mathcal{H}$, whose "eigenvalue-projection" on the "Q-space" $\mathbb{R}^n$ gives the hamiltonian dynamics of type I. Thus we get a solution

[10]The angular momentum operator arises from the Noether $SU(n)$-symmetry of the free motion in $\mathcal{H}$. Indeed, the Lagrangian $\mathcal{L} = |\dot{x}|^2 = tr(\dot{x}\dot{x}^*)$, is invariant under conjugations, $x \to uxu^{-1}$, $u \in SU(n)$, whose infinitesimal generators $\xi = \xi(x) = i[x;\xi]$ are identified with elements $\xi \in su(n)$. Since the dual space of $\mathcal{H}$ is identical to itself via pairing: $(p;x) \to \langle p \mid x \rangle = tr(px^*)$, we get the conserved Noether integral, $J = \langle p \mid \xi(x) \rangle = \langle p \mid [\xi;x] \rangle = \langle \xi \mid [p;x] \rangle = const$, for all ξ. Hence, the angular momentum $J = [x;p]$, thought of as a $su(n)$ -valued map on the phase-space $\mathcal{H} \times \mathcal{H}$ is constant. Let us remark, that $SU(n)$ is a subgroup of yet larger symmetry-group $SO(N)$, $N = n^2 - 1$, so J consists of $su(n)$ - components of the total angular momentum of $\mathcal{L} = \frac{1}{2}|\dot{x}|^2$!

$$Q(t)=(\dots q_j(t)\dots)=\left\{\text{“eigenvalues” of } (Q_0+tL_0)=\begin{bmatrix} q_1+tp_1 & \cdots & \frac{it}{q_i-q_j} \\ \vdots & \ddots & \vdots \\ \frac{it}{q_j-q_i} & \cdots & q_n+tp_n \end{bmatrix}\right\}; \tag{4.14}$$

where $\{q_i;p_j\}$ are initial values of $\{q;p\}$. In other words, the initial data $\{q;p\}$ is lifted from phase-space $\mathbb{R}^n\times\mathbb{R}^n$ to a point $\{(Q;L): L=P+iX(q)\}$ in the extended space $\mathcal{H}\times\mathcal{H}$, then the free evolution in $\mathcal{H}$ is applied: $(Q+tL;L)$, and the result is projected back to $\mathbb{R}^n$. For $n=2$ the eigenvalues of Q_0+tL_0 can be computed explicitly (problem 1).

Systems of type II and III. The role of the flat space $\mathcal{H}$ will be played now by the hyperbolic (symmetric) space $\mathcal{M}=SL(n;\mathbb{C})/SU(n)$. The latter can be realized by all hermitian positive-definite matrices of $\det=1$, via the map, $x\to x^*x=r$. The free motion on space $\mathcal{M}$ is generated by all G-invariant vector fields,

$$\xi(r)=x^*\xi_0x,$$

where $r=x^*x$, and $\xi_0\in\mathcal{H}$, identified with the tangent space of $\mathcal{M}$, at $r_0=\{I\}$. The projection consists once again of diagonalization of $r\in\mathcal{M}$,

$$r=ue^{aQ}u^{-1}.$$

Differentiating the latter we can recast it into the Lax form:

$$i\dot{L}=[M;L],$$

with matrices

$$L=P+\tfrac{i}{4a}(e^{-2aQ}Me^{2aQ}-e^{2aQ}Me^{-2aQ});\ M=-iu^{-1}(t)\dot{u}(t). \tag{4.15}$$

Solution of (4.15) satisfying natural consistency condition are given by matrix $L=P+iX$, with entries $x_{ij}=a\,coth\,a(q_i-q_j)$. We can lift up the initial data to a pair of matrices $b=e^{aQ}$ (diagonal), and matrix $\mathfrak{A}$, which solve the equation:

$$2aL_0=b\mathfrak{A}b^{-1}+b^{-1}\mathfrak{A}b.$$

Thence we find the entries of $\mathfrak{A}$, $\mathfrak{A}_{jk}=ap_j\delta_{jk}+ia^2(1-\delta_{jk})sinh^{-1}a(q_j-q_k)$.

4.6. Poisson structure on co-adjoint orbits; the Toda lattice. We remind the definition of Poisson bracket on Lie algebra $\mathfrak{G}$ (§6.3). Let $F(x);H(x)$ be two observables (functions) on $\mathfrak{G}$. Then

$$\{F;H\}(x)=\langle x\mid[\partial F;\partial H]\rangle,\text{ or (in local coordinates)}\sum_{ijk}x_kC_{ij}^k\partial_iF\partial_jH, \tag{4.16}$$

where $\{C_{ij}^k\}$ denote structure constants on algebra $\mathfrak{G}$, relative to some basis $\{e_1;e_2;\dots\}\subset\mathfrak{G}$, and $\{x_j\}$-coordinates of $x\in\mathfrak{G}^*$ relative to the dual basis $\{e_1^*;e_2^*;\dots\}$. The corresponding hamiltonian dynamics on $\mathfrak{G}$ is given by

$$\dot{x} = ad^*_{\partial F}[x]. \tag{4.17}$$

Let us remark that Poisson bracket (4.12) on $\mathfrak{G}$ is highly degenerate, as all G-invariant functions,

$$I(ad^*_g(x)) = I(x), \text{ for all } g \in G,$$

equivalently,

$$ad^*_{\partial I(x)}[x] = 0, \text{ for all } x \in \mathfrak{G}^*$$

yields trivial flows (4.17). So such $\{I\}$ form trivial integrals of (4.17), Poisson commuting with all F on $\mathfrak{G}$! The degeneracy of (4.12) can be resolved by restricting the dynamics from $\mathfrak{G}^*$ to *co-adjoint orbits* of group G,

$$\mathcal{O} = \{ad^*_g(x): x \in G\} \simeq G_x \backslash G,$$

where G_x denotes the stabilizer subgroup of x. Clearly, stabilizer $\mathfrak{G}_x$ coincides with the null-space of the Poisson structure:

$$J_x(\xi;\partial F) = \langle x \mid [\xi;\partial F(x)]\rangle = 0, \text{ for } \xi \in \mathfrak{G}_x; \text{ and all } F.$$

So J becomes nondegenerate on the quotient $\mathfrak{G}/\mathfrak{G}_x \simeq T_x(\mathcal{O})$-tangent space of $\mathcal{O}$ at $\{x\}$, and furnishes $\mathcal{O}$ a symplectic structure. It turns out that many interesting examples of hamiltonian systems (classical and newly discovered) arise in this way. To begin we remark that the ad-invariant (Killing) product on $\mathfrak{G}$ identifies Lie algebra $\mathfrak{G}$ with its dual space $\mathfrak{G}^*$ and transforms (4.17) into the Lax form,

$$\dot{x} = [M;x], \text{ with } M = \partial F.$$

There are several ways to construct Poisson-commuting integrals on $\mathfrak{G}^*$, and on co-orbits $\mathcal{O} \subset \mathfrak{G}^*$. We shall outline 2 of them.

1-st method to produce commuting hamiltonians on the Lie algebra $\mathfrak{L}$ is to look for a larger algebra $\mathfrak{G} \supset \mathfrak{L}$. We assume that $\mathfrak{G}$ is decomposed into the direct sum: $\mathfrak{G} \simeq \mathfrak{L} \oplus \mathfrak{K}$. Our goal is to build commuting integrals on $\mathfrak{L}$, starting from $\mathfrak{G}$-invariants.

Theorem 2: *Any pair of $\mathfrak{G}$-invariant functions $F, H \in I(\mathfrak{G})$, restricted on $\mathfrak{L}$, Poisson commute on $\mathfrak{L}$.*

We denote two projections from $\mathfrak{G}$ to $\mathfrak{L};\mathfrak{K}$ by $\Pi_1;\Pi_2$, and call $F' = F \mid \mathfrak{L}$, $H' = H \mid \mathfrak{L}$ - the restrictions of F, H on $\mathfrak{L}$. Their dual spaces $\mathfrak{L}^* = \mathfrak{K}^\perp$ and $\mathfrak{K}^* = \mathfrak{L}^\perp$. For any function F on $\mathfrak{G}$ its gradient ∂F is decomposed into the sum of $\mathfrak{L}$ and $\mathfrak{K}$-components, $\partial F = \partial F_1 + \partial F_2$. We take point $x \in \mathfrak{L}^*$ and observe that for any $\mathfrak{G}$-invariant F,

$$\langle x \mid [\partial_1 F; \partial_1 H]\rangle = -\langle x \mid [\partial_2 F; \partial_1 H]\rangle;$$

since $\langle x \mid [\partial F; ...]\rangle = 0$. But H is also $\mathfrak{G}$-invariant, hence

$$\langle x \mid [\partial_2 F;\partial_1 H]\rangle = -\langle x \mid [\partial_2 F;\partial_2 H]\rangle,$$

and the latter is 0, since the commutator belongs to $[\mathfrak{K};\mathfrak{K}] \subset \mathfrak{K}$, while $x \in \mathfrak{L}^*$ is orthogonal to $\mathfrak{K}$. So the Poisson bracket $\{F';H'\} = \langle x \mid [\partial_1 F;\partial_1 H]\rangle = 0$, for all $F;H$ in $I(\mathfrak{G})$, QED.

Once again the inner product on $\mathfrak{G}$ allows to recast the hamiltonian dynamics

$$\dot{x} = ad^*_{\Pi_1(\partial F)}[x],$$

on an orbit $\mathcal{O} \subset \mathfrak{L}^*$ into the Lax form,

$$\dot{L} = [L;M]; \text{ where } L = x;\ M = \partial_1 F(x).$$

The Toda lattice. As an application of Theorem we shall discuss now the *Toda lattice*. Here algebra $\mathfrak{G} = sl(n;\mathbb{R})$; $\mathfrak{L} = \mathfrak{N}_-$ consists of lower-triangular matrices and $\mathfrak{K} = so(n)$. A $\mathfrak{G}$-invariant product is given by

$$(A;B) \to tr(AB);\ A,B \in sl(n).$$

So the dual space $\mathfrak{L}^* = \mathfrak{K}^\perp$ consists of symmetric matrices, $Sym(n)$. We take hamiltonian $H = \frac{1}{2}tr(L^2)$ on $\mathfrak{G}$, and the evolution equation in the Lax form,

$$\dot{L} = [M;L] = [\partial_2 H;L],$$

where $\partial_2 H$ denotes the $\mathfrak{K}$-(antisymmetric) part of $\partial H(L) = L$. If

$$L = L_+ + L_- + D;\ (L_- = {}^{T}L_+),$$

denotes a decomposition of L into the upper, lower and diagonal parts, then

$$M = L_+ - L_-.$$

The Toda lattice corresponds to a special choice of L. Namely, we pick a point

$$L_0 = \begin{bmatrix} 0 & 1 & & \\ 1 & \ddots & 1 \\ & 1 & 0 \end{bmatrix} \in \mathfrak{L}^*; \tag{4.18}$$

and take its co-orbit $\mathcal{O} = \{L = (u^{-1}L_0 u)_{sym}: u \in N_-\} \subset \mathfrak{L}^*$, which consists of all tridiagonal matrices

$$\left\{ L = \begin{bmatrix} b_1 & a_1 & \\ a_1 & \ddots & a_{n-1} \\ & a_{n-1} & b_n \end{bmatrix} : trL = \sum b_j = 0 \right\} \tag{4.19}$$

The Lax "mate" of L is found to be

$$M = \begin{bmatrix} 0 & a_1 & \\ -a_1 & \ddots & a_{n-1} \\ & -a_{n-1} & 0 \end{bmatrix};$$

The reader can verify the Poisson bracket relation for coordinate functions $\{a_i;b_j\}$ on $\mathcal{O}$,

$$\begin{gathered}\{a_i;a_j\} = \{b_i;b_j\} = \{a_i;b_j\} = 0;\\ \{a_i;b_{i-1}\} = a_i;\ \{a_i;b_{i+1}\} = -a_i.\end{gathered} \tag{4.20}$$

Those are, indeed, the Lie bracket relations for $\mathfrak{sl}_n$-matrices, representing $\{a_i;b_j\}$:

$$a_i \leftrightarrow \begin{bmatrix} 0 & & & \\ & & 1 & \\ & 1 & & \\ & & & 0 \end{bmatrix};\ b_j \leftrightarrow \mathrm{diag}(0;\dots 1;0;\dots -1).$$

Variables $\{a_i;b_j\}$ are not canonical, but one can easily pass to a canonical set, via the coordinate change,

$$b_j = p_j;\ a_i = e^{q_{i+1} - q_i}. \tag{4.21}$$

The hamiltonian H then takes the form

$$H = \tfrac{1}{2}tr(L^2) = \tfrac{1}{2}\sum_{i=1}^{n} b_i^2 + \sum_{i=1}^{n-1} a_i^2 = \tfrac{1}{2}\sum_{i=1}^{n} p_j^2 + \sum_{i=1}^{n-1} e^{2(q_{i+1} - q_i)}. \tag{4.22}$$

It is precisely form (4.22) that the Toda lattice was first discovered, namely as a chain/ lattice of n particles[11] on $\mathbb{R}$ with the nearest-neighbor exponential (nonlinear!) interactions. Functions $\{I_k = tr(L^k): k = 2;3;\dots n\}$ make up a complete set of commuting integrals. The corresponding flows are given by the Lax operators,

$$M_k = (L^k)_+ - (L^k)_-;$$

"upper triangular" – "lower triangular" parts of L^k. This result was discovered by P. van Moerbeke.

Another general method to cook up commuting integrals involves shifts in vector space $\mathfrak{G}$ of certain G–invariants functions (polynomials). One starts with any F in the algebra of ad-invariant functions, $I(\mathfrak{G})$, and forms a shifted family

$$\{F_{\lambda, a} = F(x+\lambda a): \text{all } \lambda \in \mathbb{R}\}.$$

The shifts turn out to form a complete, involutive family on $\mathfrak{G}$, i.e. hamiltonians $F_{\lambda, a}; F_{\mu, a}$ Poisson commute,

$$\{F_{\lambda, a}; F_{\mu, a}\} = 0,\ for\ \mu,\lambda, \tag{4.23}$$

and any function, Poisson commuting with all $\{F_{\lambda,a}\}$ belongs to the algebra, generated by them. The argument ([Per], chapter 1) exploits the structure theory of semisimple Lie

[11] Notice that coordinates $\{q_j\}$ are determined by (4.21) only modulo constant shift, $q_j \to q_j + c$. Constant c can be identified with the center of mass coordinate $\overline{q} = \frac{1}{n}\sum q_j$, that undergoes a uniform constant-speed motion, $\overline{q} = at + b$, due to momentum conservation: $P = \sum p_i = \text{const}$, for hamiltonian H. The removal of the center of mass (conserved momentum) reduces the number of degrees of freedom from n to $n-1$.

algebras (§5.1-2).

The are many other modification of the shift construction, we shall skip further details, and just remark, that both methods, as well as other known constructions of commuting (integrable) hamiltonians turn out to be special cases of a general *R-matrix (recursion operator) method* (see [Per]; [Olv] for details). Let us mention that some classical problems (like "rigid body in an ideal fluid") can also be brought into such framework.

Problems and Exercises:

1. Show that the first 3 of Lax trace-integrals $\{\chi_m\}$ of (4.19) are
$\chi_1 = \sum p_j$ - momentum;
$\chi_2 = \sum p_j^2 + 2\sum v_{jk} = 2H$ – Hamiltonian (Noether integrals),
$\chi_3 = \sum p_j^3 + 3\sum_{j\leq k} (p_j - p_k) v_{jk}$.
2. Show that the only steady state (time independent) rational solutions of KdV are sums of $\left\{\frac{2}{(x-q)^2}\right\}$.

3. Solutions of Functional-differential equation:

$$\boxed{x(\xi+\eta)\{z(\xi)-z(\eta)\} = x(\xi)x'(\eta) - x(\eta)x'(\xi)} \tag{4.24}$$

We are interested in odd solutions $\{x(\xi)\}$, and even $\{z(\xi)\}$.

i) Show that x can not be regular at $\{0\}$, unless $x(\xi)=0$ (take $\eta=0$, and $z=const$, hence x=0);

ii) derive an asymptotic expansion of $x(\xi)$ at small ξ,

$$x(\xi) \sim \alpha(\frac{1}{\xi} + \beta\xi + ...).$$

Hint: changing $\eta \to -\eta$, the equation is transformed into,

$$x(\xi-\eta)\{z(\xi)-z(\eta)\} = x(\xi)x'(\eta) + x(\eta)x'(\xi).$$

Let $\xi \to \eta$, and expand both sides in powers of small $\epsilon = \xi - \eta$.

Derive small-η asymptotics of $z(\eta)$,

$$z(\eta) \sim \alpha\left(\eta^{-2} + \beta\right);$$

iii) Expand both sides of (4.24) at a fixed ξ in powers of small η (starting from η^{-2}), and show that the coefficients,
η^{-2}: $-\alpha x = -\alpha x$;
η^{-1}: $-\alpha x' = -\alpha x'$
η^0 : $(z - \alpha\beta)x - \frac{1}{2}\alpha x'' = \alpha\gamma x$
η^1 : $(z-\beta)x' - \frac{1}{6}\alpha x''' = -\alpha\gamma x$.

Here $x = x(\xi)$; $z = z(\xi)$.

The 3-rd equation (η^0) yields

$$z = \alpha\frac{x''}{2x} + \alpha(\beta+\gamma),$$

constant term can always be made 0, i.e. $\beta = -\gamma$, since z is determined up to constant, So we get

$$\boxed{z = \alpha\frac{x''}{2x}}. \tag{4.25}$$

The last equation along with (4.25) gives a 3-rd order ODE for x,

$$(\alpha\frac{x''}{2x} + \gamma)x' - \frac{\alpha}{6}x''' + \alpha\gamma x = 0. \tag{4.26}$$

iv) Multiply (4.26) by x^{-3} and integrate it to get,

$$x^{-3}x'' + 6\gamma x^{-2} + c = 0; \tag{4.27}$$

From the boundary condition at $\{0\}$, $x(\xi) \sim \alpha\xi^{-1}$, it follows that $c = -2\alpha^{-2}$.

v) Multiply (4.27) by x^3x' and integrate once again to bring it into the form

$$(x')^2 = \alpha^{-2}x^4 - 2\mu x^2 + \lambda.$$

Now the inverse function $\xi(x)$ becomes an elliptic integral

$$\xi(x) = \int_x^{\infty} \frac{dx}{\sqrt{\alpha^{-2}x^4 - 2\mu x^2 + \lambda}}. \tag{4.28}$$

In special cases integral (4.28) simplifies:

1) $\boxed{\mu = \lambda = 0} \Rightarrow \qquad x(\xi) = \alpha\xi^{-1};$

2) $\boxed{\mu = \pm b^2;\ \lambda = \alpha^2 b^4} \Rightarrow \qquad x(\xi) = \alpha b \coth b\xi;\ \alpha b\, ctg\, b\xi;$

3) $\boxed{\mu = \pm b^2/2;\ \lambda = 0} \Rightarrow \qquad x(\xi) = \frac{\alpha b}{\sin b\xi};\ \frac{\alpha b}{\sinh b\xi}.$

In all other cases integral (4.28) is expressed in terms of elliptic functions. Explicit formulae depend on roots of the quadratic equation: $\omega^2 - 2\mu\alpha^2\omega + \lambda\alpha^2 = 0$.

We skip the rest of the analysis (see [Per]), and just remark that all cases the 2-body potential has the form,

$$v(\xi) = a^2\wp(b\xi) + const.$$

4. **Lax integrals** of (I-IV), can be chosen either as traces of $\{L^m\}$, $\chi_m = \mathrm{tr}(L^m)$, or the characteristic coefficients $\{\beta_j\}$ of L. Show that

$\beta_1 = \sum_1^n p_j = P$-total momentum;

$\beta_2 = \sum_{ij} p_i p_j - \sum x_{ij}^2$ (so $\frac{1}{2}\beta_1^2 - \beta_2 = \frac{1}{2}\sum p_i^2 + \sum v_{ij}$-energy!);

$\beta_3 = \sum_{i<j<k} p_i p_j p_k - (p_i v_{jk} + p_j v_{ik} + p_k v_{ij});$

Also

$\chi_1 = \sum p_j = P;$

$\chi_2 = \sum p_j^2 + \sum_{ij} x_{ij}^2 = H$- hamiltonian/energy

$\chi_3 = \sum p_j^3 + 3\sum_{i<j} (p_i + p_j) v_{ij}$

5. Find eigenvalue $\{q_1; q_2\}$ of system (4.18) for $n = 2$, and show that $q_{1,2} = Q(t) \pm q(t)$, where $Q = \frac{1}{2}(q_1 + q_2)$; $P = \frac{1}{2}(p_1 + p_2)$ denotes the center of mass coordinates, that undergo a free motion: $Q(t) = Q_0 + Pt$, while the relative motion in the "center of mass frame",

$$q = \sqrt{(q_0 + tp_0)^2 + t^2/q^2}.$$

6. Check that co-orbit of any tridiagonal L_0 (e.g. (4.18)) under conjugations with lower-triangular subgroup N_- consists of all tridiagonal $\{L\}$ (4.19).

7. Show the Poisson bracket relation (4.28).

i) Split variable

$$x = \frac{\mu}{\mu - \lambda}(x + \lambda a) - \frac{\lambda}{\mu - \lambda}(x + \mu a) = x_1 + x_2.$$

ii) Denote by $F_1; F_2$ functions $F_{\lambda,a}; F_{\mu,a}$; by $\partial_1; \partial_2$ – gradients $\partial_{x_1}; \partial_{x_2}$, and derive

$$\{F_1; F_2\}(x) = \frac{\mu - \lambda}{\mu}\langle x_1 \mid [\partial_1 F_1; \partial F_2]\rangle - \frac{\mu - \lambda}{\lambda}\langle x_2 \mid [\partial F_1; \partial_2 F_2]\rangle.$$

iii) Use G-invariance of $F_{1,2}$ to show $ad^*_{\partial F_1(x_1)}(x_1) = ad^*_{\partial F_2(x_2)}(x_2) = 0$.

§8.5. The Kepler problem and the Hydrogen atom.

In the last section we reexamine the classical Kepler 2-body problem and its quantum counterpart: the hydrogen atom, from the standpoint of hamiltonian dynamics, symmetry and quantization.

Following Moser, we show that Kepler problem is equivalent to the geodesic flow on 3-sphere. This explains the source of the hidden (Runge-Lenz) $SO(4)$-symmetry, mentioned in §8.2, and also the high degeneracy of eigenvalues (energy levels) of the hydrogen atom. Then we suggest a direct approach to the hydrogen-spectrum problem by realizing it as a Laplacian on the 3-sphere.

Going further along these lines one can discover even larger $SO(2;4)$-symmetry of the Kepler problem. The resulting "Kepler manifold" then appears as a Weinstein-Marsden reduction of the minimal co-orbit of $so(2;4)$. The latter can be quantized according to the Dirac prescription: namely, one first canonically quantizes the extended system (co-orbit), then imposes symmetry-constraints on the quantum/ operator level. The net result is once again the exact hydrogen spectrum!

The Kepler problem in $\mathbb{R}^3$ describes the motion of a body about the fixed gravity center at the origin. Its hamiltonian,

$$H = \tfrac{1}{2}p^2 - \tfrac{\alpha}{r}; \tag{5.1}$$

has an obvious $SO(3)$-symmetry, whose Noether generators make up 3 components of the angular momentum $J = p \times q$. Furthermore, we have shown in §8.3 (two-center gravity problem) that H possesses an additional (hidden) symmetry called the *Runge-Lenz vector,*

$$L = p \times J - \tfrac{\alpha}{r}q.$$

In fact, the J and L-vectors combine together to form Lie algebra $so(4)$ (or $so(1;3)$), where J sits in the upper-diagonal $so(3)$ block, while L fills in the complementary row and column. Precisely, for any 3-vectors $\xi,\eta \in \mathbb{R}^3$ we take ξ and η components of both vectors,

$$J(\xi) = \xi \cdot J; \; L(\eta) = \eta \cdot L.$$

These scalar hamiltonians on the phase-space $\mathbb{R}^6 = \{(q;p)\}$ satisfy the following Poisson bracket relations,

$$\begin{cases} \{J(\xi);J(\eta)\} = J(\xi \times \eta); \\ \{J(\xi);L(\eta)\} = L(\xi \times \eta); \\ \{L(\xi);L(\eta)\} = -2H\,L(\xi \times \eta), \end{cases} \tag{5.2}$$

where H is the hamiltonian (5.1). Relations (5.2) imply that on any "energy shell" (level surface) $H=\lambda$, functions $\{J(\xi);L(\eta)\}$ form a 6-dimensional Poisson-Lie algebra $\mathfrak{G}$ of one of the following types: *orthogonal, pseudo-orthogonal,* or *Euclidian-motion,* depending on the sign of λ. Namely,

$$\mathfrak{G}=\begin{cases} so(4);\ \text{if } \lambda>0; \\ so(3;1);\ \text{if } \lambda<0; \\ \mathfrak{E}_4=\mathbb{R}^3 \rhd so(3),\ \text{if } \lambda=0 \end{cases}$$

The proofs of all these statements are outlined in problem 1. Of course, both J and L commute with H. Furthermore, Newton potential $V=\frac{\alpha}{r}$ is the only one among central potentials $\{V=f(r)\}$, that makes Runge-Lenz $L=p\times J+f(r)\frac{q}{r}$, a symmetry of H (problem 2).

In §8.3 we applied the J and L-symmetries to establish complete integrability of both the Kepler problem and a 2-center gravity problem. Next we would like to explore further their implications the corresponding quantum problem: the hydrogen atom. This will bring us back to the subject of *Quantization* (§6.3), and its connections to the group representations.

5.2. Quantization of classical hamiltonian systems. A classical mechanical system is determined by its *phase-space,* a symplectic manifold $\mathfrak{P}$ (e.g. cotangent bundle), with a canonical/Poisson structure $\mathfrak{J}$, or equivalently, canonical 2-form $\Omega=\mathfrak{J}^{-1}dx\wedge dx$. Points $\{x\in\mathfrak{P}\}$ describe *classical states* of the system, while functions $\{f\}$ on $\mathfrak{P}$ represent *classical observables.* One particular observable $h(x)$ represents the *energy (hamiltonian)* of the system. Any observable f defines a one-parameter flow on $\mathfrak{P}$, $\{\exp t\Xi_f\}$, generated by its hamiltonian vector-field $\Xi_f=\mathfrak{J}(\partial f)$. The flow takes any initial state x_0 into the state $x(t)$ at time t.

Quantum system is usually described by the *quantum (Hilbert) phase-space* $\mathcal{H}$, whose (unit) vectors $\{\psi\in\mathcal{H}:\|\psi\|^2=1\}$ give *quantum states,* while operators $\{A\}$ on $\mathcal{H}$ represent *quantum observables.* The act of observation consists in evaluating observable A at a state ψ,

$$A\to\langle A\psi\mid\psi\rangle.$$

Often Hilbert space $\mathcal{H}$ consists of L^2-functions $\{\psi(q)\}$ on the classical configuration space $\mathcal{M}$. One requires $\int_{\mathcal{M}}|\psi|^2dq=1$, and interprets $\int_D|\psi|^2dq$, for a region $D\subset\mathcal{M}$, as a "*probability to find the quantum system in region D*". There are two standard views of the quantum evolution. The Schrödinger picture considers *evolution of states*, generated by a one-parameter group of the *quantum hamiltonian H,*

$$\psi_0 \to \psi(t) = e^{itH}[\psi_0],$$

while the Heisenberg picture takes the corresponding evolution of observables

$$B_0 \to B(t) = e^{-itH} B_0 e^{itH}. \tag{5.3}$$

Quantization, from a geometric standpoint, amounts to constructing a Hilbert space $\mathcal{H} = \mathcal{H}(\mathcal{P})$, and assigning *quantum observables (operators)* $\{F\}$, to classical observables $\{f\}$ on $\mathcal{P}$, $f \to F$. Typically, one would like to preserve all basic symmetries (conservation laws) of the classical system in the new quantum system. So the correspondence, $f \to F$, should maintain the basic commutator (bracket) relation between observables. Precisely, given a classical hamiltonian h we denote by G its symmetry-group (of canonical transformations on $\mathcal{P}$), and by $\mathfrak{G}$ its Poisson-Lie symmetry-algebra, made of observables $\{f\}$ on $\mathcal{P}$, that commute with h. We want to construct a representation of algebra $\mathfrak{G}$ (or group G) by operators on $\mathcal{H}$, that would take all Poisson-Lie brackets on $\mathcal{P}$, into the commutator brackets of operators $\{A_f : f \in \mathfrak{G}\}$,

$$\{f;h\} \to i\hbar[A_f; A_h];\ f,h \in \mathfrak{G}.$$

Plank constant $\hbar$ in front of the commutator (a small parameter depending on the choice of physical units) will be assumed 1.

The standard example of quantization discussed in §6.1 was the Heisenberg *canonical commutation relations*, i.e. the Poisson-Lie algebra spanned by all position and momentum variables $\{q_i; p_j\}$ in $\mathbb{R}^{2n}$, with brackets,

$$\{p_i; q_j\} = \delta_{ij}.$$

Quantization of CCR led, via Stone-von Neumann Theorem, to essentially unique[12] quantum phase-space $\mathcal{H} = L^2(\mathbb{R}^n)$, where the basic observables $\{q_i; p_j\}$ were represented by the multiplications and differentiations,

$$q_i \to Q_i[\psi] = q_i\psi;\ p_j \to P_j[\psi] = i\partial_{q_j}\psi;\ \psi \in L^2. \tag{5.4}$$

Furthermore, we have shown that representation (5.4) can be extended from the Heisenberg algebra, 1-st degree (linear) operators in {q;p},

$$\mathcal{W}_1 = \{\sum a_i q_i + b_j p_j + c\},$$

to the Weyl algebra $\mathcal{W} = \mathcal{W}(\mathbb{R}^n)$, generated by all differentiations and multiplications. In particular, the 2-nd degree component

$$\mathcal{W}^2 = \{f = \sum a_i q_i^2 + b_{ij} q_i p_j + c_j p_j^2 + \dots\},$$

gave the metaplectic (oscillator) representation of the semidirect product $\mathbb{H}_n \rhd \boldsymbol{Mp}_n$.

[12]Precisely, Stone-von Neumann Theorem proves uniqueness of *irreducible representations* of CCR, so the only source of non-uniqueness are possible multiplicities of $\{T^\lambda\}$.

There are many different ways to extend the Heisenberg-CCR through a representation of $\mathcal{W}$. One of them is the standard Weyl convention, which assigns each (ordered) monomial $f = q_i p_j$ a symmetrized operator,

$$f \to A_f = \tfrac{1}{2}(Q_i P_j + P_j Q_i).$$

This convention maintains the Poisson brackets, so operators: $A_f = f(Q;P)$, $A_h = h(Q;P)$ obey the relation,

$$[A_f; A_h] = iA_{\{f;h\}}. \tag{5.5}$$

The Weyl quantization rule also extends to higher degree polynomials, for instance,

$$q^2p^2 \to \tfrac{1}{6}(Q^2P^2 + QPQP + QP^2Q + PQ^2P + PQPQ + P^2Q^2).$$

But now it fails to maintain the Poisson bracket relation (5.5) (the Poisson bracket for operators gets replaced by a more complicated *Moel bracket*, which involves higher derivatives of "*symbols*" f, h to all orders (see §2.3). In fact, a general result of Grünwald-van Howe claims that there is no consistent quantization rule that would work for the entire Weyl algebra, and would extend the Heisenberg CCR for $\{q;p\}$ (see [Ch], [GS2]).

Of course, any classical hamiltonian $h = \frac{1}{2}p^2 + V$, is still consistently quantized to a Schrödinger operator,

$$H = -\tfrac{1}{2}\nabla^2 + V(q).$$

With this in mind we turn now to the hydrogen atom.

5.3. The hydrogen atom. The real (physical) hydrogen atom consists of single proton in the nucleus and a single electron. The proton is about 2000 times heavier that electron, and densely packed at the center. So the "classical prototype" of the hydrogen atom" is the standard 2-body (Kepler) system, the role of the Newton's gravitational potential being played by the electrostatic Coulomb force (both happen to be the same $V = \frac{1}{r}$!).

So the quantum model of an electron orbiting nucleus[13] is the Schrödinger operator

[13]Our model is clearly a hybrid of the quantum and classical principles, the nucleus being treated as a fixed classical point-charge, while the electron represented by a quantum ψ-function. Yet such model makes a good first approximation to the real physical system. It accurately predicts the energy levels of hydrogen, as measure through the radiation (emission/absorption) spectra, explains chemical bonds, etc.

$$H = -\tfrac{1}{2}\Delta - \frac{1}{|x|};\ \text{in}\ \mathcal{H} = L^2(\mathbb{R}^3). \tag{5.6}$$

The evolution of quantum states $\{\psi(x) \in L^2\}$ is governed by the Schrödinger equation,

$$i\dot{\psi} = H[\psi];\ \psi(0) = \psi_0\text{-initial state},$$

whose formal solution

$$\psi(t) = e^{itH}[\psi_0].$$

As always we want to analyze $\psi(t)$ via spectral decomposition of H (chapter 2). Operator H (6) is one of the best studied "model examples" in quantum mechanics. Its spectrum is well known to consists of an absolutely continuous part: $[0;\infty)$ (of infinite multiplicity), and a discrete sequence of negative eigenvalues $\{\lambda_k = -\frac{1}{k^2}: k = 1;2;3;...\}$, accumulating to $\{0\}$. In other words, space $L^2(\mathbb{R}^3)$ is decomposed into the direct sum,

$$L^2 = \mathcal{E}_c \bigoplus_1^\infty \mathcal{E}_k;$$

of eigensubspaces $\{\mathcal{E}_k\}$, and an absolutely continuous (spectral) subspace $\mathcal{E}_c$. Eigenfunctions $\{\psi_k\}$ of H are called *bound states,* since the quantum evolution takes on a particularly simple form for such ψ,

$$\psi(t) = e^{it\lambda_k}\psi_k;$$

a time periodic motion of frequency λ_k. In contrast, states $\psi \in \mathcal{E}_c$ *scatter* in the sense that probability to find a state in any finite region $D \subset \mathbb{R}^n$ diminishes in time,

$$\int_D |\psi(x;t)|^2 d^3x \to 0,\ \text{as}\ t \to \infty.$$

The eigenvalue problem for operator (5.6) can be solved explicitly (see any Quantum mechanics text, e.g. [Bö]; [LL]). For the sake of completness we shall briefly outline the solution.

Spherical symmetry of the Coulomb potential allows one to separate variables in polar coordinates $(r;\phi;\theta)$ in $\mathbb{R}^3$, i.e. space $L^2(\mathbb{R}^3)$ is decomposed into the tensor product $L^2(\mathbb{R}^+;r^2dr) \otimes \mathcal{H}$, $\mathcal{H} = L^2(S^2)$, and operator

$$H = -\tfrac{1}{2}(\partial_r^2 + \tfrac{2}{r}\partial_r + \tfrac{1}{r^2}\Delta_S) - \tfrac{1}{r};$$

where

$$\Delta_S = \partial_\theta^2 + cot\theta\partial_\theta + \frac{1}{sin^2\theta}\partial_\phi^2,$$

denotes the spherical Laplacian on S^2. In chapter 4 (J4.4) we examined spectral theory of spherical Laplacian Δ_S, and established a decomposition $\mathcal{H} = \bigoplus_0^\infty \mathcal{H}_m$, into the sum of spherical harmonics

$$\mathcal{H}_m = Span\{Y_j^m: -m \le j \le m\};\ dim\mathcal{H}_m = 2m+1.$$

Each $\mathcal{H}_m$ constitutes an irreducible subspace of the angular momentum algebra

$so(3) = \{J = x \times i\nabla\}$[14], and the Laplacian (central Casimir element) becomes scalar on $\mathcal{H}_m$,

$$\Delta_S|\mathcal{H}_m = m(m+1).$$

Accordingly, the entire quantum space $L^2(\mathbf{R}^3)$ breaks into the direct sum of "π^m-isotropic components of J", respectively, eigenspaces of Δ_S,

$$\mathcal{L}_m = \{\psi(r;\phi,\theta): \Delta_S[\psi] = m(m+1)\psi\} \simeq L^2(\mathbf{R}^+; r^2dr) \otimes \mathcal{H}_m,$$

and the reduced operators $H_m = H \mid \mathcal{L}_m$ turn into an ordinary differential operators on $\mathbf{R}^+$,

$$H_m = -(\partial_r^2 + \tfrac{2}{r}\partial_r + \frac{m(m+1)}{r^2}\Delta_S) - \tfrac{1}{r}.$$

The eigenvalue problem $H[\psi] = E\psi$ is reduced now to an ODE for the radial part $R(r)$,

$$R'' + \tfrac{2}{r}R' + [2(E + \tfrac{1}{r}) - \frac{m(m+1)}{r^2}]R = 0.$$

Change of parameters: $E \to n = \frac{1}{\sqrt{-2E}}$, and $r \to \rho = \frac{2r}{n}$, brings it to the form

$$R'' + \tfrac{2}{r}R' + [(\tfrac{n}{r} - \tfrac{1}{4}) - \frac{m(m+1)}{r^2}]R = 0. \tag{5.7}$$

Further simplification of (5.7) results from the substitution,

$$R = \rho^m e^{-\rho/2} w(\rho).$$

Exponential $\rho/2$ comes here from the "leading part" of ODE (5.7), at $\rho = \infty$, $R'' - \frac{1}{4}R = 0$, while power ρ^m are due to the homogeneous Euler-type ODE $\partial^2 + \frac{2}{\rho}\partial - \frac{m(m+1)}{\rho^2}$ (see problem 3). Then function w solves an ODE,

$$\rho w'' + (2m+2-\rho)w' + (n-m-1)w = 0, \tag{5.8}$$

known as *confluent hypergeometric*, or *generalized Laguerre equation*. Solution of (5.8) is a generalized Laguerre polynomial, $L_{n+m}^{(2m+1)}(\rho)$, of degree $n+m$. In order to get a "regular" solution of the "singular" ODE (5.8) (points $\rho = 0;\infty$ are singular!), n must be integer and $m \le (n-1)$. So the radial components of an eigenfunction ψ is

$$R_{nm}(\rho) = \rho^m e^{-\rho/2} L_{n+m}^{(2m+1)}(\rho).$$

For a fixed eigenvalue parameter n, the angular momentum number m takes on values $m = 0, 1, \ldots (n-1)$, and we have $1 + 3 + \ldots + (2n-1) = n^2$ "hydrogen eigenfunctions" $\psi = R_{nm}(\rho)Y_k^m(\phi;\theta)$, $\{Y_k^m \in \mathcal{H}_m\}$ of eigenvalue (energy-level)

$$\boxed{E_n = -\frac{1}{2n^2}}$$

We immediately notice that the multiplicity of the n-th eigenspace, $\dim \mathcal{E}_n = n^2$, is much higher than could be expected on the basis of the apparent $so(3)$-symmetry! Indeed, a generic rotationally symmetric hamiltonian,

---

[14] Let us remark that the angular momentum observables: $J_{km} = p_k x_m - p_m x_k$, belong to the 2-nd degree (symplectic) component $\mathcal{W}^2$ of the Weyl algebra. Hence, the angular momentum can be consistently quantized: $J_{km} = x_k\partial_m - x_m\partial_k$; so $J = x \times \nabla = \nabla \times x$.

$$H = -\tfrac{1}{2}\Delta + V(|x|),$$

should have only $(2k+1)$-degeneracy of the "spherical-harmonic decomposition": $\mathcal{H} = \bigoplus_0^\infty \mathcal{H}_k$. The abnormal degeneracy of specH indicates some other (hidden) symmetries of the hydrogen problem. The previous discussion should prepare the reader to make a correct guess: the hidden symmetry of H is related to the "*Runge-Lenz vector*". However, the quantization of $\boldsymbol{L}$ is not straightforward, as $\boldsymbol{L}$ involves cubic terms:

$$p \times \boldsymbol{J} = |p|^2 q - (p \cdot q)p,$$

which can not be canonically quantized in general. It can be shown, however, (the details are left to the reader) that the correct choice, the one that maintains the classical Poisson-Lie brackets (5.2), is given by the symmetrized Weyl rule

$$\boldsymbol{L} = \tfrac{1}{2}(P \times \boldsymbol{J} - \boldsymbol{J} \times P) + \tfrac{\alpha}{r}x = \tfrac{i}{2}(\nabla \times \boldsymbol{J} - \boldsymbol{J} \times \nabla) + \tfrac{\alpha}{r}x, \tag{5.9}$$

where $\boldsymbol{J} = ix \times \nabla = i\nabla \times x$ (problem 2).

If we now fix an "energy level" of H, eigensubspace $\mathcal{E} = \{\psi : H\psi = \lambda\psi\}$, the commutator brackets (5.2) would yield a representation of Lie algebra[15] $\mathfrak{so}(4)$ in $\mathcal{E}$. But irreducible representations of $\mathfrak{so}(4) \simeq \mathfrak{su}(2) \times \mathfrak{su}(2)$, are products $\{\pi^m \otimes \pi^k : k,m = 1;2;...\}$ (chapters 4; 1), of degree mk. In particular, $\mathfrak{so}(4)$ contains a series of representations $\{\pi^k \otimes \pi^k\}$ of degree k^2, which should correspond to eigenspaces of H. So Runge-Lenz suggest a possible explanation of spectral degeneracies of H.

However, to establish the precise correspondence between the eigenspaces $\{\mathcal{E}_k\}$ of H, and irreducible representations $\{\pi^k \otimes \pi^k\}$, we need to revisit the classical Kepler problem. The analysis will reveal the true nature of the $SO(4)$-symmetry of H. We shall see that H (more precisely, its inverse H^{-1}) can be realized, as the Laplacian on the 3-sphere S^3!

But S^3 is the quotient of the orthogonal group $SO(4) \simeq SU(2) \times SU(2)/\{\pm I\}$, whose representations were analyzed in ch.4-5. In particular, we know that the regular representation R on $L^2(S^3)$ has multiplicity-free "spectrum" (true for any S^n), made of the tensor products $\pi^k \otimes \pi^k$, where $\{\pi^k\}$ are standard spin-k representations of $SU(2)$ (see problems 7,8 of §5.1). So the Laplacian Δ_{S^3} has spectrum $\{\lambda_k = k^2\}$ of multiplicities $d_k = k^2$, and we get spec(H) at once. Thus the "Runge-Lenz" explains the mystery of hydrogen spectral degeneracies.

To convert the hydrogen problem H into the S^3-Laplacian, we shall follow a

[15] assuming negative energy value $\lambda = H \leq 0$, as otherwise one should take Lorentz algebra $\mathfrak{so}(3;1)$. The assumption is fully justified, since all eigenvalues of H are indeed, negative.

geometric approach of J. Moser [Mos], based on the stereographic projection.

5.4. Kepler problem and the geodesic flow on S^3. A *stereographic map* takes sphere $S^n = \{x_0^2 + |x|^2 = 1\}$ in $\mathbb{R}^n$, with the north pole removed, onto the horizontal hyperplane,

$$\Phi:(x_0;x) \to w = \frac{x}{1-x_0}. \tag{5.10}$$

In polar coordinates $\{(\rho;\phi)\}$ on $\mathbb{R}^n$ and $\{(\theta;\phi)\}$ on S^n (fig.5),

$$\rho = \cot\frac{\theta}{2}; \text{ and } \phi \to \phi.$$

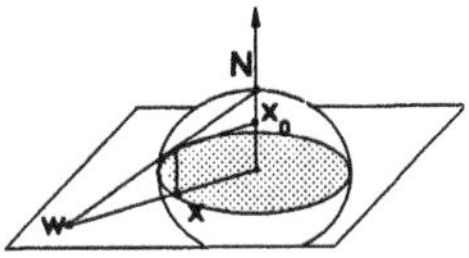

Fig.5. *Stereographic projection Φ takes a sphere punched at the north pole onto the plane (hyperplane) in $\mathbb{R}^{n+1}$.*

Since Φ is clearly a clearly a diffeomorphism (smooth 1-1 map), whose inverse

$$\Phi^{-1}:w \to (x_0;x);\ x_0 = \frac{|w|^2-1}{|w|^2+1};\ x = \frac{2w}{|w|^2+1};$$

the associated canonical map $\widetilde{\Phi} = (\Phi;{}^{\mathrm{T}}\Phi'^{-1})$ takes the natural symplectic structure (1- and 2-forms) from $T^*(S^n)$ to $T^*(\mathbb{R}^n)$.

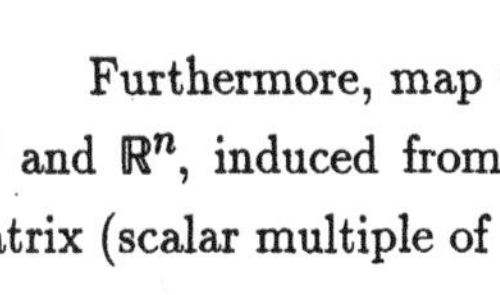

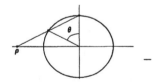

Fig.6: *Polar angle θ on sphere S^n is taken into the polar radius $\rho = \cot\frac{\theta}{2}$ on $\mathbb{R}^n$.*

Furthermore, map Φ is *conformal* with respect to natural Riemannian metrics on S^n and $\mathbb{R}^n$, induced from the ambient space $\mathbb{R}^{n+1}$. So its Jacobian Φ' is a conformal matrix (scalar multiple of the orthogonal one),

$${}^{\mathrm{T}}\Phi' \cdot \Phi' = \mu^2 I, \tag{5.11}$$

where ρ denotes the conformal factor. The matrix transposition in (5.11) refers, of course, to a choice of (Riemannian) metrics in tangent spaces of S^n and $\mathbb{R}^n$. The conformal factor can be easily computed (problem 3), and is found to be

$$\mu = \tfrac{1}{2}(1 + |w|^2)|\xi|,$$

at each point $\{(w;\xi)\}$ on the tangent bundle $T(\mathbb{R}^n)$ of the hyperplane. To check conformality and compute factor μ, one just write the Riemannian metric on $\mathbb{R}^n$ and S^n in polar coördinates:

$$ds^2 = d\theta^2 + \sin^2\theta\, d\phi^2 \text{ (on } S^n) \Rightarrow \frac{4}{(\rho^2+1)}\{d\rho^2 + \rho^2 d\phi^2\} \text{ (on } \mathbb{R}^n);$$

due to $\rho = \cot\theta/2$.

Then substitution of Φ in the kinetic energy form (Riemannian metric) on the bundle $T^*(S^n)$ yields a new kinetic energy function on $T^*(\mathbb{R}^n)$,

$$K = \tfrac{1}{8}(1+|w|^2)^2|\xi|^2.$$

Here we used a general fact: any diffeomorphisms (coordinate changes) Φ from manifold $\mathcal{M}$ to $\mathcal{N}$, $\Phi: x \to y$, induces a canonical transformation on the phase-spaces, $\tilde{\Phi}: T^*(\mathcal{M}) \to T^*(\mathcal{N})$;

$$\tilde{\Phi}: (x;\xi) \to (y;\eta) = (\Phi(x); {}^{\mathrm{T}}A^{-1}(\xi)), \tag{5.12}$$

where A in the second term of the RHS of (5.12) means the Jacobian matrix $\Phi_x{}'$ of Φ at x". In particular, any hamiltonian $f(x;\xi)$ on $T^*(\mathcal{M})$ is transformed into

$$\tilde{f}\,(y;\eta) = f \circ \tilde{\Phi} = f(\Phi^{-1}(y); {}^{\mathrm{T}}\Phi'(\eta)).$$

In the study of hamiltonian dynamics K can be replaced with any function $F = u(K)$, the corresponding hamiltonian vector fields being related by a constant (on any energy level) factor,

$$\Xi_F = u'(K)\Xi_K.$$

In particular, we can take

$$u(K) = \sqrt{2K} - 1 = \tfrac{1}{2}(1+|w|^2)|\xi| - 1.$$

The crucial observation of Moser was to notice, that the hamiltonian flow of K (the geodesic flow on the sphere!), restricted on 0-th level surface of $u(K)$, turns into our old friend, the Kepler flow. Precisely,

$$u(K)/|\xi| = H + \tfrac{1}{2},$$

where

$$H(w;\xi) = \tfrac{1}{2}|w|^2 - \frac{1}{|\xi|};$$

is the Kepler hamiltonian with the reversed position and momentum variables $(w;\xi)$. The interchange between $\{w\}$ and $\{\xi\}$ could be implemented by a canonical transformation (involution),

$$\sigma: (w;\xi) \to (-\xi; w). \tag{5.13}$$

The composition of 2 maps, $\sigma\circ\tilde{\Phi}: T^*(S^n)\to T^*(\mathbb{R}^n)$, takes the level-1 set of the kinetic energy (metric) form $\{(x;\eta): 2K = |\eta|^2 = 1\}$ - a unit cosphere bundle over S^n, into the level set $\{H(w;\xi) = -\frac{1}{2}\}$ of the Kepler hamiltonian,

$$H = H_1 = \tfrac{1}{2}|\xi|^2 - \frac{1}{|w|}.$$

Here subscript 1 refers to constant in the numerator of the Newton/Coulomb potential of H. More generally, we call $H_\alpha = \frac{1}{2}|\xi|^2 - \frac{\alpha}{|w|}$, and observe, using simple homogeneity properties of H, with respect to symplectic dilations: $(w;\xi)\to(\lambda w;\frac{1}{\lambda}\xi)$, that level sets,

$$\{H_\alpha = E\} = \{H_{\lambda\alpha} = \lambda^2 E\}.$$

In particular, cosphere bundles of different radii in $T^*(S^n)$,

$$S^*_\alpha(S^n) = \{(x;\eta): 2K(\ldots) = |\eta|^2 = \alpha^2\},$$

are taken by $\Psi = \sigma\circ\tilde{\Phi}$ into the level set,

$$\{(w;\xi):\ H_1 = -\frac{1}{2\alpha^2}\}.$$

Clearly, hamiltonians K and H are inverse one to the other, via map Ψ,

$$H = \frac{1}{2K}\circ\Psi.$$

Following [Sou], we call the union of all nonzero covectors over S^n, $T^+(S^n) = \{(x;\eta): \eta \neq 0\}$, the *Kepler manifold.* Let us summarize the foregoing discussion in the following

Theorem 1: *The stereographic map* (5.10) *composed with symplectic involution* σ (5.13), $\Psi = \sigma\circ\Phi$, *takes the Kepler manifold,* $T^+(S^n)$, *onto the negative energy shell* $\mathcal{P}_- = \{(w;\xi): H_1 < 0\}$ *of the Kepler hamiltonian, and transforms the kinetic (metric) form* $K = \frac{1}{2}|\eta|^2$ *on* $T^+(S^n)$ *into the inverse* $1/H$ *on* $\mathcal{P}_-$.

Finally, it remains to quantize both problems and to compare their spectra. The geodesic flow of hamiltonian, $K = |\eta|^2$ on S^n (in fact, on any manifold $\mathcal{M}$) yields the Laplacian[16] Δ.

A straightforward approach to relate the "negative part of Schrödinger operator H" to Δ_S^{-1} would be to *quantize* the canonical map $\Psi = \sigma\circ\tilde{\Phi}$, i.e. assign a unitary

[16] On symmetric spaces $\mathcal{M}$, like the n-sphere, the "quantized Laplacian" can be derived from symmetry considerations. Lie algebra elements must be represented by invariant vector fields on $\mathcal{M}$, generators of point-transformations $g: x\to x$, $g\in G$. Classically, invariant vector fields $\{X\}$ on $\mathcal{M}$, are in 1-1 correspondence with invariant hamiltonians $\{F_X(x;\xi) = X\cdot\xi;\ \xi\in T^*_x\}$ on the phase-space $T^*(\mathcal{M})$. Quantizing algebra $\{F_X\}$ we get a family of 1-order differential operators $\{i\partial_X\}$, acting on $L^2(\mathcal{M})$. But the hamiltonian K is the sum of squares of invariant vector fields, which upon quantization yields the invariant Laplacian.

operators $U: L^2(S^n)\to L^2(\mathbb{R}^n)$, to the stereographic (symplectic) map $\widetilde{\Phi}$, and to quantize involution σ (the latter is clearly given by the Fourier transform $\mathfrak{F}: L^2(\mathbb{R}^n)\to L^2(\mathbb{R}^n)$).

Then one has to verify that the product $W = \mathfrak{F}U$ takes $L^2(S^n)$ into the (negative) discrete component of H^{-1}, and intertwines both operators,

$$\boxed{W\Delta_S W^{-1} = H^{-1}}$$

Since, spectrum of the Laplacian Δ on S^3 is well known (chapter 5):

$$\boxed{\lambda_k = k^2} \text{ with multiplicity } \boxed{d_k = k^2};$$

this would yield the result, as spec$H$ (its discrete part) is the inverse of specΔ!

Another formal "derivation" would be

i) to take the "stereographed" spherical Laplacian

$$K = (1+|w|^2)^2\Delta; \text{ on } L^2\Big\{\frac{d^n w}{(1+w^2)^n};\mathbb{R}^n\Big\};$$

ii) to show that the eigenvalue problem

$$K[\psi] = E^2\psi;$$

is reduced to a "square root problem",

$$\sqrt{-\Delta}(1+|w|^2)[\psi] = E\psi;$$

iii) to observe that the latter turns into a hydrogen eigenvalue problem:

$$(w^2 - \frac{E}{\sqrt{-\Delta}})\psi = \psi; \tag{5.14}$$

with interchanged position and momentum $\{w; i\partial\}$.

Then (5.14) becomes an eigenvalue ($\lambda = 1$) problem for the Schrödinger operator H_E. Using the obvious scaling properties of the hydrogen hamiltonian:

$$\text{spec}(H_E) = \frac{1}{E^2}\text{spec}(H_1)$$

would yield the requisite relation between the k-th eigenvalue of H_1 and eigenvalue E of Δ_S:

$$\boxed{\lambda_k E^2 = 1}$$

Once one could get the discrete hydrogen spectrum from spec(Δ_S)! Neither of 2 *direct arguments*, has yet been establlished rigorously. The known methods involve an extension of the Kepler problem to a higher dimensional manifold, $\mathbb{R}^4\times\mathbb{R}^{4*}$ (Kepler represents a $U(1)$-symmetry reduction of the latter), and the Dirac formalism for imposing constraints in quantum systems with symmetries. The Dirac's procedure [Dir] involves quantizing first the extended (large) system, then imposing "quantum constraints" as operators on the extended quantum Hilbert space $\mathcal{H}$, so that the "physical states" become null-vectors of quantum constraints. This procedure could be implemented in many different ways (see [Simm]; [Kum]; [Sou1-2]; [Mla]; [GS2]; [Hur]). We shall outline an approach, due to [Simm]; [Kum] (see also [GSp]), where the Kepler problem is "conformally regularized", embedded into a system of 4 constraint harmonic oscillators.

5.5. Conformal regularization and the Dirac quantization. We consider a system of 4 oscillators described by the hamiltonian $h_0 = \frac{1}{2}\sum |\alpha_j|^2$ in the phase-space $\mathbf{C}^4$ equipped with the natural symplectic 2-form:

$$\omega = \frac{i}{2}\sum d\alpha_j \wedge d\bar{\alpha}_j; \tag{5.15}$$

and impose the constraint,

$$|\alpha_1|^2 + |\alpha_2|^2 = |\alpha_3|^2 + |\alpha_4|^2. \tag{5.16}$$

Notice that (5.16) defines a *twistor-space* Π, made up of all complex null-lines in $\mathbf{C}^4$, quotient modulo the one-parameter subgroup $\{e^{it}:\alpha \to e^{it}\alpha\}$ - a flow of h_0. So the twistor-space Π (a 6-D manifold with the canonical form (5.15)) arises from the Weinstein-Marsden reduction of the constraint oscillator system. Let us also remark that the conformal group $G = SU(2;2)$ acts naturally on Π (since G preserves the (2;2)-indefinite form (5.16) on $\mathbf{C}^4$, hence it takes null-lines of (5.16) into null-lines). In fact, Π represents a minimal co-adjoint orbit of $SU(2;2)$, and the constraint system can be obtained by geometrically quantizing orbit Π in the sense of §6.3.

2°. **Embedding.** We shall embed the Kepler problem,

$$\{T^*(\mathbf{R}^3\backslash\{0\});\omega = dp\wedge dq; h = \tfrac{1}{2}p^2 - \tfrac{1}{q}\}$$

into the constraint oscillator problem. We fix energy-levels $h = -E$; $(E > 0)$ (for the Kepler hamiltonian), and $h_0 = \frac{1}{\sqrt{8E}}$ (for the oscillator). The embedding procedure (of "Kepler" on the energy shell, $h_0 = -E$ to the "oscillator"), will exploit the familiar conserved integrals of h,

$$J = q \times p \text{ - angular momentum; and } L = J \times p + \frac{q}{|q|}.$$

We remind the basic Poisson-bracket relations for $\{h; J; L\}$,

$$\begin{gathered}\{h; J\} = \{h; L\} = 0;\\ \{J_i;J_j\} = -\epsilon_{ijk}J_k;\ \{J_i;L_j\} = -\epsilon_{ijk}L_k;\ \{L_i;L_j\} = \epsilon_{ijk}2hJ_k\end{gathered} \tag{5.17}$$

in other words h commutes with J and L, and the latter form a Lie algebra $so(4)$ (for negative energies $h < 0!$). Alongside (5.17) we shall use 2 other geometric relations (problem 4):

$$J\cdot L = 0;\ |L|^2 - 2h|J|^2 = 1. \tag{5.18}$$

We set $\rho = \sqrt{-2h}$ (constant for a fixed energy level), and define a new set of vector-variables

$$x = \rho J + L;\ y = \rho J - L.$$

Relations (5.17-18) show that $|x|^2 = |y|^2 = \rho^2|J|^2 + |L|^2 = 1$, so map $\Psi:(q;p)\to(x;y)$, takes a negative (fixed) energy shell $\mathcal{P}_E = \{h(p;q) = -E\}$ of the Kepler phase-space $T^*(\mathbf{R}^3)$ into the product of 2-spheres,

$$\Psi:\mathcal{P}_E\to S^2\times S^2 = \{(x;y)\} \simeq \mathbf{CP}^1\times\mathbf{CP}^1.$$

Furthermore, the symplectic structure (5.15) on the shell $\mathcal{P}_E$ goes into the sum of the natural (volume) forms on the spheres (problem 4),

$$\omega_E = \frac{1}{\sqrt{8E}}\left(\frac{dx_1\wedge dx_2}{x_3} + \frac{dy_1\wedge dy_2}{y_3}\right). \tag{5.19}$$

One could check that variables $\{x;y\}$ do, indeed, Poisson-commute (problem 4). The latter could also be written in terms of the complex (stereographic) parametrization of S^2:

$$z\to x = \frac{1}{1+|z|^2}(z;\ |z|^2-1);\ \text{similarly } w\to y = \dots;\ z,w\in\mathbf{C}.$$

Then

$$\omega_E = \frac{i}{\sqrt{8E}}\left(\frac{dz\wedge d\bar z}{(1+|z|^2)^2} + \frac{dw\wedge d\bar w}{(1+|w|^2)^2}\right). \tag{5.20}$$

Thus the reduced Kepler problem yields the product of two projective lines $\mathbf{CP}^1$, with a 1-parameter family of symplectic forms $\{\omega_E\}$, depending on the energy-level E. Next we remark that the projective space $\mathbf{CP}^1$ gives the energy-reduced phase-space for a pair of oscillators. Indeed, restricting the hamiltonian $h_{12} = |z_1|^2 + |z_2|^2 = 1/\sqrt{2E}$ (fixed energy), and factoring the energy-shell modulo its flow, $z = (z_1;z_2)\to e^{it}z$, we do recover space $\mathbf{CP}^1$ with the 2-form[17] (5.20).

3°. Now we turn once again to the constraint 4-oscillator (5.16) and fix its energy-shell, $\sum_1^4 |z_j|^2 = \frac{1}{\sqrt{2E}}$. Along with (5.16) this relation gives the same pair of projective spaces and 2-forms, as the above energy-reduced Kepler system,

$$|z_1|^2 + |z_2|^2 = |z_3|^2 + |z_4|^2 = \frac{1}{\sqrt{8E}}.$$

So 2 hamiltonian systems: Kepler (on the negative energy region), and the constraint 4-oscillator produce the identical (symplectomorphic) family of reduced energy-shells. Hence, both systems are equivalent (better to say the Kepler system is embedded in the constraint-oscillator, as the latter includes the "collision Kepler trajectories", so it gives a conformal regularization of the Kepler manifold).

4°. Finally, the constraint oscillator problem can be quantized according to the Dirac

[17] By the same pattern projective space $\mathbf{CP}^n$ gives $(n+1)$ oscillators, energy-constraint oscillators, $|z_0|^2 + \dots + |z_n|^2 = \text{Const}$.

prescription (see [Hur]). Namely, we first quantize free oscillators (see §6.2) and get 4 commuting operators $\{\alpha_j = -\partial_j^2 + x_j^2 : j = 1;2;3;4\}$ in the Fock-space of symmetric tensors (polynomials) in 4 variables, $\mathcal{H} \simeq \bigoplus_0^\infty \mathcal{H}_k$, $\mathcal{H}_k$ - spanned by the products of the Hermite functions $\{\psi_i(x_1)\psi_j(x_2)\psi_m(x_3)\psi_l(x_4): i+j+m+l = k\}$ (§6.2). Each of oscillators $\{\alpha_j\}$ has an $\frac{1}{2}$-integral spectrum of eigenvalue (§6.2), $\mathrm{spec}[\alpha_j] = \{k + \frac{1}{2}: k = 0;1;...\}$. Following the Dirac procedure we impose constraint on the "quantum level", i.e. consider solution of the joint eigenvalue problem,

$$\alpha_1 + ... + \alpha_4 = \lambda;\ \alpha_1 + \alpha_2 = \alpha_3 + \alpha_4.$$

The latter boils down to a solution of a simple Diophantian system,

$$\begin{cases} k_1 + ... + k_4 = 2n - 2 \\ k_1 + k_2 = k_3 + k_4 \end{cases};\ k_j \geq 0 \tag{5.21}$$

Here $2n$ represents the n-th eigenvalue of the quantized constraint oscillator. Thus we get $k_1 + k_2 = k_3 + k_4 = n - 1$. In other words, the $2n$–th eigenspace of the "constraint oscillator problem (quantized Kepler)" is obtain from the n-th eigenspace $\mathcal{H}_n(12)$ of the 2-D oscillator $\alpha_1 + \alpha_2$, tensored with the n-th eigenspace $\mathcal{H}_n(34)$ of $\alpha_3 + \alpha_4$: $\mathcal{H}_{2n} = \mathcal{H}_n(12) \otimes \mathcal{H}_n(34)$. Spin variable n for a 2-D oscillator $(\alpha_1 + \alpha_2)$ takes on all integral and $\frac{1}{2}$-integral values. Indeed, in §6.2 we learned that eigenspaces of the n-D oscillator coincide with irreducible representations $\{\pi^m\}$ of $SU(n)$, so 2-D oscillators correspond to $SU(2)$. Thus we get the dimension of eigenspace $\mathcal{H}_{2n}$ to be $n \cdot n = n^2$, the "constraint-oscillator-eigenvalue" $\lambda_n = 2n$, and the corresponding "Kepler eigenvalue",

$$\boxed{E_n = -\frac{1}{\sqrt{2\lambda_n}} = -\frac{1}{4n^2}};\ \text{QED.}$$

Additional results and comments:

The material of §8.1 is fairly standard and could be found mo... ...on classical mechanics [AM]; [Arn]. The same applies to §8.2 (the Marsden-Weinstein reduction appeared first in [MW]). Classical problem from the standpoint of Lie symmetries are treated in many sources ([Arn], [Olv]). The recent book [Per] by A. Perelomov contains a comprehensive survey of the classical results, as well the recent developments in "integrable hamiltonians" (see also [FM]). This book along with the review article [OP] was our main source in §8.4. The "hydrogen quantization problem" goes to the very onset of quantum mechanics. W. Pauli (1926) and V. Fock (1935) first discovered the $SO(4)$-symmetry of the hydrogen hamiltonian on the Lie algebra level (see [LL]). J. Souriau [Sou] and J. Moser [Mos] reviewed the classical Kepler problem, and applied the "stereographic projection" method to it. The higher $SO(2;4)$-symmetries were analyzed in

[Simm];[Sou];[Kum];[Hur];[GSp], and more recently in [GS2]. The general review of geometric quantization and the references could be found in [Sn], [Hur].

Problems and Exercises:

1. Verify that commutation relations (5.2) with constant H define one of 6-dimensional Lie algebras: $so(4)$; $so(3;1)$ or $\mathfrak{E}_3 = \mathbf{R}^3 \triangleright so(3)$, depending on sign of H.
2. Check that the "combined angular momentum"-"Runge-Lenz" quantum vectors (5.9) obey the commutation relations (5.2), and both operators commute with H.
3. **Laguerre polynomials and the hydrogen hamiltonian.** Generalized Laguerre polynomial $L_n^\alpha(x)$, of degree n and order α, can be be defined as a regular solution of the ODE,
$$xy'' + (\alpha+1 - x)y' + ny = 0.$$
They form an orthogonal family on $\mathbf{R}_+$ with weight $w(x) = x^\alpha e^{-x}$, and can be generated by the Rodrigues formula,
$$L_n^\alpha(x) = Const\, x^{-\alpha} e^x [x^{\alpha+n} e^{-x}]^{(n)}.$$
Show that the reduced hydrogen hamiltonian,
$$R'' + \tfrac{2}{r}R' + [(\tfrac{n}{r} - \tfrac{1}{4}) - \tfrac{m(m+1)}{r^2}]R = 0. \qquad (5.22)$$
could be brought to the Laguerre form. Steps:

   1° Verify the following commutation relations for differential operators: $L = \partial^2 + b\partial + c$, and $M = x^2\partial^2 + bx\partial + c$ (Euler-type);

   (i) $L[e^{\lambda x}...] = e^{\lambda x}[(\partial + \lambda)^2 + b(\partial + \lambda) + c] = e^{\lambda x}[L + 2\lambda\partial + (b\lambda + \lambda^2)]$

   (ii) $M[x^s...] = x^s[M + 2sx\partial + s^2 + (b-1)s]$

   2° Take the reduced "hydrogen operator" (5.22), and apply (i) to the product $R = e^{-r/2}u(r)$, to get an ODE,
$$M[u] = \partial^2 u + (\tfrac{2}{r} - 1)\partial u - (\tfrac{n-1}{r} + \tfrac{m(m+1)}{r^2})u = 0.$$
Break M into the sum of 2 Euler-type operators, $M_0 + M_1$, and show that
$$(M_0 + M_1)[r^s v] = r^s\{\partial^2 + (\tfrac{2+2s}{r} - 1)\partial - \tfrac{n-1-s}{r} + \tfrac{s(s+1) - m(m+1)}{r^2}\}[v].$$
For the singular term $(...)r^{-2}$ in the potential to cancel out index s must be equal m, and the resulting ODE becomes a generalized Laguerre equation!
4. i) Verify relations (5.18) for $\{h; J; L\}$;
   ii) show that the canonical 2-form on $\mathcal{P}_E \subset T^*(\mathbf{R}^3)$ is taken into the form ω_E (5.19) by map Ψ;
   iii) check that variables $\{x_i\}$, and $\{y_j\}$ Poisson-commute, $\{x_i; y_j\} = 0$. Find the commutation relations among $\{x_i\}$.

Appendix A: *Spectral decomposition of self adjoint operators.*

Any selfadjoint operator A in Hilbert space $\mathcal{H}$ (bounded or unbounded) admits a *spectral decomposition,* which generalizes the notion of an eigen-expansion of a symmetric matrix,

$$A \sim \begin{pmatrix} \lambda_1 & & \\ & \dots & \\ & & \lambda_n \end{pmatrix}$$

Here $\{\lambda_1; \dots \lambda_n\} = spec(A)$ denotes the eigenvalues of A, and the k-th diagonal entry/block corresponds to the k-th eigensubspace of A. In other words space $\mathcal{H}$ is decomposed into the direct sum of eigenspaces: $\mathcal{H} \quad \bigoplus_k E_k$; and operator $A \mid E_k = \lambda_k I$, i.e.

$$\boxed{A = \bigoplus_k \lambda_k P_k}, \tag{A.1}$$

where P_k means the orthogonal projection from $\mathcal{H}$ to E_k. Diagonal matrices can be thought of as multiplication operators $\Lambda: (f_k) \to (\lambda_k f_k)$, on spaces of (scalar/vector) n-tuples $\{(f_k): 1 \le k \le n\}$. In other words there exists a unitary map $\mathcal{U}: \mathcal{H} \to \oplus E_k$, that conjugates A to a multiplication operator,

$$\boxed{\mathcal{U} A \mathcal{U}^{-1} = \Lambda: f(\lambda) \to \lambda f(\lambda)} \tag{A.2}$$

Both results: spectral decomposition (A.1) and diagonalization (canonical form) (A.2), can be generalized for all selfadjoint Hilbert space operators.

Precisely, spectrum of a selfadjoint operator, $spec(A)$, is a closed subset of $\mathbb{R}$. The role of eigen-projections $\{P_k\}$ is played in general by the family of *spectral projections* $\{P(\Delta)\}$, associated to closed subsets of $\Delta \subset \mathbb{R}$, or equivalently by the *spectral measure (resolution)* $dP(\lambda)$. Projections $\{P(\Delta)\}$ form a *commuting, monotone* $(P(\Delta_2) < P(\Delta_1)$, for any pair $\Delta_2 \subset \Delta_1)$, and *disjoint* $(P(\Delta_1 \bigcup \Delta_2) = P(\Delta_1) \oplus P(\Delta_2)$, for any pair $\Delta_1 \bigcap \Delta_2 = \emptyset)$, *family of orthogonal projections.*

The analog of the direct sum spectral decomposition (A.1) of A becomes the *direct integral decomposition,*

$$A = \oint \lambda dP(\lambda).$$

Spectral subspaces $\{E(\Delta)\}$ of A are images of spectral projections $\{P(\Delta)\}$. If operator A has discrete spectrum $\{\lambda_k\}$ (e.g. finite rank or compact), then

$E(\Delta) = \bigoplus_{\lambda_k \in \Delta} E_k$ - direct sum of eigensubspaces sitting "inside Δ".

The proper analog of the diagonal matrices (A.2) are multiplication operators

$$\Lambda: f(\lambda) \to \lambda f(\lambda), \tag{A.3}$$

on spaces $L^2(\mathbb{R}; d\mu)$ of scalar/vector valued functions on $\mathbb{R}$, with a Borel measure $d\mu$.

It turns out that multiplications provide canonical models of all selfadjoint operators. The basic result of spectral theory of selfadjoint operators states that any such A is unitarily equivalent to the direct sum of multiplications (A.3). Namely,

Theorem A1: *There exists unitary operator* $\mathcal{U}: \mathcal{H} \to \bigoplus_m L^2(d\mu_m)$, *such that*

$$\boxed{A = \mathcal{U}\Lambda\mathcal{U}^{-1} \sim \Lambda}$$

In the canonical realization of A spectral subspace $E(\Delta)$ consists of all functions vanishing outside Δ, $E(\Delta) = \{f: supp(f) \subset \Delta\}$, whereas spectral projection $P(\Delta)$ becomes a multiplication with the indicator function $\chi_\Delta: f \to \chi_\Delta f$.

Canonical realization also yields some natural properties of spectral subspaces, e.g. *spectral subspace* $E(\Delta)$ *is invariant under the commutator of* A, $\mathrm{Com(A)} = \{B: AB = BA\}$. Indeed, any B, that commutes with a multiplication by λ on L^2-space of scalar/operator functions on $\mathbb{R}$, must itself be a multiplication with a scalar/operator valued function, $B: f(\lambda) \to B(\lambda)[f(\lambda)]$.

The **proof** of spectral decomposition (spectral measure) exploits some basic harmonic analysis on commutative groups, $\mathbf{R}$ and $\mathbf{Z}$. We shall demonstrate it for a single unitary operator to give the reader a gist of the arguments involved. Function $\phi(\xi)$ on $\mathbf{R}$ or $\mathbf{Z}$ is called *positive-definite*

$$\sum_{jk} \phi(\xi_j - \xi_k) a_j \bar{a}_k \geq 0,$$

for all n-tuples $\{\xi_1; \dots \xi_n\} \subset \mathbf{R}$ and all coefficients $\{a_1; \dots a_n\}$ in $\mathbf{C}$.

Bochner's Theorem : *Any positive-definite sequence* $\{\phi_k\}$ *on* $\mathbf{Z}$ *coincides with Fourier coefficients of a positive measure* $d\mu$ *on the dual group* $\mathbf{T}$, i.e.

$$\boxed{\phi_k = \frac{1}{2\pi}\int_0^{2\pi} e^{-ik\theta} d\mu(\theta)} \qquad (A.4)$$

The proof is outlined in problem 12 of §2.1 (chapter 2).

Now given a unitary operator U in Hilbert space $\mathcal{H}$, we pick any vector $\psi_0 \in \mathcal{H}$, restrict U onto a cyclic subspace $\mathcal{H}_0$ spanned by ψ_0,

$$\mathcal{H}_0 = \mathrm{Span}\{U^k(\psi): k = 0; \pm 1; \pm 2; \dots\},$$

and observe that the sequence of matrix-entries $\{\phi_k = \langle U^k\psi \mid \psi\rangle\}$ is positive-definite on $\mathbf{Z}$. Hence, by Bochner's Theorem

$$\phi_k = \frac{1}{2\pi}\int_0^{2\pi} e^{-ik\theta} d\mu_0(\theta), \qquad (A.5)$$

for some measure $\mu = \mu_0$. Let us remark that any spectral measure $\{dP(\lambda)\}$ in Hilbert space $\mathcal{H}$ defines a family of scalar measures $\{d\mu_\psi\}$ associated to vectors $\{\psi \in \mathcal{H}\}$,

$$d\mu_\psi(\lambda) = \langle dP(\lambda)\psi \mid \psi\rangle.$$

We claim that Bochner measure μ_0 (A.5) is the spectral measure of $U \mid \mathcal{H}_0$.

Indeed, space $\mathcal{H}_0$ is identified with $L^2(\mathbf{T};d\mu_0)$, via map

$$\psi \to \hat{\psi}(x) = \sum b_k e^{ikx};\ b_k = \langle U^k \psi_0 | \psi \rangle;$$

and operator U acts by multiplication on L^2-functions,

$$U: \hat{\psi}(\theta) \to e^{i\theta}\hat{\psi}(\theta). \tag{A.6}$$

Indeed, space $\mathcal{H}_0$ contains a dense subspace of vectors

$$\psi = \sum a_j U^j[\psi_0] = f(U)[\psi_0],\ \text{where } f = \sum a_k e^{ikx},$$

and operator U acts by the shift, $\{b_j\} \to \{b_{j-1}\}$ on the space of Fourier coefficients.

It follows now from (A.6) that any "function of U", $M = f(U) = \sum a_j U^j$, also acts by multiplication on L^2,

$$M: \hat{\psi}(x) \to f(x)\hat{\psi}(x).$$

The latter holds for all polynomials $\{f(U)\}$, hence all continuous and even bounded (L^∞) functions $\{f\}$ on $\mathbf{T}$. Hence spectral projections $\{P(\lambda)\}$ are nothing, but multiplications in $L^2(d\mu)$ with *indicator-functions*,

$$\chi_\lambda(x) = \begin{cases} 1; x \in [0;\lambda] \\ 0; x \notin [0;\lambda] \end{cases},$$

i.e.

$$P(\lambda): \hat{\psi}(x) \to \chi_\lambda(x)\hat{\psi}(x),\ \psi \in L^2.$$

This yields spectral measure of U on a cyclic subspace $\mathcal{H}_0$. To get it for the entire space $\mathcal{H}$, we take an orthogonal complement $\mathcal{H}' = \mathcal{H} \ominus \mathcal{H}_0$, find a cyclic subspace $\mathcal{H}_1 \subset \mathcal{H}'$, resolve it into $L^2[d\mu_1]$, and continue the process. Eventually entire space $\mathcal{H}$ is resolved into the direct sum of L^2-spaces,

$$\mathcal{H} \simeq \bigoplus_0^\infty \mathcal{H}_m;\ \mathcal{H}_m \simeq L^2[d\mu_m],$$

and we establish spectral decomposition for any unitary operator U.

Remark 1: After spectral decomposition is proven for unitary $\{U\}$ one can easily pass to self-adjoint operators $\{A\}$ (bounded or unbounded), via the Caley transformation: $A \to U = (A-i)^{-1}(A+i)$, that relates both classes. Thence one proceeds to more general cases, e.g. *joint spectral resolution for a commuting family (representation) of selfadjoint operators*, or *unitary groups*, etc.

Remark 2: Spectral resolution allows to construct "functions of self-adjoint operators" $\{F(A)\}$, used often in differential equations (see §2.3). We define

$$F(A) = \oint F(\lambda)dP(\lambda),$$

a clear generalization of $F(A) = \bigoplus_k F(\lambda_k)P_k$, for a matrix A.

Our last result concerns spectral decomposition of *compact* selfadjoint operators.

Theorem A2: *Any compact selfadjoint operator A in Hilbert space $\mathcal{H}$ has purely discrete spectrum of real eigenvalues $\lambda_k \to 0$; all eigenspaces $E_k = \{\xi : A\xi = \lambda_k \xi\}$ $(\lambda_k \neq 0)$ are finite-dimensional, and the system of eigenvectors is complete, i.e.*

$$\bigoplus_{\lambda \in specA} E(\lambda) = \mathcal{H}.$$

Proof: We recall that compact operators are characterized by the property: (*i*) *A-image of the unit ball* $\mathbf{B} = \{\|\xi\| \leq 1\}$, $K = closure\{A(\mathbf{B})\}$, *is compact in* $\mathcal{H}$. Let us also remark that (ii) *compact projections* P *are always finite-rank* (since a unit ball $\mathbf{B}$ in any Banach space $\mathfrak{B}$ is compact iff $\dim \mathfrak{B} < \infty$).

Discreteness of *specA* is proven by taking its upper bound $\lambda_0 = \text{supp}\{\langle A\xi \mid \xi\rangle : \|\xi\| \leq 1\}$, and using (i) to show that the bound is attained at some vector ξ_0. Such ξ_0 is clearly an eigenvector of the eigenvalue λ_0. If $\lambda_0 > 0$, the corresponding eigenspace E_0 is finite-dimensional by (ii).

Next we peel off subspace E_0, i.e. restrict operator A (respectively. quadratic form $\langle A\xi \mid \xi\rangle$) onto the orthogonal complement of E_0, $\mathcal{H}_1 = \mathcal{H} \ominus \mathrm{E}_0$, and repeat the process. This yields the next eigenvalue

$$\lambda_1 = \sup\{\langle A\xi \mid \xi\rangle : \xi \in \mathcal{H}_1 ; \|\xi\| \leq 1\} < \lambda_0;$$

and its eigenspace E_1. The process continues until the entire Hilbert space $\mathcal{H}$ is exhausted, i.e. resolved into the direct sum of eigenspaces: $\bigoplus_{\lambda} E_\lambda$, *QED*.

Appendix B: *Integral operators.*

Throughout the book we often encounter integral operators,

$$Kf(x) = \int_X K(x;y)f(y)dy,$$

acting in various function-spaces: $L^2; L^p; \mathbb{C}(X)$, etc., and need to estimate their operator-norms, and analyze other properties, like *positivity, compactness, Hilbert-Schmidt, trace-class*. In this section we shall briefly outline some basic facts of the integral-operator theory.

Integral operators could be thought of as "continuous matrices", with "entries" $\{K(x;y)\}$ labeled by points of X. Many natural features of matrices extend to integral operators fairly straightforward, for example, self-adjointness of K means $K(x;y) = \overline{K(y;x)}$ - kernel of the adjoint operator K^*; positive definiteness $\langle Kf \mid f\rangle \geq 0$, for all f, is equivalent to positive definiteness of any matrix $\{(a_{ij}): a_{ij} = K(x_i;x_j);\ x_1;...x_n \in X\}$.

The operator-norm of $K: L^p \to L^q$ could be estimated via *Minkowski inequality*,

$$\left(\int \left| \int F(x;y)dy \right|^q dx\right)^{1/q} \leq \int \left(\int \mid F(x;y) \mid^q dx\right)^{1/q};\ \text{for any } F(x;y),$$

whence follows,

$$\| Kf \|_q \leq \left(\int \left(\int \mid K(x;y) \mid^q dx \right)^{p'/q} dy \right)^{1/p'} \| f \|_p; \tag{B.1}$$

where $p' = \frac{p}{p-1}$ - the dual Hölder index.

The constant in the RHS of (B.1) called *mixed* $L^{q;p'}$-norm of $K(x;y)$, is not optimal, but sufficient for many purposes. For L^2-spaces ($q = p = p' = 2$) we get

$$\| Kf \|_2 \leq \| K \|_2 \| f \|_2;$$

the mixed norm of K becomes its $L^2(dxdy)$-norm.

Compact operators in Hilbert/Banach spaces take bounded sets $\{\| f \| \leq C\}$ into compact (the latter could be verified in L^p-spaces by *Ascolli-Arzella Theorem*[1]).

Self-adjoint compact operators are easily described in terms of their eigenvalue spectrum: $specA = \{\lambda_k\}$ - purely discrete, $\lambda_k \to 0$, as $k \to \infty$, and all nonzero λ's have finite multiplicities, $dim E(\lambda) < \infty$, for any $\lambda \neq 0$ in $specA$. In general (non self-adjoint) case a convenient way to describe compact operators is terms of their

[1]We remind that according to Ascolli-Arzella: *compactness of a subset* $\mathcal{M} \subset L^p$; or $\mathbb{C}$ is equivalent to *boundedness + equicontinuity*.

modulus,

$|A| = (AA^*)^{1/2}$- a positive self-adjoint operator. Then

$$\underbrace{A\text{ - compact}} \Leftrightarrow \underbrace{|A|\text{ - compact}} \Leftrightarrow \underbrace{\lambda_k(|A|) = s_k(A) \to 0, \text{ as } k \to \infty}.$$

The eigenvalues of modulus $|A|$ are called *s-numbers* of A, and the class of all compact operators $\mathfrak{S}_\infty(\mathcal{H})$ is characterized by asymptotic vanishing of s-numbers, $s_k(A) \to 0$.

Among all compact operator we distinguish two important subclasses (ideals of the algebra $\mathcal{B}(\mathcal{H})$):

- *Hilbert-Schmidt:* $\mathfrak{S}_2 = \{A : \|A\|_{HS}^2 = \sum s_k(A)^2 = \sum \lambda_k(AA^*) < \infty\}$;
- *Trace-class:* $\mathfrak{S}_1 = \{A : \|A\|_{tr} = \sum s_k(A) = \sum \lambda_k(|A|) < \infty\}$;

eigenvalues counted with their multiplicities.

To describe both we choose an orthonormal basis $\{e_1; ... e_k; ...\}$ in $\mathcal{H}$ and represent any bounded operator A by an infinite matrix $(a_{ij})_{ij=1}^\infty$: $a_{ij} = \langle Ae_i | e_j \rangle$. We define the trace of A, $trA = \sum a_{jj}$ - sum of diagonal entries (provided the sum converges).

Proposition 1: *Operator A is Hilbert-Schmidt iff $\sum_{ij} |a_{ij}|^2 < \infty$, furthermore the sum of squares of matrix entries gives the Hilbert-Schmidt-norm of A,*

$$\|A\|_{HS}^2 = \sum_{ij} |a_{ij}|^2 = tr(AA^*).$$

The statement is easily verified by diagonalizing $B = AA^*$ and using invariance of *trace* under all conjugations: $B \to U^{-1}BU$, with invertible U. This also shows that positive self-adjoint operators with finite sum $\sum a_{ij} < \infty$, belong to the trace-class, and $\|A\|_{tr} = trA = \sum \lambda_k(A)$.

Let us remark that Hilbert-Schmidt-norm defines the inner product on $\mathfrak{S}_2$

$$\langle A | B \rangle = tr(AB^*);$$

and the (operator) product of two Hilbert-Schmidt-operators belong to the trace-class. Space $\mathfrak{S}_2$ equipped with such product turns into the Hilbert space, isomorphic to the tensor product $\mathcal{H} \otimes \mathcal{H}$.

Next we want to study compactness, trace-class and Hilbert-Schmidt properties of integral operators. It is difficult to completely characterize compact and trace-class operators in terms of the kernel $K(x;y)$, although there are many sufficient conditions. The Hilbert-Schmidt class proves to be more manageable.

Proposition 2: (i) *Operator* $K \in \mathfrak{S}_2$ *iff* $\int\int |K(x;y)|^2 dxdy < \infty$, *and the* L^2-*norm of the kernel is exactly the Hilbert-Schmidt-norm of* K.
(*ii*) *Trace of an integral operator:* $trK = \int K(x;x)dx$ - *integral over the diagonal of* $X \times X$.

The proof easily follows by choosing an orthonormal basis of functions $\{\psi_k(x)\}$ and writing Dirac $\delta(x;y) = \sum \psi_k(x)\bar{\psi}_k(y)$, or in the physicist bra-ket convention: $\delta = \sum |\psi_k(x)\rangle\langle\psi_k(y)|$. In the mathematical convention we represent the identity operator in $\mathcal{H} = L^2(X)$, as the sum of orthogonal projections: $I = \bigoplus_0^\infty \langle ...|\psi_k\rangle\psi_k$. Then

$$trK = \sum \langle K\psi_k | \psi_k\rangle = \int K(x;y)\delta(x;y)dxdy = \int K(x;x)dx,$$

and (i) follows from (ii) by writing $\|K\|_{HS}^2 = tr(KK^*)$.

It follows from Proposition 2, that continuous integral kernels $K(x;y)$ on compact spaces X yield compact (even Hilbert-Schmidt) operators. In fact, one has,

Proposition 3: *Integral operators with continuous kernels* $K(x;y)$ *on compact manifolds* X *are trace-class with* $trK = \int K(x;x)dx$.

It suffices to check that self-adjoint operators with positive kernels $K(x;y) = K(y;x) \geq 0$ are trace-class. But any such K is the square of an Hilbert-Schmidt operator M with continuous kernel $M(x;y) = \sqrt{K(x;y)}$.

As a corollary, we immediately get compactness of the convolution-operators, $R_f: u \to f*u$, on torus $\mathbb{T}^n$, or any compact group G.

Appendix C: *A primer of Riemannian geometry: geodesics, connection, curvature.*

Riemannian manifolds typically come equipped with 2 structures. One of them is *metric-tensor* $\{g_{ij}(x)\}$ on tangent spaces $\{T_x : x\in\mathfrak{X}\}$, along with the dual metric $\{(g^{ij})=(g_{ij})^{-1}\}$ on cotangent spaces. Metric brings in a host of other interesting geometric structures.

C1. Geodesics is defined as a shortest path $\gamma=\{\mathrm{x(t)}:0\le \mathrm{t}\le 1\}$, that connects 2 given points $x,y\in\mathfrak{X}$. So γ solves the variational problem, minimizes the *length functional*

$$L[\gamma]=\int_0^1 \sum g_{ij}(x)\dot{x}^i\dot{x}^j dt=\int_0^1 \mathcal{L}(x;\dot{x})dt.$$

Hence γ satisfies the Euler-Lagrange equations (see §8.1),

$$\frac{\delta L}{\delta\gamma}=\partial_x\mathcal{L}-\frac{d}{dt}\partial_{\dot{x}}\mathcal{L}=0,$$

or

$$\frac{d}{dt}\Big(\sum_{i\neq k} g_{ki}\dot{x}^i+2g_{kk}\dot{x}^k\Big)-\sum_{ij}\partial_k(g_{ij})\dot{x}^i\dot{x}^j=0. \tag{C.1}$$

The latter can be written in terms of the so called *Christoffel symbol* of metric g,

$$\Gamma_{ij}^k=\sum_p g^{kp}\Big\{\frac{\partial_i(g_{jp})+\partial_j(g_{pi})-\partial_p(g_{ij})}{2}\Big\}, \tag{C.2}$$

by differentiating the LHS (C.2), using symmetry of metric tensor g_{ij} in ij, and multiplying the result by the inverse matrix $\{g^{ij}\}$ (see Proposition below). This yields

$$\ddot{x}_k+\sum_{ij}\Gamma_{ij}^k\dot{x}^i\dot{x}^j=0. \tag{C.3}$$

As a simple illustration we shall obtain geodesics on the Euclidian space, Poincare half-plane $\mathbb{H}$, and the 2-sphere.

Examples. 1) Geodesics in the Euclidian space $\mathbf{R}^n$ are straight lines: $\ddot{\mathrm{x}}=0\Rightarrow \mathrm{x(t)}=\mathrm{a}+\mathrm{bt}$.

2) The Poincare plane $\mathbf{H}=\{x+iy: y>0\}$ carries metric $ds^2=\frac{dx^2+dy^2}{y^2}$, so tensor $(g_{ij})=\begin{bmatrix}1 & \\ & y^{-2}\end{bmatrix}$; its dual $(g^{ij})=\begin{bmatrix}1 & \\ & y^2\end{bmatrix}$. The Euler-Lagrange equations then become:

$$\begin{aligned}&(\mathrm{I})\ 2\frac{d}{dt}\Big(\frac{\dot{x}}{y^2}\Big)=0\\ &(\mathrm{II})\ 2\frac{d}{dt}\Big(\frac{\dot{y}}{y^2}\Big)=\partial_y\Big(\frac{\dot{x}^2+\dot{y}^2}{2}\Big)\end{aligned} \tag{C.4}$$

To solve system (C.4) we note that $\dot{x}/y^2=a$ - const[2]. Introducing an auxiliary variable

[2]In chapter 8 the reader will see the meaning and reason for conservation of $\dot{x}/y^2$. This quantity represents the p_x-momentum (conjugate to the x–variable) of a hamiltonian system associated to Lagrangian $\mathcal{L}$. Its conservation reflects a translational symmetry of $\mathcal{L}$ in the x–direction.

$u = \dot{y}/y^2$, we rewrite (I)-(II) as a system

$$\begin{cases} \frac{du}{dt} = y(a^2-u^2) \\ \frac{dy}{dt} = uy^2 \end{cases}; \frac{dx}{dt} = ay^2. \tag{C.5}$$

Variables y and u in the first 2 equations (C.5) separate and we get

$$\int \frac{udu}{a^2-u^2} = \int \frac{dy}{y};$$

whence follows

$$u = \frac{\sqrt{C-a^2y^2}}{y}. \tag{C.6}$$

On the other hand the last 2 equations yield

$$\frac{dy}{dx} = \frac{u}{a};$$

and substitution function $u = u(y)$ from (C.6) results in

$$\int \frac{ydy}{\sqrt{C-a^2y^2}} = -\int \frac{dx}{a}.$$

But the latter has a general solution

$$\boxed{\sqrt{C-a^2y^2} = a(x-x_0)}$$

that represents a family of all semicircles in the upper half-plane perpendicular to the real axis!

3) Sphere S^2 carries metric $ds^2 = d\phi^2 + sin^2\phi d\theta^2$, in polar angles ϕ,θ. The reader is asked to show that geodesics are great circles: $C = C(\phi_0;\theta_0)$, that make angles ϕ_0 with z-axis, and θ_0 with x-axis. Circle $C(\phi_0;\theta_0)$ can be parametrized in terms of the arc-length $\{\phi(s);\theta(s)\}$, as

$$\begin{cases} \tan(\theta-\theta_0) = \frac{1}{sin\phi_0}\tan s \\ \cos\phi = cos\phi_0 \cos s \end{cases}$$

Geodesics on a Riemannian manifold $\mathfrak{M}$ allow to introduce two other important structures: *exponential map*, and *parallel transport*.

C2. Exponential map $exp{:}T_x{\to}\mathfrak{M}$, takes tangent vectors at a point x onto the manifold (in a neighborhood of $\{x\}$). Point $x_0 = x(0)$ gives the initial position for a 2-nd order system (C.3), while tangent vector $\xi = \dot{x}(0)$ plays the role of initial velocity. The combined initial data $(x_0;\xi)$ produces a unique solution-curve, $\gamma = x(t;x_0;\xi)$ - a geodesics issued from x_0 in the direction ξ. Map $\xi{\to}x(t;\xi)$ is called exponential and denoted, $exp_{x_0}(t\xi)$, sends ray $\{t\xi\} \subset T_{x_0}$ into a geodesics γ. One can show that exp takes a neighborhood of $\{0\}$ in T onto a neighborhood of $\{x_0\}$ in $\mathfrak{M}$ in a 1-1 way, its inverse is often called *log*.

C3. Covariant derivative and parallel transport. Parallel transport on surfaces consists of moving tangent vectors $\{\xi(t)\}$ along a geodesics $\{\gamma = x(t)\}$, in a way that preserves their length $|\xi(t)|$, and the angle between ξ and $\dot{x}$. In other words we get a family of linear maps $\{U(t)\}$ on tangent spaces $\{T_t = T_{x(t)}\}$ along γ, $U_t: T_0 \to T_t$, which preserve metric $g \mid T_t$. Clearly, U_t is nothing, but a Jacobian matrix $\phi' = (\frac{\partial \phi}{\partial x_0})$ of the map,

$$\phi_t: x_0 \to x(t; x_0; \xi) = exp_{x_0}(t\xi).$$

Covariant derivative then represents an infinitesimal generator of U_t. The formal definition of *covariant derivative* (*affine connection*) involves a family of linear transforms $\{\nabla_X: X \in T_x\}$ on tangent spaces, , labeled by vectors $X \in T_x$, that satisfy

$$\nabla_X(fY) = (\partial_X f)Y + f\nabla_X Y, \tag{C.7}$$

for any pair of tangent vectors $X, Y \in T_x$, and any function f (here $\partial_X f$ denotes the derivative of f along X). Covariant derivative (C.7) is uniquely determined by its values on basic vector fields $\{\partial_j\}$, for a chosen set of local coordinates $\{x_j\}$. Namely,

$$\nabla_{\partial_i}(\partial_j) = \sum \Gamma_{ij}^k \partial_k,$$

where coefficient $\{\Gamma_{ij}^k\}$ form the so called *Christoffel symbol* of connection ∇. Alternatively we can describe ∇ as a linear map that takes vector-fields $\{X\}$ on $\mathfrak{M}$ to 1-st order differential operators on sections of the *tangent-bundle*,

$$X \to \nabla_X = \partial_X + \Gamma(X;...): Y \to [X;Y] + \Gamma(X;Y),$$

where $\partial_X(Y) = [X;Y]$ means *Lie derivative* of field Y along X (or commutator of 2 vector fields), and $\Gamma(X;...)$ – the Christoffel operator (matrix) on T_x.

Covariant derivative gives rise to the *parallel transport* of tangent vectors/fields along curves $\gamma \subset \mathfrak{M}$. Namely, field X is called *parallel along* γ, if its derivative

$$\nabla_{\dot{\gamma}}(X) = 0, \tag{C.8}$$

where $\dot{\gamma} = \frac{d\gamma}{ds}$ denotes a unit tangent (velocity) vector. The geometric content of (C.8) becomes apparent when one first looks at a linear OD system,

$$\frac{d}{ds}X + A(s)X = 0;\ X(0) = X_0; \tag{C.9}$$

whose trajectories $\{X = X(s; X_0)\}$ define a family of "parallel lines". In other words, the fundamental matrix-solution $\{U(s;t): (\frac{d}{ds} + A)U = 0\ (s \geq t),\ U(s,s) = I\}$, implements the parallel transport of initial states (vectors) $\{X_0\}$, at time t to terminal states $\{X(s, X_0) = U(s,t)[X_0]\}$, at time s. In our geometric setup we move tangent vectors along γ, via ODS $\frac{d}{ds} + A(s)$, where ds – arc-length along γ, and $A = \Gamma(\dot{\gamma}(s);...)$ – Christoffel operator in the direction of $\dot{\gamma}(s)$. An interesting situation arises when structures (the metric and the connection) are consistent, i.e. parallel transport preserves metric (norms, angles between tangent vectors). So

$$\langle U(s)\xi | U(s)\eta\rangle = \langle \xi|\eta\rangle, \text{ for any pair } \xi,\, \eta \in T_x,$$

equivalently,

$$\tfrac{d}{ds}\langle X \mid Y\rangle = \langle \nabla_{\dot\gamma}(X) \mid Y\rangle + \langle X \mid \nabla_{\dot\gamma}(Y)\rangle, \text{ for any pair } X, Y, \qquad (C.10)$$

along any path $\gamma \subset \mathfrak{S}$. Here $\langle | \rangle$ denotes the Riemannian product on tangent spaces.

We introduce *torsion* of connection ∇ as a vector-valued bilinear form on T_x,

$$T(X;Y) = \nabla_X(Y) - \nabla_Y(X) - [X;Y].$$

In local coordinates T represents an anti-symmetric part of Christoffel tensor Γ in lower indices,

$$T(\partial_i;\partial_j) = \sum (\Gamma_{ij}^k - \Gamma_{ji}^k)\partial_k.$$

Proposition: *There exists a unique torsion-free covariant derivative (parallel transport) ∇ on $\mathfrak{S}$, consistent with metric $\{g_{ij}\}$, whose Christoffel symbol can be written explicitly in terms of the metric tensor.*

We take a coordinate system $\{x_j\}$ and compute the covariant derivative $\nabla_i = \nabla_{\partial_i}$ of the product $\langle \partial_j | \partial_k\rangle = g_{jk}$, via (C.10). This yields

$$\partial_i(g_{jk}) = \sum_m (\Gamma_{ij}^m g_{mk} + \Gamma_{ik}^m g_{mj}). \qquad \text{(I)}$$

Then we write 2 similar relations for all cyclic permutations of the triple $\{ijk\}$,

$$\partial_j(g_{ki}) = \sum_m (\Gamma_{jk}^m g_{mi} + \Gamma_{ji}^m g_{mk}); \qquad \text{(II)}$$

$$\partial_k(g_{ij}) = \sum_m (\Gamma_{ki}^m g_{mj} + \Gamma_{kj}^m g_{mi}). \qquad \text{(III)}$$

Using symmetry of tensor Γ in lower indices (torsion-free), we observe that first term of (I) is equal to the 2-nd term of (II), 1-st of (II) is equal to 2-nd of (III), and $2-$nd of (I) is the first of (III). So 3 equations (I-III) are of the type

$$\begin{cases} A + B = p \\ C + A = q \\ B + C = r \end{cases} \Rightarrow \begin{cases} A = \frac{p+q-r}{2} \\ B = \frac{p+r-q}{2} \\ C = \frac{q+r-p}{2} \end{cases}.$$

Here A (resp. B, C) represents $\sum_m \Gamma_{ij}^m g_{mk}$ (resp. its permutations in $\{ijk\}$), and we get the Christoffel tensor, contracted with the metric tensor $\{g_{jk}\}$ expressed in terms of $\{g_{jk}\}$ and its derivatives. Multiplying (contracting) it with the inverse metric tensor $\{g^{jk}\}$, we get the final formula

$$\Gamma_{ij}^k = \sum_p g^{kp}\left\{\frac{\partial_i(g_{jp}) + \partial_j(g_{pi}) - \partial_p(g_{ij})}{2}\right\}.$$

So we recover the Christoffel symbol of the Riemannian metric (C.2) determining the equation of geodesics, $\ddot{x} + \Gamma(\dot{x};\dot{x}) = 0$.

Parallel transport along a closed path $U(t;0)$, where $x(t) = x_0$, gives a linear map in T_{x_0} called *holonomy*. In case of Riemannian connection operators $\{U(t)\}$ clearly preserve the norm $\|\xi\|_g$ in T_x, so they define isometries of the tangent space.

C4. Curvature is closely related to the parallel transport and covariant derivative. We start with 2-D manifolds. Given such $\mathfrak{S}$ we pick a tangent vector $\xi = \xi_0$ at point $x_0 \in \mathfrak{S}$, and transport it along a small loop $\gamma =$ boundary of "small region D". On return to the initial point vector ξ will rotate by an angle $\phi = \phi(D)$ (angle ϕ gives the holonomy matrix $U_\phi = \begin{bmatrix} \cos\phi & \sin\phi \\ -\sin\phi & \cos\phi \end{bmatrix}$). Functional $\phi(D)$ is easily seen to satisfy the relations:

$$\phi(D_1 \cup D_2) = \phi(D_1) + \phi(D_2);\ \text{for any pair of regions } D_1;\ D_2,$$

while the change of orientation: $D \to -D$ (γ traversed in the opposite direction) yields

$$\phi(-D) = -\phi(D).$$

Such ϕ clearly represents a differential 2-form (*curvature-form*) integrated over D,

$$\phi(D) = \int_D \omega.$$

The value of ω on a pair of tangent vectors $\xi;\eta$ can be found from the following procedure. We take a small parallelogram P_ϵ in space T, spanned by $\xi;\eta$ compute holonomy angle ϕ along P_ϵ, then set

$$\omega(\xi;\eta) = \lim_{\epsilon \to 0} \frac{\phi(P_\epsilon)}{\epsilon^2}.$$

The reader is asked to compute ω on the 2-sphere of radius R and the Poincare plane (the former has: $\omega = R^{-2}dS$, the latter: $\omega = -dS$, where dS is the natural Riemann surface area element). For a general Riemann surface $\omega = KdS$, with respect the (Riemannian) area element, where function K gives the *Gauss (Riemann) curvature* of the surface (on S^2 and $\mathbb{H}$, function K is constant ± 1).

In higher dimensions the role of angle ϕ is played by the holonomy matrix $U = U(\gamma)$. As above any pair of tangent vectors ξ,η defines an infinitesimally-small parallelogram $P_\epsilon = P_\epsilon(\xi;\eta)$. We take the holonomy $U_\epsilon = U(P_\epsilon)$ in the orthogonal group $SO(T_x)$. Also higher-D ($n \geq 3$) orthogonal groups are non-commutative (so $U(D_1 \cup D_2)$ is no more a simple product of two U's), we can still expand $U_\epsilon(\xi;\eta)$ in small ϵ and find,

$$U_\epsilon(\xi;\eta) = I + \epsilon^2 \Omega(\xi;\eta) + O(\epsilon^3),$$

where Ω is a 2-form with values in the Lie algebra $so(n)$ of skew-symmetric matrices in the tangent space T_x, called the ***Riemann curvature tensor*** of $\mathfrak{G}$. The curvature tensor gives rise to sectional curvatures of $\mathfrak{G}$ in all ***geodesic planes,*** passing through x. The latter means all 2-D surfaces Σ in $\mathfrak{G}$ (near x_0), spanned by the exponential (geodesic) images of all 2-planes $\mathcal{S}$ in T, $\Sigma = exp(\mathcal{S})$. The Gaussian curvature of Σ is expressed in terms of the Riemann tensor, as

$$K_{\xi\eta} = \langle R(\xi;\eta)\xi | \eta \rangle.$$

These are sectional curvatures of $\mathfrak{G}$.

The formal definition of the ***Riemann curvature tensor*** is given in terms of the connection ∇, namely,

$$R(X,Y) = \nabla_X \nabla_Y - \nabla_Y \nabla_X - \nabla_{[X;Y]};$$

an operator-valued bilinear form on tangent space $\{T_x\}$. Here $[X;Y]$ denotes the Lie bracket of vector-fields X, Y. We can also write it as a quatric form, $\langle R(X;Y)\cdot V \mid W \rangle$, or rank-4 tensor of type (1,3),

$$R^m_{ijk} = \langle (\nabla_i \nabla_j - \nabla_j \nabla_i)\partial_k | \partial_m \rangle.$$

Riemann curvature tensor is shown to satisfy the relations:

- $R(X,Y) = -R(Y,X)$;
- $g(R(X,Y)\cdot Z; V) = -g(R(X,Y)\cdot V; Z)$.

In other words operators $R(X,Y)$ on T_x are skew-symmetric relative to metric g, and the bilinear operator-valued map $(X,Y) \to R(X,Y)$, is anti-symmetric.

- If torsion $T = 0$, then

$$R(X,Y)\cdot Z + R(Y,Y)\cdot X + R(Z,X)\cdot Y = 0 \text{ (Bianchi identity)}$$

Remark: The Christoffel, torsion and curvature tensors are often convenient to recast in the language of differential forms. We choose a basis of vector fields $\{X_1;...X_n\}$, and consider the corresponding dual set of 1-forms: $\omega = (\omega^j)$, e.g. (dx^j) for $\omega = (\partial_j)$. Then one can introduce a *connection 1-form:* $\Omega = \Gamma \cdot \omega$, i.e.

$$\Omega^i_j = \sum_k \Gamma^i_{kj} \omega^k.$$

One can think of ω, Ω as vector-valued and operator-valued 1-forms. Clearly, Ω determines Γ, hence covariant derivative ∇. On the other hand both can be described in terms of the torsion and curvature, via the *Cartan structural equations:*

$$\boxed{\begin{array}{l} d\omega = \Omega\wedge\omega + \tfrac{1}{2}T\omega\wedge\omega \\ d\Omega = \Omega\wedge\Omega + \tfrac{1}{2}R\omega\wedge\omega \end{array}}$$

Cartan's equations generalize the Frenet equations for the orthogonal moving frame $\{\mathbb{T};\mathbb{N};\mathbb{B}\}$ (tangent, normal, binormal), along curve γ,

$$\frac{d}{dt}\begin{bmatrix} \mathbf{T} \\ \mathbf{N} \\ \mathbf{B} \end{bmatrix} = \begin{bmatrix} & \kappa & \\ -\kappa & & \tau \\ & -\tau & \end{bmatrix}\begin{bmatrix} \mathbf{T} \\ \mathbf{N} \\ \mathbf{B} \end{bmatrix}$$

in terms of curvature κ and torsion τ of γ. For 2-sphere they give the standard relations between derivatives $\{\partial_\phi;\partial_\theta\}$ of the orthogonal frame $\{\boldsymbol{u};\boldsymbol{v};\boldsymbol{w}\}$,

$$\partial_\phi\begin{bmatrix} \boldsymbol{u} \\ \boldsymbol{v} \\ \boldsymbol{w} \end{bmatrix} = \begin{bmatrix} & 1 & \\ -1 & & \\ & & 0 \end{bmatrix}\begin{bmatrix} \boldsymbol{u} \\ \boldsymbol{v} \\ \boldsymbol{w} \end{bmatrix}; \quad \partial_\theta\begin{bmatrix} \boldsymbol{u} \\ \boldsymbol{v} \\ \boldsymbol{w} \end{bmatrix} = \begin{bmatrix} & & \sin\phi \\ & & \cos\phi \\ -\sin\phi & -\cos\phi & \end{bmatrix}\begin{bmatrix} \boldsymbol{u} \\ \boldsymbol{v} \\ \boldsymbol{w} \end{bmatrix}$$

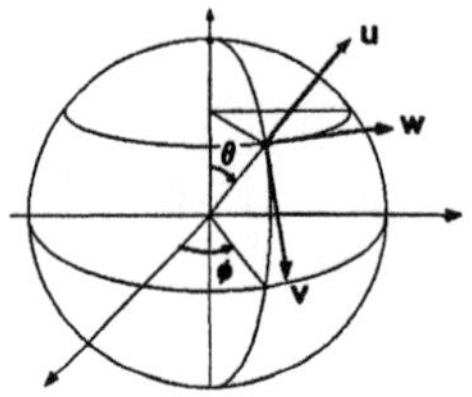

Fig.1: Polar angles $(\phi;\theta)$ on the 2-sphere S^2 and the orthogonal frame $\{\boldsymbol{u};\boldsymbol{v};\boldsymbol{w}\}$.

C5. Geodesics and curvature on Lie groups and homogeneous spaces. A Lie group G always possesses a right (or left) invariant Riemannian metric[3] g: it suffices to fix g at the identity $\{e\}$, and then translate it by right/left multiplications, $\phi_x: y \to yx$ (or $x^{-1}y$), that maps tangent spaces $T_e \to T_x$. Lie algebra $\mathfrak{G}$ of G can be identified with the tangent space T_e, and also with the algebra of left-invariant fields on G (see §1.4). Such fields are considered parallel, as left translates on G form parallel transports. So all geometric structures: metric, covariant derivative, curvature tensor become left-invariant. In particular, geodesics through the identity are all one-parameter subgroups $\{\gamma = x(t)\}$, holonomies are Jacobians of left translates $\{\phi'_{x(t)}: \mathfrak{G} = T_e \to T_x\}$, and curvature will measures a "degree of non-commutativity of G". It suffices to compute

[3]Let us remark that G typically may not possess a bi-invariant Riemannian metric. In chapter 5 we show that all simple/semisimple Lie groups G do have a bi-invariant (Killing) form $B(X;Y)$, but this form is positive-definite only for compact G.

them at $\{e\}$. The Riemannian product $\langle|\rangle$ on tangent spaces defines a bilinear map $B:\mathfrak{G}\times\mathfrak{G}\to\mathfrak{G}$,

$$\langle B(X;Y)\,|\,Z\rangle = \langle [X;Y]\,|\,Z\rangle;\ \text{all } X,Y,Z\in\mathfrak{G},$$

which plays the role of the Riemann curvature tensor. The covariant derivative at $\{e\}$ then becomes,

$$\nabla_X(Y)_e = \tfrac{1}{2}([X;Y] - B(X;Y) - B(Y;X)),$$

while sectional curvatures takes on somewhat more complicated form,

$$K_{\xi\eta} = \left\langle \frac{B(\xi;\eta)+B(\eta;\xi)}{2}\middle|\frac{B(\xi;\eta)+B(\eta;\xi)}{2}\right\rangle + \frac{1}{2}\left\langle [\xi;\eta]\middle|\frac{B(\xi;\eta)-B(\eta;\xi)}{2}\right\rangle - \\ -\frac{3}{4}\langle[\xi;\eta]\,|\,[\xi;\eta]\rangle - \langle B(\xi;\xi)|B(\eta;\eta)\rangle;\ \xi,\eta\in\mathfrak{G}.$$

The latter simplifies for bi-invariant Riemannian metric on compact Lie groups,

$$K_{\xi\eta} = \tfrac{1}{4}\langle[\xi;\eta]\,|\,[\xi;\eta]\rangle.$$

Next we turn to homogeneous space $\mathfrak{X} = K\backslash G$. Space $\mathfrak{X}$ has a G-invariant metric, iff the stabilizer subgroup K of point x_0, acts by isometries on the tangent space $T_0 = T_{x_0}$, i.e. Jacobians, $\{\phi'_u : u\in K\}$ preserve metric $g_0 = g(x_0)$ on T_0. The latter is always the case, when stabilizer K-compact (there any metric g_0 on T_0 could averaged over K (see §3.1) to make it K-invariant). In particular, symmetric spaces $K\backslash G$ (semisimple Lie group, modulo maximal compact subgroup) possess a G-invariant metric. Manifolds $\mathbb{R}^n$; S^2; $\mathbb{H}$ provide examples of symmetric spaces, as we explained in §1.1,

$$\mathbb{R}^n = \mathrm{M}_n/SO(n);\ S^2 = SO(3)/SO(2);\ \mathbb{H} = SL_2(\mathbb{R})/SO(2).$$

Obviously all geometric structures (covariant derivative, Riemann tensor) on symmetric spaces are G-invariant, and could be computed in terms of the so called Cartan decomposition, discussed in §5.7. Also geodesics on $\mathfrak{X}$ are transformed one to the other by symmetries (isometries) of $\mathfrak{X}$.

There are plenty of references to the basic differential geometry, of which we mention only a few, that were used most often: [Arn]; [Sp]; [DFN]; [Car]; [Hel,2]; [Cha].

References.

For the sake of convenience we have roughly divided the reference list into a few basic subjects.

Differential Geometry and Topology:

[ABP] *M. Atiyah, R. Bott, V. Patodi,* On the heat equation and the index theorem, *Inv. Math.* 19, **1973**, 279-330.

[AB] *M. Atiyah, R. Bott,* A Lefschetz fixed point theorem for elliptic complexes, *Ann. Math.* 86 (2), **1967**, 374-407.

[Cha] *I. Chavel,* Eigenvalues in Riemannian geometry, *Pure and Appl. Math. vol. 115, Acad. Press,* **1985**.

[Car] *H.Cartan,* Differential calculus; differential forms, *Houghton Mufflin, Boston,* **1971**.

[DFN] *Dubrovin, Fomenko, S.P. Novikov,* Modern differential geometry, *Moskwa, Nauka,* **1984**.

[Gil] *P.B. Gilkey,* Invariance theory, the heat equation and Atiyah-Singer index theorem, *Publish or Perish (Boston),* **1985**.

[Gun] *R. Gunning,* Lectures on Riemann surfaces, *Princeton Univ. Press,* **1967**.

[KB] *A. Katok, K. Burns,* Manifolds with nonpositive curvature, *Ergod. Th. of Dyn. Sys. 5* (**1985**), 307-317.

[Mil] *J. Milnor,* Hyperbolic geometry; the first 150 years, *Bull. AMS, 6,* **1982**, 9-24.

[Sp] *M. Spivak,* Compehensive introduction to differential geometry, *Publish or Perish,* **1979**.

[Wei] *A. Weinstein,* Lectures on symplectic manifolds, *CBMS Regional Conf. Series in Math. AMS, Providence,* **1977**.

Groups and algebras; Lie theory:

[BH] *M.Born, K.Huang,* Dynamical theory of crystal lattices, *Oxford Univ. Press, London,* **1956**.

[Che] *C. Chevalley,* Theory of Lie groups, *vol. I, II, III, Princeton Univ. Press,* **1946**; *Hermann, Paris,* **1955**.

[Cox] *H.S.M. Coxeter,* Regular polytopes, *Collier-Macmillan,* **1963**.

[HC] *D. Hilbert, S. Cohn-Vossen,* Geometry and imagination, *Chelsea, NY,* **1952**.

[Jac] *N. Jacobson,* Lie algebras, *Wiley (Interscience), NY* **1962**.

[Kac] *V. Kac,* Infinite-dimensional Lie algebras, *Birkhauser,* **1983**.

[Pon] *L. Pontrjagin,* Topological groups, *Princeton Univ. Press,* **1958**.

[Ser] *J.P. Serre,*
1) Lie algebras and Lie groups, *Benjamin, NY-Amsterdam,* **1965**;
2) Algebres de Lie semisimple complexes, *Benjamin, NY,* **1966**.
3) Linear Representations of finite groups, *Springer,* **1968**.

[Spr] *T.A. Springer,* Invariant theory, *Springer,* **1977**.

[Schw] *R. Schwarzenberger,* N-dimensional crystallography, *Pitman,* **1980**.

[Wel] *H. Weyl,* Symmetry, *Princeton Univ. Press,* **1952**.

Harmonic and functional analysis:

[DS] *N. Dunford, J.T. Schwartz,* Linear operators, *I-III, Wiley-Interscience,* 1971.

[DM] *H. Dym, H.P. McKean,* Fourier series and integrals, *Academic Press, NY,* 1972.

[Ed] H. M. Edwards, Riemann Zeta function, *Acad. Press, NY-London,* 1974.

[HR] *E. Hewitt, K. Ross,* Abstract harmonic analysis, *vol.I,* 1963; vol. II, 1970, *Springer Verlag, Berlin-NY.*

[Hof] *K. Hoffman,* Banach spaces of analytic functions, *Prentice-Hall,* 1962.

[Lo] *L.H. Loomis,* Abstract harmonic analysis, *Van Nostrand Reinhold, NY,* 1953.

[PW] *R. Paley, N. Wiener,* Fourier transforms in the complex domains, *AMS, Providence, RI* 1934.

[St] *E. Stein,* Singular integrals and differentiability properties of functions, *Princeton Univ. Press,* 1970.

[SW] *E. Stein, G. Weiss,* Introduction to Fourier analysis on Euclidian space, *Princeton Univ. Press,* 1971.

[Tit] *Titchmarch,* The theory of the Zeta-function, *Oxford, Claredon,* 1951.

[WW] *E. Whittaker, G.N. Watson,* A course of Modern analysis, *Cambridge Univ. Press, London-NY,* 1927.

[Zy] *A. Zygmund,* Trigonometric series, *vol. I, II, Cambridge Univ. Press, London-NY,* 1959.

Differential equations; Spectral theory; Special functions:

[BB] *Berard, G. Besson,* Spectres et groupes cristallographiques, I: Domaines euclidiens, *Inven. Math. 58,* 1980, 179-199; II: Domaines spheriques, *Ann. Inst. Fourier, Grenoble, 30,* 1980, 237-248.

[CQ] *A. Cormack, E. Quinto,* A Radon transform on spheres through the origin in $\mathbb{R}^n$ and applications to the Darboux equation, *Tran. AMS, 260,* 1980, 575-580.

[CH] *R. Courant, D. Hilbert,* Methods of mathematical Physics, *Wiley (Interscience), NY,* 1953.

[DG] *J. Duistermaat, V. Guillemin,* The spectrum of positive elliptic operators and periodic bicharacteristics, *Inv. Math. 29,* 1975, 39-79.

[DKV] *J. Duistermaat, J. Kolk, V. Varadarajan,*
1) Spectra of compact locally symmetric manifolds of negative curvature, *Inv. Math. 52,* 1979, 27-93.

2) Functions, flows and oscillatory integrals on flag manifolds and conjugacy classes of real semisimple Lie groups, *Comp. Math. 49,* 1983, 309-398.

[Erd] *A. Erdelyi et al.* Higher transcendental functions, *v. I-III (Bateman Manuscript Project), McGraw-Hill, NY,* 1953.

[FP] *L.D. Faddeev, B.S. Pavlov,* Scattering theory and automorphic functions, *J. Soviet Math. 3, no.4,* 1975.

[GK] *V. Guillemin, D. Kazdan,* Some Inverse spectral results for negatively curved manifolds, *Proc. Sym. Pure Appl. Math.36,* 1980, 153-180.

[Hö] *L. Hörmander,* The analysis of Linear partial differential operators, *I-IV, Springer,* 1985.

[Ka] *M. Kac,* Can one hear the shape of the drum? *Amer. Math. Monthly 73,* 1966, 1-23.

[KF] *F. Klein, R. Fricke,* Vorlesungen über die Theorie der Elliptischer Modularfunktionen/Automorphenfunktionen, *G.Teubner, Leipzig,* 1896/1912.

[LP] *P.Lax, R. Phillips,* Scattering theory for automorphic functions, *Princeton Univ. Press,* 1976.

[Leb] *Lebedev,* Special functions and their applications, *Dover,* 1972.

[Mc] *H.P. McKean,* Selberg trace formula as applied to a compact Riemann surface, *Comm. Pure Appl. Math. 25,* 1972, 225-246.

[McS] *H.P. McKean, I.M. Singer,* Curvature and the eigenvalues of the Laplacian, *J. Diff. Geom. 1,* 1967, 43-69.

[Mil] *W. Miller, Jr.*
1) Lie groups and special functions, *Acad. Press, NY,* 1968.
2) Symmetry and separation of variables, *Addison-Wesley, Reading, MA.* 1977.

[MP] *S. Munakshisundaram, A.Pleijel,* Some properties of the eigenfunctions of the Laplace operator on Riemannian manifolds, *Can. J. Math. 1,* 1949.

[Un] *A. Untenberg,* Oscillateur harmonique et operateurs pesudodifferentiels, *Ann. Inst. Fourier, Grenoble, 29,* 1979, 201-221.

[Ur] *A. Uribe,* A symbol calculus for a class of pseudodifferential operators on S^n, *J. Funct. Anal. 59,* 1984, 535-556.

[Olv] *P. Olver,* Applications of Lie groups to differential equations, *Graduate Text in Math. Springer,* 1988.

[Ovs] *L.V. Ovsiannikov,* Group analysis of differential equations, *Moskwa, Nauka,* 1978.

[Tal] *M. Taylor,* Pseudo-differential operators, *Princeton Univ. Press,* 1981.

[Ven] *A. Venkov,* Spectral theory of automorphic functions and its applications, *Kluwer,* 1990.

[Vi] *M. Vigneras,* Varietes riemanniennes isospectrales et non isometriques, *Ann. Math. 112,* 1980, 21-32.

[We2] *H. Weyl,* Ramifications, old and new, of the eigenvalue Problem, *Bull. AMS 56,* 1950, 115-139.

[Wi] *H. Widom,* Szego's Theorem and a complete symbolic calculus for pseudo-differential operators, *Ann. Math. Studies, 91, Princeton Univ. Press,* 1979.

[Wo] *S. Wolpert,* The eigenvalue spectrum as moduli for compact Riemann surfaces, *Bull. AMS 83,* 1977, 1306-1308.

Group representations:

[At] *M.F.Atiyah,* Elliptic operators and compact groups, *Lect. Notes in Math. 401, Springer,* **1974**.

[AS] *M. Atiyah, W. Schimdt,* A geometric construction of the discrete series for semi-simple Lie groups, *Inven. Math. 42,* **1977**, 1-62.

[Bar] V. Bargmann, Irreducible unitary representations of the Lorentz group, *Ann. Math.* **1947**, 48, 568-640.

[Ber1] *F. Berezin,* Laplace operators on semi-simple Lie groups, *AMS Transl. 2, 21,* **1962**, 239-339.

[Bla] *R. Blattner,* Harmonic analysis and representation theory of semi-simple Lie groups, *D. Reidel, Boston,* **1980**.

[CT] *R. Cahn, M. Taylor,* Asymptotic behavior of multiplicities of representations of compact groups, *Pac. J. Math., 84 (1),* **1979**, 17-28.

[GN] *I. Gelfand, M. Naimark,* Unitary representations of classical groups, *Trans. MIAN,* **1950**, 36.

[GGP] *I.M. Gelfand, M.I. Graev, I.I. Piatecki-Shapiro,* Generalized functions, vol. 6: Group representations and automorphic functions, *Moskwa, Nauka,* **1966**.

[GGV] *I.M. Gelfand, M.I. Graev, N.Ya. Vilenkin,* Generalized functions, vol. 5: integral geometry and the related problems of the representations theory, *Moskwa, Nauka,* **1966**.

[GMS] *I.M. Gelfand, R. Milnos, Z. Shapiro,* Representations of the rotation and Lorentz groups and their applications, *Pergamon Press, NY,* **1963**.

[Geo] *H. Georgi,* Lie algebras in particle Physics, *Benjamin, Reading, MA.* **1982**.

[God] *R. Godement,* A theory of spherical functions, *Trans. AMS,* **1952**, 73, 496-556.

[Gr] *K.I. Gross,* On the evolution of noncommutative harmonic analysis, *Amer. Math. Mon.* **1978**, 85, 525-548.

[GK] *K.I. Gross, R.A. Kunze,* Bessel functions and representation theory, *J. Func. Anal.* (I) 22, **1976**, 73-105; (II) 25, **1977**, 1-49.

[GS1] *V. Guillemin, S. Sternberg,* The metaplectic representation, Weyl operators, and spectral theory, *J. Func. Anal. 42,* **1981**, 128-225.

[Har] *Harish-Chandra,*
1. Representations of semisimple Lie groups, *Trans. AMS, I, 75,* **1953**; II-III, 76, **1954**; IV, *Amer. J. Math. 78,* **1956**.
2.Plancherel formula for complex semisimple Lie groups, *Proc. NAS USA 37,* **1951**;
3. Spherical functions on semisimple Lie groups, *I-II, Amer. J. Math,* **1958**, 80;
4. Harmonic analysis on semisimple Lie groups, *Bull. AMS,* **1970**, 78, 529-551.
5. Discrete series for semisimple Lie groups, *I,II, Acta Math. 113,* **1965**, 241-318; 116, **1966**.

[Hej] *D.A. Hejhal,* The Selberg trace formula for *PSL*(2;ℝ), Lect. Notes Math. no. 548, 1001, Springer, **1976, 1983**.

[Hel] *S. Helgason,*
1) Differential geometry, Lie groups and symmetric spaces, *Acad. Press,* **1978**;
2) Groups and geometric analysis: integral geometry, invariant differential operators and spherical functions, *Acad. Press, NY,* **1984**.

3) The Radon transform, *Birkhauser,* **1980.**

[HST] The Selberg trace formula and related topics, *D.Hejhal, P.Sarnak, A.Terras, ed. Contem. Math. 53, AMS,* **1986.**

[Hir] *T. Hirai,*
On irreducible representations of the Lorentz group of n-th order, *Proc. Japan Acad. 38,* **1962,** 258-262;
The characters of irreducible representations of the Lorentz group of n-th order, *Proc. Japan Acad. 41,* **1965,** 526-531;
The Plancherel formula for the Lorentz group of n-th order, *Proc. Japan Acad. 42,* **1965,** 323-326.

[JL] *H. Jaquet, R. Langlands,* Automorphic forms on $\mathbf{GL}_2$, *Lect. Notes in Math. 114, Springer,* **1970.**

[Kar] *F. Karpelevich,* The geometry of geodesics and the eigenfunctions of the Laplace-Beltrami operator on symmetric spaces, *Trans. Moscow Math. Soc. 14,* **1965,** 48-185.

[Kir] *A.A. Kirillov,*
1) Elements of representation theory, *Springer, NY,* **1976.**

2) Unitary representations of nilpotent groups, *Russian Mat. Surveys (Uspekhi), 17,* **1962,** 57-101.

[KS] *A. Knapp, B. Speh, Status of classification of unitary representations, Lect. Notes Math. 908, Springer,* **1982,** 1-38.

[Kn] *A. Knapp,* Representation theory of semisimple groups: an overview based on examples, *Princeton Univ. Press,* **1986.**

[Ko] *B. Kostant,* Quantization and unitary representations, *Lect. Notes in Math.* **170, 1970,** 87-207.

[Lan] *S. Lang,* $\mathbf{SL}_2(\mathbb{R})$, *Addison-Wesley, Reading, MA.* **1975.**

[Lip] *R. Lipsman,* Group representations, *Lect. Notes in Math. 388,* **1974.**

[Mac] *G.W. Mackey,*
1) The theory of unitary group representations, *Univ. of Chicago Press, Chicago,* **1976;**
2) Unitary group representations in Physics, probability and number theory, *Benjamin-Cummings Publ. Reading, Mass.* **1978.**

3) Unitary representations of group extensions, *I, Acta Math. 99, 3-4,* **1958,** 265-311;

4) Imprimitivity for representations of locally compact groups, *Proc. NASci. USA, 35,* **1949,** 193-221.

5) Harmonic analysis as the exploition of symmetry: a historical survey, *Bull. AMS,* **1980,** 3, 543-698.

[NZ] *M.A. Naimark, D. Želobenko,* Completely irreducible representations of semisimple complex Lie groups, *Izv. USSR Acad. Sci. 34, 1,* **1970,** 57-85.

[Par] *R. Parthasarathy,* Dirac operators and discrete series, *Ann. Math, 96,* **1972,** 1-30.

[Pu] *L. Pukanszky,* On the unitary representations of exponential groups, *J. Func. Anal. 2,* **1968,** 73-113;
Representations of solvable Lie groups, *Ann. Sci. Ecole Norm. Sup. 4,* **1971,** 464-608.

[Ros] *J. Rosenberg,* A quick proof of Harisch-Chandra Plancherel formula for spherical functions on a semisimple Lie group, *Proc. AMS, 63,* **1977**, 143-149.

[Sch] *W. Schmid,* On the characters of the discrete series, *Inven. Math. 30,* **1975**, 47-144
Representations of semisimple Lie groups, *London Math. Soc. Lect. Notes, 34,* **1979**, 185-235.

[Seg] *I. Segal,* An extension of Plancherel's formula to separable unimodular groups, *Ann. of Math. 52,* **1950**, 272-292.

[Sel] *A. Selberg,* Harmonic analysis and discontinuous groups in weakly symmetric Riemannian spaces with applications to Dirichlet series, *J. Ind. Math. Soc. 20,* **1956**, 47-87.

Discontinuous groups and harmonic analysis, *Proc. Intern. Congress of Math. Stockholm,* **1962**, 177-189.

[Sp] *B. Speh,* The unutary dual of $GL_3(\mathbb{R})$ and $GL_4(\mathbb{R})$, *Math. Ann. 258,* 1981, 113-133.

[SŽ] *A.I. Stern, D.P. Želobenko,* Representations of Lie groups, *Moskwa, Nauka,* **1983**.

[Sug] *M. Sugiura,* Unitary representations and harmonic analysis, *Addison-Wiley, NY,* **1975**.

[Ta2] *M. Taylor,* Noncommutative harmonic analysis, *AMS, Providence,* **1986**.

[Ter] *A. Terras,* Harmonic analysis on symmetric spaces and applications, *Springer,* **1985**.

[Va] *V.S. Varadarajan,*
1) Lie groups, Lie algebras and their representations, *Springer,* **1984**;
2) Harmonic analysis on real reductive groups, *Springer,* **1977**.

[Vi] *N. Ya. Vilenkin,* Group representations and special functions, *AMS, Providence,* **1968**.

[Vo] *D. Vogan,* Representations of real reductive groups, *Birkhauser, Boston,* 1982.

[Wal] *N. Wallach,*
1) Harmonic analysis on homogeneous spaces, *Dekker, NY,* **1973**;
2) Symplectic geometry and Fourier analysis, *Math. Sci. Press, Brookline,* **1977**.

[War] *G. Warner,* Harmonic analysis on semisimple Lie groups, *vol. I, II, Springer-Verlag, Berlin-NY,* **1972**.

[We3] *H. Weyl,* The classical groups, their invariants and representations, *Princeton Univ. Press, Princeton, NJ.* **1939**.

[Wig] *E. Wigner,*
1) Group theory and its applications to the quantum mechanics of atomic spectra, *Acad. Press,* **1959**.

2) On unitary representations of the inhomogeneous Lorentz group, *Ann. Math. 40,* **1939**, 149-204.

[Wol] *J. Wolf,* Foundations of representation theory for semisimple Lie groups, *D. Reidel, Boston,* **1980**.

[Ze] *D. Želobenko,* Compact Lie groups and their representations, *AMS,* **1973**.

Classical and quantum mechanics; Quantization:

[AM] *R. Abraham, J.E. Marsden,* Foundations of mechanics, *2-nd ed. Benjamin-Cummings, Reading, Mass.* **1978**.

[Arn] *V.I. Arnold,* Mathematical methods of classical mechanics, *Springer, NY,* **1987**.

[Be] *F.A. Berezin,* 1) General concept of quantization, *Comm. Math. Phys. 40,* **1975**, 153-174.
2) Quantization, *Math. USSR Izv. 8,* **1974**, 1109-1163.

[BL] *L. Biedenharn, J. Louck,* Angular momentum in quantum Physics, the Racah-Wigner algebra in quantum theory, *vol. 8,9, Encyclop. of Math. and its Applic. Addison-Wesley,* **1981**.

[Ch] *P. Chernoff,* Mathematical obstruction to quantization, *Hadr.J. 4,* **1981**, 879-898.

[Dir] *P.A.M. Dirac,* Lectures on quantum mechanics, *Belfer Grad. School, Yeshiva Univ. NY,* **1964**.

[FM] *A. Fomenko, A. Mischenko,* Euler equations on finite-dimensional Lie groups, *Izv. AN USSR,* **1978**, 42, 2, 396-415.

[GSp] *G. Gaeta, M. Spera,* Remarks on geometric quantization of the Kepler problem, Lett. Math. Phys., 16, **1988**, 187-197.

[GSW] *M.Green, J. Schawarz, E. Witten,* Superstring theory, *I, II, Cambridge Univ. Press,* **1987**.

[GS2] *V. Guillemin, S. Sternberg,* Variations on a theme by Kepler, **1991**.

[How] *R. Howe,* On the role of the Heisenberg group in harmonic analysis, *Bull. AMS, 3, 2,* **1980**, 821-843.
Quantum mechanics and partial differential equations, *J. Func. Anal. 32,* **1980**, 188-254.

[Hur] *N. Hurt,* Geometric quantization in action, *D. Reidel, Amsterdam,* **1983**.

[Kum] M. Kummer, On the regularization of the Kepler problem, Comm. Math. Phys., 84, **1982**, 133-152.

[LL] *L.D. Landau, E.M. Lifshitz,* Quantum mechanics, *Moskwa, Nauka,* **1989**.

[MF] *P. Morse, H. Feshbach,* Methods of Theoretical Physics, *I-II, McGraw-Hill,* **1953**.

[MW] *J. Marsden, A. Weinstein,* Reduction of symplectic manifolds with symmetry, *Rep. Math. Phys. 5,* **1974**, 121-130.

[Mla] *I. Mladenov,* Geometric quantization of the MIC-Kepler problem via extension of the phase-space, *Ann. Inst. H. Poincare (Phys. Theor.) 50,* **1989**, 183-191.

[Mos] *J. Moser,* Regularization of Kepler's problem and the averaging method on a manifold, *Comm. Pure Appl. Math. 23,* **1970**, 609-636.

[OP] *M.A. Olshanetski, A.M. Perelomov,* Classical integrable finite-dimensional systems related to Lie algebras, *Phys. Rep. 71 (313),* **1981**;
Quantum integrable systems related to Lie algebras, *Phys. Rep. 94, 6,* **1983**.

[Per] *A.M. Perelomov,* Integrable systems of classical mechanics and Lie algebras, *Birkhauser, Boston,* **1991**.

[RS] *M. Reed, B.Simon,* Methods of modern mathematical Physics, I-IV, Ac. Press, **1972**

[Sn] *J. Sniatycki,* Geometric quantization and quantum mechanics, *Springer,* **1980**.

[Sou] *J.-M. Souriau,*

1. Structure des system dynamiques, ***Dunod Univ. Paris,*** **1970.**
2. Sur la variete de Kepler, Symp. Math., vol. XIV, Acad. Press, **1974.**

[Simm] *D.J. Simms,* Geometric quantization of energy levels in the Kepler problem, Symp. Math., vol. XIV, Acad. Press, **1974.**

[Sim] *B. Simon,* The classical limit of quantum partition functions, *Com. Math. Phys. 71,* **1980**, 247-276.

[To] *M. Toda,* Theory of nonlinear lattices, *Springer, NY,* **1981.**

List of frequently used notations:

$\mathbb{R};\mathbb{C};\mathbb{Q};\mathbb{Z}$ - reals; complex numbers; quaternions; integers

Manifolds/domains

$\mathcal{M};\mathfrak{S}$ - manifold

$\mathbb{H}$ -Poincare half plane $\{\Im z > 0\}$;

$\mathbb{D}$ -Poincare disk $\{\,|z| < 1\}$

$\mathbb{M}^4;\ \mathbb{M}^n$ - Minkowski space

$\mathbb{H}^n$ - rank-one hyperbolic space of dim $= n$

$\mathbb{P}^1;\ \mathbb{P}^k$ - projective spaces of dim 1,k; also

$\mathbb{RP}^1;\ \mathbb{CP}^k$ - projective spaces over reals, complex spaces

S^n - the n-sphere $\{\|x\| = 1\}$ in $\mathbb{R}^{n+1}$

$\Sigma;\Sigma_+$ - root system (positive roots) of a semisimple Lie algebra

$\Lambda;\mathfrak{H}_+$ - Weyl chamber

Spaces

$\mathcal{V},\ \mathcal{H}$ - vector/Hilbert space

$\mathcal{T}^m(\mathcal{V})$ - tensors of rank m over $\mathcal{V}$; also $T^{\otimes N}(\mathcal{V})$.

$\mathcal{S}^m(\mathcal{V})$ - symmetric tensors of rank m

$\mathcal{P}_m$ - polynomials of degree m

$Mat_d; Mat_{n\times m}$ - algebra/space of $d\times d$, $n\times m$ matrices

$\mathcal{H}_m$ - harmonic polynomials/spherical harmonics of degree m

$\Lambda^m(\mathcal{V})$ - m-th anti-symmetric tensor power of space $\mathcal{V}$

SM_n - symmetric $n\times n$ matrices

Groups and algebras

$G;\ \widehat{G}$ - group and its dual group (dual object)

$\mathbb{T}^n$ - torus

$\mathbb{W}_n$ - permutation group of n elements;

$\mathbb{A}_n$ - alternating group (of even permutations)

$\mathbb{W}$ -Weyl group of a semisimple Lie algebra (root system)

The standard notations in the upper-case bold are reserved for the classical Lie groups: $\boldsymbol{SO}(3)$ $\boldsymbol{SO}(p;q)$; $\boldsymbol{SO}(n)$; $\boldsymbol{SU}(n)$; $\boldsymbol{SU}(2)$; $\boldsymbol{SL}_2(\mathbb{C})$; $\boldsymbol{SL}(n)$; $\boldsymbol{SL}(3)$; $\boldsymbol{GL}_n$ etc., and the corresponding lower-case bold denote their Lie algebra

$\mathbb{E}_n = \mathbb{R}^n \rhd \boldsymbol{SO}(n)$ - Euclidian motion group; $\mathfrak{E}_n$ - its Lie algebra

$\mathbb{H}_n$ - Heisenberg group

$H;\mathfrak{H}$ - Cartan (diagonal) subgroup/subalgebra of a semisimple Lie group

$B_{\pm}; \mathfrak{B}_{\pm}$	- Borel subgroup/subalgebra of upper/lower triangular matrices in SL_n
$\mathfrak{G} = \mathfrak{K} \oplus \mathfrak{P}$	- Cartan decomposition
	Function spaces; algebras; distributions
$\mathcal{C}(\mathcal{M})$	- space of continuous functions
$\mathcal{C}^m(\mathcal{M})$; $\mathcal{C}^\infty(\mathcal{M})$	- m-smooth (∞-smooth) functions
$\mathcal{C}_c$; ${\mathcal{C}^m}_c(\mathcal{M})$	- compactly supported functions
$L^1; L^p; L^2$	- Lebesgue spaces of integrable functions
$\mathcal{H}_s$	- Sobolev space of order s
$\mathcal{D}$; $\mathcal{D}^m$	- distributions
$\mathcal{B}(\mathcal{V})$	- algebra of all (bounded) operators of a Hilbert/Banach space $\mathcal{V}$
$\mathcal{W} = \mathcal{W}_n$	- Weyl algebra on $\mathbb{R}^n$, generated by all multiplications and differentiations
	Transforms; Operators
$\mathcal{F}$, $\mathcal{F}^{-1}$	- Fourier transform and its inverse
$\mathcal{R};\hat{\mathcal{R}};\widetilde{\mathcal{R}}$	- Radon transform and its dual
Sym_m	- operation of symmetrization, $\mathcal{T}^m(\mathcal{V})\to\mathcal{S}^m(\mathcal{V})$
$\wedge$	- wedge product of vectors, tensors, differential forms
Δ	- Laplacian (Laplace-Beltrami operator)
$M^r f(x)$	- spherical mean of function f at $\{x\}$, of radius r
	Representations
$\mathrm{Int}(S;T)$	- intertwining space of representations S and T
$\mathrm{Com}(T)$	- the commutator algebra of T
$T \otimes S$	- tensor (Kronecker) product of representations T and S
$\oplus T^j$	- direct sum of representations $\{T^j\}$
$\oint T^x d\mu(x)$	- direct integral of representations $\{T^x : x \in \mathfrak{X}\}$
ad_X; Ad_X	- adjoint representation of Lie algebra/group

Index.

A

B

C

D

E

F

G

H

I

M

N

O

Y

Z